Lecture Notes in Networks and Systems

Volume 274

The series "Lecture Notes in Networks and Systems" publishes the latest developments in Networks and Systems—quickly, informally and with high quality. Original research reported in proceedings and post-proceedings represents the core of LNNS.

Volumes published in LNNS embrace all aspects and subfields of, as well as new challenges in, Networks and Systems.

The series contains proceedings and edited volumes in systems and networks, spanning the areas of Cyber-Physical Systems, Autonomous Systems, Sensor Networks, Control Systems, Energy Systems, Automotive Systems, Biological Systems, Vehicular Networking and Connected Vehicles, Aerospace Systems, Automation, Manufacturing, Smart Grids, Nonlinear Systems, Power Systems, Robotics, Social Systems, Economic Systems and other. Of particular value to both the contributors and the readership are the short publication timeframe and the world-wide distribution and exposure which enable both a wide and rapid dissemination of research output.

The series covers the theory, applications, and perspectives on the state of the art and future developments relevant to systems and networks, decision making, control, complex processes and related areas, as embedded in the fields of interdisciplinary and applied sciences, engineering, computer science, physics, economics, social, and life sciences, as well as the paradigms and methodologies behind them.

Indexed by SCOPUS, INSPEC, WTI Frankfurt eG, zbMATH, SCImago.

All books published in the series are submitted for consideration in Web of Science.

More information about this series at http://www.springer.com/series/15179

Stefan Trzcielinski · Beata Mrugalska ·
Waldemar Karwowski ·
Emilio Rossi · Massimo Di Nicolantonio
Editors

Advances in Manufacturing, Production Management and Process Control

Proceedings of the AHFE 2021 Virtual Conferences on Human Aspects of Advanced Manufacturing, Advanced Production Management and Process Control, and Additive Manufacturing, Modeling Systems and 3D Prototyping, July 25–29, 2021, USA

Springer

Editors
Stefan Trzcielinski
Poznan University of Technology
Poznan, Poland

Waldemar Karwowski
University of Central Florida
Cocoa Beach, FL, USA

Massimo Di Nicolantonio
Department of Architecture
University of Chieti-Pescara
Pescara, Pescara, Italy

Beata Mrugalska
Faculty of Engineering Management
Poznan University
Poznan, Poland

Emilio Rossi
Lincoln School of Design
University of Lincoln
Lincoln, UK

ISSN 2367-3370 ISSN 2367-3389 (electronic)
Lecture Notes in Networks and Systems
ISBN 978-3-030-80461-9 ISBN 978-3-030-80462-6 (eBook)
https://doi.org/10.1007/978-3-030-80462-6

This Springer imprint is published by the registered company Springer Nature Switzerland AG
The registered company address is: Gewerbestrasse 11, 6330 Cham, Switzerland

Advances in Human Factors and Ergonomics 2021

AHFE 2021 Series Editors

Tareq Z. Ahram, Florida, USA
Waldemar Karwowski, Florida, USA

12th International Conference on Applied Human Factors and Ergonomics and the Affiliated Conferences (AHFE 2021)

Proceedings of the AHFE 2021 Virtual Conferences on Human Aspects of Advanced Manufacturing, Advanced Production Management and Process Control, and Additive Manufacturing, Modeling Systems and 3D Prototyping, July 25–29, 2021, USA.

Advances in Neuroergonomics and Cognitive Engineering	Hasan Ayaz, Umer Asgher and Lucas Paletta
Advances in Industrial Design	Cliff Sungsoo Shin, Giuseppe Di Bucchianico, Shuichi Fukuda, Yong-Gyun Ghim, Gianni Montagna and Cristina Carvalho
Advances in Ergonomics in Design	Francisco Rebelo
Advances in Safety Management and Human Performance	Pedro M. Arezes and Ronald L. Boring
Advances in Human Factors and Ergonomics in Healthcare and Medical Devices	Jay Kalra, Nancy J. Lightner and Redha Taiar
Advances in Simulation and Digital Human Modeling	Julia L. Wright, Daniel Barber, Sofia Scataglin and Sudhakar L. Rajulu
Advances in Human Factors and System Interactions	Isabel L. Nunes
Advances in the Human Side of Service Engineering	Christine Leitner, Walter Ganz, Debra Satterfield and Clara Bassano
Advances in Human Factors, Business Management and Leadership	Jussi Ilari Kantola, Salman Nazir and Vesa Salminen
Advances in Human Factors in Robots, Unmanned Systems and Cybersecurity	Matteo Zallio, Carlos Raymundo Ibañez and Jesus Hechavarria Hernandez
Advances in Human Factors in Training, Education, and Learning Sciences	Salman Nazir, Tareq Z. Ahram and Waldemar Karwowski

(continued)

(continued)

Advances in Human Aspects of Transportation	Neville Stanton
Advances in Artificial Intelligence, Software and Systems Engineering	Tareq Z. Ahram, Waldemar Karwowski and Jay Kalra
Advances in Human Factors in Architecture, Sustainable Urban Planning and Infrastructure	Jerzy Charytonowicz, Alicja Maciejko and Christianne S. Falcão
Advances in Physical, Social & Occupational Ergonomics	Ravindra S. Goonetilleke, Shuping Xiong, Henrijs Kalkis, Zenija Roja, Waldemar Karwowski and Atsuo Murata
Advances in Manufacturing, Production Management and Process Control	Stefan Trzcielinski, Beata Mrugalska, Waldemar Karwowski, Emilio Rossi and Massimo Di Nicolantonio
Advances in Usability, User Experience, Wearable and Assistive Technology	Tareq Z. Ahram and Christianne S. Falcão
Advances in Creativity, Innovation, Entrepreneurship and Communication of Design	Evangelos Markopoulos, Ravindra S. Goonetilleke, Amic G. Ho and Yan Luximon
Advances in Human Dynamics for the Development of Contemporary Societies	Daniel Raposo, Nuno Martins and Daniel Brandão

Preface

Contemporary enterprises aim to deliver a great number of consumer products and systems, assuring at the same time friendly and satisfying working environments for people involved in manufacturing services. Human-centered design factors, which strongly affect manufacturing processes and the potential end-users, are crucial for achieving continuous progress in this respect. These topics have been extensively discussed at the AHFE 2021 conference, and selected papers are included in this book.

Researchers around the world attempt to improve both the quality of consumer products and working environments. The advanced production management and process control (APMPC) focuses on exchanging ideas and recent developments in production, sustainability, life cycle, innovation, development, fault diagnostics, and control systems. It also addresses a spectrum of theoretical and practical topics.

Furthermore, this book covers topics in additive manufacturing, advanced modeling systems, and 3D prototyping. Digital modeling systems, empowered by tangible and intangible tools that apply technologies such as ICTs, virtual environments, cloud computing, intelligent user interfaces, interconnected tools, and hybrid systems, enable us to conceive, model, analyze, virtualize, and simulate sophisticated solutions. Additive manufacturing (AM) refers to all production processes made with 3D printers, rapid prototyping systems, and methods for remote fabrication, which use raw materials to create three-dimensional objects. Such fabricated solutions become pervasive in many environments, decreasing the cost of the entire manufacturing process.

We believe that the up-to-date knowledge presented in this book can inspire and support researchers in manufacturing and process control to advance their designs and implement them into practice. Therefore, this book is addressed to both researchers and practitioners.

The papers presented in this book have been arranged into seven sections. The first two sections focus mainly on topics in additive manufacturing, while the remaining six sections focus on topics related to production management, process control, and manufacturing.

Additive Manufacturing, Modeling Systems, and 3D Prototyping

1. Advanced Technologies and Manufacturing Processes for 3D Printing
2. 3D Printing Design Applications

Advanced Production Management and Process Control

3. Production Management and Process Control

The Human Aspects of Advanced Manufacturing: Managing Enterprise of the Future

4. Production Management in Industry
5. Human Capital Management in Industry
6. Applications in Industry
7. Lean and Agility and Enterprise

The content of this book is the outcome of the dedicated effort of many people. We would like to thank the authors, whose research and development efforts are published here. Finally, we also wish to thank the following editorial board members of the respective AHFE 2021 affiliated conferences for their diligence and expertise in selecting and reviewing the presented papers:

Advanced Manufacturing

Production Management and Process Control

S. Ávila Filho, Brazil
M. Gibson, USA
A. Goto, Japan
A. Hamrol, Poland
M. Macias, Mexico
J. Niemann, Germany
T. Ota, Japan
E. Ramos, Peru
C. Raymundo, Peru
S. Simani, Italy
Y. Tansel İÇ, Turkey
M. Wyrwicka, Poland

Additive Manufacturing, Modeling Systems, and 3D Prototyping

P. Barcarolo, Italy
M. Calvano, Italy
A. D'Onofrio, Italy
J. Keist, USA
F. Medola, Brazil
R. Narayan, USA
E. Pei, UK
P. Rao, USA
F. Sandnes, Norway

Stefan Trzcielinski
Beata Mrugalska
Waldemar Karwowski
Emilio Rossi
Massimo Di Nicolantonio

Contents

Applications in Industry

Lean and Agility and Enterprise

Advanced Technologies and Manufacturing Processes for 3D Printing

Determining the Optimal Orientation of AM-Parts Based on Native 3D CAD Data

Slim Krückemeier(✉), Georg Staudter, and Reiner Anderl

Department of Computer Integrated Design,
Technische Universität Darmstadt, 64287 Darmstadt, Germany
{krueckemeier,staudter,anderl}@dik.tu.darmstadt.de

Abstract. Due to the separation of part design and pre-processing in the industrial context of additive manufacturing, design changes caused by part orientation determination lead to time-consuming iterations. This work provides an approach to determine the optimal part orientation and post-processing surfaces in CAD-Systems, based on native 3D CAD data, to overcome these shortcomings. The authors identified relevant parameters being influenced by part orientation and propose a method to solve the multi-criteria optimization problem of the part orientation. An application example is implemented using the programming language C# and the NX Open API of the CAD-system Siemens NX 12 by vendor Siemens PLM Software is used for validation purposes.

Keywords: Additive manufacturing · Part orientation · Additive process chain optimization · Software application

1 Introduction

Additive manufacturing (AM) technologies have developed rapidly over the last decades. AM-technologies are no longer limited to the rapid manufacturing of prototypes but can produce highly complex, functional, and end-use parts for various applications. Besides the significant advantages over traditional manufacturing technologies, the comparatively low surface quality is associated as a significant disadvantage with AM. This results from the unique process of layered manufacturing and highly dependent on the built orientation of the surface. The manufacturer typically does the determination of the build orientation during the pre-processing. The AM pre-process encompasses the steps between design and printing and comprises the definition of part position, orientation, the setting of manufacturing parameters, and the toolpath generation. One main objective of the pre-processing is to ensure that the surface quality requirements are met. Due to the overall lack of AM-specific knowledge, parts are often manufactured by additive manufacturing service providers, whose experts perform the pre-processing for their clients. Therefore, part design and pre-processing are typically separated in an industrial context. If it cannot be determined with certainty, that the surface quality requirements can be achieved with the existing production equipment and the pre-processing settings, either faulty parts are produced, or additional material must be added to the surfaces

S. Trzcielinski et al. (Eds.): AHFE 2021, LNNS 274, pp. 3–9, 2021.
https://doi.org/10.1007/978-3-030-80462-6_1

for post-processing. Both cases lead to time and cost-intensive design adaptation and communication loops.

The design of a part has a significant influence on its orientation in the printing volume. In turn, the orientation primarily determines surface quality, whether the surfaces require support structures and whether surfaces are accessible for post-processing. Furthermore, most quality characteristics are also mainly influenced by the part orientation. Due to these strong dependencies, part orientation is of crucial importance for part design. Thus, the determination of an optimal built orientation has been the subject of research for many years, and several approaches and software applications are available. The majority of these approaches are developed to support AM pre-processing and are therefore based on the de facto AM standard for data exchange – the STL (Standard Tessellation Language) data format. Due to the separation from design, quality-maintaining adaptations must still be communicated to the design department, which is why the problem of cost-intensive design- and communication loops is not yet solved. This work provides an approach to determine the optimal part orientation and post-processing surfaces in CAD-Systems based on native 3D CAD data to overcome these shortcomings. The sections of this paper are organized as follows: Sect. 2 gives an overview of related work, in Sect. 3 the concept is presented and Sect. 4 describes the implemented application example. Finally, Sect. 5 draws a conclusion and presents an outlook for future work.

2 Related Work

Part orientation is a crucial aspect in additive manufacturing, which has no trivial solution. There is a large number of published scientific papers and commercial software applications dealing with the topic. The solution approaches can be divided according to two distinctive features: the optimization parameters and the optimization procedures.

The most commonly used optimization parameters in the literature are surface quality, the volume of support structures, and build time and cost [1–9]. Besides that, the surface area in contact with support structure [10], the surface are needing post-processing [11], the quality of critical features [1], the structural performance [12] as well as the removability of the support structures [5] are also considered in individual approaches. The second distinctive feature is the optimization procedure. The most widespread approach is using a genetic algorithm for solving the optimization problem [3, 4, 6, 7, 9, 11]. On the other hand, many publications use customized algorithms or adapt different algorithms from other fields [1, 2, 8, 10, 13]. The algorithms evaluated by *Rocha et al.* in [10], for example, are based on completely different principles. They investigate whether the electromagnetism-like (EM) algorithm and the stretched simulated annealing (SSA) algorithm are suitable for the optimization of the part orientation. For a further in-depth review of publications in the field of the orientation of AM-parts, see the work of *Di Angelo et al.* [14].

The majority of these approaches are based on the STL data format. Due to STL-data being an approximation of geometry, design changes for requirement satisfaction caused by the part orientation have to be conducted in the CAD-system. After the design change a new STL-file needs to be exported and pre-processed again. There are two approaches for tackling these time-consuming iterations. The first option is to integrate the determination of the part orientation into the design domain and perform design changes based on native CAD data. The second option is to use alternative file formats for data exchange, such as *STEP ISO 10303*, which enable the transfer of the geometric representation instead of a mere approximation. Furthermore, when using STL as data exchange format, the surface roughness can only be considered in respect to a global requirement equally valid for all surfaces. This is caused by the loss of information about individual part surfaces due to the global approximation of the entire part via tessellation.

This work presents a solution for the first approach by determining the optimal part orientation and post-processing surfaces in CAD-Systems based on native 3D CAD data.

3 Concept

Part orientation has a significant influence on various process and part properties. To compare different orientations in order to find the best solutions, it is necessary to define parameters that describe the quality of individual orientations. The entire set of quality parameters should represent all aspects of the part orientation to perform a target-oriented determination of the optimal orientation. Within the scope of this work, seven quality parameters are used, which are to be minimized. The parameters used are:

- Support Surface
- Support Volume
- Surface quality
- Build time
- Post-processing-area
- The factor for important faces
- The factor for load direction

In this work, the post-processing area is defined as all surfaces in contact with the support structure and whose surface quality requirements are not met, so that these surfaces must be post-processed. *Frank and Fadel* already pointed out in 1995 that for most applications, the requirement for the surface quality of the entire part is not as critical as the surface quality of some specific geometric features [15]. Therefore, analogous to [16], a factor for important faces is introduced, which describes the quality of these specific surfaces. The factor for load direction has its origin in the consideration of the anisotropic part properties of additively manufactured parts. Parts have the best properties parallel to the building plane, such as the highest stiffness or strain at break [17]. Therefore, a part orientation is preferred in which the load direction is parallel to the building plane.

The quality parameters are described by mathematical models, which rely on machine parameters such as layer thickness, hatch distance or velocities and geometric information such as part volume, surface area or face orientation. To provide the geometric information, this work aims to extract these information from the integrated product model of the native 3D CAD data [18]. The integrated product model divides all product data into three information subsets: Product definition, product representation and product presentation. The machine-readable product representation is accessed in this concept. The following equation shows the mathematical model for calculating the support area of a cylindrical surface:

$$A_{S,cylinder} = 2 * \pi * r_{cyl} * h_{cyl} * \delta_Z * \frac{180^\circ - \theta_{crit}}{180^\circ}, \tag{1}$$

$$\delta_Z = \begin{cases} 1, & \gamma_i + 90^\circ > \theta_{crit} \\ 0, & \gamma_i + 90^\circ \leq \theta_{crit} \end{cases} \tag{2}$$

$A_{S,cylinder}$: support area of a cylindrical surface
r_{cyl}: cylinder radius
h_{cyl}: cylinder height
δ_z: indicator function whether support is needed
θ_{crit}: critical overhang angle
γ_i: angle between cylinder axis and build direction

Each quality parameter is calculated in a separate model to ensure an easy update for more accurate calculation models and adaptations for other additive manufacturing machines. The quality parameters are combined into a description vector to represent the quality of a specific orientation.

To solve the part orientation problem, it is necessary to deal with two subproblems: The generation of a selection of different orientations and the determination of the best orientation from this selection. The determination of the set of orientations is the first substantial building block of the developed concept. As already stated in [15], the requirements for a subset of specified surfaces are more important than the requirements of the entire part. Therefore, it is likely that the optimal orientation of the entire part is equal to the optimal orientation of one of these crucial surfaces. Consequently, the user must specify such surfaces, usually functional surfaces, mating surfaces, or other surfaces with special quality requirements. For each specified part surface, all orientations are calculated, by which the surface can be produced in optimal quality. Thus, a set of orientations is created, which is examined for its quality in the next step. For determining the optimal orientation, this paper adopts the procedure presented by *Zhang and Bernard* in [16, 19] and adapts it to the defined quality parameters. This algorithm enables multi-attribute decision making (MADM) by identifying the best candidate from a given set of alternatives. This approach leads to valid results when combined with an application and problem oriented methodology that generates the set of orientations [13].

The MADM model used is called a distance-similarity-model. For each orientation, the description vector of a given orientation is compared with the target vector, which consists of the best entries of the quality parameters of all orientations. For this purpose, the distance and the similarity of both vectors are calculated in two submodules and merged into an overall index. When calculating the distance, the influence of the different optimization parameters is not considered equally, but according to a relative weighting set by the user. This work aims to provide a solution, which allows the ranking of the orientations by their overall index. In the end, the user shall choose between the orientations and select the optimal orientation for the problem, based on the information provided.

4 Application Example

For validation purposes, an application example was implemented using the programming language C# and the NX Open API of the CAD-system Siemens NX 12 by the vendor Siemens PLM Software. Due to the complexity of freeform surfaces, this implementation only uses planar and cylindrical. Therefore, the models for the calculation of the quality parameters focus on those types of faces. The part selected for validation (Fig. 1a) was introduced in [20], successfully used in different publication and is therefore established for the validation of part orientation approaches.

The program structure is oriented to the phases of data input, calculation, and result preparation. The input phase is divided into five steps. First, the user has to select the part to be oriented. Second, a set of machine parameters has to be chosen from a list or a new set has to be created. Third, the user has to specify which quality parameters are considered by defining their relative weights. For validation three sets of parameters where used, one of which is for optimal surface quality using the parameters roughness, support area, post-processing-area and factor for important faces (Fig. 1b). Further on, only parameters with a weight greater than zero are considered. Therefore, the last two steps are only conducted if the corresponding parameters are enabled. During the fourth step, a vector or plane is defined to represent the load direction within the part. In the last step, the implementation enables the selection of crucial part surfaces and the specification of local, face-specific target surface roughness values. A global target value valid for all faces can be set as well.

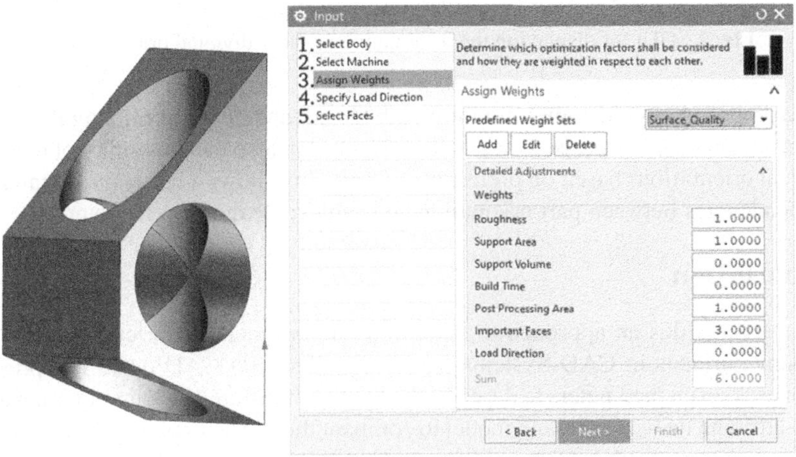

(a) (b)

Fig. 1. (a) Part used for validation (b) GUI for setting the relative weights for the quality parameters

During the calculation phase, the set of orientations is created, the selected quality parameters for every orientation are calculated, the distance-similarity-model is applied and the overall index is computed. In the last step of the calculation phase, the best orientation for each quality parameter as well as the three orientations with the best overall indices are preselected. Those preselected results are prepared for presentation and displayed as a list (Fig. 2). Two functionalities support the visual inspection of the part. They enable the coloring of faces that meet their surface roughness requirements in green and faces that do not meet their requirements in red. Furthermore, areas that require post-processing can be colorized as well. For more detailed information, the user can select a table for the estimated roughness value of each face. Eventually, the user has to select one orientation and the part is rotated into this orientation automatically.

The validation shows that the quality parameters can be significantly improved compared to those of the initial orientation. The roughness is reduced by 61%, support area by 64%, post-processing-area by 55% and factor for important faces by 33%.

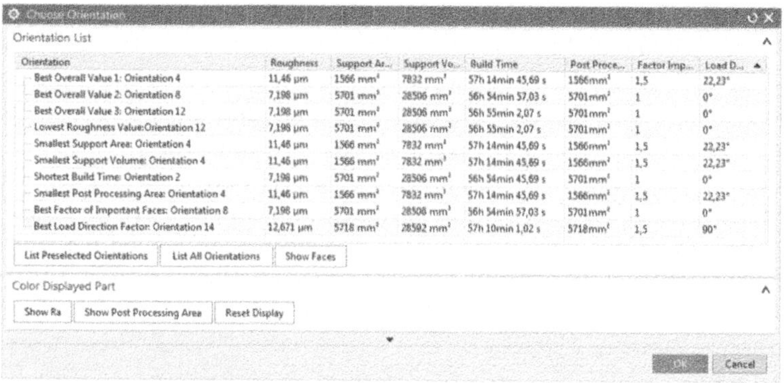

Fig. 2. GUI for displaying the information of the individual orientations.

The proposed system comes not only with the advantage of optimizing the part orientation and providing a pre-selection of best results but also the possibility of inspecting individual orientations based on preferred criteria. This allows a better understanding of the dependencies between part orientation and achievable quality characteristics.

5 Conclusion

This paper provides an approach to determine the optimal part orientation and post-processing surfaces in CAD-Systems, based on native 3D CAD data. The presented concept uses seven parameters to describe the quality of different orientations and adopts a multi-attribute decision making model to compare these orientations. The concept was implemented using the NX Open API for the CAD-system NX 12 by Siemens PLM and a validation proofed the feasibility of the concept. This validation shows improvements of quality parameters up to 64% when comparing the optimal orientation to the initial orientation.

References

1. Pham, D.T., Dimov, S.S., Gault, R.S.: Part orientation in stereolithography. Int. J. Adv. Manuf. Technol. **15**, 674–682 (1999)

2. Alexander, P., Allen, S., Dutta, D.: Part orientation and build cost determination in layered manufacturing. Comput. Aided Des. **30**, 343–356 (1998)
3. Pandey, P.M., Thrimurthulu, K., Reddy, N.V.: Optimal part deposition orientation in FDM by using a multicriteria genetic algorithm. Int. J. Prod. Res. **42**(19), 4069–4089 (2004). https://doi.org/10.1080/00207540410001708470
4. Canellidis, V., Giannatsis, J., Dedoussis, V.: Genetic-algorithm-based multi-objective optimization of the build orientation in stereolithography. Int. J. Adv. Manuf. Technol. **45**, 714–730 (2009)
5. Danjou, S.: Mehrzieloptimierung der Bauteilorientierung für Anwendungen der Rapid-Technologie. Dissertation, Universität Duisburg-Essen (2010)
6. Danjou, S., Köhler, P.: Ermittlung optimaler Bauteilorientierung zur Verbesserung der Prozessplanung in der CAS/RP-Kette. RTejournal 6 (2009)
7. Khodaygan, S., Golmohammadi, A.H.: Multi-criteria optimization of the part build orientation (PBO) through a combined meta-modeling/NSGAII/TOPSIS method for additive manufacturing processes. Int. J. Interact. Des. Manuf. (IJIDeM) **12**(3), 1071–1085 (2017). https://doi.org/10.1007/s12008-017-0443-7
8. Qin, Y., Qi, Q., Shi, P., et al.: Automatic determination of part build orientation for laser powder bed fusion. Virtual Phys. Prototyping **16**(1), 29–49 (2020). https://doi.org/10.1080/17452759.2020.1832793
9. Brika, S.E., Zhao, Y.F., Brochu, M., et al.: Multi-objective build orientation optimization for powder bed fusion by laser. J. Manuf. Sci. Eng. **139** (2017)
10. Rocha, A.M.A.C., Pereira, A.I., Vaz, A.I.F.: Build orientation optimization problem in additive manufacturing. In: Gervasi, O., et al. (eds.) ICCSA 2018. LNCS, vol. 10961, pp. 669–682. Springer, Cham (2018). https://doi.org/10.1007/978-3-319-95165-2_47
11. Ahn, D., Kim, H., Lee, S.: Fabrication direction optimization to minimize post-machining in layered manufacturing. Int. J. Mach. Tools Manuf. **47**, 593–606 (2007)
12. Ulu, E., Korkmaz, E., Yay, K., et al.: Enhancing the structural performance of additively manufactured objects through build orientation optimization. J. Mech. Des. **137** (2015)
13. Zhang, Y., Bernard, A., Gupta, R.K., et al.: Feature based building orientation optimization for additive manufacturing. Rapid Prototyping J. **22**, 358–376 (2016)
14. Di Angelo, L., Di Stefano, P., Guardiani, E.: Search for the optimal build direction in additive manufacturing technologies: a review. JMMP **4**, 71 (2020)
15. Frank, D., Fadel, G.: Expert system-based selection of the preferred direction of build for rapid prototyping processes. J. Intell. Manuf. **6**, 339–345 (1995)
16. Zhang, Y., Bernard, A.: Using AM feature and multi-attribute decision making to orientate part in Additive Manufacturing. In: da Silva Bártolo, P., Lemos, A., de Pereira, A., et al. (eds.) High Value Manufacturing: Advanced Research in Virtual and Rapid Prototyping, vol. 15, pp. 411–416. CRC Press (2013)
17. Kirchner, K., Jäschke, H., Franke, H.-J., et al.: Mechanisch-technologische Eigenschaften generativ gefertigter Bauteile in Abhängigkeit von der Bauteilorientierung. RTejournal 7 (2010)
18. Grabowski, H., Anderl, R., Polly, A.: Integriertes Produktmodell, 1. Aufl. Entwicklungen zur Normung von CIM. Beuth, Berlin (1993)
19. Zhang, Y., Bernard, A.: An integrated decision-making model for multi-attributes decision-making (MADM) problems in additive manufacturing process planning. Rapid Prototyping J. **20**, 377–389 (2014)
20. Cheng, W., Fuh, J., Nee, A., et al.: Multi-objective optimization of part-building orientation in stereolithography. Rapid Prototyping J. **1**, 12–23 (1995)

Node-Based Shape Optimization and Mechanical Test Validation of Complex Metal Components and Support Structures, Manufactured by Laser Powder Bed Fusion

Aditya Ghantasala[1](✉), Johannes Diller[2], Armin Geiser[1], David Wenzler[3], Dorina Siebert[2], Christina Radlbeck[2], Roland Wüchner[1], Martin Mensinger[2], and Kai-Uwe Bletzinger[1]

[1] Chair of Structural Analysis, Technical University of Munich, Arcisstr. 21 80333, Munich, Germany
{aditya.ghantasala,armin.geiser,wuechner,kub}@tum.de

[2] Chair of Metal Structures, Technical University of Munich, Arcisstr. 21 80333, Munich, Germany
{johannes.diller,dorina.siebert,c.radlbeck,mensinger}@tum.de

[3] Institute for Machine Tools and Industrial Management, Technical University of Munich, Boltzmannstr. 15, 85748 Garching b. München, Germany
david.wenzler@iwb.tum.de

Abstract. Vertex morphing parametrization in combination with node-based shape optimization has proven to be invaluable in improving the initial engineering designs as it offers the largest possible design space. In this contribution, these optimization procedures were applied to additively manufactured components with the aim to improve their mechanical properties. The components present the nodes of a tensegrity-tower, which is designed and realized by the Technical University of Munich and will be built at the Deutsches Museum in Munich in 2021. The nodes are highly complex connections between the compression rods and cables of the tower. They are made of aluminum and manufactured using laser powder bed fusion of metals (PBF-LB/M/AlSi10Mg). After the optimization, the nodes were printed and tested mechanically to validate and verify the numerical optimization results. Finally, an attempt to optimize the support structure required for the additive manufacturing process is presented.

Keywords: Node-based shape optimization · Additive manufacturing · Powder-based fusion of metals · Tensile testing · Strain measurement · Tensegrity-tower

1 Introduction

Design optimization has been an indispensable part of the engineering design process across many disciplines. Prominent fields include aerospace and airplane design. Topology optimization procedures are prominently used to improve the designs and automate

S. Trzcielinski et al. (Eds.): AHFE 2021, LNNS 274, pp. 10–17, 2021.
https://doi.org/10.1007/978-3-030-80462-6_2

the design process in the above disciplines. However, the application of these techniques is limited as they result in complex free-form geometries that can hardly be built by conventional manufacturing processes such as milling. Additive manufacturing techniques such as laser-based powder bed fusion of metals (PBF-LB/M) have shown the potential to change this. PBF-LB/M allows to economically produce topology-optimized components. In the course of these advancements, the construction sector is also making increasing use of optimization methods and additive manufacturing techniques such as PBF-LB/M. Manufacturing topology-optimized components using PBF-LB/M results in considerably more support structure, as the topology optimization process typically removes material, creating holes and overhangs [1].

In contrast to that, shape optimization modifies the shape of the component without changing the topological features. Thus, the shape optimization process will neither create new nor remove existing holes. Shape optimization, especially in combination with the vertex-morphing technique can further improve the engineering designs. In this context, [2–4] presented some interesting applications of shape optimization in practice. Previous works together with the original paper of [5] and the works [6, 7] introducing vertex morphing regularization in the context of node-based shape optimization have shown promising results in different applications.

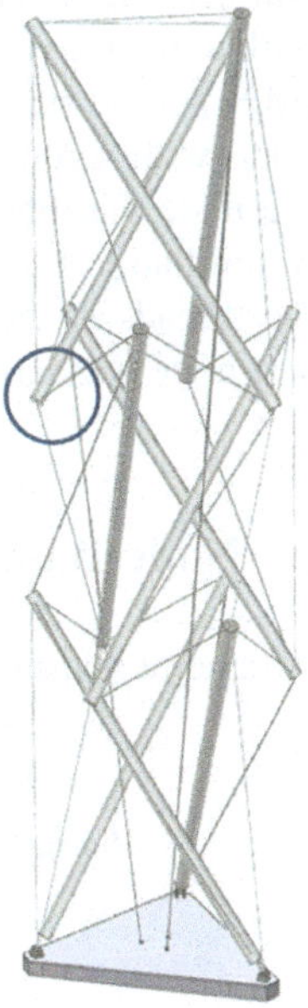

Fig. 1. Tensegrity-tower

In this paper, the workflow of applying shape optimization to a connecting node of a tensegrity-tower is shown. In Fig. 1 the global design of the tensegrity-tower is presented. This tower is to become an exhibit in the Deutsches Museum in Munich. The tower is divided in three sections, with a total height of 5 m. Each compression bar is connected to multiple tension cords (see Fig. 2). The tower contains 18 connecting nodes with four different designs.

The initial design of the nodes was done manually, followed by the shape optimization. Subsequently, the nodes were produced using the PBF-LB/M technique out of AlSi10Mg material. Figure 7 shows the node printed through PBF-LB/M. To validate the results obtained by the numerical shape optimization, the printed optimal node was subjected to mechanical testing and the experimental results were compared to the solution from numerical analysis.

2 Tensegrity-Structure Design

Tensegrity is a structural principle based on the use of isolated components in compression inside a net of continuous tension. This is achieved in such a way that the compressed members (usually bars or struts) do not touch each other and the pre-stressed tensioned members (usually cables or tendons) delineate the system spatially. The connections between the compression rods and the tension cables require nodes with a complex geometry [7].

The respective nodes of this tower require multiple physical constraints. To connect the node to the compression rod, a clearance hole with a diameter of 24.5 mm had to be kept, such that a threaded bar (M24) could be inserted and fixed to the compression rod and the node with screw nuts (see Fig. 2). The cables were connected to the node via threads (M12). Hence, a minimum wall thickness of 3 mm had to be excluded from the design space. For pre-stressing the threads of the cables, a flat support surface at the end of every sub-node was required to apply screw nuts.

Fig. 2. Close-up of the connecting node with cables attached.

3 Node-Based Shape Optimization and Vertex Morphing

A general optimization problem can be expressed as

$$\begin{array}{lll} \min\limits_{s_\Gamma} & J(X_\Gamma(s), Q(X_\Gamma)) & \\ \textit{Subjected to} & A_{i,j}s_j - X_{\Gamma,\mathrm{i}} = 0 & i = 1, \ldots, m_\Gamma, j = 1, \ldots, m_s \\ & g(X_\Gamma(s), Q(X_\Gamma)) \leq 0 & j = 1, \ldots, m_g \\ & h_j(X_\Gamma(s), Q(X_\Gamma)) = 0 & j = 1, \ldots, m_h \end{array} \tag{1}$$

where, $\Gamma \subset d\Omega$ is the design boundary of the computational domain Ω; A is a generic mapping operator which performs transformation between the design variables X_Γ and the control variables s; J, g, and h denote the objective function to be minimized, the vector of inequality constraints, and the vector of equality constraints, respectively.

For node-based shape optimization, the design variables X_Γ are the nodal coordinates of the discretization itself. This choice using nodal coordinates as the design variables has two major advantages: First, it provides the largest possible design space for optimization, and second, this approach does not require an explicit parametrization of the design boundary's geometry (Γ). This reduces the effort required to set up the optimization problem and eliminates the limitations introduced with the parametrization used. The solution of the optimization problem in Eq. (1) using gradient-based first order methods with respective shape gradients will result in not manufacturable shapes for the components. This is because of the noisy nature of the gradients. To deal with this, the *vertex-morphing* technique was applied. This defines the mapping operator A as

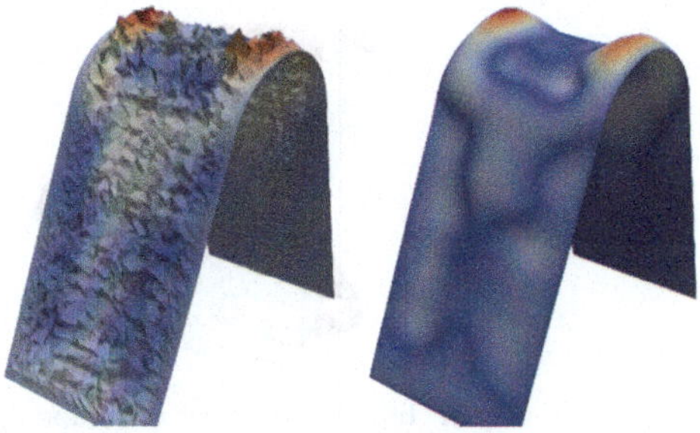

Fig. 3. Optimal shape without and with vertex morphing.

$$A = \int_{\Sigma(s,r)} F(X, X_0) s \, d\Sigma \tag{2}$$

where F is any possible filter function, r is the filter radius, and Σ is the part of Γ, which is in the radius of r to the point X_0. This filters the sensitivities producing a smooth optimal shape of the component. Figure 3 displays a comparison of optimal shapes generated with and without vertex morphing applied.

For an in-depth discussion about the properties and behavior of this technique, readers are referred to [5] and [6]. This technique is implemented in the open-source framework KratosMultiphysics [10]. In this work, this framework was used for both finite element analysis (FEA) and shape optimization procedures.

4 Shape Optimization of the Tensegrity-Node

The shape optimization of the tensegrity-node was performed with mass as objective and compliance as a constraint. This means that the shape of the node was optimized to reduce the mass by keeping the compliance of the node below the value at the beginning of the optimization process. The choice of the compliance as a constraint was motivated by the fact that it is opposing to the mass objective and is an aggregated quantity on the entire domain. This choice increases the stress in some locations, but care was taken that it stayed below the yield stress of the material. The set-up of the optimization together with the geometrical constraints applied is depicted in Fig. 3. Here, the surface subjected to optimization is shown in grey. Since the ends of each sub-node and the holes inside them had to comply with the rest of the tensegrity-tower (see Fig. 2) these surfaces were not changed during the optimization process. These surfaces can be seen in red in Fig. 3. The numerical optimization was conducted with simultaneous loading on all the sub-nodes, which was derived from the tension in the connected cables (Fig. 4).

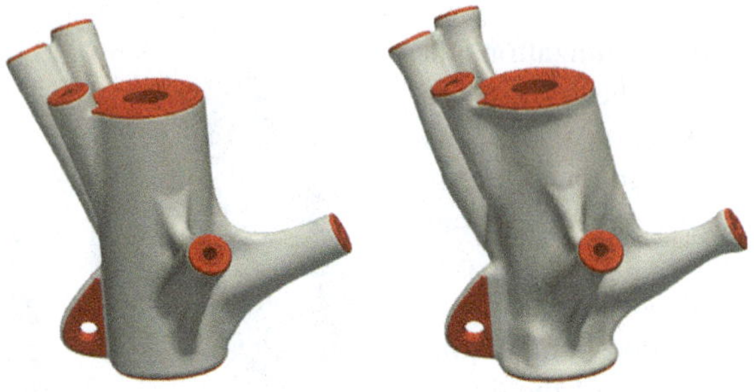

Fig. 4. Initial tensegrity-node and optimized node (25% mass reduction)

The optimization used adjoint surface sensitivities and a gradient projection algorithm. The internal volume mesh was adapted to the updated surface mesh using a pseudo-structural mesh motion. This avoided the collapse of the internal volume elements and thus a remeshing during the optimization process. Figure 3 shows the shape of the node after 12 optimization iterations. At this point, a convergence in the objective was observed with an improvement, i.e., decrease, of 25% in mass and a marginal decrease in compliance.

5 Validation of the Optimization by Mechanical Testing

After optimization, a mass reduction of 25% was achieved, while maintaining the compliance of the node as constant. To validate the applied structural optimization methods, mechanical tests were executed. This was necessary, since the results of standardized tensile specimens may differ from complex geometries. In this study, the optimized node was subjected to tensile loading (see Fig. 5). However, testing all the sub-nodes at once is not possible, as the tensile testing machine can only apply one load magnitude at a time. Thus, a spherical bearing was applied to the top to ensure a continuous tensile load on the sub-node (see Fig. 5). The lower part was tightened with load slings on a shackle, resulting in a fully hinged connection, only transmitting axial loads. Two strain gauges were applied on the curvature to measure the elastic strains. The resulting stresses were calculated and compared to the results of the numerical simulation.

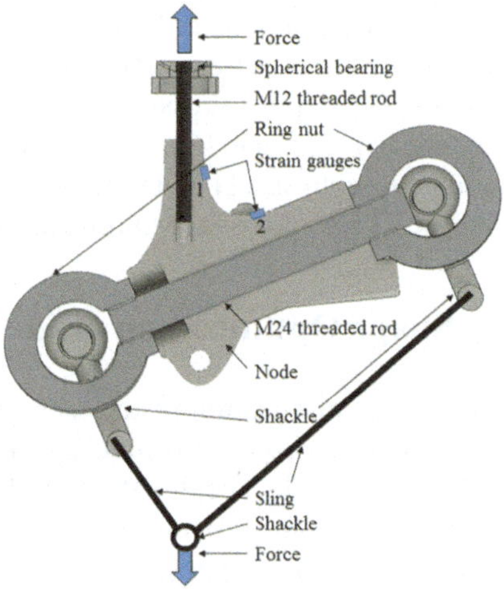

Fig. 5. Half-section of testing setup of the optimized tensegrity-node

6 Numerical Results and Comparison with Test Results

The stress values at the locations of the strain gauges were calculated by multiplying the measured strain with the material's Young's modulus. This equals 85000 N/mm^2 according to previously conducted standardized tensile tests. Though the maximum design load on this sub-node was 9.8 kN, to investigate the safety reserves, a maximum load of 40 kN was applied to the threaded rod (M12, see Fig. 6), which resulted in stresses of 45.6 N/mm^2 and 37.5 N/mm^2 on strain gauge 1 and 2. The location of the strain gauges is marked in Fig. 5.

Fig. 6. von-Mises stress distribution obtained from numerical simulation.

A numerical model was set up to simulate the conditions and forces applied in the test set-up (see also Fig. 5). Zero Dirichlet boundary conditions were applied on the connecting surfaces colored in red in Fig. 3. A cumulative force of 40 kN was applied on the nodes belonging to the surface of the hole inside the sub-node. The test set-up also restricted the bending of the sub-node and thus the bending related stresses. To realize this in the numerical model, the nodes on the surface of the hole inside the sub-node were constrained to move only in the direction of the applied force.

This resulted in the Von-Mises stress distribution shown in Fig. 5. The values corresponding to the locations of the strain gauges 1 and 2 were 58 N/mm^2 and 44 N/mm^2, respectively. These are in good agreement with the values obtained in the tests. The small differences observed in the numerical results can be attributed to modelling simplifications, particularly, exclusion of threads in the hole and resulting distribution of the applied force. The results show, that with a mass reduction of 25%, an increase of stress is measured compared to the initial node. However, the stresses were still in the acceptable range, as they were below the yield strength of 290 N/mm^2. Since the strain energy was optimized (see Sect. 4), the increase of stresses was expected. The results also strongly suggest that the safety factor must be higher than four, since no visible damage was detected after applying a load of 40 kN, which is four times higher than the design load.

Fig. 7. Printed optimized node

7 Summary and Conclusion

- A novel application of shape optimization method to additively manufactured parts was introduced.
- The method was applied to a tensegrity-tower with complex connection nodes, which was designed by TUM and will be realized as an exhibit in the Deutsches Museum.
- The nodes were shape optimized by using vertex morphing and manufactured by PBF-LB/M/AlSi10Mg, resulting in a reduction of mass by 25%.
- The optimized tensegrity-nodes were subjected to tensile loading in a fully hinged testing set-up.
- FE simulations of the testing set-up were performed with the optimized tensegrity-nodes.
- The results of the tensile testing were compared to the FE simulations.
- Though the optimized node had considerably less mass compared to the original, the mechanical properties of the optimized node were well below the safety limit.

This work establishes a workflow to link the numerical simulations and mechanical testing environments. This feedback mechanism from the tensile testing results to the numerical simulations can be used to verify and improve the shape optimization process.

In conclusion, the viability of shape optimization confirmed that it is a promising alternative to conventional (topology) optimization methods. It has the advantage that less support structure is created, because in contrast to the topology optimization, the shape optimization does not cause any holes. It was also shown that the numerical results from the optimization are in good and safe agreement with the experimental values.

8 Outlook

To further reduce the use of support structures, an algorithm is being developed to reduce overhanging structures, which are greater than 45°. Furthermore, shape optimization of the support structures using the thermal dissipation properties of conventional support structure will be investigated. The aim of this investigation is to reduce the amount of necessary support structures and to improve the thermal dissipation simultaneously.

Acknowledgements. The authors thank the Deutsche Forschungsgemeinschaft (DFG, German Research Foundation) – Project number 414265976 – TRR 277, for funding this research project.

References

1. Mirzendehdel, A., Suresh, K.: Support structure constrained topology optimization for additive manufacturing. Comput. Aided Des. **81**, 1–13 (2016)
2. Najian Asl, R., Shayegan, S., Geiser, A., Hojjat, M., Bletzinger, K.-U.: A consistent formulation for imposing packaging constraints in shape optimization using Vertex Morphing parametrization. Struct. Multidiscip. Optim. **56**(6), 1507–1519 (2017). https://doi.org/10.1007/s00158-017-1819-9

3. Nonogawa, M., Takeuchi, K., Azegami, H.: Shape optimization of running shoes with desired deformation properties. Struct. Multidiscip. Optim. **62**(3), 1535–1546 (2020). https://doi.org/10.1007/s00158-020-02560-0
4. Othmer, C.: Adjoint methods for car aerodynamics. J. Math. Ind. **4**(1), 1–23 (2014). https://doi.org/10.1186/2190-5983-4-6
5. Bletzinger, K.-U.: A consistent frame for sensitivity filtering and the vertex assigned morphing of optimal shape. Struct. Multidiscip. Optim. **49**(6), 873–895 (2014). https://doi.org/10.1007/s00158-013-1031-5
6. Hojjat, M.: Node-based parametrization for shape optimal design. Ph.D. thesis, Technical University of Munich, Munich (2014)
7. Sultan, C., Skelton, R.: Deplyment of tensegrity structures. Int. J. Solids Struct. **40**, 4637–4657 (2003)
8. Ghantasala, A., Asl, R.N., Stahl, S., Shayegan, S., Hojjat, M., Bletzinger, K.U.: Node based shape optimization for higher productivity in additive manufacturing. In: II International Conference on Simulation for Additive Manufacturing - Sim-AM 2019, Pavia, Italy (2019)
9. Jáuregui, V.: Tensegrity Structures and Their Application to Architecture. PUbliCan - Ediciones de la Universidad de Cantabria (2010)
10. Dadvand, P., Rossi, R., Oñate, E.: An object-oriented environment for developing finite element codes for multi-disciplinary applications. Arch. Comput. Methods Eng. **17**, 253–297 (2010)

A Semi-skilled Fabrication Approach of Shape-Changing Interfaces through Fused Filament Fabrication

Robin Vandormael[1](✉), Kamie Leten[1,2], Marieke Van Camp[1], Jouke Verlinden[1,2], and Regan Watts[1]

[1] Department of Product Development, Faculty of Design Sciences, University of Antwerp, Antwerp, Belgium
{robin.vandormael,kamie.leten,marieke.vancamp,jouke.verlinden, regan.watts}@uantwerpen.be

[2] Faculty of Industrial Design Engineering, Delft University of Technology, Delft, Netherlands

Abstract. The common additive manufacturing techniques like fused filament fabrication (FFF) routinely produce physical, rigid structures. But using this production technique for manufacturing flexible structures with high-end materials such as thermoplastic polyurethane (TPU) is more difficult. Because of its difficulty, the fabrication of these structures requires higher-end machinery, time-intensive fabrication, and skilled users. Therefore, we focus on the malleable dynamics of a rigid thermoplastics with mid-range FFF technology to expand the design-space of shape-changing interfaces and propose a fabrication approach for it. As a result, the intended user, for example a creative designer, can also integrate shape-changing interfaces of rigid thermoplastics in their designs, much sooner than if constrained by an FFF printing platform. In a first phase, we experiment with different materials through an iterative design-based process. In a second phase, we perform an explorative design-case study to test the material's flexibility and the fabrication approach. The research is concluded with an approach proposal, discussion and future work.

Keyword: Fused filament fabrication · Shape-changing interfaces · 3D printing · Additive manufacturing · Interaction design · Design experience

1 Introduction

We aim to investigate the fused filament fabrication (FFF) technology of rigid thermoplastics for shape-changing interfaces through an iteration-based study. FFF techniques facilitate the production of complex design artefacts with unique aesthetics and functional properties [1]. In the field of shape-changing systems, the concept of 4D printing is also discussed [2]. It permits 3D printed structures to include a dimension of time. As a result, it can change their form under the influence of environmental factors such as temperature, illumination, pressure etc. [2]. But 3D printed materials and structures mostly lack mechanical properties for 4D printing. To overcome these limitations, researchers

S. Trzcielinski et al. (Eds.): AHFE 2021, LNNS 274, pp. 18–26, 2021.
https://doi.org/10.1007/978-3-030-80462-6_3

have investigated the flexible capabilities of FFF, for example with printing layers as thinly as possible. The garments designed by the Iris van Herpen atelier, Catherine Wales and Michael Smidt in collaboration with architect Frances Bitonti are examples hereof which demonstrate the flexible capabilities of FFF technology for computational design [2, 3]. The textile properties of these woven structures with FFF printed structures have already been determined [3–5]. Fabricating these garments require specific skills, a time-intensive production process and a considerable amount of manual labor. The Foliage Dress by Iris van Herpen for the Paris Fashion Week 2018 took 260 h to 3D print, only for the parts, excluding the assembly [1, 6]. Therefore, we developed and propose a semi-skilled fabrication approach based on primarily using the malleable dynamics of FFF technology for printing upon woven fabrics to expand the design space of shape-changing interfaces for the creative intended user.

We try make this technique accessible for the early majority of the intended users who own mid-range 3D printers and could fabricate the proposed hybrid material with low effort. As an opportunity, the hybrid material will consist of rigid thermoplastics conventional materials to increase its flexibility and its appeal. In this study, we demonstrate the following. In a first phase, we investigated the malleable dynamics, materials and fabrication opportunities of FFF by performing an iterative prototyping process in five phases. After the iteration process, a fabrication approach is proposed with a prototyped hybrid material as a result. In a second phase, we conducted an explorative design case study to demonstrate the potential of this hybrid material.

2 Methods

An iterative-design process suited this type of research best, due to the trial-and-error approach and low-end to high-end prototyping of the hybrid material. In total, five iterations were performed. Each iteration focuses on a different aspect. Several tools and machinery were used to cut or pre-fabricate each material. To speed up prototyping with each iteration, fabrication was initially done using handheld tools, but once the design prototype was finalised, we switched to using machinery such as laser cutters and 3D printers. The first iterations were performed without FFF printing to ensure quick prototyping. Wooden veneer was used as replacement material for the PLA for the FFF tiles in these iterations.

2.1 Experimental

The materials in all five iterations can be divided in three main categories: fabrics, wood, and thermoplastics. A transparent, cellulose nitrate-based adhesive, applicable to all wood-based materials and most types of plastic, is not included. Cotton fabric and nylon tulle were used in the fabric category. They were all cut manually using scissors. Three types of wooden veneer were included in the wood category: oak, bamboo and meranti, each with a thickness of 1 mm, cut with a laser cutter. In the final category, only the thermoplastic polylactic acid (PLA) is used as filament to 3D print structures. The 3D models of the structures were designed with CAD software, namely Solidworks 2019. Once modelled, the structures were exported as Standard Template Library (STL) files and imported in slicer software. The slicer software used is PrusaSlicer, because one of the mid-range FFF printer, here the Prusa MK3s (Prusa Research, Czechia). The second mid-range FFF printers was the Makerbot Replicator 2 (United States). The parameters of both 3D printers were the extruder temperature, set at 215 °C, and the printing bed temperature, set at 60 °C.

2.2 Iterative Design Solutions

To make the hybrid material as applicable as possible for the case study described in Sect. 3, the test set-up resembles a cylindrical shape with the same dimensions. The fabrics, which are fully flexible, form the base of the hybrid material to which either the wooden or PLA tiles are glued. The design of the tiles, formerly 3D structures, determines the foldability and the look of a corrugated effect. This design is based on earlier research on foldable structures, such as origami, the art of making objects by folding sheets of paper into shapes, miniature origami robots and tessellation mathematics [7, 8]. The foldability parameter in this experiment is measured as the Z-length reduction, as seen in Fig. 1. As objective for the iterations, the Z-length reduction must be at least 50% compared to the original Z-length.

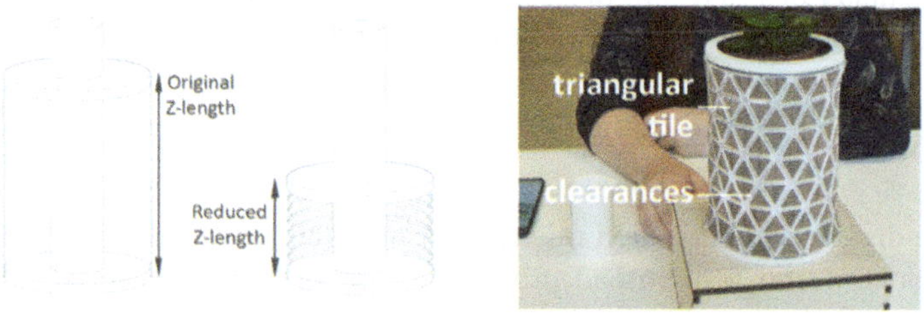

Fig. 1. Z-length reduction – tessellated triangular design

We decided on a triangular tessellation for following reasons. Firstly, the two strongest shapes possible are a circle, on the first place, and a triangle, on the second place. However, the advantage of a triangle is that the folding lines, seen in origami, are already defined. Secondly, a triangle possesses nine degrees of freedom; six corner translations and three corner normal rotations. With each corner added, rectangles, pentagons, hexagons etc., the number of degrees of freedom increases. And thirdly, as a linear force is applied to stretch or compress the hybrid material in the Z-direction, all force is distributed on the tiles down the sides to the triangular joints, similar to principles in truss design. For instance, when using a rectangular tessellation, the linear force could also be distributed across the diagonals, resulting in in-plane bending. In-plane bending occurs even more frequent when polygons with more corners are used.

Iteration One. The first iteration focuses on the exploration of the low-end materials, cotton fabric and oak veneer. The hybrid material resulting from this iteration is used as a baseline. ***Design process:*** The size of one triangular tile was set at 20 mm to 17 mm. Seven rows and five columns of these tiles results in a size of 100 mm to 119 mm, the size of the cotton fabric as well. ***Prototype:*** The tiles were laser cut out of oak, meranti and bamboo veneer. Doing so, an intricate design led to a more appealing look. Within 2 min, 100 tiles could be laser cut, called the cut-to-tile ratio further on. The tiles, 63 in total, were glued on the cotton fabric, one by one. This assembly took one hour. The clearance between the tiles, as can be seen in Fig. 1, was 2 mm. ***Experiment:*** The targeted Z-length reduction was 0.34 equal to 34% (x).

$$(\text{Original Z-length} - \text{Z-length reduction})/\text{Original Z-length} = \text{x}. \qquad (1)$$

Analysis: The small tiles and large clearances set a Z-length reduction baseline of 34%. However, the assembly time took too long. Using three wooden veneer types created a more appealing look but will only be an option as the user sees fit.

Iteration Two. Here, a reduction of the assembly time is most pressing. This is more specifically done by adjusting the laser cutting process. ***Design process:*** The width of the clearance increased, hoping to improve the Z-length reduction. Seven rows and six columns of oak veneer tiles changed the hybrid material's size from 120 mm to 123 mm. The tiles, however, were still connected after cutting due to nodes between them. From this point on, the triangular tile and hybrid material sizes remain the same in each iteration. ***Prototype:*** The tiles were laser cut in oak veneer. The cut-to-tile ratio increased slightly. The entire piece was glued on the cotton fabric and each node was broken. This assembly took 15 min, a reduction by 300% compared to iteration one. ***Experiment:*** The targeted Z-length reduction was only 0.11 equal to 11%. ***Analysis*:** Introducing nodes reduced the assembly, but at the cost of the Z-length reduction, because it prevented the foldability. The adhesive also did not adhere to all tiles. Nodes are therefore not the solution.

Iteration Three. Although cotton fabric is flexible, it proved to be too stiff for a hybrid material. Therefore, it is replaced with nylon tulle and 3D printed tiles. Inspiration for this technique can be found in the work of Iris van Herpen [3]. The goal is to decrease labor and assembly time by snap fitting the oak veneer tiles without adhesive. ***Design process:*** The design of the 3D printed tile consisted of a full printed base and a rim acting as a holder for the oak veneer tiles. ***Prototype:*** The nylon tulle was attached to the acrylic build plate using paper tape. The tiles were printed directly on the nylon tulle. The PLA melted around the nylon mesh threads, encasing it. The printing time was 124 min at a 0.1 mm layer thickness. The oak veneer tiles were laser cut and fitted in place. The assembly took six minutes. ***Experiment:*** The targeted Z-length reduction was 0.23 equal to 23%. ***Analysis*:** Using FFF printing, the positioning of the triangular tiles and the clearances were much more accurate. The triangular tiles did not adhere good enough to the nylon tulle. The absence of adhesive and too little friction caused some oak veneer tiles to fall out the triangular tiles (Fig. 2).

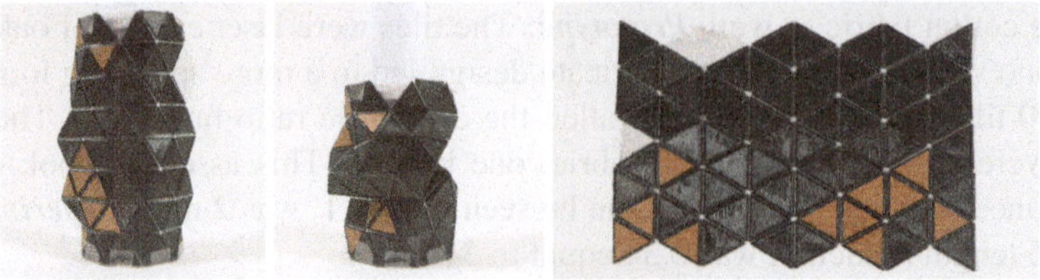

Fig. 2. Iteration Four: Experiment and top view hybrid material

Iteration Four. The triangular tiles must adhere better to the nylon tulle by increasing the friction between the triangular tiles and the oak veneer tiles. ***Design process:*** The center of 3D printed tile was removed in the design to decrease printing time. The rim was still 1 mm wide, but had a smaller tolerance between the rim and the oak veneer tile. ***Prototype:*** A new mid-range printer was used, a Prusa MK3s. As for the fabrication process, firstly, the bases of the triangular pieces were 3D printed. Secondly, midway, the print was paused and the nylon tulle was clamped on the metal build plate using eight magnets. Thirdly, the rim was 3D printed on top, adhering to the bases below with the nylon tulle in between. The printing time was 89 min. The oak veneer tiles were laser cut and fitted in place. The assembly took six minutes. ***Experiment:*** The targeted Z-length reduction was 0.34 equal to 34%. ***Analysis*:** The triangular tiles adhered much better to the nylon, as can be seen below. Some of oak veneer tiles however did still fall out. The clearances must still be larger to reach a targeted Z-length reduction of 50%.

Iteration Five. The last problems to be solved were increasing the Z-length reduction to 50% and preventing loose oak veneer tiles. ***Design process:*** The rim needed to be thicker, to create more friction for the oak veneer tiles. Therefore, the base was printed thinner, so that the number of layers remained the same. The clearances were 2 mm. ***Prototype:*** As with iteration five, magnets were used to hold the nylon tulle in place. The printing time was again 89 min. The oak veneer tiles were laser cut and fitted in place with adhesive on the back. The assembly took twelve minutes. ***Experiment:*** The targeted Z-length reduction was 0.58 equal to 58%. ***Analysis*:** The clearances were large enough to achieve the 50% Z-length reduction and the oak veneer tiles did not fall off because of the friction and adhesive (Fig. 3).

Fig. 3. Iteration Five: Experiment - top view hybrid material - fabrication

2.3 Results

An overview of the parameters of each iteration can be found below. Afterwards the fabrication process of the hybrid material is explained in detail as a guide for the intended end-users (Table 1).

Table 1. Iteration overview

Iteration	Materials	Shape	3D Print time (minutes)	Assembly time (minutes)
1	Cotton; oak, bamboo and meranti veneer; adhesive		/	60
2	Cotton; oak veneer; adhesive		/	15
3	PLA; nylon tulle; oak veneer		124	6
4	PLA; nylon tulle; oak veneer		89	6
5	PLA; nylon tulle, oak veneer; adhesive		89	12

Fixed Dimensions of Materials. To ensure a repeatable experiment with the hybrid material, following dimensions must be maintained. Nylon tulle: 122 mm to 123 mm. 56 triangular tiles in PLA: 20 mm to 17 mm to 1.1 mm. The diameter of the PLA is 1.75 mm. 56 wooden veneer tiles: 19 mm to 16 mm to 1 mm.

Hybrid Material Fabrication. The fabrication is done in four steps with a combination of PLA and nylon tulle. The 3D printer used here is the Prusa MK3s.

1. Generate the CAD model consisting of 56 triangular tiles spaced in seven rows and four columns, like the design in iteration five. Each tile consists of a base, 0.3 mm high, and a rim, 0.8 mm high. The clearances between the tiles are 2 mm.
2. Use PrusaSlicer to convert the model to Gcode and select PLA as material. "Print three layers at 0.1 mm layer height. Then, pause the print.
3. Place the nylon tulle on top and secure it with eight magnets (four magnets on the corners and four magnets at the centre of each side).
4. Print the remaining eight layers at 0.1 mm layer height on top of the nylon tulle and so encasing it between the prints.

Hybrid Material Customization. This process is optional. Multiple wooden veneer types can be used to customize the hybrid material to the users taste. The customization is done in three steps. The machinery used, is a laser cutter BRM 6090 (BRM Lasers, Netherlands), but can be done by hand.

5. Convert only the inner edges of the rim of the CAD model to a DXF-file. The BRM 6090 software sends the DXF-file to the laser cutter. Set the speed to 100 mm/s and the power to 60%. It takes one minute to laser cut 56 wooden veneer tiles or 30 min to cut them by hand.
6. Once done, snap fit the wooden veneer tiles in the triangular tiles. The rim creates enough resistance to hold the wooden veneer tiles in place.
7. Flip the hybrid materials and use transparent cellulose nitrate adhesive to glue the wooden veneer pieces to the nylon tulle. The gap in the base of the triangular tiles allows for an area large enough for the adhesive to be applied.

In total, the fabrication time is 102 min; 89 min to 3D print, one minute to laser cut and 12 min to assemble. The handwork, however, only takes 12 min.

3 Design Exploration

3.1 Case Study

In the second phase, we conducted an explorative design-case study with the objective to test the material's flexibility and the fabrication approach. Here, the hypothesis was to test the aesthetics of the hybrid material as a shape-changing interface, implemented in an intuitive flowerpot design. When water evaporates and nutrients are cut short, the plant starts to shrink, corrugate and eventually collapse. The experiment was conducted with six participants, all schooled in Interaction Design. They were asked to evaluate three interactive motions and if the hybrid material was suitable here for.

The hybrid material for this experiment had a size of 133 mm to 268, with seven rows of 20 triangular tiles on it. With this length, a full cylindrical shape could be made with a diameter of 150 mm. Inside the flowerpot, a linear mechanism moved the hybrid material vertically between its stretched and the corrugated position. A servo motor, controlled by an Arduino, powered the actuated motion in a Wizard of Oz test-up. In real-life, the actuated motion would be controlled by environmental sensors. LED's inside the flowerpot aided the interactive motions through colour changes. Using a variable resistor, operated by the observer, the servo could be controlled (Fig. 4).

Fig. 4. Test set up (left: stretched position - right: slightly corrugated position)

3.2 Experiment and Verification

The three interactive motions are a stretched position, a corrugated position and a pulsating motion. We try to anthropomorphise the flowerpot through these motions.

Stretched Position. The flowerpot finds itself in its normal state. A white LED-light implicates a natural feeling. The purpose of this interaction is to communicate that the plant is healthy; enough water, nutrients and sunlight.

Corrugated Position. The flowerpot finds itself in a corrugated position. It has shrunk, thereby creating a corrugated effect indicating that the plant has dried-out because one of nutrients is missing. The red LED-light implicates danger. The participant needs to replenish the growth factor.

Pulsating Motion. The flowerpot pulses between the stretched position and the corrugated position of 25% instead of 58%. This interactive motion simulates joy. It occurs when the plant wants attention or when one or multiple growth factors of the plant is replenished. This pulsating motion tested the hybrid material to its limits.

4 Conclusion and Future Work

The fabrication approach of this hybrid material can be used by creative intended user, but it is only applicable on small scale. With larger clearances, the Z-length reduction, can even be greater than 58%, but this results in the hybrid material being more see-through. Field of application are, for example, the fashion industry as proven by Iris van Herpen, engineering or advanced shape-changing interfaces [3]. Experiments on different specifications, increasing size, communicative properties etc. of this hybrid material does however need further research. Moreover, measurable specifications such as robustness, yield strength, e-modulus etc. also need to be determined.

References

1. Doubrovski, Z., Stokhuijzen, A.N., van Herpen, I., Schuddeboom, P., et al.: Fabricating the foliage dress: computational design in couture. In: Proceedings of SCF 2018, ResearchGate, New York, pp. 1–2 (2017). https://doi.org/10.1145/nnnnnnn.nnnnnnn
2. Cambridge Dictionary: 'Origami', Cambridge Dictionary, Cambridge Dictionary, Cambridge (n.d.)
3. Miyashita, S., Guitron, S., Lidersdorfer, M., Sung, R.C., Rus, D.: Fabricating the foliage dress: computational design in couture. In: Proceedings of 2015 IEEE International Conference on Robotics and Automation (ICRA), Seattle, WA, USA. IEEE (2015)
4. Partsch, L.N., Vassiliadis, S., Papageorgas, P.: 3D printed textile fabrics structureres. In: Proceedings of 5th International Istanbul Textile Congres 2015: Innovative Technologies "Inspire to Innovate", TEI Piraeus, Athens, Greece, p. 1 (2015)
5. Pernet, D.: The process behind the making of the Iris Van Herpen Couture Collection 'Ludi Naturae' (2018). ASVOF. https://ashadedviewonfashion.com/2018/01/29/236151/. Accessed 02 Nov 2019

6. Sabantina, L., Kinzel, F., Ehrmann, A., Finsterbusch, F.: Combining 3D printed forms with textile structures - mechanical and geometrical properties of multimaterial systems. In: Proceedings of 2015 Global Conference on Polymer and Composite Materials, IOP, vol. 8, no.1 (2015). https://doi.org/10.1088/1757-899X/87/1/012005
7. Tibbits, S., Linor, S., Dikovsky, D., Hirsch, S.: 4D printing: multi-material shape change. In: Proceedings of TED conference 2013, pp. 1–6. MIT, Cambridge (2013). https://doi.org/10.1002/ad.1710
8. Vanderploeg, A., Lee, S., Mamp, M.: The application of 3D printing technology in the fashion industry. Int. J. Fashion Des. Technol. Educ. **10**(2), 170–179 (2017). https://doi.org/10.1080/17543266.2016.1223355

Printing Small: Measuring the Resolution Limits of CLIP Technology

Addis Tessema[1(✉)], Justin Mecham[1], Kimon Iliadis[1], Sue J. Mecham[1], Robert Pinschmidt[1], and Joseph M. DeSimone[1,2]

[1] DeSimone Material Group, Department of Chemistry, University of North Carolina, Chapel Hill, USA
{addis,sjmecham,pinschrk}@email.unc.edu, joe@carbon3d.com
[2] Carbon Inc., Redwood City, CA, USA

Abstract. This study seeks to understand the impact of process variables, especially UV light distribution, on the 3D fabrication of micro features at the optical resolution scale. In CLIP 3D technology, a Digital Micromirror Device (DMD) is used to create a pattern of projected UV light by controlling each pixel independently. However, as the light outputs of the projected pixels are partially overlapping, there are challenges in obtaining the targeted size and shape of microscale features. From our study, 1. the applied exposure has a direct effect on the size of small printed positive features; 2. smaller sized negative features are even more sensitive and require a careful tuning of the applied exposure to successfully print desired fine structures. Overall, the applied exposure, particularly the UV dosage is critical in the accurate fabrication of micro features that incorporate negative and positive features.

Keywords: Digital light processing · 3D printing · Micro-fabrication · Photo polymerization · Print resolution

1 Introduction

Additive Manufacturing (AM) has shown fascinating growth in the last couple of decades [1, 2]. Currently, 3D printing is considered as the first choice for rapid prototyping in the AM sector. 3D printing technology gives extreme design freedom, greatly reduced fabrication times, and cost reductions in manufacturing of complex structures and micro devices [2]. Currently, 3D printing technologies can be categorized into four main groups [2]: fused deposition manufacturing (FDM), selective laser sintering (SLS), inkjet 3D printing [3], and stereolithography (SLA) [4–10].

SLA follows a layer-by-layer build scheme to construct a 3D object using photopolymerizable liquid resin. The printing process starts from designing the 3D model of the part in STL format. The model is then converted to sliced (very thin) 2D layers (typically 1 um–100 um) and these slices are printed gradually into 3D objects. Light is projected continuously for each slice onto the surface of a liquid photo-curable resin and solidifies the resin in a desired pattern. Continuous Liquid Interface Printing (CLIP) is a novel and

S. Trzcielinski et al. (Eds.): AHFE 2021, LNNS 274, pp. 27–34, 2021.
https://doi.org/10.1007/978-3-030-80462-6_4

powerful method relative of SLA, where the normally problematic oxygen inhibition during the polymerization process is turned to an advantage [11]. Since its development, CLIP 3D technology has been used for a variety of applications, from medical device fabrication [12, 13], sports equipment, and aerospace to automotive industries, for both rapid prototyping and direct industrial floor production. In the CLIP system, the bottom side of a resin vat is designed with an oxygen permeable window, and the light projected from underneath passes through this window (see Fig. 1 (a)). Due to the uniquely permeable window, there is abundant oxygen in the resin near the window, even during UV exposure, but the concentration of oxygen falls rapidly as the distance from the window increases. A high concentration of oxygen inhibits polymerization of the resin near the window to maintain a very thin layer of liquid resin, also known as the Dead Zone (DZ), between the window and curing resin. The CLIP process provides two main advantages, firstly, it preserves a continuous interface between printed layers, providing isotropic cure and properties in printed objects. Secondly, it eliminates sticking of printed parts on the window and enhances flow of fresh resin towards the center of the window from the side as the build platform is raised for the next layer. The result is the dead zone reduces printing time by 25–100% compared to conventional printing techniques and improves final part properties.

The DLP [digital light processing] device used in CLIP technology is a Digital Micromirror Device (DMD), an array of independently controlled reflective aluminum plates. The resolution limit of a CLIP printer is defined by the light projected from a single pixel (mirror) of the optical setup. Using available high quality large micromirror arrays, CLIP is well suited for fabrication of microscale parts and complex structures. CLIP has been applied to produce parts with micro featured architecture such as microneedles, micro-lattice structures and micro membranes.

However, it can be a challenge to obtain the targeted size and shape for micro scale features at the scale of its defined resolution. In theory, there are two major factors that can undermine the precise fabrication of small features: the cure-through effect and the scattered light from adjacent pixels. In the DMD operation, light reflected from a single pixel mirror is not perfectly sharp and confined within the pixels' size but expands beyond the desired projection area (following a Rayleigh Gaussian distribution). This is known as pixel bleeding [7, 9]. Thus, in cases where there are multiple adjacent active pixels, there is bleeding of light from one pixel into the neighboring pixels as shown in Fig. 1(b). The cumulative light from multiple pixels shows a significant discrepancy from the theoretical output of individual pixels, that induces over-curing of certain areas (grey region in Fig. 1(b)) and through-cure into prior layers. These circumstances impact the accurate fabrication of micro sized features (mainly features sized between 1–5 pixels). Therefore, it is vital to understand and take into consideration the effect of the cumulative light distribution during accurate microscale printing.

In the case where negative features (like holes, channels or dents) are printed, a cure-through effect is a major issue to address. The cure-through effect occurs during a layer-by-layer curing where a layer with a hole (an intentionally uncured region) is followed by a solid slice layer, the light projected to cure the solid slice layer can penetrate to the prior slice and solidify the uncured resin in the hole region. As a result, it can completely or partially change the geometric shape of layers with holes.

This study is aimed to understand how 3D fabrication of microscale features at the scale of the print resolution will be affected by the process challenges, mainly flux bleeding (scattered light) and cure-through, in the case of CLIP 3D printing technology. In this work, both positive and negative micro-features are considered.

The study designated rectangular thin bars as positive features and rectangular channels as negative features. Further, a chessboard structure was designed to combine the effects of both features. Printing parameters were considered as experiment variables and expected to affect the magnitude of bleeding flux (scattered light) and cure-through. With this intention, print parameters: applied UV-intensity (I) and printing speed (s) in continuous mode, are taken under consideration. The measured feature sizes obtained for the different print parameters are presented.

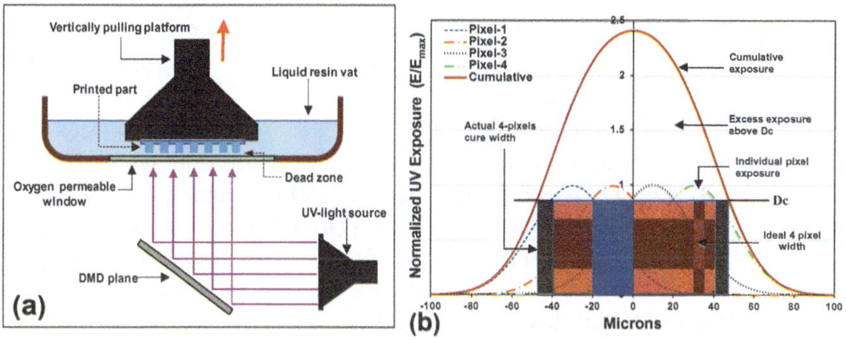

Fig. 1. (a) Schematic diagram of CLIP 3D printing technology, and (b) A graph that shows the light/exposure distribution in each pixel and their cumulative/total distribution

2 Material and Methods

A custom photon-initiated resin with controllable resin photochemistry characteristics was formulated. 350 MW poly-ethylene glycol di-methacrylate (PEGdMA-350), was mixed (by weight) with 3% of diphenyl(2,4,6-trimethylbenzoyl) phosphine oxide (TPO), 2% of 2-isopropylthioxanthone (ITX) and 1% benzotriazole basic light stabilizer (BLS)), using a plate shaker. BLS materials are light absorbing compounds that have been used to reduce light scatter and cure-through within the resin. TPO and ITX are intended to serve as photo-initiators and also enhance photo absorption in the resin. The final formulation has UV absorption (α) = 0.01/um and curing dosage (Dc) = 20 mJ/cm^2, where the curing dosage (Dc) is the (minimum) amount of photon energy required to accurately cure a many pixel spot of the resin to a desired depth, and UV absorption (α) is the measure of UV light penetration through the resin.[13]

A prototype 3D printer (S1-Lassen) from Carbon, Inc. is used in this study and this printer is designed to have a pixel resolution of 20 um. The printer is operated using user-machine interface software developed by Carbon, Inc. This software is used to upload the 3D model STL files and generate the 3D slice model. Further, during printing, the software enables the user to adjust the printing parameters (printing speed (S), UV light intensity (I), and slice thickness (S_t)). For these experiments, three categories of designs were created and studied: a positive feature model (slender bars as shown in Fig. 2(a)), a negative feature model (long rectangular channels as shown Fig. 3(a)), and a combination of both (checkerboard structure (i.e. pattern of solid squares placed next to square holes) as shown in Fig. 5(a)). Each of these designs are printed at different printing speeds (S) and UV intensities (I). Following the completion of the print, the part is removed from the printer and washed thoroughly using IPA, either before or after removing from the build platform, to eliminate residual resin from the surface. Finally, the printed sample is post-cured using a FireJet© UV lamp (UV wavelength of 345–385) at 50% intensity for 3 min. The samples were evaluated by laser confocal microscopy for quality and feature size measurement.

3 Results

3.1 Positive Features (Solid Parts)

The For positive features at resolution scale, a slender bar/line with width/thickness of 3-pixels (~60 um), 5-pixels (~100 um), 8-pixels (~160 um) and 10-pixels (~200 um) are considered (see Fig. 2(a)). These single lines are printed using different printing parameters, UV intensity and printing speed. In this experiment, the printing speed (S) and UV intensity (I) are varied, so the overall UV energy increased correspondingly. The width/thickness of the printed bars was measured using a confocal laser microscope, shown in Fig. 2(b).

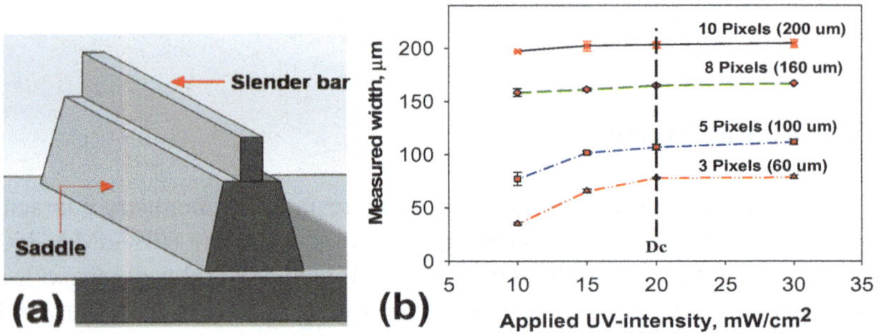

Fig. 2. (a) Schematic diagram of positive features, and (b) Measured width of bars with different width vs printing exposure

It was noted that, for exposure (E = I × t) below the resin curing dosage (Dc = 20 mJ/cm^2), the final print had a smaller width than its design, which indicates under-curing. On the other hand, when the applied exposure was higher than Dc, less significant changes in the width were observed as exposure was increased. For instance, for the 3-pixel width sample, the printed lines were seen to experience under-cure (size of 40 um), however, samples were consistent with their design size of 60 um in the over-exposure region (exposure > 20 mJ/cm^2). Furthermore, it appeared that samples with thinner width are much more sensitive to change in applied exposure. The thickness variation with exposure was much less for the thicker bar (10-pixels) than for the thin bar (3-pixels). Overall, this phenomenon is directly related to the cumulative UV light distribution across multiple pixels. For a large number of adjacent pixels, the cumulative effect dominates and the exposure to cure will be $\geq$Dc, while smaller clusters of exposed pixels are under-cured. However, in the case where there are a small number of adjacent pixels, the cumulative effect of adjacent pixels is much less, and severe under-curing can result. In summary, the lower cumulative exposure of individual or small clusters of exposed pixels defines the printed size (under-cure) for small features ($\leq$5-pixels). Whereas, for large size features, the dominant cumulative effect explains the width consistency in 8 and 10 pixel-wide samples showing small under- and over-curing.

3.2 Negative Features

In order to study micro negative features, plates with multiple microchannels (shown in Fig. 3(a)) were printed at ranges of applied UV intensity (I) and exposure ($E = I \times t$). Samples were intended to have through-channel widths of 3-pixels, 4-pixels and 5-pixels (60, 80 and 100 um). The channels and the plate were printed vertically to minimize the cure-through effect, so the scattered light issue could be addressed separately. Because of scattered light from pixels in the plate walls, the channels were frequently over-cured, leaving them narrower than planned and often closed. The normalized channel depth (measured depth of channels into the plate (d) divided by measured thickness of the base plate (t_p)) of channels printed at different applied UV intensities (I) (while the printing speed was maintained the same) are presented in Fig. 3(b). Here, it should be noted that some of the printed channels showed through-channels (completely open across the thickness, normalized depth = 1), while others showed discontinuity (solid bridge at the middle of the channel normalized depth < 1, see Fig. 4(b)). For the samples printed at uniform speed, the channel depth was seen to decrease with increasing UV-intensity, which clearly illustrated the effect of scattered light. For smaller channels (where the walls are closer to each other), more than half the depth of the channel was solidified, even though the total light was significantly less than Dc for all but the 20 mW/cm^2 exposure. This shows that smaller gap/negative features are more sensitive to changes in applied intensity than wider channels.

Figure 3(c) depicts the normalized depth of samples printed at constant exposure (the speed and intensity were varied proportionally). Despite a constant total exposure, the 3- and 4-pixels channels still decreased in channel depth (extent of over-cure) with increases in the applied UV intensity. However, channels with widths $\geq$ 5-pixels showed full through-plate thickness channels regardless of increases in intensity (for the record, samples with channels wider than 5-pixels were printed and all of them showed a through-(normalized depth = 1) channel). Explanations for the increasing cure with higher intensity/faster speed at constant total exposure include 1. changes in the kinetics of resin chain propagation and termination with intensity, and 2. reduced time for diffusion of oxygen into actively polymerizing areas from near-by under-illuminated areas, causing less oxygen inhibition.

To evaluate the plate thickness effect on over-curing due to active pixels along the thickness, the design was remodeled with different plate thickness. Figure 4(a) shows

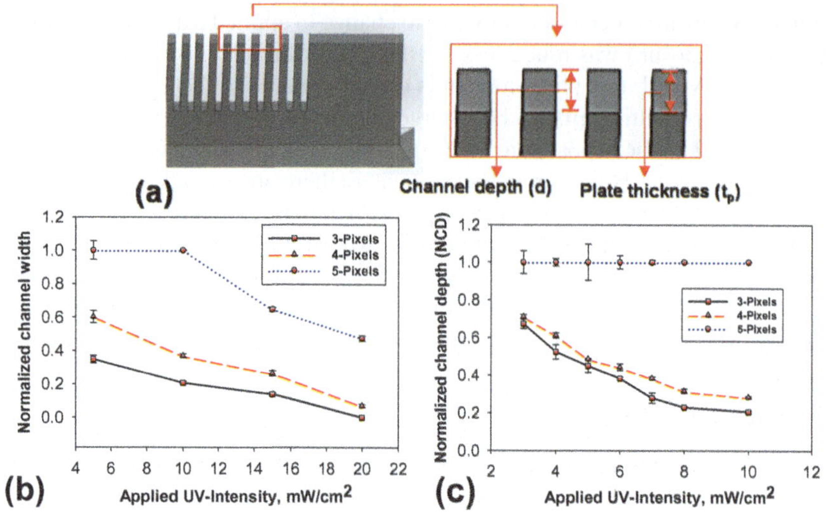

Fig. 3. (a) Schematic diagram of negative feature/micro-channels, (b) Normalized channel depth vs applied UV intensity (I) at constant printing speed (4mm/hr), and(c) Normalized channel depth vs applied UV intensity (I) at constant printing exposure

the normalized depth of micro channels obtained for 200, 300, 400 and 500 um thickness plates. it was observed that the normalized depth of the channel decreased with increasing thickness of the plate. This result was consistent for two different values of applied UV-intensity. It is presumed that reducing the plate thickness reduces the number of active pixels, which will lead to a decrease in the cumulative scattered peak intensity and facilitate over-cure in the micro channels (see Fig. 4(b)).

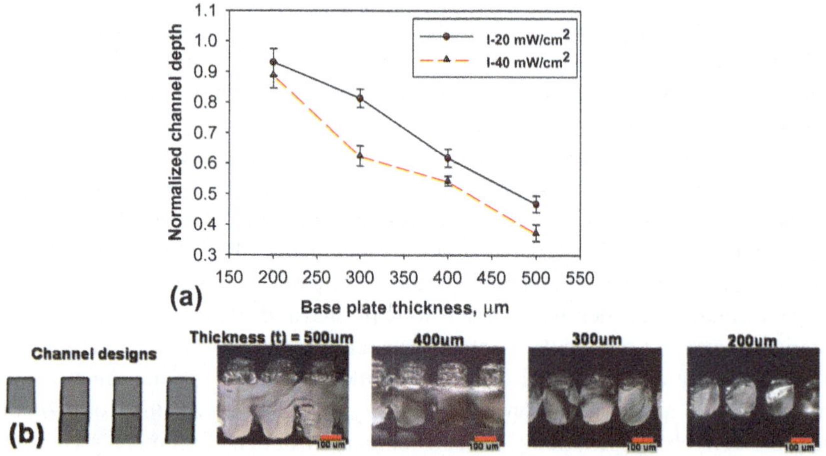

Fig. 4. a) Variation in depth of the channel along with the thickness of the plate, and (b) Schematic and microscopic images of cut micro channels with different plate thicknesses

3.3 Combined Chessboard Structure

In the third group of experiments, checkerboard designs featuring hole (negative) and block (positive) features (as shown in Fig. 5(a)) were printed. Checkerboards were built with unit size (square positive and negative features) of 3-pixels (60 um), 4-pixels (80 um), 5-pixels (100 um), 8-pixels (160 um) and 10-pixels (200 um) and a thickness of 200 um and printed at a series of UV intensities (same printing speed of 4 mm/hr). At the lowest applied UV intensity of 4 mW/cm^2 (exposure ~20% of Dc for this resin), the samples came out with through-holes (normalized depth of 1), except the 3-pixel samples. As the applied intensity increased gradually, samples with unit size 3-pixels and 4-pixels showed a decrease in normalized depth of the negative feature, while the other samples showed through-holes. The high sensitivity of the 3 and 4-pixel samples to the change in applied UV intensity is similar to the result obtained in the negative feature experiments. Samples with unit size of 5-, 8-, and 10-pixels showed through-holes for most of the applied UV intensity range. Therefore, during printing of geometries with combined features, special consideration should be given to the scattered light around the negative features.

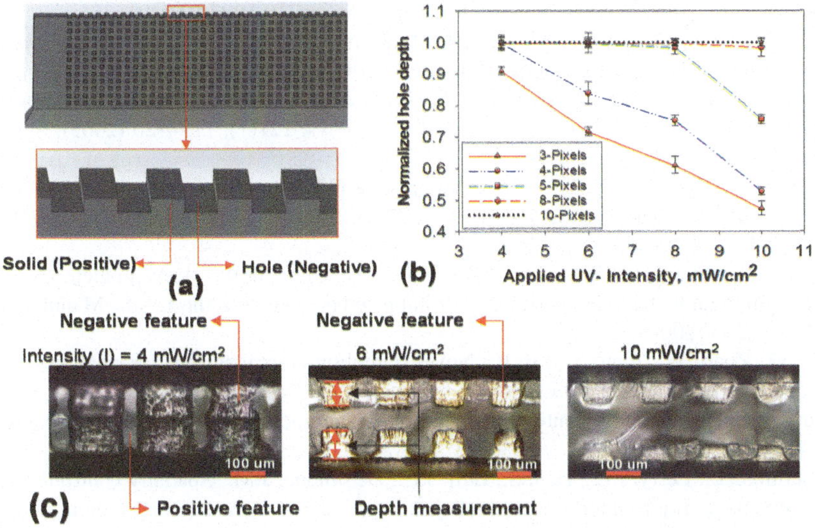

Fig. 5. (a) Schematic diagram of combined negative and positive features, (b) Normalized hole depth vs applied UV intensity (I) at constant printing speed (s), and (c) Microscopic images of cut 4-pixel checkerboard sample with different applied intensity (I)

4 Conclusion

Experimental work on an S1 CLIP 3D printer for printing small, high resolution objects revealed that the optical quality of light projection is very critical in accurate micro-printing. The scattering of projected light from nearby pixels induces a significant cumulative effect that causes resin curing that leads to inaccurate features. Thus, calibrating light distribution in a given photo-curing system is crucial to the precise printing of designed micro-structured parts. Furthermore, printing parameters which influence the overall energy exposure, light intensity, and exposure time, need to be adjusted to reduce the effect of the scattered light and cure-through effects.

References

1. Wang, B.: The future of manufacturing: a new perspective. Engineering **4**, 722–728 (2018)
2. Ngo, T.D., Kashani, A., Nguyen, K.T.Q., Hui, D.: Additive manufacturing (3D printing): a review of materials, methods, applications and challenges. Com. Part B: Eng. **143**, 172–196 (2018)
3. Yan, Q., Dong, H., Su, J., Song, B., Wei, Q., Shi, Y.: A review of 3D printing technology for Medical applications. Engineering **4**, 729–742 (2018)
4. Bertsch, A., Jezequel, J.Y., Andre, J.C.: Study of the spatial resolution of a new 3D microfabrication process: the microstereo photolithography using a dynamic mask-generator technique. J. Photoch. Photobio. A **107**, 275–281 (1997)
5. Choi, J.W., Ha, Y.M., Lee, S.H., Choi, K.H.: Design of Microstereolithography system based on dynamic image projection for fabrication of three-dimensional microstructures. J. Mech. Sci. Technol. **20**(12), 2094–2104 (2006). https://doi.org/10.1007/BF02916326
6. Han, D., Yang, C., Fang, N.X., Lee, H.: Rapid multi-material 3D printing with projection micro-stereolithography using dynamic fluidic control. Addit. Manuf. **27**, 606–615 (2019)
7. Sun, C., Fang, N., Wu, D.M., Zhang, X.: Projection micro-stereolithography using digital micro-mirror dynamic mask. Sens. Actuators A: Phys. **121**(1), 113–120 (2005). https://doi.org/10.1016/j.sna.2004.12.011
8. Zhang, X., Jiang, X.N., Sun, C.: Micro-stereolithography of polymeric and ceramic microstructures. Sens. Actuators A: Phys. **77**(2), 149–156 (1999). https://doi.org/10.1016/S0924-4247(99)00189-2
9. Choi, J.S., Kang, H.W., Lee, I.H., Ko, T.J., Cho, D.W.: Development of micro-stereolithography technology using a UV lamp and optical fiber. Int. J. Adv. Manuf. Technol. **41**, 281–286 (2009)
10. Xu, G., Zhao, W., Tang, Y., Lu, B.: Novel stereolithography system for small size objects. Rapid Prototype J. **12**, 12–17 (2006)
11. Tumbleston, J.R., et al.: Continuous liquid interface production of 3D objects. Science **347**, 1349–1352 (2015)
12. Caudill, C.L., Perry, J.L., Tian, S., Luft, J.C., DeSimone, J.M.: Spatially controlled coating of continuous liquid interface production microneedles for transdermal protein delivery. J. Controlled Release **284**, 122–132 (2018)
13. Johnson, A., et al.: Single-step fabrication of computationally designed microneedles by continuous liquid interface production. PLOS ONE **11**(9), e0162518 (2016). https://doi.org/10.1371/journal.pone.0162518

Development of an Elastic Inflatable Actuator for Active Seating Systems

Peter Gust, Sebastian Kampa(✉), Marco Kuhlmeier, Christian Sauder, and Alina Sersch

Chair of Engineering Design, University of Wuppertal, Gaußstraße 20, 42119 Wuppertal, Germany
{peter.gust,sebastian.kampa,marco.kuhlmeier,christian.sauder, alina.sersch}@uni-wuppertal.de

Abstract. In this paper, the development of an elastic inflatable actuator (EIA) for the generation of unconscious sitting discomfort to encourage users towards dynamic sitting is discussed. The EIA is printed with fused filament fabrication (FFF) of flexible thermoplastic polyurethane (TPU). In order to adapt the actuator to the requirements, a finite element analysis (FEA) is built up, which is based on tensile tests of the TPU. The FEA is then compared with force-displacement measurements of the printed EIA. It is shown that the simulation represents the experiment, as long as similar conditions are ensured. Due to the given use case, an almost static load can be assumed, which is represented in the simulation. In this context, the simulation model can thus be utilized for a subsequent optimization of the actuator for the application in intelligent seating systems.

Keywords: Soft robotics · Finite element analysis · Dynamic sitting · Intelligent seating system · Human factors · 3D printing

1 Introduction

Static sitting is one of the main reasons of musculoskeletal disorders. The first physiological effects can already be felt after a duration of 15 to 30 min [1]. Seated workplaces should therefore be designed to promote frequent changes in posture, which is called dynamic sitting. One approach is the introduction of strain through automatic and prospective actuator movement within the seat surface dependent on the user's individual sitting posture and behavior [2, 3]. The aim is to create subconscious sitting discomfort, that leads to a change of posture.

Pneumatic actuators are already successfully used in active seating systems, e.g. in the automotive industry, for individual adjustment of the seat and backrest, as lumbar support or for massage applications. In addition, there are already some health maintenance systems that promote dynamic sitting [4, 5].

The possibilities for executing the movement of soft robotics can generally be categorized into two classes. On the one hand, fluidic actuators are operated with hydraulic

S. Trzcielinski et al. (Eds.): AHFE 2021, LNNS 274, pp. 35–42, 2021.
https://doi.org/10.1007/978-3-030-80462-6_5

and pneumatic drives [6, 7]. On the other hand, electric actuators use the material properties of smart materials to exert a force [6]. EIA, as part of soft robotics, are suitable for integration into seats or generally for human-machine interaction, since they do not interfere with the natural movement of the human body [8]. EIA are compliant and therefore intrinsically suitable for contact with soft tissue [9]. They are characterized by running on a pressurized gas or liquid [10]. Their variable stiffness ensures inherently safe operation [10, 11], because the material can adapt to the human body and does not lead to painful stress concentrations or point contacts [8, 10]. It is also of great advantage that there are no electric fields, currents or heat interacting with the human body [9, 12].

2 Methods

2.1 Requirements

In previous papers, tests were carried out in which the required displacement and force of the actuator were determined by using a load cell, a linear guidance and a displacement transducer [1, 2]. A fully automated test bench was built on the basis of an office chair [2]. Since the subconscious sitting discomfort cannot be consciously indicated, the threshold value at which the strain could be felt and localized by the participant was recorded. The determined data are thus to be understood as maximum force and displacement required for the actuator. The tests were carried out using three different velocities: 25, 50 and 75 mm/min [2]. The results of these measurements show that an average of 32.78 N and 20.20 mm are required [2].

2.2 Design of a Soft Actuator

According to [8], soft actuators can be classified into four groups depending on their direction of movement: contracting, expanding, twisting and bending actuators. In this paper, monolithic expanding actuators are investigated due to the given use case. The four types of EIA differ in terms of their boundary conditions [3]. Linear actuators have a symmetrical cross-section, whereby the pressure contracts or expands the material and thus creates unidirectional movement of the actuator [3, 4]. The EIA are basically consisting of bellows- or balloon-like structures, which vary in arrangement and size. Depending on the desired direction of movement or number of degrees of freedom, several motion mechanisms can be combined with each other in addition to the use of materials with different stiffnesses or thicknesses [5]. The motion path is thus predefined in the actuator design [3]. To enable motion, both positive and negative pressure can be used. An example of negative pressure actuators are linear soft vacuum actuators as used by [6], where a monolithic rotationally symmetrical bellows is used, which contracts under negative pressure. If positive pressure is applied, the folds of the bellows of the actuator open and thus cause a linear expansion. Other inflatable actuators have an non-symmetrical cross-sections, which allows bending or twisting [5]. The asymmetry causes a bending when the structure is inflated [3]. An inextensible, but flexible bottom layer is needed to enable the bending due to the different compliance of the material [7].

In the use case a high strain on the material is required. Furthermore, in order to reduce the fatigue and failure of a soft robot, it is important to limit the deformation and load to 70% of the maximum permissible load of the material [8]. To achieve the required displacement, materials with large elastic deformability are required [4, 9]. According to [9], five types of materials are suitable for 3D printing: elastomers, flexible/elastic materials, smart composites, shape memory polymers and shape memory alloys. Since 3D printing will be done using FFF, TPU is used, which is already established for soft robotics [6, 9–12].

2.3 Material Tests

In the material tests the three materials NinjaTek NinjaFlex, Recreus FilaFlex and BASF Ultrafuse are being tested. The properties of the material are determined in a tensile test according to ISO 37 [13]. For this purpose, five type 2 samples each with 0°/90° and 45°/135° filling were punched out of printed mats. To optimize the print quality, the material is dried at 40 °C for at least 24 h before printing. The samples are printed on a Prusa i3 MK3S. Prior to testing, the samples are stored at room temperature for at least 24 h. In the tests all materials are tested on a Zwick Z010 tensile testing machine. Since NinjaFlex shows the lowest stress, it is used for further tests. It can be seen that the material orientation has only a minor effect on the characteristic curve and so the mean values can be used. Furthermore, the test speed has a significant influence on the overall results. The slower the tests are performed, the lower the stress is at the same strain. Figure 1 shows the characteristic curve of the mean values of NinjaFlex at 200 mm/min. Figure 2 shows the Mooney-Rivlin parameters determined from the tensile tests.

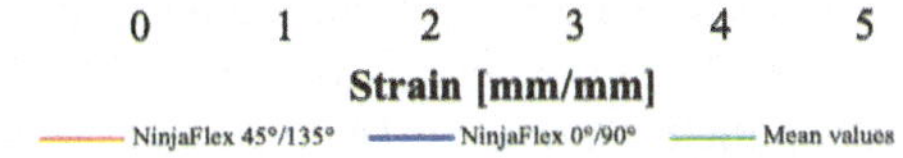

Fig. 1. Means of the tensile tests of NinjaTek NinjaFlex at 200 mm/min

	Test Speed in mm/min		
	100	200	500
C10 (MPa)	-0,233	-2,194	-4,896
C01 (MPa)	2,562	5,3953	9,2449
C20 (MPa)	0,116	0,0302	0,0715
C11 (MPa)	-0,561	-0,129	-0,365
C02 (MPa)	0,9	0,8004	1,9541
D1 (MPa^{-1})	0	0	0

Fig. 2. Mooney-Rivlin parameters of NinjaTek NinjaFlex

2.4 FEA Model

In order to predict the individual characteristic of the EIA, a parameterized CAD model is built up in Catia V5, which is then transferred to Ansys Workbench via the bidirectional CAD-Nexus interface. The nominal geometry of the EIA is shown in Fig. 3 and is designed to meet the requirements for the intelligent seating system as well as for 3D printing. In consideration of these requirements the EIA has a given height $h = 35$ mm, an outer diameter $D = 60$ mm and base diameter $d = 58$ mm. The folds of the bellows

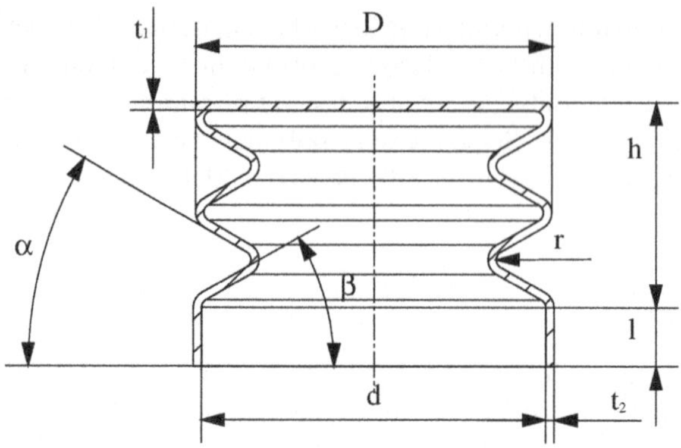

Fig. 3. CAD model of the EIA

have an angle $\alpha = \beta = 30°$ to the horizontal. In addition, the EIA has a constant wall and top thickness $t_1 = t_2 = 1{,}4$ mm, radius $r = 1{,}5$ mm and a base length $l = 10$ mm.

The simulation of the EIA is done by a FEA in Ansys Mechanical. The model is connected to the ground by a fixed support on the inside and outside of the base section of the EIA. In the first load step, a constant pressure p is applied to all surfaces inside the EIA. Subsequently, a rigid body is moved down 60 mm with a constant velocity to compress the EIA. The initial position of the rigid body is 40 mm above the top surface of the EIA under atmospheric pressure. While the rigid body is moved, the force reaction in longitudinal direction is calculated. Due to the different initial expansions caused by varying the applied pressure from 10 kPa to 100 kPa – with a step size of 10 kPa – the location of the first contact shifts. Therefore, the total displacement is hold fixed so that the change in length Δs can be compared between the different pressure settings.

Due to the design of the EIA the model is reduced to a quarter of the full geometry and a cyclic symmetry condition is implemented to reduce the effort of calculation. The model is discretized by applying a mesh to the geometry that covers the model evenly. To achieve this structured mesh along the perimeter of the EIA hexahedral and wedge elements alongside the MultiZone method are used. The maximum length of the elements is 0.02 mm in radial direction to ensure that the results are independent from the mesh. Due to the design as well as the application of the EIA, different nonlinear characteristics have to be considered in the simulation. In this context the material behavior is described by the nonlinear stress-strain curve which is determined in 2.3. This approach requires a hyperelastic material model. As in [6], the approach of Mooney-Rivlin represents the stress-strain curve best and is therefore used to describe the material [14]. For the interaction of the EIA and the rigid body as well as the individual folds of the bellows frictionless contacts are defined by using the pure penalty algorithm for each contact region individually. In addition, the large elastic deformation is another characteristic, which requires the consideration of geometric nonlinearity. The evaluation of the simulation results is carried out by evaluating the maximum displacement in longitudinal direction, the maximum Von-Mises stress as well as the maximum elastic strain for the first load step. During the second load step the maximum Von-Mises stress, the maximum elastic strain and the force reaction on the rigid body in longitudinal direction along its travel is evaluated.

2.5 Experiment

The validation of the simulation model is done by a test where the characteristic curve of the 3D printed EIA is recorded and compared with the simulation results. The tests are carried out on a fixed surface and an ABB IRB 4600 industrial robot for the linear movement. The force is measured with an ABB load cell at 100 Hz and a resolution of 0.1 N. The load cell is also used as contact surface to ensure that the actuator is in full contact with the load cell. During the experiment, the actuator is fixed on an aluminum base with a hose clamp. The base is equipped with a 1/8″ Festo pneumatic hose connection to which a 6 mm Festo pneumatic hose is attached. The pressure is set manually by using a Mader digital pressure gauge type 60 with an accuracy of ±2% FS which is mounted behind a compressed-air maintenance unit. The force-displacement measurements are performed under the same conditions as for the simulation. The load cell is moved down with a constant velocity. As soon as an increasing force is recorded, contact to the EIA is assumed. The correlating displacement value is used for the determination of the expansion due to the applied pressure. The load cell is moved down until the actuator is compressed to the bottom dead center located 60 mm below the initial position and then returned to the starting position of the measurement. In contrast to the static simulation, the force-displacement measurements in the experiment are recorded for both compression and expansion to describe the characteristic curve of the EIA. Figure 4 shows the test setup.

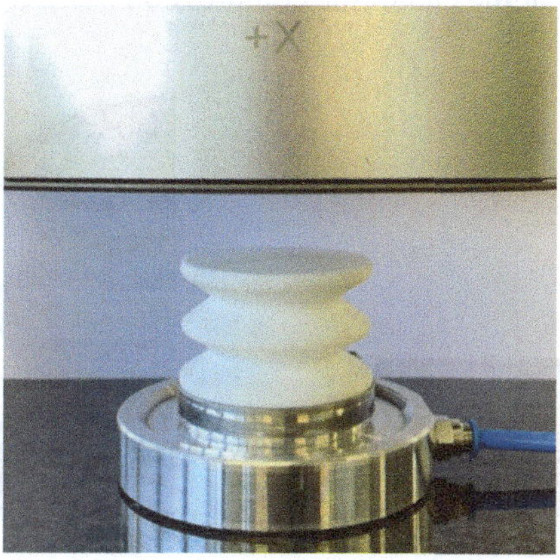

Fig. 4. Experimental force-displacement measurement of the EIA

3 Results and Discussion

Figure 5 shows a comparison of the measurement results and the simulation results for the maximum displacement, depending on the applied pressure. The use case requires low forces associated with low pressures. Considering the results, it can be seen that the simulation with the material data of tests with 200 mm/min best matches the measurement results in the range of 10 – 50 kPa.

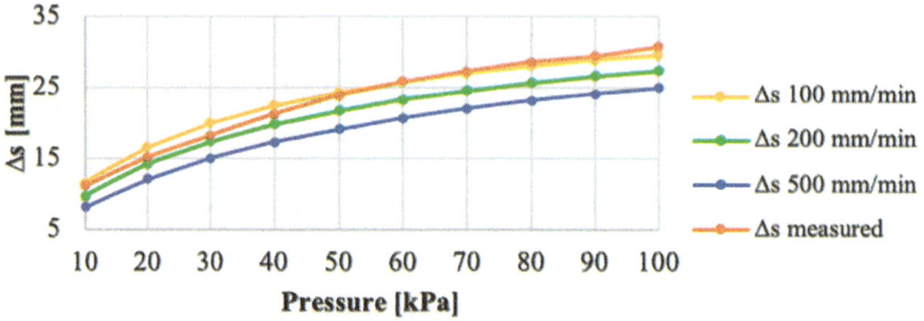

Fig. 5. Comparison of the sumulation results and the measurements

The results of the FEA are described for an applied pressure of 50 kPa exemplary. Figure 6 shows the Von-Mises stress for the inflation and compression of the EIA. The maximum longitudinal elongation is 22.50 mm at the top center. The maximum Von-Mises stress during the inflation of 4.20 MPa occurs at the outside of the top inner bend. After the compression the maximum Von-Mises stress in this location is 4.83 MPa. Figure 7 shows the results for the elastic strain. The maximum elastic strain during the inflation is 23.51% at the inner radius at the top of the EIA. The maximum elastic strain after compression can be found at the outside of the lower inner bend with 28.50%. The maximum force reaction at the end of the compression is 133.04 N.

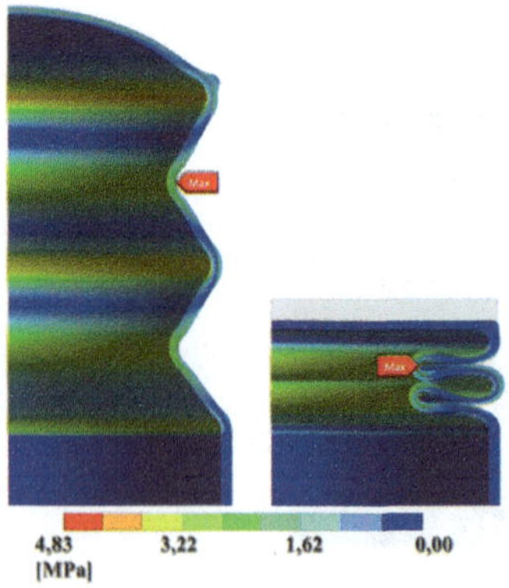

Fig. 6. Von-Mises Stress after inflation and compression with an applied pressure of 50 kPa

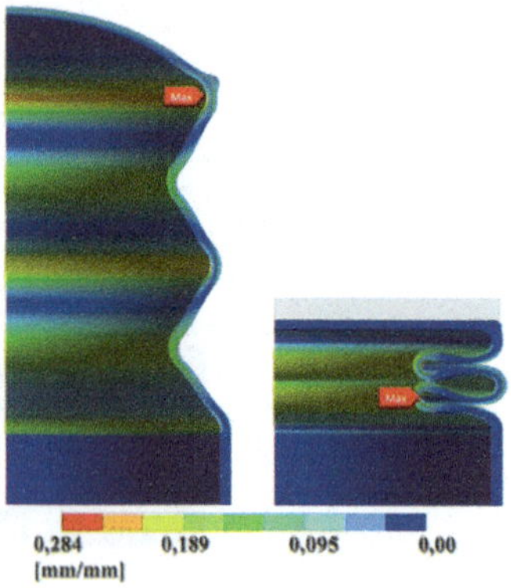

Fig. 7. Elastic strain after inflation and compression with an applied pressure of 50 kPa

As a result of the resolution, eleven characteristic curves are compared, first the simulation in Fig. 8, then the test in Fig. 9. The simulation and the experiment are structured in such a way that the same experiment is represented. Due to the setup, the displacement is shown negatively. The zero point is to be understood at the top dead center. The greatest displacement and force occur at the bottom dead center. At the beginning, a steep rise of the curves can be seen, which becomes increasingly flattened in the wide course. In the upper area, relative force increase is small before the folds of the EIA reach block. Reaching the block force is represented by a steep rise of the curve in the end. Since the pressure inside the EIA remains constant throughout the simulation, a uniform characteristic curve is obtained for both compression and expansion. The characteristic curves from the experiment show a counterclockwise progression and are very smooth in a range from 30 kPa to 80 kPa. Starting at 90 kPa, the curves show a sequential behavior during compression, which is more intense with increasing pressure. Thereby the upper fold is compressed at the beginning and afterwards the lower fold. Furthermore, the characteristic curves show a decreasing force of the EIA at high pressures at −40 mm and −55 mm, which corelates to the sequential folding.

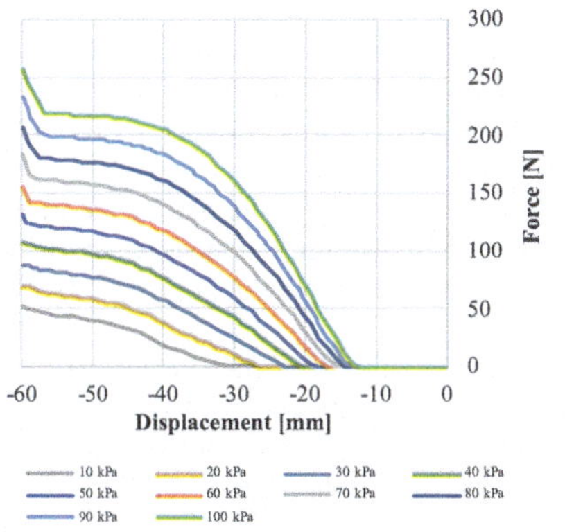

Fig. 8. Calculated force-displacement diagram

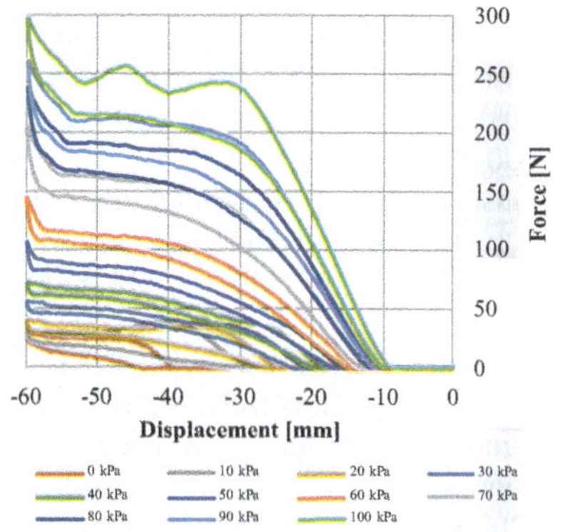

Fig. 9. Measured force-displacement diagram

In contrast to the static simulation, the characteristic curves of the tests show a hysteresis, whereby during compression, a higher force acts in vertical direction than during expansion of the actuator. This hysteresis is related to a pressure increase in the test, which is caused by a non-regulated pneumatic system when the actuator respectively the air in the actuator is compressed. The observation of the experiment shows that the hysteresis behavior as well as the block force increases with increasing pressure. During compression, the values of expansion are lower due to an almost constant pressure inside the EIA. This effect can be reduced by decreasing the speed in the experiment.

4 Conclusion

The results of this work show that monolithic bellows-like structures with symmetrical boundary conditions are basically suitable for use in active seating systems. Depending on the scaling and the geometry of the actuator, the required characteristic behavior can be achieved. The mechanical behavior of an EIA shows a hysteresis, which is caused by changes in inner pressure during compression and expansion. Due to the material used and its properties, an increasing deformation of the top surface can be observed depending on the applied pressure, whereby plastic deformations already occur after a few load changes. As the tests have shown, EIA exhibit a dynamic behavior which depends not only on the type of load but also on the test setup. These influences are not considered in a static analysis. For this reason, the tests must be carried out under quasi-static conditions to correspond to the FEA results. The material data must also be determined in an almost static test, which correlates to slow speeds. If this is ensured, the behavior of an EIA can be sufficiently predicted with a static FEA. With the results obtained in this paper, a design of experiments (DOE) will then be carried out to conduct a sensitivity study. In a further step, an optimization an endurance tests of the actuator are planned.

References

1. Gust, P., Kampa, S., Feller, N., Vom Stein, M., Haase, I., Virzi, V.: Intelligent seating system with haptic feedback for active health support. In: Arai, K., Bhatia, R., Kapoor, S. (eds.) FTC 2018. AISC, vol. 880, pp. 856–873. Springer, Cham (2018). https://doi.org/10.1007/978-3-030-02686-8_64
2. Gust, P., Kampa, S., Mersch, F., Kuhlmeier, M.: Determination of the perception threshold between conscious and subconscious sitting discomfort. In: Ahram, T., Falcão, C. (eds.) AHFE 2020. AISC, vol. 1217, pp. 815–821. Springer, Cham (2020). https://doi.org/10.1007/978-3-030-51828-8_108
3. Gorissen, B., Reynaerts, D., Konishi, S., Yoshida, K., Kim, J.W., De Volder, M.: Elastic inflatable actuators for soft robotic applications. Adv. Mater. **29**(43), 1–14 (2017)
4. Zentner, L., Böhm, V., Minchenya, V.: On the new reversal effect in monolithic compliant bending mechanisms with fluid driven actuators. Mech. Mach. Theory **44**(5), 1009–1018 (2009)
5. Gorissen, B., Chishiro, T., Shimomura, S., Reynaerts, D., De Volder, M., Konishi, S.: Flexible pneumatic twisting actuators and their application to tilting micromirrors. Sens. Actuators A Phys. **216**, 426–431 (2014)
6. Tawk, C., Alici, G.: Finite element modeling in the design process of 3D printed pneumatic soft actuators and sensors. Robotics **9**(3), 52 (2020)
7. Mosadegh, B., et al.: Pneumatic networks for soft robotics that actuate rapidly. Adv. Funct. Mater. **24**(15), 2163–2170 (2014)
8. Alici, G.: Softer is harder: what differentiates soft robotics from hard robotics? MRS Adv. **3**(28), 1557–1568 (2017). https://doi.org/10.1557/adv.2018.159
9. Yap, Y.L., Sing, S.L., Yeong, W.Y.: A review of 3D printing processes and materials for soft robotics. Rapid Prototyp. J. **26**(8), 1345–1361 (2020)
10. Yap, H.K., Ng, H.Y., Yeow, C.H.: High-force soft printable pneumatics for soft robotic applications. Soft Robot. **3**(3), 144–158 (2016)
11. Hu, W., Li, W., Alici, G.: 3D printed helical soft pneumatic actuators. In: 2018 IEEE/ASME International Conference on Advanced Intelligent Mechatronics (AIM), vol. 2018-July, pp. 950–955 (2018)
12. Gul, J.Z., et al.: 3D printing for soft robotics – a review. Sci. Technol. Adv. Mater. **19**(1), 243–262 (2018)
13. International Organization for Standadization (ISO): ISO 37:2017-11 - Rubber, vulcanized or thermoplastic - Determination of tensile stress-strain properties (2017)
14. Mooney, M.: A theory of large elastic deformation. J. Appl. Phys. **11**(9), 582–592 (1940)

3D Printing Design Applications

Reusable Kit for 3D Printable Sustainable Cutlery

Massimo Di Nicolantonio[1], Emilio Rossi[2(✉)], Antonello Vespoli[1], and Giovanni Annese[3]

[1] Department of Architecture, University of Chieti-Pescara, Viale Pindaro 42, 65127 Pescara, Pescara, Italy
m.dinicolantonio@unich.it, antonello.vespoli@outlook.it

[2] Lincoln School of Design, University of Lincoln, Brayford Pool, Lincoln LN6 7TS, UK
erossi@lincoln.ac.uk

[3] COESUM, Idea & Sviluppo Srl, Via Erasmo Piaggio 75, 66100 Chieti, Chieti, Italy
design@coesum.com

Abstract. The relationship between plastics and disposable domestic products like cutlery has a long history in Design. For long time, the use of thermoplastics has been seen as the most cost-effective solution to meet the market demands with an affordable solution 'for all'. Only in last decades the society got the awareness that this way of producing products is unsustainable for many reasons: environmental wastes, ecological threats, pollution and lack of lifecycle strategies. An alternative creative approach for the sustainable design and manufacturing of cutlery is needed to overcome this unsustainable business model. In this study, we present the research and design results that have allowed to proposing an original sustainable 3D printed solution for mitigating the use of disposable plastic in cutlery sector.

Keywords: Sustainable 3D printing · Human centered design · Sustainable design · Sustainable production · Cutlery · Design experimentation

1 Introduction

3D Printing process, technologies and tools are rapidly becoming pervasive both in industrial and in informal contexts. However, the risk to have new unsustainable printing processes and production's behaviors is high and, potentially, can increase the environmental threats. On the other hand, Design for Sustainability is a discipline that works, since late 80's, on the mitigation of productions-related environmental footprint and, recently, on the development of socio-technical systems and distributed solutions empowering both environmental aspects and socio-economic ones [1].

The recent market demands have significantly reshaped the current approaches on sustainable productions; this aspect is today crucial for the Design discipline, which is intrinsically links end-users' needs, production technologies and entrepreneurial competitiveness. Therefore, 3D Printing can be considered an opportunity to test and verify new sustainable business models aimed to mitigating the use of disposable plastics.

S. Trzcielinski et al. (Eds.): AHFE 2021, LNNS 274, pp. 45–52, 2021.
https://doi.org/10.1007/978-3-030-80462-6_6

Today, a large part of the 3D Printing sector uses plastics. Consequently, one of the related environmental threats is the extremization of the objects' obsolescence, as a reflection of the idea of 'disposable' products. Three additional roots have fed the production and the disposal of plastic waste, generating the current 'plastic problem': 1) Quantity: a disproportionate amount of plastic is produced every year; the world production moved from 15 million tons in 1964 to over 310 million tons today [2]. 2) Overall environmental dispersion and contamination: the amount of plastic dispersed in the ecosystem [3]. 3) The 'throwaway culture' as a reflection of the idea of modernity. These factors, along with the plastics' slow degradability and its reduced reuse connote disposable products as the main cause of environmental problems to which the future society will be asked to look for an alternative solution.

Today it is very hard to separate consumerism behavior from the 'plasticulture', since this relationship has become an intrinsic aspect of people's lifestyle. Single-dose objects have brought to light real needs that, however, plastics can no longer respond; this is generating more problems than solutions. Our society needs to look carefully to valid alternatives that will not necessarily be limited to replacing plastics with bioplastics, bamboo or any other eco-material. Addressing the cultural root to this problem is the only solution [4] to make in practice the slogan 'producing less and better'.

In line with the discussion started within the Design for Sustainability scientific domain, which aims to propose alternative solutions to mitigate the environmental threats whilst increasing the social and manufacturing awareness on the design of creative eco-efficient solutions, this paper proposes the research and design results for the cutlery sector, which is considered as one of the most representative examples of everyday use products. In particular, by recognizing the need of an alternative approach for the sustainable design and manufacturing, this paper presents a sustainable 3D printed solution for mitigating the use of disposable plastic in cutlery sector. This paper is part of the studies started in 2018 on 'Sustainable 3D Printing' [5] and on the 'Integration between Human Centered Design (HCD) and 3D Printing' [6].

2 Strategies and Opportunities for Sustainable 3D Printing

The development of new sustainable 3D printed products needs a new approach [5]. By combining new research models matching different levels for sustainable innovations, which are composed by primary and secondary levels of interaction, it is possible to develop a new scenario for Sustainable 3D Printing; this scenario defines a new set of design opportunities for Sustainable-oriented 3D Printing Systems [6].

These data have been later combined with the Human Centered Design approach (HCD) [7]. The combination with the so-called 'human factor' – the human dimension of any design intervention – into the scenario for Sustainable-oriented 3D Printing Systems is significant for supporting the development of novel sustainable-oriented business models, promising transition studies, aware market demands and wellbeing-based sustainable lifestyles [6]. This integration generated a number of Research Topics and Promising Design Opportunities for 'HCD-oriented Sustainable 3D Printing', which are linked to the current advances in Design for Sustainability.

The result of this interpolation process is the 'Framework of Design Opportunities for Sustainable 3D Printing Systems' [6], which portraits a new complete set of research and design topics that can be used by designers, researchers, entrepreneurs and policymakers interested to work with 3D printed solutions to support the transition toward sustainable ways of design, production and consumption[1].

3 Design Scenario

Waste management and consumer awareness are fundamental to change the way disposable products are used. New design angles focused on the reduction of plastic products must be introduced to exceed the meanings of technological obsolescence, psychological obsolescence, and planned obsolescence, as well as reintroducing environmentally sustainable products that culturally stimulate aware productions and uses. The main purpose of this stage is exploring valid substitutes to disposable products.

Food, likewise medical and personal care sectors, predisposes a wide use of disposable products; this is due to the high levels of hygiene required. As reported by Legambiente [8], the use of disposable products is deeply rooted in the European food sector, with 40% of production destined for food products, including packaging or tableware. The scenario analysis aims to explore the plastics' mitigation strategies in the production of FMCGs for the food sector, which consequently intents to understand how to reduce the impact deriving from the production and the use of products. This assumption is essential to improve the linear economy model, represented by the 'Take-Make-Dispose' process, toward the new 'Use-Use-Reuse' pattern belonging to the circular economy model (Fig. 1).

Fig. 1. The sustainable scenario used to support the 'Use-Reuse-Reuse' circular patterns.

4 Design Experimentation

The design solution here proposed is a reusable 3D printable kit for sustainable cutlery self-production, which can be used both in formal and informal food-related sectors. The concept has been designed using the 'Use-Use-Reuse' model (Fig. 2).

[1] The analysis of the Scenario is discussed in the work of Rossi and Di Nicolantonio [6].

Forks, plates, glasses, and all related cutlery items represent the majority of disposable domestic products. Therefore, a basic cutlery set (fork, spoon and knife), was used to verify the research hypothesis. People consuming food away from home every day: workers, students, travelers, tourists, professionals, etc. are the potential consumers. In this study, HE students were considered as the final end-users of this solution; this because the student population shows a specific attitude on the consumption of disposable products. To collect the research data on habits, preferences and market trends, a questionnaire has been distributed within a targeted student population.

Fig. 2. Functions generated in the early design explorations: guarantee highest levels of hygiene, transportability, resistance and sustainability).

4.1 Survey and Data Analysis

To outline the needs of potential consumers, a questionnaire composed by eighteen questions has been submitted to 180 students[2]. The questions aimed to investigate:

1. The students' understanding about Sustainability.
2. Possible alternatives to the use of disposable products (cutlery).
3. The students' competence on additive manufacturing technologies.
4. The needed features of reusable products for the consumption of food.

The analysis demonstrated that students:

- Prefer to use bioplastic rather than traditional plastics.
- Want to purchase recyclable solutions with a longer product life cycle.
- Have an interest in all Sustainability-related issues.

[2] Students across all levels in the BA Design program at the University of Chieti-Pescara, Italy.

- Desire to mitigate the problem of disposable items.
- Want to design, use and implement alternative solutions to cope with the problems caused by disposable use / led by throwaway culture.
- Packaging, portability and ergonomic features are fundamental aspects.
- Hygiene is one of the main characteristics that any product must have.
- Customization is an important parameter to stimulate the product's purchase, though this is perceived as not essential in the context of Sustainability.

4.2 Concept Design

The survey has shown some important aspects, which have been later used for the design of the reusable 3D printable kit for sustainable cutlery self-production: hygiene, transportability, and ergonomic features. These fully meet the criterion of having, always, a clean and uncontaminated product.

The shape of the kit contributes to keep high the level of hygiene, supporting all cleaning processes. An elegant case contributes to achieve this parameter, whilst preventing the contamination as well as simplifies the transportability, the storage and the filth removal. The case is also made with lightweight material, either thermoplastics or metal. About ergonomics, the solution is easy to be used, suitable for repeated uses as well as adaptable, assuring good mechanical performances.

4.3 3D Design Thinking

Autocad, Rhinoceros and SolidWorks have been used to design the kit. The technical design has been conducted with the collaboration of COESUM, an Italian company specialized in 3D modeling and rapid prototyping [9].

Three phases characterized the technical design (Fig. 3).

- In the first phase, 2D design software was used to originate the main product's geometries, which have been made by translating raster photos of handmade prototypes into vector files (CAD).
- In the second phase, three-dimensional modeling tools were used to create virtual models and to test geometries, shapes and solution's ergonomic features.

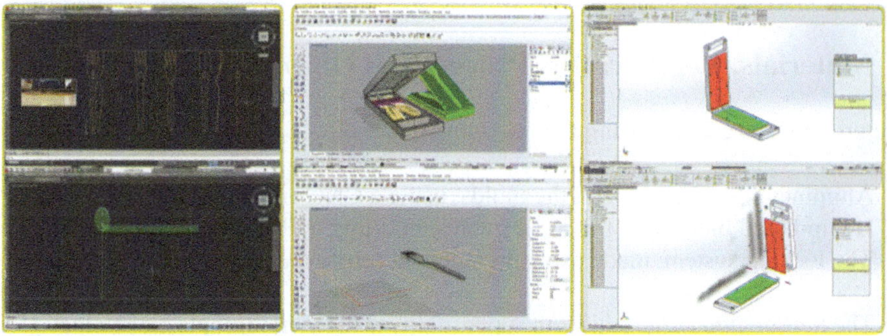

Fig. 3. From left to right, the virtual modeling in phases 1, 2 and 3.

- In the third phase, the use of 3D parametric engineering software allowed designing the complex mechanical models; in this phase, all project elements have been tested to comply with the production criteria.

4.4 Final Design

The final design – *ReBio-Cutlery* kit – is a customizable, portable and reusable 3D printable kit for sustainable cutlery, containing all basic items for consuming food away from home: fork, spoon, and knife (Fig. 4). It can be 3D printed from scratch, whose main components are: 1) a case made in molded aluminum; 2) moulds in aluminum; 3) a tool holder in hemp-based bioplastics; 4) some tools (i.e. separators) in hemp-based bioplastics; 5) a hex key.

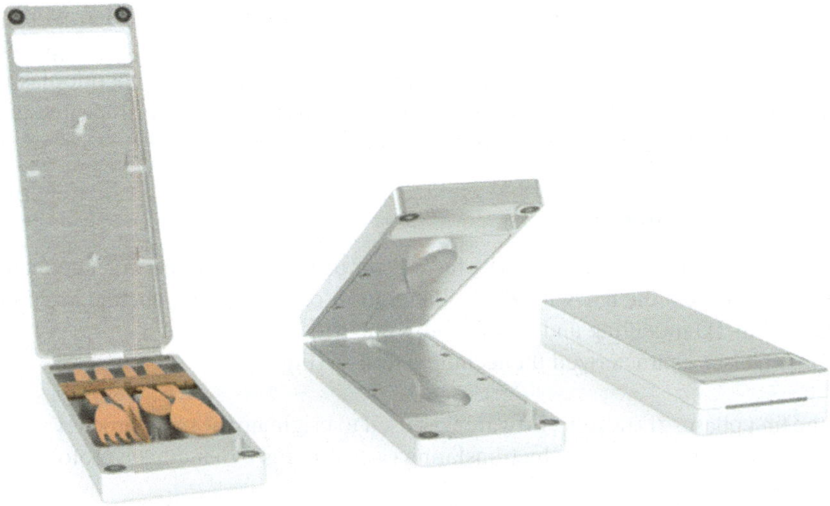

Fig. 4. *ReBio-Cutlery* kit: concept visualization.

5 Materials, Regeneration Process and Obsolescence Strategy

The prototype was made using the following materials:

1. Aluminum, for the case and the molds, using a CNC cutter.
2. Hemp-based bioplastic, for cutlery and supporting tools, using 3D printer.
3. The locking system, made using commercial ferrite magnets.

The *ReBio-Cutlery* introduces the opportunity to regenerate bioplastic products (fork, spoon, and knife), if damaged. This is possible by reconverting bioplastics through thermoplastic cycles. Hemp-based bioplastics, having organic origin, do not produce toxic substances during the thermal cycles. This feature allows consumers to regenerate the damaged parts of their products at home (i.e. using the kitchen oven). At the end of the product life cycle, hemp-based bioplastics can be macerated and reused, reproducing a new filament that can be used for 3D printing, or to be used with different technologies and processes.

Aluminum, instead, is considered an efficient sustainable alternative to plastics. It is 100% recyclable and reusable for multiple applications as well as it guarantees better endurance than plastics and a lower lightness than glass.

The proposed solution also complies with the idea of planned obsolescence for products with a relatively short life cycle. This condition is often caused by a voluntary 'bad design', which often is linked to the use of poor or unsuitable materials. In this way, the *ReBio* process is in countertendency with current consumerist thinking, which over the years has substantially contributed to the 'plastics issue' worldwide.

6 Conclusions

This study introduces important advances in the sector of Sustainable 3D Printing as well as in the Product Design domain by proposing a scalable process that can be used for the production, and/or the self-production, of a wide range of sustainable domestic products (i.e. FMCGs). According to the recent theories within the Design for Sustainability scientific domain, this can be seen as an effective and proactive strategy to mitigate the use of disposable plastics in domestic contexts of use and beyond.

The will to design new sustainable products that reduce the impact on the environment, whilst supporting the economic growth, denotes the main factor that connects circular economy and contemporary consumerism. Although in last decades the consumer dynamics have been characterized by the excessive use of plastics, the transition to renewable energy and aware business models could be considered one of the possible approaches to connect consumerism with circular economy. A new idea of sustainable consumerism can be promoted in the transition from the linear economy model, the one responsible of the global problem of waste disposal and summarized through the 'Take-Make-Dispose' process, to a circular one characterized by 'Use-Use-Reuse' processes.

Consumerism, per se, is independent from the use of plastic. It depends by markets, by consumers and by products' expectations, standards and needs. For this reason it will be very difficult to reshape the consumers' attitude using vertical approaches; this means that the sole economy- or market-driven strategies are no longer sufficient for this purpose. Accordingly, Sustainable 3D Printing can be considered a cultural revolution. In line with this cultural shift, *ReBio-Cutlery* is an example of how this technology can be applied to mitigate the problem of disposable products, introducing new sustainable perspectives on the use and the reuse of products. *ReBio-Cutlery* aims to sensitize the users by making them part of the recycling process as well as giving them the opportunity to learn from recycling processes, even from the ones performed in the domestic environments. In this way, consumers can start operating a more conscious use of the products they use everyday, outdoing the concept of disposable products.

Acknowledgements. This paper shows the result of a research carried out in the Antonello Vespoli's M.Arch. dissertation conducted at the 'G. d'Annunzio' University of Chieti-Pescara, in Italy (tutors: Massimo Di Nicolantonio, Emilio Rossi and Giovanni Annese). The design results here presented have allowed to proposing an original solution for mitigating the use of disposable plastic in cutlery sector.

The various paragraphs of this paper are the result of a collective dialogue between authors on the current role of Sustainable 3D printing and its implication for the Industrial Design domain. However, the writing of the various sections of this paper can be attributed to Massimo Di Nicolantonio for Paragraphs 3, 4, 5 and 6, and to Emilio Rossi for Abstract, Paragraphs 1 and 2.

References

1. Ceschin, F., Gaziulusoy, I.: Evolution of design for sustainability: from product design to design for system innovations and transitions. Des. Stud. **47**, 118–163 (2016)
2. WWF Italia. https://www.wwf.it/news/notizie/?uNewsID=37160
3. Ocean Motion. http://oceanmotion.org/
4. Vezzoli, C., Manzini, E.: Design for Environmental Sustainability. Springer, London (2008)
5. Rossi, E., Di Nicolantonio, M., Barcarolo, P., Lagatta, J.: Sustainable 3D printing: design opportunities and research perspectives. In: Di Nicolantonio, M., Rossi, E., Alexander, T. (eds.) AHFE 2019. AISC, vol. 975, pp. 3–15. Springer, Cham (2020). https://doi.org/10.1007/978-3-030-20216-3_1
6. Rossi, E., Di Nicolantonio, M.: Integrating human-centred design approach into sustainable-oriented 3D printing systems. Hum.-Intell. Syst. Integr. **2**, 57–73 (2020)
7. Giacomin, J.: What is human centred design? Des. J. **17**(4), 606–623 (2014)
8. Legambiente, #usaegettanograzie. https://shop.usaegettanograzie.it/
9. COESUM. https://coesum.it/en/

A Systematic Review of Differences Between Conventional Orthoses and 3D-Printed Orthoses

Ana Cláudia Tavares Rodrigues[1(✉)], Letícia Vasconcelos Morais Garcez[1], Fausto Orsi Medola[1], Luciana Ramos Baleotti[2], Frode Eika Sandnes[3,4], and Atiyeh Vaezipour[5]

[1] Graduate Program in Design – PPGDesign-UNESP, São Paulo State University (UNESP), Bauru, Brazil
{act.rodrigues,leticia.vm.garcez,fausto.medola}@unesp.br
[2] Graduate Program in Human Development and Technologies, São Paulo State University (UNESP), Rio Claro, Brazil
luciana.r.baleotti@unesp.br
[3] Oslo Metropolitan University (OsloMet), Oslo, Norway
frodes@oslomet.no
[4] Kristiania University College, Oslo, Norway
[5] RECOVER Injury Research Centre, The University of Queensland, Brisbane, Australia
a.vaezipour@uq.edu.au

Abstract. Rapid prototyping and 3D printing technologies have been innovative options for orthotic development, that are assistive devices applied to any part of the body to different disabilities or injuries. The aim of the study was to investigate scientific evidence about the effectiveness of using 3D printed orthoses in relation to conventional orthoses, through a systematic literature review in the years 2010 to 2020. Scientific evidence identifies that there are improvements associated with the development and use of rapid prototyped and 3D printed orthoses however the publications found present were small in sample sizes and lack comprehensive end-user evaluation.

Keywords: Orthosis · 3D printing · Assistive technology · Rapid prototyping

1 Introduction

Currently, more than one billion people worldwide live with some form of disability, of which around 200 million experience considerable functional difficulties [1]. With the aim of improving the functioning of people with disabilities, Assistive Technology (AT) resources are individually customized to assist the independence and autonomy of users. Successful AT initiatives often require interdisciplinary practices among professionals from the areas of Rehabilitation, Health Sciences, Product Design and Engineering, with a view of sharing knowledge and ideas in the developing of products that can have a positive impact on the quality of life, participation and social inclusion for people with disabilities [2]. Interdisciplinarity approaches have given positive results in minimizing

S. Trzcielinski et al. (Eds.): AHFE 2021, LNNS 274, pp. 53–60, 2021.
https://doi.org/10.1007/978-3-030-80462-6_7

gaps in the development and improved usability of Assistive Technology, favoring quality in research and innovation and exploring the different perspectives of the interface between the auxiliary device and the user [3, 4].

Orthoses are assistive devices applied to any part of the body, alone or covering more than one joint [5]. With the aim of meeting the individual needs of each person, orthoses may perform several functions including maintaining or promoting the range of articular movement, preventing or correcting deformities, protecting against injuries, assisting in rehabilitation, and maximizing function [6]. Orthotic devices are essential components of a rehabilitation program. Although the use of such devices has increased over the years, the range of materials used in the manufacture is still limited and costly [7].

An orthosis requires proper adjustment to the limb, since inappropriate fit can lead to discomfort, pain and skin lesions [8]. Such problems can result in early abandonment of the orthosis by the user before the rehabilitation program is completed. Shaping the device on the subject's limb requires extensive experience of the rehabilitation professional [9]. Moreover, the manufacturing process is laborious and time-consuming [10].

To reduce costs and increase user satisfaction, 3D printing technologies have been indicated as innovative options for orthotic development [10]. Although the literature presents studies about the process of creation and development of customized prostheses through 3D printing [11], there are still few studies that have compared their effectiveness to the orthotic devices made by the conventional process.

Therefore, this study reports a systematic review aimed to answer the following questions: Are the 3D printed orthoses better than conventional orthoses? What evaluation measurement have been used to compare these types of orthoses?

2 Method

A systematic review of the literature was conducted through several steps. First, a search for scientific articles published in the last ten years in the Scopus and Pubmed databases was conducted. We used the query terms "Orthosis; 3D printing", "Splint; 3D printing", "Orthosis; Rapid Prototyping", "Orthosis; Assistive Technology; 3D printing", "Orthosis; Assistive Technology; Rapid Prototyping", "Splint; Assistive Technology; 3D printing", plus the Boolean operator "and".

Next, the abstracts of the studies found were read to verify which ones met the inclusion criteria, namely scientific articles reporting the manufacture of orthoses with rapid prototyping and 3D printing technologies; comparative evaluation of orthotic devices made through conventional methods and 3D printing and rapid prototyping technologies.

After discarding duplicate studies and studies that did not meet the inclusion criteria, article data were tabulated in a Microsoft® Excel spreadsheet. The tabulated data included the year of publication, the type of orthosis addressed, the number of participants of the study and the evaluation and comparison measures used.

3 Results and Discussion

Study Selection

A total of 87 articles were found using Pubmed and 117 articles were found using Scopus, totalling 204 hits (Fig. 1). The descriptor pair that returned the highest number of results was "Orthosis and 3D Printing", with 77 results.

Of the 204 studies, 147 studies were excluded as they reported (a) literature reviews, (b) other types of Assistive Technologies such as prostheses; (c) prostheses and dental implants, (d) items not considered assistive technology resources, (e) internal prostheses and surgical procedures, (f) reported only the software and processes used to develop the final product, or (g) no evaluation tests. Thus, 57 studies were classified as being within the scope of this study. However, after the detailed reading, only 9 were finally selected for discussion herein, as they met all inclusion criteria.

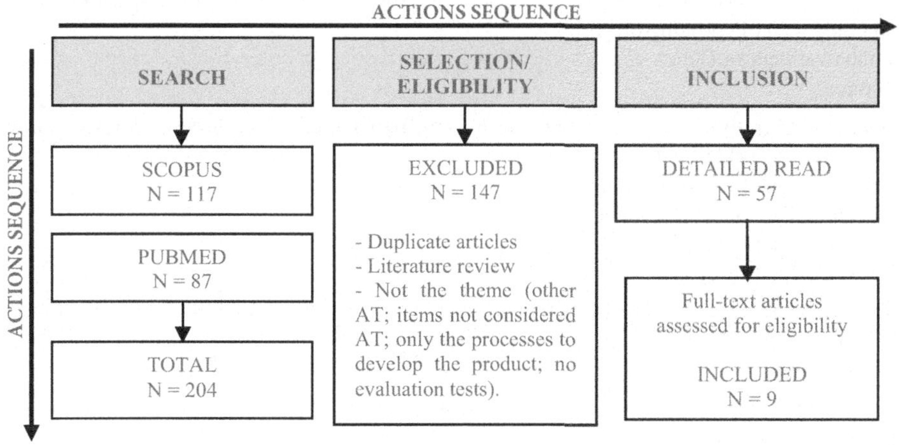

Fig. 1. Procedure.

Study Characteristics

The characteristics of the selected articles were that 3D printing was used in the development of orthoses. Moreover, tests were reported comparing orthoses made by conventional methods (prefabricated (canvas or thermoplastic) or made to measure) of upper or lower limbs with orthoses by means of 3D printing (see Table 1).

According to the year of publication (see Fig. 2) there was an increase in publications addressing the comparison of 3D printing orthoses with conventional orthoses from 2017 and onwards. In 2019, 44,4% of the publications [12–15] were on this topic. Expectations for 2020 are there the number of publications on these topics will increase. Note that the data collection presented herein was carried out during January 2020, and only one publication [16] was included in the graph. However, an increase in research related 3D printed orthoses was identified.

Table 1. Studies that met the inclusion criteria.

Title	Authors (year), country	Type of orthoses
Effects of a 3D-printed orthosis compared to a low-temperature thermoplastic plate orthosis on wrist flexor spasticity in chronic hemiparetic stroke patients: a randomized controlled trial	Zheng, Y. et al. (2020), China	Wrist, hand and fingers orthosis (Upper limb)
Personalized assistive device manufactured by 3D modelling and printing techniques	Lee, K. H. et al. (2019), Republic of Korea	Cock-up (Upper limb)
The biomechanical difference between running with traditional and 3D printed orthoses	Mo, S. et al. (2019), China	Ankle-foot orthosis (Lower limb)
Small splint external fixation combined with 3D printing brace for the treatment of Colles fractures	Zeng, T. et al. (2019), China	Cock-up (Upper limb)
Effect of 3D Printing Individualized Ankle-Foot Orthosis on Plantar Biomechanics and Pain in Patients with Plantar Fasciitis: A Randomized Controlled Trial	Xu, R. et al. (2019), China	Ankle-foot orthosis (Lower limb)
Fabrication and stress analysis of ankle foot orthosis with additive manufacturing	Banga, H. et al. (2018), India	Ankle-foot orthosis (Lower limb)
Effect of personalized wrist orthosis for wrist pain with three-dimensional scanning and printing technique: A preliminary, randomized, controlled, open-label study	Kim, S. J. et al. (2018), Republic of Korea	Cock-up (Upper limb)
Analysis and comparison of wrist splint designs using the finite element method: Multi-material three-dimensional printing compared to typical existing practice with thermoplastics	Cazon, A. et al. (2017), United Kingdom	Cock-up (Upper limb)
Patient specific ankle-foot orthoses using rapid prototyping	Mavroidis, C. et al. (2011), United States	Ankle-foot orthosis (Lower limb)

Fig. 2. Publications in chronological order.

Regarding the type of orthoses (see Fig. 3), we observed that there was a balance in publications related to the target limbs, where 55,5% of the papers addressed upper limb orthoses and 44,4% of the papers addressed lower limb orthoses. Additionally, we found few variations as all the lower limb orthoses studies ($n = 4$) covered ankle-foot orthoses (AFO) [13, 15, 19, 20]. The other 5 studies addressed upper limb orthoses. The majority ($n = 4$) of these discussed the wrist orthoses, called "cock-up" [12, 14, 17, 18] and one study documented a wrist, hand and fingers orthosis (WHFO) [16].

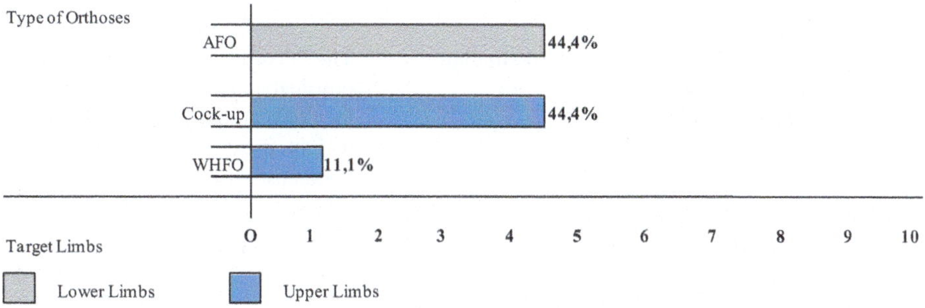

Fig. 3. Orthosis types.

Evaluation Measurement

The evaluation measurement methods used varied greatly. Table 2 lists details about the evaluation and comparison measures used, as well as the number of participants involved and the summary of findings.

Table 2 reveals that only a handful of studies compared 3D printed orthoses and conventional orthoses. In addition, several studies did not evaluate the orthoses with participants or end-users [18, 19]. Others relied on case studies [12, 20].

Regarding the measures of evaluation and comparison of upper limb orthoses, it was possible to observe better results in goniometry, Modified Asworth Scale, Fugl-Meyer Assessment, Edema Level [16]; Jebsen-Taylor Hand Function Test, Quebec User Evaluation of Satisfaction with Assistive Technology (QUEST 2.0) [12]; Cooney Modification

Table 2. Evaluation and comparison measures used and the number of participants.

Evaluation and comparison measures	Participants	Summary of findings
Modified Ashworth Scale, Fugl-Meyer Assessment, Visual Analogue Scale, Goniometry and Edema level	40	3D orthosis showed greater changes in reducing spasticity and motor function
Jebsen-Taylor Hand Function Test and Quebec User Evaluation of Satisfaction with Assistive Technology	1	3D printed devices showed better results in the JHFT score and in the most of QUEST items than ready-made assistive devices
Measurement of the Hindfoot Eversion Angle and Comfort level	13	No significant differences between the two types of orthosis
Visual Analogue Scale, Edema level, Cooney Modification of the Green and O'Brien Score and Patient-Rated Wrist Evaluation	60	VAS and edema level did not show significant differences; the Green and O'Brien Score and PRWE had better results with the 3D orthosis
Footscan® recorded Maximum Pressure, Maximum Force and Contact Area at specific points, in addition to Visual Analogue Scale and Comfort Level	60	Higher level of comfort in the 3D orthosis
Biomechanical Tests (Deformation of material, Constant Load, Mechanical Properties and Custom Product)	0 (4 simulations)	The 3D orthosis corresponded to the physical and mechanical aspects of the conventional; presented less weight, greater thermal comfort and good cost-wise
Patient-Rated Wrist Evaluation, Jebsen-Taylor Hand Function Test and Orthotics and Prosthetics Users' Survey	22	The 3D cock-up was better in relation to wrist pain and 2 OPUS tasks
Performance Level in displacement and Stress values	0 (40 simulations)	The 3D orthosis remains at the same or even better performance level
Gait Analysis	1	The 3D orthosis provided a better fit of the individual's anatomy

of the Green and O'Brien Score, Patient-Rated Wrist Evaluation [14] and Orthotics and Prosthetics Users' Survey (OPUS) [17] with the use of 3D orthoses, pointing out benefits in such orthoses with regard to the range of motion, manual function, spasticity, satisfaction and performance in Activities of Daily Living (ADLs).

The Visual Analogue Scale, Edema Level (in one of the studies) and biomechanical tests (tension/resistance) did not present statistically significant differences between the two types of orthoses [14, 18].

As for publications about lower limb orthoses, it was possible to identify improvements with 3D printed orthoses over conventional orthoses in items of adjustment of the equipment to the user [20], comfort [15] and cost-wise [19].

4 Conclusions

With the worldwide increase in life expectancy and increase of individuals who experience disabilities or reduced functioning, it is expected that an increasing number of people will need orthoses. There has been increasing use of Rapid Prototyping and 3D printing technologies in the field of Assistive Technologies, with a great potential for the development of personalized devices that are more likely meet the users' needs and preferences. It is therefore important to investigate the effectiveness and potential benefits of these devices.

The results of this systematic review indicate that there are improvements associated with the development and use of rapid prototyped and 3D printed orthoses compared to conventional orthoses in terms of upper limb range of motion, manual function, spasticity, satisfaction, performance in activities of daily life, comfort and low cost. Although the number of publications were low, there has been an increase in recent years, with a sharp increase expected in the coming years. Also, some studies found no significant differences between the two orthosis types.

In conclusion, the field of orthosis 3D printing is rapidly evolving and has presented important benefits for the users of such resources. However, there is insufficient evidence in support for the claim that 3D printed orthoses are better than conventional orthoses, due to the low number of publications on this topic. In addition, the studies included in this review were small in sample sizes and lack comprehensive end-user evaluation which would be critical for the uptake of Assistive Technologies.

References

1. World Health Organization (WHO). World Report on Disability (2011)
2. Sandnes, F.E., Medola, F.O., Berg, A., Rodrigues, O.V., Mirtaheri, P., Gjøvaag, T.: Solving the grand challenges together: a Brazil-Norway approach to teaching collaborative design and prototyping of assistive technologies and products for independent living. The Design Society. (2017)
3. Medola, F.O., Sandnes, F.E., Silva, S.R.M., Rodrigues, A.C.T.: Improving assistive technology in practice: contributions from interdisciplinary research and development and collaboration. Assistive Technol. Outcomes Benefits **12**(1), 1–10 (2018)
4. Medola, F.O., Sandnes, F.E., Rodrigues, A.C.T., Paschoarelli, L.C., Silva, L.M.: Rehabilitation and product design: towards the inclusion of people with disabilities through interdisciplinary collaboration. In: Machado, J., Soares, F., Veiga, G. (eds.) HELIX 2018. LNEE, vol. 505, pp. 185–191. Springer, Cham (2019). https://doi.org/10.1007/978-3-319-91334-6_26
5. Ferrari, A.L.M., Medola, F.O., Sandnes, F.E.: How do orthoses impact ease of donning, handwriting, typewriting, and transmission of manual torque? A study of three prefabricated wrist-hand orthoses. J. Prosthet. Orthot. (2020, in press)
6. Deshaies, L.D.: Upper extremity orthoses. In: Trombly, C.A., Radomski, M.V. (Org.) Occupational Therapy for Physical Dysfunction, 6th edn, pp. 421–464. Lippincott Williams & Wilkins (2008)

7. Rodrigues Jr., J.L.: Claw hand: a proposal of Occupational Therapy Intervention for leprosys. Master's dissertation. Federal University of Pará. Post Graduate Programme in Tropical Diseases (2013)
8. Edelstein, J.E., Bruckner, J.: Introduction to orthotics. In: Edelstein, J.E., Bruckner, J. (eds.) Orthotics: A Comprehensive Clinical Approach, vol. 1, pp. 1–16. SLACK Incorporated (2002)
9. Martinez, L.B.A., Elui, V.M.C., Martinez, R.A., Agnelli, J.A.M.: Elaboration of standard instrument for the test of thermoplastic materials for orthoses. Rev. Interinst. Bras. Ter. Ocup. Suppl. **1**(4), 518–525 (2017)
10. Cha, Y.H., et al.: Ankle-foot orthosis made by 3D printing technique and automated design software. Appl. Bionics Biomech. (2017)
11. da Silva, L., Medola, F., Rodrigues, O., Rodrigues, A., Sandnes, F.: Interdisciplinary-based development of user-friendly customized 3D printed upper limb prosthesis. In: Ahram, T.Z., Falcão, C. (eds.) AHFE 2018. AISC, vol. 794, pp. 899–908. Springer, Cham (2019). https://doi.org/10.1007/978-3-319-94947-5_88
12. Lee, K.H., Kim, D.K., Cha, Y.H., Kwon, J.Y., Kim, D.H., Kim, S.J.: Personalized assistive device manufactured by 3D modelling and printing techniques. Disabil. Rehabil. Assist. Technol. **14**(5), 526–531 (2019)
13. Mo, S., et al.: The biomechanical difference between running with traditional and 3D printed orthoses. J. Sports Sci. **37**(19), 2191–2197 (2019)
14. Zeng, T., Gao, D.W., Wu, Y.F., Chen, L., Zhang, H.T.: Small splint external fixation combined with 3D printing brace for the treatment of Colles fractures. Zhongguo gu shang = China J. Orthop. Traumatol. **32**(6), 513–518 (2019)
15. Xu, R., Wang, Z., Ma, T., Ren, Z., Jin, H.: Effect of 3D printing individualized ankle-foot orthosis on plantar biomechanics and pain in patients with plantar fasciitis: a randomized controlled trial. Med. Sci0 Monit. Int. Med. J. Exp. Clin. Res. **25**, 1392–1400 (2019)
16. Zheng, Y., et al.: Effects of a 3D-printed orthosis compared to a low-temperature thermoplastic plate orthosis on wrist flexor spasticity in chronic hemiparetic stroke patients: a randomized controlled trial. Clin. Rehabil. **34**(2), 194–204 (2020)
17. Kim, S.J., Kim, S.J., Cha, Y.H., Lee, K.H., Kwon, J.Y.: Effect of personalized wrist orthosis for wrist pain with three-dimensional scanning and printing technique: a preliminary, randomized, controlled, open-label study. Prosthet. Orthot. Int. **42**(6), 636–643 (2018)
18. Cazon, A., Kelly, S., Paterson, A., Bibb, R., Ian Campbell, R.: Analysis and comparison of wrist splint designs using the finite element method: multi-material three-dimensional printing compared to typical existing practice with thermoplastics. Proc. Inst. Mech. Eng. Part H: J. Eng. Med. **231**(9), 881–897 (2017). https://doi.org/10.1177/0954411917718221
19. Banga, H., Belokar, R., Kalra, P., Kumar, R.: Fabrication and stress analysis of ankle foot orthosis with additive manufacturing. Rapid Prototyping J. **24**(2), 301–312 (2018)
20. Mavroidis, C., et al.: Patient specific ankle-foot orthoses using rapid prototyping. J. Neuroeng. Rehabil. **8**(1) (2011)

Product Design and Development by Additive Manufacturing for Automotive Systems

Álvaro M. Sampaio[1,2(✉)] and António J. Pontes[2]

[1] Lab2PT, School of Architecture, University of Minho, Campus de Azurém, 4800-058 Guimarães, Portugal
amsampaio@arquitetura.uminho.pt
[2] Institute for Polymers and Composites IPC/I3N, University of Minho, Campus de Azurém, 4800-058 Guimarães, Portugal
pontes@dep.uminho.pt

Abstract. Developing new products through additive manufacturing is becoming imperative in today's industry, mainly because this technology enables the production of products in just one step. For the Automotive industries, additive manufacturing is centered on prototypes to accelerate the product design phase. The concept evaluation phase needs to be optimized to provide mechanical and kinematic evaluation but not overloading the design teams with assembly tasks. This paper presents ongoing research focusing on producing head-up displays kinematic concept solutions by selective laser sintering technology using part consolidation additive manufacturing principle. To achieve this, several production tests were performed with one particular kinematic automotive system concept solution, changing design concepts. The performed tests permitted to better understand the limitations and capabilities of the laser sintering technology process.

Keywords: Design rules · Additive manufacturing · Selective laser sintering · Head-up displays

1 Introduction

Additive Manufacturing (AM) is the process of building a 3D object in a layer-by-layer fashion. Each successive layer of material is fused with the preceding layer by the application of thermal energy, binders, or curing agents. In the automotive industry beyond the customization of tools for the assembly lines, AM is widely used to shorten product development or, by the development of physical models in the iteration stages, to choose design concepts or, to perform tests with the combination of simulation to evaluate form and fit [1]. Therefore, AM in the automotive sector is centered on prototypes and not final parts, but those prototypes play an important role in accelerating the product design phase.

In recent years, head-up display (HuD) systems have come into use in the automotive sector. As information requested by a driver for driving increases and is more diversified, HuD systems display critical information as a reflected image in the immediate viewing

S. Trzcielinski et al. (Eds.): AHFE 2021, LNNS 274, pp. 61–67, 2021.
https://doi.org/10.1007/978-3-030-80462-6_8

area of the driver, so that drivers do not need to take their eyes from the road to read the information .since the information is delivered in a superimposed manner. These automotive systems are divided into four sub-group solutions, which are the Kinematic, Housing, Image Projection, and Assembly. The start point of the development process for these systems is the work volume, which tends to be smaller and smaller, therefore, design concepts are being optimized to develop small scale solutions achieved by implementing smaller parts but, at the same time, by diminishing the number of parts that the system needs, mainly in the kinematic group. In AM technology one way to achieve this is by part consolidation, one of the several advantages of AM [2]. Therefore, this paper presents ongoing research focusing on producing HuD kinematic concept solutions by Selective Laser Sintering (SLS) technology, where the mechanism is produced on the machine i.e., pre-assembled. The importance of the research was to verify if it was possible to produce a fully functional mechanism pre-assembled right out of the machine, thus avoiding the use of fixation parts/methods (e.g., Screws, retaining rings, glue), and enabling a fast development process through the iteration stages.

2 Additive Manufacturing Approach

A variety of AM technologies are currently available for different types of polymers, metals, alloys, composites, ceramics, and resins [3, 4], and these AM processes allow easy and convenient creation of complex geometries [4, 5]. SLS is the main processes currently being used and further investigated, being the most feasible process for this approach in the short term because parts built with this technology do not use support so this allows the operator to arrange parts more freely in a build chamber and similar parts can be nested, therefore yielding higher productivity while lowering the associated cost [6]. Using SLS, designers have the opportunity to fabricate almost any shape or topology, including complex internal structures such as cellular or lattice structures [7], since the part is directly built up from CAD data by selectively sintering powder materials layer by layer [8]. In spite of its great potentials and benefits, AM's industrial establishment for end-use part production purposes is still limited. One reason for this is the insufficient availability of comprehensive design rules for AM [9]. It can be difficult for experienced or novice engineers to design for AM. The expansive capabilities of AM lift the Design for Manufacturing constraints of conventional manufacturing processes, thereby expanding the design freedom of the designer, but they also leave the designer with new sets of process-specific design rules that are often poorly understood and quantified [10]. Therefore, Design for AM (DFAM) was introduced in order to improve the productions of AM parts and products and to overcome some manufacturing issues regarding this technology [10]. When considering part consolidation several DFAM rules were already prescribed although there is not a common understanding and proper specification. Although DFAM varies regarding the AM technology, for example for the case of SLS clearance (i.e., the gap between two parts) several manufacturing AM worldwide providers specify a 0.5 mm of clearance as a rule of thumb [11–14]. Nevertheless, 'clearance' is not used to describe the same type of interaction (i.e., the geometry of the parts involved) or even interface (e.g., moving, sliding), for example for [11] clearance is for interlocking and moving parts, for [12] it is defined for hinges and gears and for [13, 14] it is general and not specified. What is common to all is the need to design a clearance measure between parts that are manufactured together and to evaluate the mass of the parts, since it could provide the over sintering effect due to overexposed the parts to the laser power.

3 HuD's Product Design and Manufacturing

To evaluate if it was possible to manufacture and HuD System with part consolidation, experimental tests were performed using a vertical kinematic HuD concept (Fig. 1). Since this strategy will enhance the development process by avoiding the use of fixation parts and methods such as screws, retaining rings and glue, and it will permit to reduce the total number of parts of the mechanism, this experimental investigation was centered in determining the best clearance to achieve a fully functional mechanism.

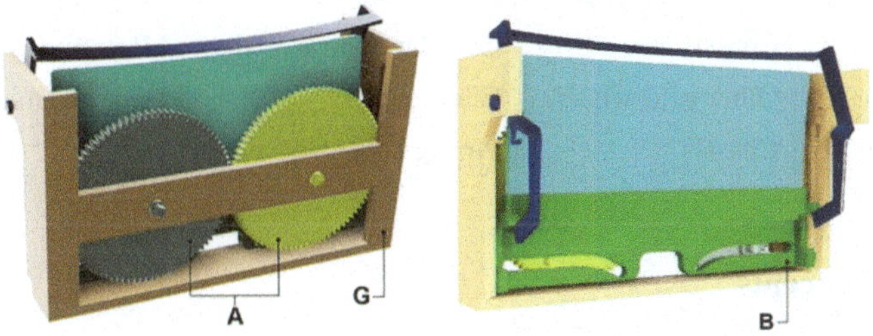

Fig. 1. Vertical Kinematic HUD concept used to perform the experimental tests.

In this mechanical concept, a set of gear wheels (A in Fig. 1) is responsible to (i) trigger the vertical movement of the combiner (B in Fig. 1), (ii) the angle adjustment of the combiner and, (iii) for the opening/closing movement of the cover.

The combiner is moved through the rotation movement of the two gear wheels. Each gear wheel possesses a pin built-in, and these pins (C in Fig. 2) goes throw two paths of the combiner support (D in Fig. 2). When the gear wheels rotate, the pins move in these paths, promoting vertical movement to the combiner support. The combiner support possesses four guide pins (E in Fig. 2), two in each side, that move in guide paths (F in Fig. 2), built-in the chassis (G in Fig. 1), to conduct the movement. Figure 2 gives a detail view of the mechanism designed.

The HuD kinematic mechanism had four different types of part interfaces, that were present individually or combined. Additionally, parts had interactions with each other through the interfaces and by different contact features that promote or facilitate those interfaces. Table 1 presents parts involve with interactions by contact features and type of interface. For each interaction and interface, it was necessary to establish clearance values to be able to produce a functional kinematic concept with total part consolidation.

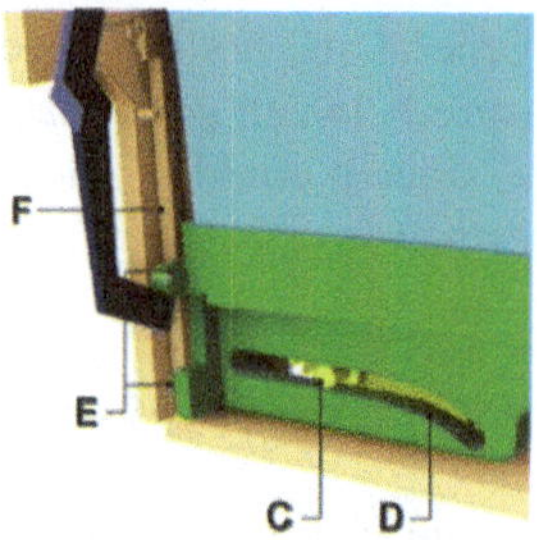

Fig. 2. Detail view of the kinematic mechanism designed.

Table 1. Part consolidation matrix

Part	Interaction with	Contact feature	Type of interface
Gear wheels	Combiner support	Pins & rails	Rotating and sliding
	Chassis	Circular feature & shaft	Rotating
	Other gear wheel	Gear teeth	Connection and friction
Combiner support	Chassis	Pins & Rails	Sliding

3.1 Design Concepts

Before performing the production of the mechanisms, it was necessary to establish some design considerations to successfully produce the HuD's. The first consideration was to define the clearance between two or more moving parts to avoid the excess of friction between them, which could turn hard the movement between the parts, and to avoid the melting of the material that compose the parts, thus merging the parts. The second consideration states that all the parts must possess some kind of feature that does not allow them to move out of their assembly position during the manufacturing process or during the movement of the mechanism, thus avoiding the usage of fixation methods/parts. For the experimental tests, in the first phase, it was designed two different mechanisms with different clearance between the walls of two adjacent parts. A mechanical concept was designed with a clearance defined for 0.5 mm, as prescribed by several AM manufacturing companies (i.e., Materialise, 3d Hubs, Shapeways), and a second concept was designed with 0.8 mm since other experimental studies were already developed that suggested a more conservative approach [15]. Since some features of the kinematic concept present different mass and contact interfaces, it was necessary to develop a second phase, to overcome the possible over sintering issue that could occur in some features that have a larger width. Therefore, a third concept was developed with a combination of clearances of 0.5 mm and 0.8 mm after evaluating the two first concepts and their performance. Moreover, all parts were "locked in place" through a circular feature in case of shafts or were fitted between parts, thus not allowing displacement movement by the components.

3.2 Production and Evaluation

The equipment used for the research was the EOS P396 and the material was an unfilled PA12 powder (PA2200 from EOS). The process parameters used, which remained constant for all builds, were the EOS Balance Standard Parameters.

First and Second Experimental Tests. After applying the post-processing operations required (i.e., powder removal), it was possible to observe that the experimental test with 0.5 mm of clearance did not work. Although some part interfaces had been manufactured correctly, the kinematics mechanism was not possible since the support powder required to build the clearances between two or more parts, due to the high thermal load that the sintered material has, melted and consequently, merged the parts. This was true for all the part interfaces regardless of the features. The second experimental test with 0.8 mm of clearance had the same overall result. The kinematic mechanism produced did not work also. The only improvements in comparison with the first mechanism were achieved in the rotation and sliding of all the pins and shafts of the gear wheels and the sliding of pins of the combiner support. In fact, in this mechanism the gear wheels were able to rotate, promoting the kinematic of the mechanical concept, but due to the high clearances applied, the interaction between the two wheels and other parts move out of its assembly position and rotate freely, thus not working as designed.

Third Experimental Test. Attending to the result obtained in the first two experimental trials, it was decided to do a third attempt to produce a fully functional mechanism. In this attempt, it was decided to use a combination of clearances values between the parts of 0.5 mm and 0.8 mm. Therefore, for this production, the clearance between the gear wheels and the chassis were design with a 0.8 mm value to avoid the over sintering effect, since these parts had higher mass and that were nearby during a large surface. The same clearance was defined for the rotating interface between the chassis and the gear wheels since that interface relays on a feature that secures the shaft-containing parts in the place. Other clearances were designed with 0.5 mm. After producing the kinematic mechanism, it was possible to evaluate that this concept worked properly. The 0.5 clearance was defined for the core of the kinematic mechanism where the tolerances needed to be narrow. Thus, gear wheels and their teeth, rotating and sliding interfaces, not geometrical constrained, were manufactured with the lower clearance value possible. Although the clearance designed in the first experiment was not suitable to achieve a full mechanism it was decided to go back to a value of 0.5 mm since it was verified in the second experimental test that the increase of this value led to malfunctions in the movement of the parts, moreover, the interfaces with locking features and a large mass were analyzed as the ones that contribute to the increase of the thermal load thus 'gluing' the parts. Therefore, these last features were produced with a 0.8 clearance value.

4 Conclusions

The main concerns to establish part consolidation is, clearance of parts and, the mass of the parts involve. Therefore, to evaluate if it was possible to manufacture and HuD

System with only one manufacturing cycle and with parts assembled in their right places, three experimental tests were performed using a kinematic HuD concept. The first two experimental parts with mechanism design with clearances of 0.5mm, in the first one and, 0.8mm in the second one, did not work, so it was done a third experimental test with a combination of clearances. The third experimental tests revealed that important kinematic features and rotation and sliding interfaces that are not locked can be manufactured with 0.5mm of clearance and that parts with locking features and large mass needed a bigger clearance value of 0.8mm. The overall result showed that is possible to manufacture kinematic mechanisms with part consolidation principle in AM improving thereby the development process in terms of efficiency and time.

Acknowledgments. This work is supported by: European Structural and Investment Funds in the FEDER component, through the Operational Competitiveness and Internationalization Programme (COMPETE 2020) [Project nº 039334; Funding Reference: POCI-01-0247-FEDER-039334].

References

1. Klahn, C., Leutenecker, B., Meboldt, M.: Design for additive manufacturing - supporting the substitution of components in series products. Procedia CIRP **21**, 138–143 (2014)
2. Renjith, S.C., Park, K., Okudan Kremer, G.E.: A design framework for additive manufacturing: integration of additive manufacturing capabilities in the early design process. Int. J. Precis. Eng. Manuf. **21**(2), 329–345 (2019). https://doi.org/10.1007/s12541-019-00253-3
3. Frazier, W.E.: Metal additive manufacturing: a review. J. Mater. Eng. Perform. **23**, 1917–1923 (2014)
4. Wong, K.V., Hernandez, A.: A review of additive manufacturing. ISRN Mech. Eng. **2012**, 1–10 (2012). https://doi.org/10.5402/2012/208760
5. Barr, J.: Makers: the new industrial revolution by Chris Anderson. J. Bus. Finan. Librariansh. **18**(3), 276–278 (2013)
6. Soe, S.P.: Quantitative analysis on SLS part curling using EOS P700 machine. J. Mater. Process. Technol. **212**(11), 2433–2442 (2012)
7. Seepersad, C.C., Govett, T., Kim, K., Lundin, M., Pinero, D.: A designer's guide for dimensioning and tolerancing SLS parts. In: 23rd Annual International Solid Freeform Fabrication Symposium, pp. 921–931 (2012)
8. Yan, M., Zhou, C., Tian, X., Peng, G., Cao, Y., Li, D.: Design and selective laser sintering of complex porous polyamide mould for pressure slip casting. Mater. Des. **111**, 198–205 (2016)
9. Gausemeier, J., Echterhoff, N., Wall, M.: Thinking ahead the Future of Additive Manufacturing – Analysis of Promising Industries. University of Paderborn Direct Manufacturing Research Center, vol. 103 (2012)
10. Adam, G.A.O., Zimmer, D.: On design for additive manufacturing: evaluating geometrical limitations. Rapid Prototy. J. **21**(6), 662–670 (2015)
11. Materialise. https://www.materialise.com/en/manufacturing/materials/pa-12-sls/design-guidelines
12. Shapeways. https://www.shapeways.com/materials/versatile-plastic
13. Sculpteo. https://www.sculpteo.com/en/3d-learning-hub/design-guidelines/3d-printing-tolerances/

14. Redwood, B., Schoffer, F., Garret, B.: The 3D Printing Handbook: Technologies, Design and Applications. 3D Hubs (2017)
15. Sampaio, Á.M., et al.: Design for additive manufacturing of mechanical connections toward hybrid products. In: Di Nicolantonio, M., Rossi, E., Alexander, T. (eds.) AHFE 2019. AISC, vol. 975, pp. 418–427. Springer, Cham (2020). https://doi.org/10.1007/978-3-030-20216-3_39

3D Printing and Social Inclusion: A Design Research Framework

Emilio Rossi[1](✉), Massimo Di Nicolantonio[2], and Alessio D'Onofrio[2]

[1] Lincoln School of Design, University of Lincoln, Brayford Pool, Lincoln LN6 7TS, UK
erossi@lincoln.ac.uk

[2] Department of Architecture, University of Chieti-Pescara, Viale Pindaro 42, 65127 Pescara, Pescara, Italy
{m.dinicolantonio,alessio.donofrio}@unich.it

Abstract. 3D Printing is a fast-growing disruptive technology that is changing the manufacturing sector in many ways, improving both the design of products and the production processes. Whilst the most traditional research contributions tackle the role of the technology in manufacturing, the implications on the social and inclusive dimensions are less considered and explored, although it is proven that the social instances of contemporary society can motivate the development of bottom-up radical innovations at the small and large scales. This paper aims to identify the design relations between 3D Printing and Social Inclusion. This study proposes an original and promising design research framework composed by four macro design-oriented research lines linking Design discipline with 3D Printing and Social Inclusion. This study can be considered significant because it proposes a convergence between 3D printing and Social Inclusion using a Design Research processes.

Keywords: 3D printing · Social inclusion · Design research · Research framework · Design topics

1 Introduction

According to the recent trends and market data, the 3D printing sector – also known as Additive Manufacturing – is in constant expansion [1–3], with a strong concentration in industrial domains like aerospace, precision manufacturing, automobile and software engineering. Accordingly, the so-called heavy industry plays an important role in feeding the researches and the advanced experimentations in the sector. Moreover, thanks to intrinsic advantages like flexible design, CNC research, lower resource requirement, savings on tooling and faster production cycle, 3D Printing has become one of the most disruptive technologies of present times [4], able to significantly evolve the way companies and designers conceive, produce and sell new products and solutions [5, 6].

The advances in the 3D Printing sector are well documented and explored by many studies, which portrait a complete and complex picture. It is clear that the majority of studies address vertical and problem-solving issues, which contribute to increase the

S. Trzcielinski et al. (Eds.): AHFE 2021, LNNS 274, pp. 68–73, 2021.
https://doi.org/10.1007/978-3-030-80462-6_9

skills and the knowledge framework in the sector, for the sector. Only a limited number of studies face interdisciplinary and proactive scenarios (i.e. Sustainable Development, Circular Economy, Design, Heritage promotion). Consequently, it is possible to affirm that the 3D Printing sector is largely driven by technology-push actions (i.e. technology-driven innovations).

One of the research fields where the potentialities of 3D Printing technologies are still unexplored is Social Inclusion, although it is proven that the social instances of contemporary society can motivate the development of bottom-up radical innovations at the small and large scales. Referring to this context, the experiments performed mainly concern the so-called 'Design for Disability' domain (i.e. design of prosthesis and anthropomorphic parts) [7]. The duality between competitiveness and missed opportunities is partially due to two aspects: firstly, Social Inclusion is a very complex research framework, defined as 'the process of improving the terms on which individuals and groups take part in society' [8], and it is intrinsically linked with concepts like wealth, prosperity, human wellbeing and environmental protection [9], which are consequently broad and complex to be faced through vertical explorations; secondly, Design for Disability is often overlapped with the idea of Design for Social Inclusion [10], whilst it represents only a minor aspect of this broader design approach. Altogether these aspects contribute to limit the development of synergic links between 3D Printing and Social Inclusion.

Considering this current lack within the literature in the Design field, it is therefore possible to conceive a scenario for new research opportunities and design experimentations made by generating a convergence between Social Inclusion, Design and 3D Printing.

2 Aims and Methodology

This paper aims to identify the design relations between 3D Printing and Social Inclusion as well as the design opportunities for later implementations and experimentations. Accordingly, the main aim of this work is to define a consistent link between social issues, which are part of the complex scenario of Social Inclusion, and manufacturing opportunities, properly developed by 3D Printing sector. The interdisciplinary connections will be made using the Design Research.

The specific aim of this work is to propose a promising design research framework composed by 'design-oriented research lines' and 'specific design topics' linking 3D Printing and Social Inclusion. Design-oriented research lines will address the attention on the macro-domains where the convergence between 3D Printing and Social Inclusion can generate a significant impact; specific design topics are narrowed areas where practical explorations can generate immediate 'inclusive' impacts.

The methodology used by this work is line with the idea of design scenario building. In particular, the idea of inclusive-oriented design scenario is consistent with the one discussed by Ezio Manzini, et al. [11]: 'a designable vision of something complex and articulated based on a clear motivation – what is the aim? – and a practicality – the actions to undertake to favour its implementation'. The use of the design scenario building methodology can generate future – likely – ideas for inclusive living conditions as well as technology-driven opportunities for the 3D Printing sector to intercept bottom-up instances and top-down design strategies. In terms of motivation, the design scenario here presented aims to imagine inclusive visions, while the practicality focuses on the range of enabling solutions that will be possible to generate in the future convergence between 3D Printing and Social Inclusion, both at the personal and social dimension.

3 Results: Design Research Framework

The research iterations produced through the scenario building methodology allowed defining important design-oriented insights linking AM and Social Inclusion, which later have been used develop the '3D Printing for Social Inclusion' design research framework. It is composed by four 'macro design-oriented research lines', linked to a number of 'design topics' usable by researchers and designers to test the viability of these notions as well as refine the knowledge framework here presented. The design research framework is not only focused on the design aspects (i.e. product design, use of materials with enhanced haptic characteristics, etc.); some remarkable insights can be found in the cross-sectorial and cross-scalar relations between Design, 3D Printing and Social Inclusion.

In this section, the macro design-oriented research lines and the related design topics are synthetically presented and discussed. These examine all dimensions of Social Inclusion: a) inclusivity aspects (endogenous features), b) accessibility to technological systems, c) methodological impacts on communities, and d) the role of networks.

3.1 Inclusive 3D Printing Systems

Inclusive 3D Printing Systems gathers important data and insights aimed to set the main target strategy for combining Design, Social Inclusion and 3D Printing. This first macro design-oriented research line addresses the most traditional and known issues around the theme of Social Inclusion, both at the strategic dimension (i.e. system(ic) design) and the human centred one (i.e. product design dimension). The research line also integrates useful data that meet the social dimension of the Sustainable Development (i.e. community-centred actions, human ecology, etc.). Therefore, related design topics for *Inclusive 3D Printing Systems* are:

- Diversity-inspired 3D Printing systems operating on Global-Local scenarios (i.e. using diversity as a resource).
- Community-centred 3D printing systems based on context-based scenario and bottom-up needs.
- Need-based economies supported by 3D printing systems.
- Socially inclusive 3D printing systems.
- 3D printing systems 'for all'.
- Affordable 3D printing systems to prevent child labour.
- Human-centred 3D printing systems.
- Ergonomically performing 3D printing systems.

3.2 Accessible 3D Printing Systems

Accessible 3D Printing Systems focuses on the accessibility to technological systems considering both the technology-centered dimension and the physical one (i.e. human-product interaction). The explicit and hidden relationships between people, economy, services and productions systems are interpreted and translated into manageable design topics that can be implemented with simple – accessible – and complex technological iterations. Accordingly, the research and design attentions revolve around the broader notion of 'accessibility to services', rather than only to the human-product interactions; this new vision expands the design opportunities by integrating the design topics with new holistic insights for the Design of Social Inclusion domain. Therefore, related design topics for *Accessible 3D Printing Systems* are:

- Accessible 3D printing systems (and their sub-parts: products, services, distribution networks, supply chains, etc.).
- Usable 3D printing systems (i.e. solutions making users and sustainable processes closer).
- Research and design of low-cost 3D printing systems.
- Research and design of DIY 3D printing systems.
- 3D printing systems based on collective intelligence and values sharing.
- 3D printing supporting the launch on new entrepreneurships.
- Development of economically affordable new business models (transitions from unsustainable/semi-sustainable business models into promising sustainable ones).

3.3 Community-Oriented 3D Printing Systems

Community-oriented 3D Printing Systems examines the impacts of 3D Printing technology into large-scale contexts of social uses, with a particular focus on the community dimension. At this scale, the research and design interventions aim to explore, and later reinforce, the methodological impacts driven by technology-push innovations into formal communities and/or informal groups of people (i.e. spontaneous groups of stakeholders) to conceive new socially sustainable ways of living, co-producing and using resources. Therefore, related design topics for *Community-oriented 3D Printing Systems* are:

- 3D printing systems for Communities of Practice (CoP).
- 3D printing systems for Communities of Interest (CoI).
- Community-oriented 3D printing solutions for the production of goods and mixed solutions (i.e. PSSs).
- Community-oriented new business models based on the use of 3D printing solutions.

3.4 Networked 3D Printing Systems for Personal and Collective Health

Networked 3D Printing Systems for Personal and Collective Health support the collective intelligence of people and actors involved in large-scale 3D Printing platforms, which intercept the demand for proactive and socially inclusive distributed solutions, with an emphasis on the personal and collective health. Accordingly, the switch from the personal dimension to the collective one is one of the topics that, better than others, can be tackled by distributed inclusive solutions. Beyond that, the human centered dimension is still a core topic of this research line, with interesting applications and potential experimentations linking biological systems, behavioral patterns and intelligent networks. Therefore, related design topics for *Networked 3D Printing Systems for Personal and Collective Health* are:

- Development of platforms supporting human healthcare systems.
- Application of networked 3D printing systems to support housing-related issues in emerging and developing countries.
- Autopoietic networks for fair collaborations and sharing of know-how (i.e. CoP, Wikis, etc.).
- Networked 3D printing systems for sanitary purposes.
- Networked innovations in emerging and developing countries.

4 Conclusions

Whilst the 3D Printing sector is constantly growing, continuing in the development of new technology-driven results aimed to increase both the market competitiveness and the manufacturing processes, it is however possible to reflect on the future social role of this disruptive technology, which could be even oriented toward the issues and the dimensions of Social Inclusion. As demonstrated by this study, the opportunity to work on Social Inclusion is a turning point for the 3D Printing sector, despite a design-driven approach must be used to coordinate the research actions as well as to think about proactive ideas and promising applications.

This paper has proposed a design framework composed by four design-oriented research lines and a number of related design topics linking 3D Printing and Social Inclusion. Although these early data need to be further refined through proper explorations and practical experimentations to give consistency and scientific legitimacy, from the methodological point of view it is possible to affirm that the use of design-oriented methodologies can stimulate an interdisciplinary discussion around the use of 3D Printing technology beyond the mere performance and the precision of manufacturing processes. Accordingly, new ways of thinking are needed to explore alternative scenarios for proactive uses of this technology.

The design framework here proposed and synthetically discussed can be seen as an initial reflection – methodological outcome – around the will to look for new design opportunities, which will always consider the technological advances produced by vertical studies; therefore, this papers open up a new scenario for lateral developments – i.e. finding new opportunities by proposing a convergence between 3D printing and Social Inclusion using the Design Research processes.

Finally, this study has demonstrated that the issues related to Social Inclusion can be used to develop a broad scenario for new design opportunities, evolving the restricted idea of using 3D Printing 'only' for 'designing medical prostheses' toward a more holistic concept of '3D Printing for Social Inclusion'.

Acknowledgements. This paper is based on the work developed by Rossi and Di Nicolantonio on the theme of Sustainable 3D Printing [12, 13] and proposes a new vision for the Design discipline that contributes to the activities of the AHFE International Affiliated Conference on Additive Manufacturing, Modeling Systems and 3D Prototyping (AHFE-AMMSP) [14].

All authors have equally contributed to the study design and implementation, including material preparation, data collection and results development. In particular, Rossi authored Sections 3, 3.1, 3.2 and 4; Di Nicolantonio wrote Sections 1 and 2; D'Onofrio composed Abstract, Sections 3.3 and 3.4.

References

1. SmarTech Analysis: 2020 Additive Manufacturing Market Outlook and Summary of Opportunities. SmarTech Publishing (2019)
2. Research and Markets: Additive Manufacturing Market Analysis By Material Type, By Metal Type, By Polymer Type, By Ceramics Type, By Process, By End-use, And Segment Forecasts To 2027. Research and Markets (2020)
3. Statista: Projected global additive manufacturing market size between 2020 and 2028. https://www.statista.com/statistics/284863/additive-manufacturing-projected-global-market-size/
4. Campbell, T.A., Ivanova, O.S.: Additive manufacturing as a disruptive technology: implications of three-dimensional printing. Technol. Innov. **15**(1), 67–79 (2013)
5. Kurfess, T., Cass, W.J.: Rethinking additive manufacturing and intellectual property protection. Res. Technol. Manag. **57**(5), 35–42 (2014)
6. Ford, S.: Additive manufacturing technology: potential implications for U.S. manufacturing competitiveness. J. Int. Commer. Econ. 1–35 (electronic) (2014)
7. Jin, Y., Plott, J., Chen, R., Wensman, J., Shih, A.: Additive manufacturing of custom orthoses and prostheses – a review. Procedia CIRP **36**, 199–204 (2015)
8. The World Bank: Social Inclusion. https://www.worldbank.org/en/topic/social-inclusion
9. European Commission: Europe 2020: A Strategy for Smart, Sustainable and Inclusive Growth. Brusselles. European Commission, Bruxelles (2010)
10. Rossi, E., Barcarolo, P.: Design for the mediterranean social inclusion. PAD **16**, 13–34 (2019)
11. Manzini, E., Jégou, F., Meroni, A.: Design oriented scenarios: generating new shared visions of sustainable product service systems. In: Crul, M.R.M., Diehl, J.C., Ryan C. (eds.) Design for Sustainability: A Step-by-Step Approach, pp. 15–32. United Nations Environment Programme, Delft University Technology (2009)
12. Rossi, E., Nicolantonio, M., Barcarolo, P., Lagatta, J.: Sustainable 3D printing: design opportunities and research perspectives. In: Di Nicolantonio, M., Rossi, E., Alexander, T. (eds.) AHFE 2019. AISC, vol. 975, pp. 3–15. Springer, Cham (2020). https://doi.org/10.1007/978-3-030-20216-3_1
13. Rossi, E., Di Nicolantonio, M.: Integrating human-centred design approach into sustainable-oriented 3D printing systems. Hum.-Intell. Syst. Integr. **2**, 57–73 (2020)
14. AHFE-AMMSP. http://ahfe.org/board.html#ammsp

Design of Custom Breast Prosthesis for Additive Manufacturing Production

Carlos G. Helguero(✉), Jorge L. Amaya, Fausto Maldonado, Jocelyne A. Acosta, Maria Bravo, and Junior Garzón

Advanced Machining and Prototyping Laboratory CAMPRO, Faculty of Mechanical and Production Sciences Engineering, ESPOL Polytechnic University, Campus Gustavo Galindo Km 30.5 Vía Perimetral, P.O. Box 09-01-5863, Guayaquil, Ecuador
{chelguer,jlamaya,famaldon,jocaacos,marebrav, jagarzon}@espol.edu.ec

Abstract. By taking advantage of additive manufacturing techniques and Computer-Aided Design software, the present work proposes modeling a personalized external breast prosthesis through three-dimensional (3D) scanning and photopolymerization 3D printing. Modeling and validation of the external prosthesis were performed using specialized software, considering dynamic and mechanical variables such as flexibility. The results showed that the prosthesis developed in this study has approximately six times higher contact percentage with the affected area than a commercial prosthesis, besides revealing propitious mechanical properties. Additive manufacturing technologies are potential tools to produce breast prostheses and maximize comfort levels while keeping up typical sales rates. The proposed design could develop an alternative that may increase the quality of mastectomized women's lives by offering a solution that focuses on returning what cancer has taken away from them.

Keywords: 3D scanning · 3D printing · Custom design · External breast prosthesis · Photopolymerization

1 Introduction

According to the World Health Organization (WHO) study, breast cancer hit two million women from the female world population in 2018 [1]. Most cases detected in later stages require a mastectomy, *id est* a surgical procedure that involves total or partial breast removal [2, 3]. Although mastectomy increases survival chances, it may also leave physical and psycho-social scars in breast cancer patients too challenging to bear [4]. Resources such as external breast prostheses are typically used to restore the bosom's visual appearance instead of a breast reconstruction operation.

Big-scale manufacturing of external prostheses arises with this trend among breast cancer survivors in the current market. However, some usability issues have been identified with commercial prostheses; besides, user choice is strongly biased by the high costs of acceptable quality models. Prostheses currently marketed do not meet size and

S. Trzcielinski et al. (Eds.): AHFE 2021, LNNS 274, pp. 74–81, 2021.
https://doi.org/10.1007/978-3-030-80462-6_10

weight requirements, resulting in backaches, neckaches, among other common conditions. For patients who have received a mastectomy, their well-being is also affected by the generic designs, which do not consider each patient's anatomy features, such as the scar's shape and size due to surgery [5].

Previous studies have developed customized medication through hydrogels and drug-loaded implants as one of the main applications of three-dimensional (3D) printing in cancer treatments [6]. Designing a new breast prosthesis has also been proposed using the biomimetic technique for total mastectomy cases where a coordinate measuring machine (CMM) is used as a tool to 3D model the prosthesis [7].

3D printing and scanning techniques have proved to contribute to the optimization of medical procedures due to the advantages it offers by materializing real images of the human body's anatomy [8–10]. By taking advantage of them and computer-aided design (CAD) software, this work proposes modeling a personalized external breast prosthesis through 3D scanning and photopolymerization 3D printing. For this procedure, anatomical data of healthy patients were obtained with 3D scanning tools. Modeling and validation of the external prosthesis were performed using specialized software, adapting the customized external breast prosthesis for additive manufacturing purposes, and considering dynamic variables and mechanical properties such as adherence, flexibility, conformation, surface topology, contact percentage, and the bosom's density. To the best of our knowledge, no prior work on this topic has been published.

2 Materials and Methods

Both the exterior and interior design of the custom external breast prosthesis depends on different features that characterize each patient. Essential requirements a prosthesis must meet to solve previously identified problems are listed in Table 1.

Table 1. Breast prosthesis design required features.

Health issue	Prosthesis must reach the healthy breast's size and weight
Comfort issue	Prosthesis should be appropriately attached to the post-mastectomy scar, feel safe, and provide women's confidence
Socio-cultural issue	Prosthesis must have a natural appearance
Economical issue	Prosthesis must be commercially competitive
Environmental issue	Selected material must be biodegradable

The present work develops a customized prosthesis design through four phases, as shown in Fig. 1. In the first phase, the topological data is collected from the mastectomized surface and the healthy breast, processed and modeled in the second phase as an external breast prosthesis. Subsequently, the percentage of contact between the modeled prosthesis and the mastectomized surface is simulated by finite elements analysis (FEM) software to validate the design.

Within the third phase, the interior structure is designed based on the computed weight required to resemble the healthy breast in terms of density. Finally, to validate the latter, a finite element analysis is performed. Information on each phase is detailed in the sections below.

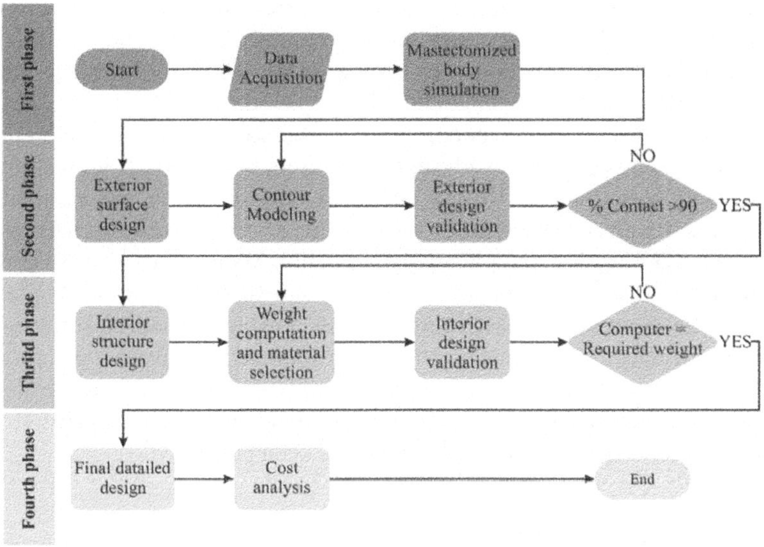

Fig. 1. External breast prothesis design flow chart with the detailed steps belonged to each phase of design.

2.1 Data Acquisition

The females' body 3D images were obtained employing a portable 3D scanner (Structure Sensor by Occipital, Inc). This technology has been developed to three-dimensionally replicate objects and structures through a mobile device whose size is smaller than a smartphone. It can be adapted to the size of the smart device to which it is mounted. Its laser sensors take billions of points from surfaces and record their topological information, creating the 3D image. The applications are miscellaneous, from virtual reality apps to shape prototyping projects.

2.2 Exterior Prosthesis Layout

3D modeling software, Meshmixer (Autodesk Inc., San Rafael, CA), was used to edit the prosthesis's outer surface in two stages. First, the anterior form of the artificial breast was obtained by mirroring the remaining healthy one. Later, the posterior prosthesis section was obtained by replicating the mastectomized surface depicted in the 3D scanned file. Finally, the two bodies are joined and refined. The final design is exported to be used as a CAD file.

The percentage of contact between the prosthesis's posterior surface and the body's mastectomized area was determined to evaluate the prosthesis's conformation to the patient's body. This step was carried out by analyzing the deviation between the modeled and the scanned surfaces, *id est.* STL and IGE files, respectively, using the Geomagic Control X 3D (3D Systems, Inc., Santa Clarita, CA) software comparison tool, as shown in Fig. 3 below in Sect. 3.

This tool requires one of the bodies to be analyzed in IGES format, so the Geomagic Design X (3D Systems, Inc., Santa Clarita, CA) software was used to convert the STL file to IGES.

2.3 Interior Prosthesis Layout

Prosthesis performance is directly linked to the selected material properties. A solid interior would increase the material's influence on the design's mechanical properties and limit the 3D printing options. On the other hand, a hollow interior design would not meet the weight requirements needed to comfort women. It would lack mechanical strength and could easily deform.

The optimized internal mesh is well shown in Fig. 3. Despite representing greater complexity, this configuration offers some advantages: 1) It allows variation of the prosthesis density until it meets the required value, 2) reduces the amount of material used compared to the solid filling, and 3) cut costs. The standard bra with deposits was defined as a support system for the prosthesis.

2.4 Detailed Design and Finite Elements Analysis (FEA)

Bra wearing is essential to keep the prosthesis attached to the body. Hence, its effects through a static and dynamic analysis were also studied. For the horizontal axis, static analysis was developed. A critical situation was considered where the bra force's vertical component does not keep the prosthesis adhered to the skin. The static model is shown in Fig. 2 (a) where the bra clamping force's horizontal component, F_n is intended to be found.

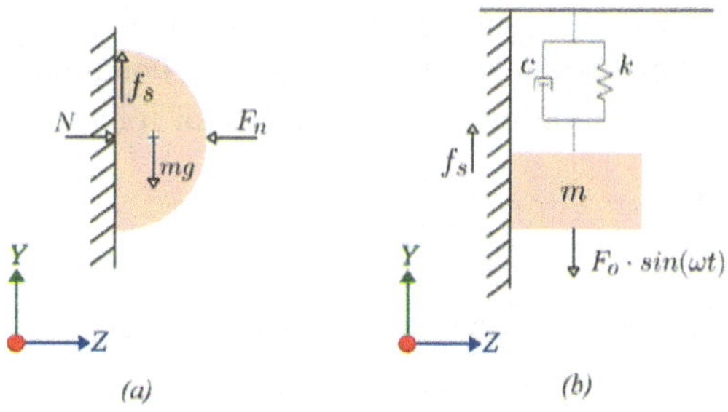

Fig. 2. Mathematical models with free-body diagrams considered for analysis: (a) static mathematical model for a critical situation where the bra clamping force's vertical component is null. (b) Dynamic mathematical model of forced damped vibration system.

For the vertical axis, dynamic analysis was proposed to calculate the prosthesis's oscillations in a chain of events where the user does some sport activity such as jogging. Figure 2 (b) exhibits the model expected to be a forced damping vibration system, considering friction with skin subtracting the mechanical system's energy.

A production cost analysis to determine the retail price of the prosthesis was also performed. The prosthesis response to daily stimuli, for example, a direct hit, was evaluated using FEA software to determine if it resembled a healthy breast's behavior. Cases were raised for the bra's pressure (distributed force) and the stress applied at a single point (punctual force). The displacement and efforts associated with each case were also obtained. The three simulated punctual forces were set in different areas: in the upper part of the breast, over the nipple, and in the lower part of the breast.

3 Results

The external breast prosthesis's final interior design is presented in Fig. 3. The exterior design change with each patient since it depends on the topological data obtained from the 3D scan. The generated layout preserved the healthy breast's characteristics and an adaptable region to the post-mastectomy scar geometry.

Contact percentage was determined from the 3D Comparison tool of the Geomagic Control X software. Figure 4 shows the contact between the prosthesis's posterior surface in Fig. 4 (a), the generic prosthesis in Fig. 4 (b), and the mastectomized area. The green surface in the colors map indicates the contact areas, while red indicates the non-contact ones. Table 2 shows the tabulated results of the contact percentage. These results reveal that the prosthesis fits the patient's body better than a generic prosthesis does.

Table 3 presents a summarized report of results from the mechanical analysis. The dynamic analysis was performed to determine the prosthesis behavior in motion and evaluate the possible effects caused by friction. Prosthesis dynamic study denoted that the bra clamping force's vertical component is negligible, so it doesn't result in shear stress. The static analysis exposed that the required horizontal component performed by bra must be above 6,9 N.

After validating the design with FEA, Table 3 also shows the maximum values obtained through the simulation. Bra Pressure application in the prosthesis with internal mesh structure featured a more significant deformation than the prosthesis with a solid interior. The stress distribution caused by the Punctual Force presented the maximum value at the point of application of the force. Punctual force in the lower part of the

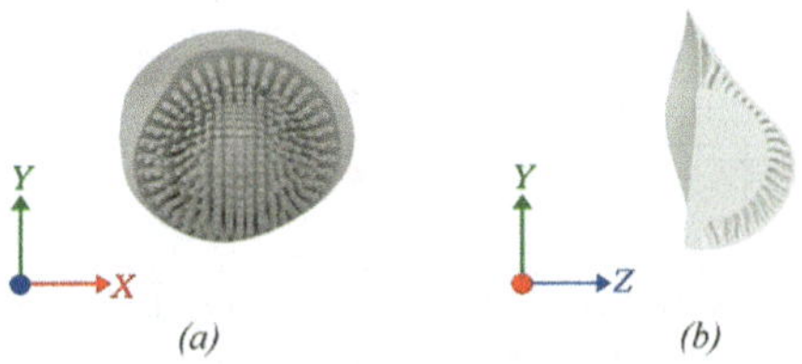

Fig. 3. External breast prosthesis with the interior mesh design; (a) back cutaway view (b) lateral cutaway view

prosthesis case was selected since it generated the maximum values of both stress and strain.

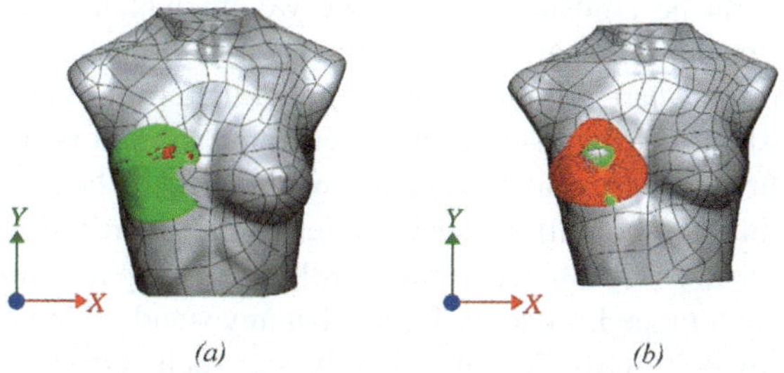

Fig. 4. Geomagic Control X 3D comparison software analysis with a green display contact surface and red non-contact surface: (a) evaluating the surface between the mastectomized area and customized prosthesis. (b) Assess the surface between the mastectomized region and the commonly marketed generic prosthesis.

Table 2. Percentage of contact comparative chart.

Prosthesis	Percentage of contact
Customized	96%
Generic	10%

Table 3. Mechanical analysis results: Dynamic model, Static model, and FEA

	Dynamic analysis (vertical *Y*-axis)	Static analysis (horizontal *X*-axis)	FEA: Bra Pressure (15 kPa)	FEA: Punctual Force (65 N)
Bra Clamping Force [N]	0,03	6,90	–	–
Maximum Strain [mm]	–	–	1,60	9,67
Maximum Stress [MPa]	–	–	0,32	2,57

4 Conclusions

Prosthesis developed in this study has approximately nine times higher contact percentage with the mastectomized surface than a commercial prosthesis. It means they adapt better to the mastectomized body and provide a natural appearance similar to the healthy breast. It could meet the minimum requirements to solve the comfort issue detailed in Table 1.

Curly tubes in the internal mesh structure provide flexibility to the design, and their diametrical length can be adjusted to meet the weight which each patient requires. Stresses generated in the external breast prosthesis by external stimuli such as punctual and distributed forces are within the acceptable range for ordinary conditions. The dynamic analysis showed that no friction is generated by deformation between both surfaces and that the prosthesis moves together with the chest; hence, it is expected not to propitiate harmful skin conditions such as dermatitis or irritations. Static analysis established the necessary force to act on the prosthesis to keep it adequately attached to the chest. It is within a range limited by forces that any standard bra can provide.

Besides revealing propitious mechanical properties such as isotropy, elasticity, resistance, and rigidity, the prosthesis reached a weight like the healthy breast, and therefore it would avoid injuries in patients. Additive manufacturing technologies are potential tools to produce breast prostheses and maximize comfort levels while keeping up typical sales rates.

Weight is a factor related to the percentage of adipose tissue in the breast that varies from one person to another, so it is recommended to obtain this value through a patient's mammogram to determine the external breast's required weight prosthesis correctly. 3D printing materials, especially the elastic type ones, do not have stress-strain engineering curves that allow a non-linear finite element analysis to be computed; future work could explore this field.

Acknowledgments. The authors would like to thank the Advanced Machining and Prototyping Laboratory CAMPRO, from ESPOL Polytechnic University, and HELGUERO 3D Inc., for their contribution to this work.

References

1. World Health Organization. https://gco.iarc.fr/today/data/factsheets/cancers/20-Breast-fact-sheet.pdf
2. Ferguson, D.F.: The actual extent of mastectomy: a key to survival. Perspect. Biol. Med. **30**, 311–323 (1987)
3. Harness, J.K., Willey, S.C.: Operative Approaches to Nipple-Sparing Mastectomy. Indications, Techniques, & Outcomes. Springer, Heidelberg (2017)
4. Andersen, B.L., Golden-Kreutz, D.M.: Depressive symptoms after breast cancer surgery: relationships with global, cancer-related, and life event stress. Psychooncology **13**, 211–220 (2004)
5. Lalani, S., Gul, R.B., Jetha, Z.A.: Women experiences of using external breast prosthesis after mastectomy. Asia Pac. J. Oncol. Nurs. **4**, 250–258 (2017)
6. Serrano, D.R., Terres, M.C., Lalatsa, A.: Applications of 3D printing in cancer. J. 3D Printing Med. **2**, 115–127 (2018)
7. Pedro Cruz, F., Hernandez, M., Zuñiga, J., Figueroa, R., Vertiz, A., Pineda, Z.: A biomimetic approach for designing a full external breast prosthesis: post-mastectomy. Appl. Sci. **8**(3), 357 (2018). https://doi.org/10.3390/app8030357
8. Helguero, C., et al.: Trabecular scaffolds' mechanical properties of bone reconstruction using biomimetic implants. Procedia CIRP **65**, 121–126 (2017)

9. Helguero, C.G., et al.: Biomechanical properties of 3D-printed bone scaffolds are improved by treatment with CRFP. J. Orthop. Surg. Res. **12**(1), 195 (2017). https://doi.org/10.1186/s13018-017-0700-2
10. Helguero, C., Castro, J., Ochoa, C., Maldonado, F., Ramírez, E.A., Amaya, J.L.: Improving cutting path on custom 3D-printed surgical guides for bone-tumor resection. Biomedical and biotechnology engineering. In: Proceedings of the ASME 2019 International Mechanical Engineering Congress and Exposition. ASME, Salt Lake City (2019)

Advanced Production Management and Process Control

Increasing Agility: A Tangible XR for Rapid Prototype Development, Design Exploration and Usability Assessment

Ronald Meyer[1(✉)], Michael Preutenborbeck[1], Marcel Usai[1], Kristof Briele[2], André Witzel[3], and Frank O. Flemisch[1]

[1] Institute of Industrial Engineering and Ergonomics, Systemergonomie, Eilfschornsteinstr. 18, 52062 Aachen, Germany
{r.meyer,m.preutenborbeck,m.usai,f.flemisch}@iaw.rwth-aachen.de
[2] Laboratory for Machine Tools and Production Engineering (WZL), RWTH Aachen University, Campus-Boulevard 30, 52074 Aachen, Germany
k.briele@wzl.rwth-aachen.de
[3] Institute for Technology and Innovation Management, Kackertstr. 7, 52072 Aachen, Germany
witzel@time.rwth-aachen.de

Abstract. Agile processes demand for increasing flexibility in product development and shorter development periods. Especially when it comes to product assessments, products have to run through a complex and often expensive development cycle before first user tests can be conducted. We are developing a method where a tangible XR (EXtended Reality) is used for product assessment in an agile development process with evolving the product from a virtual to a physical state. This approach is illustrated in a prototypical example of the implementation of an automotive interior design of a tangible XR car door opening mechanism.

Keywords: Tangible · Agile processes · Virtual reality · Augmented reality · Product development · Product assessment · Human systems exploration

1 Introduction

The exploration of new product ideas and new concepts from design, engineering and production of prototypes towards user assessments is an essential process to bring products to the market. Different market analysis methods exist to generate new products or improve product ideas by analyzing trends using tools like data mining of social media platforms. Whatever method is used, new ideas need to be implemented and tested before market release to ensure market readiness. In software development, agile development methods are applied to ensure altering cycles of implementation and testing a digital product until market readiness, nevertheless agile methods for the testing of physical products are not established. Hence, many ideas and prototypes are frequently discarded after running through a complete physical development processes or after a user assessment with fully developed prototypes. Furthermore, more and more development processes become agile, especially in the transformation process to the

S. Trzcielinski et al. (Eds.): AHFE 2021, LNNS 274, pp. 85–91, 2021.
https://doi.org/10.1007/978-3-030-80462-6_11

Fourth Industrial revolution, and may shorten production time and therefore increase productivity.

The Cluster of Excellence "Internet of Production" at RWTH Aachen University focusses on research and development of new methods in the field of the Fourth Industrial Revolution. The interaction of automation, the application of smart methods and the usage of huge datasets are object of investigation in finding new intelligent models and processes to support industrial engineering and production, as well as improved methodologies such as agile development.

The involvement of product users and/or usability experts in the development process can additionally improve development at early stages but may also be difficult since the products are not in a usable state yet in an early development phase. At those early stages, prototypes need to be balanced between cost, effectiveness and readiness for usability assessment. A possible solution to reduce costs and assess products or interaction methods before their physical development is tangible virtual reality, or tangible XR. A complex virtual environment enables the integration of a new product idea as a virtual entity in its intended context – in an interactive way for product and user assessment. Methods for explorative user assessment and design explorations have already been applied in so-called Exploroscopes [4, 7], in which experts and users assess and explore new ideas and interaction methods in a playful, creative way.

The paradigm of exploration can be a solution to a more agile development and product assessment process and is investigated the "Internet of production" in a DFG cluster of excellence established at RWTH Aachen University. Our goal is to develop a tangible XR for exploration purposes to enable both users and experts to explore tangible virtual products obtained from a digital shadow. Increasing tangibility is achieved through augmenting the virtual world with real tangible objects, i.e., buttons, handles or surfaces that are augmented in XR.

2 Related Work

Work on multimodal mixed reality prototypes has been focused by various researchers recently, with and without considering the haptic, i.e., tangible, channel. Both ways of virtual interaction may allow users and other stakeholders to be involved in the product development process at an early stage in order to capture and take into account their needs, the so-called Voice of Customer. To achieve this, there are several strategies for developing mixed reality or tangible XR prototypes under usage of different technology.

Bodergoni et al. develop a framework to structure mixed reality prototypes [1]. They identify the characteristics of the prototype, the user and the interaction as dimensions to classify mixed reality prototypes appropriately. Prototypes and users can be either real or virtual. As another dimension, the authors consider the characteristics of the interaction. The interaction of users with the prototype can be either direct, through visual and haptic feedback, or only indirect through a visual one. Depending on the prototype's characteristics, the authors suggest different methods and techniques for the evaluation of products. Based on this framework, we will now focus on approaches where a real user can directly interact with a virtual prototype via haptic elements.

Stark et al. describe the integration of mixed reality elements as Smart Hybrid Prototyping (SHP) [2]. The authors describe SHP as a concept for multimodal and cross-domain assessment of virtual prototypes in early phases of product development. This is based on methods of mixed reality and extends them with mechatronic input devices as a bridge between physical reality and digital virtuality. In this way, the user can gain haptic impressions in addition to visual and auditory ones.

In a similar way, Carulli et al. show in their work the potential of multimodal virtual prototypes to capture the Voice of Customers [3]. As a use case, the authors use a new user interface design for a washing machine. Potential customers could adapt the interface directly by interacting with the virtual prototype. The study's positive feedback confirms the use of multimodal virtual prototypes in the development process to determine the needs and ideas of potential customers.

Flemisch et al. describe the concept of multimodal virtual prototypes under the term tangible XR [4]. An initial concept of a tangible XR application has been implemented in a defense project EnUSi, which deals with innovative vision systems in armored vehicles. With a tangible XR Simulator, questions about the ergonomic design and use of vision systems (camera monitor systems) were investigated by constructing a simplified driver workplace mock-up. By including soldiers in a participatory design process, this simulator has the ability to quickly evaluate new systems since the modular design of the hardware and software allows for a faster implementation.

In addition to a tangible XR, Flemisch et al. develop the concept of Exploration, a pre-cursor of the development phase, and complement of tests and experiments, as an underlying principle providing a systematic way to explore the design and use space of potential systems [5]. The core of the methodology is an iterative and participative design approach that allows new human-machine systems to be explored and improved with stakeholders. By integrating stakeholders at an early stage in the development process, it is possible to balance different, sometimes contradictory product and process qualities and therefore improve agility.

3 A Tangible XR for Agile Development

The concept of tangible XR is meant to speed up processes and reduce costs in many ways. The implementation of the haptic channel, thus adding force feedback or physical surfaces with texture to virtual objects, facilitates combining creativity with engineering skills. However, we see the application of a tangible XR not exclusively as a creative process but as an agile method to reach goals or milestones in a product development and engineering cycle with the abilities to apply market development methods.

The usage and creation of data relates on the one hand side to the usage of the product, i.e. the user perspective, and on the other hand side on the engineering of the product, i.e. the production of a prototype. Our idea of an iterative process with the application of a tangible XR is implemented in a way, where the product evolves from a virtual to a more and more physical state.

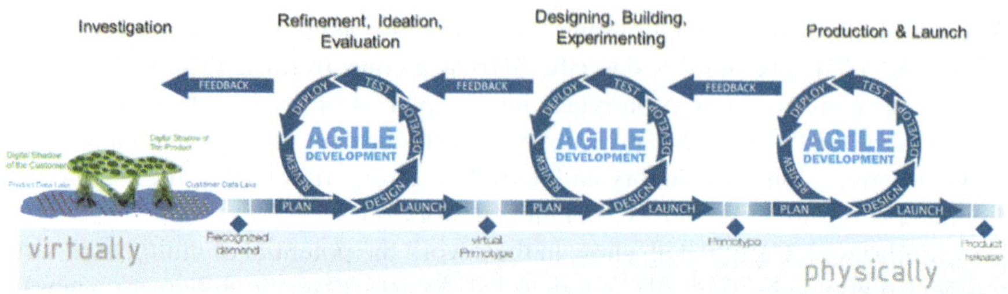

Fig. 1. Agile Process of a Tangible XR from virtual to physical state with iterations and development steps with feedback loops.

Figure 1 shows the sequence of our agile development process with its iteration steps. Each step may be implemented by its own agile development method or cycle, i.e. Scrum or Feature Driven Development. The process starts with the recognition of a demand. A demand may have different origins – the improvement or extension of an existing product or the invention of a completely new product. In both cases a data set is stored in a so-called data lake, which is a collection of databases containing a digital shadow of the customer and a digital shadow of the product. The terminology 'digital shadow' has its origin in the discussion on the debate about privacy and was first used in the context of production in 2015 with the term 'digital twin' referring to Liebenberg and Jarke [5]. In their description, the 'digital twin' or digital shadow is a dynamic digital view or trace on a physical process or a simulation, where only those aspects are represented which are necessary for a specified purpose. Liebenberg and Jarke mention the digital shadow to contain e.g. subsets of data from production processes or mapping data that is linked to other data that are usable in certain scenarios where fast reaction is crucial. Furthermore, we see the digital shadow as a twin of a physical object, with properties that are related to its composition, its use or application or its relation to other objects, either physical or virtual. We extend our understanding of the digital shadow by not solely being a digital counterpart to a physical object but also being a virtual product with a certain set of properties that may further evolve to a physical product at later stages over development iterations as shown in Fig. 1. In addition to the digital shadow of the product our described process considers a digital representation of the user, i.e. customer, whose usage data relates to the usage of a product stored in the data lake linked to the digital shadow. The shown process is designed to keep data accessible over the iterations and is meant to be altered and complemented over the process with each development step.

The iteration steps, where each of them contains the planning, design, development, testing, reviewing and deploying the product, involves user or expert groups (or the customer) for intense testing of the product. At early stages of the process a product or a demand may be handled entirely virtual, i.e. the intended improvement, redevelopment, or new product is designed as a virtual prototype, which is meant to be tested and deployed completely in a virtual environment. Conclusions and results from the testing phase flow back as new information into the data lake and are added to the digital shadow of the customer and the digital shadow of the product. At this early stage of product development where the product idea may or may not be at market readiness we implement a tangible XR to explore the usage of the product and test the developed product ideas. Therefore, we implement haptic devices replicating certain haptic impressions that are related to the product. These haptic impressions can be e.g. imitation of surfaces or handles, the implementation of force feedback devices imitating forces on controls and actuators or an implementation of frames with a combination of surfaces and forces to simulate large physical objects. Adding these impressions are meant to generate a higher degree of immersion and therefor increase testing capabilities of users' or experts' (customers) on products or ideas that are not at a ready-developed state yet.

With the progression of the process the virtual idea and virtual development evolves to a more and more physical primotype, where each step in development enables a more and more physical implementation of the newly developed idea or product. The iterative process provides product testing and exploration in each step and contains so-called design explorations as described by Flemisch et al. [4]. Design explorations allow thematically connected activities and tools to invent, design, prototype and assess the effects of different options of human machine systems. The combination of design explorations and agile development increases the speed of implementation, development and evaluation of new innovations [6, 8]. The generation of feedback to the data lake and to other iterations at each agile development step is implemented as a mechanism in the process itself. It ends with the production and launch of a physical product that has been explored and tested over several stages from a virtual to physical state.

4 First Application Scenario

As a first implementation of our described process, we developed an application scenario with a prototype from the domain of automotive interior design incorporating the improvement of the haptic quality of an opening mechanism of a car door. Our chosen step of development in the process shown in Fig. 1 is located after the 'recognition of the demand' at a first test of the 'virtual primotype'. Figure 2 shows the setup of our prototypical system. For the assessment process both user and a so-called "explorer" are present in our laboratory, where the user is wearing a pair of virtual reality glasses. Our physical laboratory setting incorporates a frame of extendible Bosch-Rexroth aluminum profiles, here, extended to a frame of a car door shape. The car door shape can be rotated to simulate the opening mechanism from the interior of the car door. The car door shape is physically connected to an actuator, which is capable of generating force to the car door shape through an axle when the car door is rotated/opened. The forces of the actuator given on the car door rotation are perceived as passive resistance by the user, when opening the car door. The intensity and the profile of the forces may be altered by the 'explorer' by manipulating force profiles on a mobile device, our so-called "exploration panel". This allows the 'explorer' to test different settings of the opening forces of the car door and therefore evaluate different force settings spontaneously in cooperation with the user. The user visually perceives an augmented version of the car door in the virtual reality environment, as shown in Fig. 2 (top left corner).

The motor control is implemented using a linux/Ubuntu-based system which provides force and angle data of the motor (SensoDrive System) to a ROS interface which sends data to a Windows based Unity System, which provides the virtual reality environment with the augmentation of the car door. A mobile device, also connected via

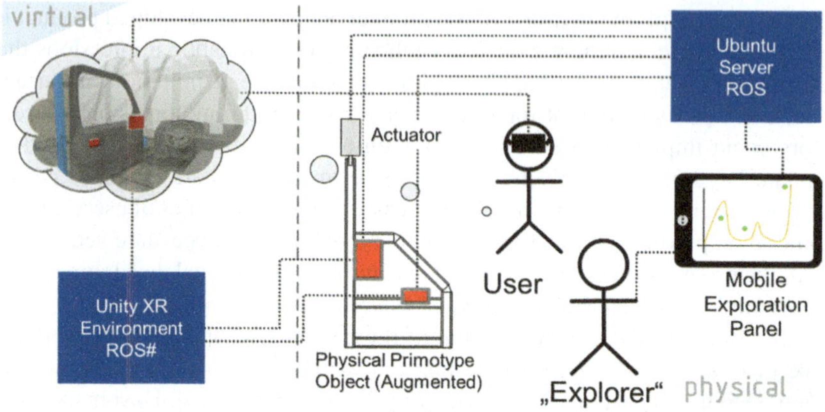

Fig. 2. Scenario of car door opening mechanism as tangible XR in a laboratory setting: Explorer with mobile exploration panel adjusts physical forces on actuator and thus explores different opening forces of a car door with augmented visualizations in virtual reality for the user. The users' haptic impression is generated by a car door frame connected to an actuator – visuals are fully virtual and mapped on real physical objects

ROS, enables the manipulation of the force/resistance of the actuator connected to the car door frame via a graphical representation.

The implementation of the scenario enables the 'explorer' to assess different types of force/resistance in collaboration with the user with immediate feedback without implementing different types of car doors. The user, i.e. the subject that opens the car door from the inside of the virtual car, may give direct feedback on the desired improvement or demand of the stakeholder (voice of the customer) and may explore extended scenarios in collaboration with the explorer.

5 Conclusion

We have described a new process that enables the creation of products or ideas that may be tested before market readiness at a virtual state by implementing a tangible extended virtual reality, or tangible XR. The process is presented as an agile method with digital representations of the customer and the product in a global data set which can be seen as a data lake. The process incorporates the implementation of design explorations, which enable a temporally and thematically connected series of activities, techniques and tools to invent, design, prototype and assess the effects of different options of human-machine systems. Our first implemented scenario shows the capabilities of the newly developed process and may be extended to future scenarios.

6 Future Work

While the process we describe represents an entire chain of agile development, including repetitive steps such plan, design, develop, test, deploy and review, we have implemented only an intermediate step with our prototypical scenario. The generation of user and customer data with a feedback loop to the data lake can be executed as a proof of concept with our prototype state. Anyway, further steps to higher physical states of development need to be implemented and connected. The development of the process has led us to the following research questions, which we will answer with further development of our prototypical scenario and user/expert assessments: Can the 'explorer' assess the "voice of the user/customer" reliably and realistically by following our process? What are the limitations of our environment and which scenarios can be implemented for assessment? How does the virtual reality influence the perception of the tangible object? Addressing these research questions is planned for future research.

Acknowledgement. This work is funded by the Deutsche Forschungsgemeinschaft (DFG, German Research Foundation) under Germany's Excellence Strategy – EXC-2023 Internet of Production – 390621612.

References

1. Bordegoni, M., Cugini, U., Caruso, G., et al.: Mixed prototyping for product assessment: a reference framework. Int. J. Interact. Des. Manuf. **3**, 177–187 (2009). https://doi.org/10.1007/s12008-009-0073-9
2. Stark, R., et al.: Competing in engineering design - the role of virtual product creation. In: Proceedings of the 19th CIRP Design Conference, Cranfield Univ. Press, Cranfield (2010)
3. Carulli, M., Bordegoni, M., Cugini, U.: An approach for capturing the voice of the customer based on virtual prototyping. J. Intell. Manuf. **24**(5), 887–903 (2012). https://doi.org/10.1007/s10845-012-0662-5
4. Flemisch, F., et al.: Towards a balanced human systems integration beyond time and space: exploroscopes for a structured exploration of human–machine design spaces. In: HFM-231 SYMPOSIUM On "Beyond Time and Space"; NATO-STO Human Factors and Medicin Panel, Orlando (2013)
5. Liebenberg, M., Jarke, M.: Information systems engineering with digital shadows: concept and case studies: an exploratory paper. In: Dustdar, S., Eric, Y., Salinesi, C., Rieu, D., Pant, V. (eds.) Advanced Information Systems Engineering: 32nd International Conference, CAiSE 2020, Grenoble, France, June 8–12, 2020, Proceedings, pp. 70–84. Springer, Cham (2020). https://doi.org/10.1007/978-3-030-49435-3_5
6. Bielecki, K., Bloch, M., Schmidt, R., Hernández, D. L., Baltzer, M., Flemisch, F.: Tangible virtual reality in a multi-user environment. In: Proceedings of the 11th International Conference on Automotive User Interfaces and Interactive Vehicular Applications, Adjunct Proceedings, pp. 76–80 (2019)
7. Flemisch, F., Baltzer, M., Sadeghian, S., Meyer, R., Hernández, D., Baier, R.: Making HSI more intelligent: human systems exploration versus experiment for the integration of humans and artificial cognitive systems. In: Karwowski, W., Ahram, T. (eds.) IHSI 2019. AISC, vol. 903, pp. 563–569. Springer, Cham (2019). https://doi.org/10.1007/978-3-030-11051-2_85
8. Flemisch, F., et al.: Let's get in touch again: tangible AI and tangible XR/AR/VR for a more tangible, balanced HIS. In: International Conference on IHSI, Modena (2020)

Applying Packing Problems to Optimize Throughput Time and Human Ergonomics in Project Shops

Thomas Henke[1(✉)] and Jochen Deuse[1,2]

[1] Institute of Production Systems, TU Dortmund University, Leonhard-Euler-Straße 5, 44227 Dortmund, Germany
{Thomas.Henke,Jochen.Deuse}@ips.tu-dortmund.de

[2] Advanced Manufacturing, School of Mechanical and Mechatronic Engineering, University of Technology Sydney, Ultimo, Australia

Abstract. Special Machine Manufactures (SMM) assembling in project shops have to ensure a high delivery performance to be competitive, while the demographic change reinforces the importance of human ergonomics. Since the degree of standardization in project shops is low, the assembly and logistical processes are planned and controlled insufficiently. Particularly the material supply shows a great potential for improvement. Instead of provoking high non-value adding shares and ergonomic risks by supplying the material for an order completely and randomly around the mounting area at the start of assembly, a structured material supply enables SMM to reduce and stabilize the throughput time in order to increase the delivery performance and helps to decrease the ergonomic risks for the workers. In this paper a promising approach for SMM is presented and validated.

Keywords: Project shop manufacturing · Packing problem heuristics · Throughput time · Value adding times · Special machine manufacturing

1 Introduction

The increasing globalization accompanied with the findings in human ergonomics result in two major challenges for special machine manufacturers (SMM) [1, 2]. Firstly, they are confronted with new competitors and therefore strive to fulfil the agreed delivery dates [1]. Secondly, the demographic change in western countries increases the work age average that reinforces the importance of human ergonomics in workplaces [2]. To master these challenges SMM need innovative approaches.

As their competitive feature SMM are focused on the production of unique large scale products that are assembled in project shops [3]. Associated with this the assembly and logistical processes are characterized by a low degree of standardization. Hence all components belonging to an order are supplied randomly around the assembly area at the start of the assembly. By that the work content of the workers is essentially determined by non-value adding walking, searching and handling times. Therefore not

S. Trzcielinski et al. (Eds.): AHFE 2021, LNNS 274, pp. 92–99, 2021.
https://doi.org/10.1007/978-3-030-80462-6_12

only the throughput time of an order spreads depending on the non-value adding times and increases the risk of delivery delays, but also the ergonomic stress for the workers increases, since their work contains preventable non-ergonomic handlings to relocate components as well as avoidable walking routes [3–5].

To meet the requirements of human factors and ergonomics as well as to lower and stabilize the throughput time to prevent delivery delays, an innovative approach for project shops is presented in this paper. Based on the technical background and related research (Sect. 2) the approach is presented in Sect. 3. It translates the depicted challenges into a mathematical optimization problem and solves them by adapting and applying heuristics for packing problems. In Sect. 4 the industrial applicability of the approach is validated by a case study. The paper concludes with a summary.

2 Technical Background and Related Research

In this section, the relevant background on project shop manufacturing and packing problems as well as existing approaches are presented.

2.1 Project Shop Manufacturing

Products assembled in project shops are characterized by a high level of uniqueness, large dimensions and high weights, leading to an assembly in fixed positions. Consequently, SMM are assigned to the engineer- or make-to-order principle [3]. The most important competitive priority is the delivery performance that is calculated from the difference between the actual and the agreed delivery date. It therefore essentially depends on the throughput time and the prevention of deviations [6].

A particular challenge in project shop manufacturing of special machines is the low repetition frequency of the products, as a result of which the degree of standardization must be classified as very low. The assembly planning and control are mainly based on experience and lead to suboptimal planning results [4, 5].

A representative example is the material supply. Due to the low degree of standardization the material of an order is mainly provided completely and randomly on a provisioning area at the start of production. This leads to a high non-value adding time share for the workers, since their assembly tasks go along with high search times, long walking and transporting distances as well as preventable material handlings [7]. By this not only the throughput times spread depending on the orders and increase the risk of delivery delays, but also the ergonomic risks for the workers rises. To avoid searching, transporting and handling efforts in order to minimize ergonomic risks as well as to stabilize the throughput time in order to ensure the delivery date, a standardized and structured allocation of the supplied components is essential [8].

A promising approach to determine the optimal arrangement of components and modules during the material supply is the application of heuristics for packing problems that are presented in the following section.

2.2 Packing Problem Heuristics

Packing problems belong to the combinatorial optimisation problems that are frequently used to solve manufacturing and logistical problems [9]. The packing problems addressed in the paper are the two dimensional Strip Packing Problem (SPP) and the Rectangle Packing Area Minimization Problem (RPAMP), that are known to be NP-hard and therefore need to be solved by heuristics [10, 11].

In the SPP a set of n non-rotatable rectangular *items* i_j with $j \in \{1,\ldots,n\}$, each having a width w_j and a height w_h have to be allocated on a strip S with a given width W so that the resulting height H is minimized. A solution is feasible, if the items do not overlap, their edges are parallel to those of the strip and the strip's width is not exceeded [12]. While the SPP aims to minimize the height of a given area, the RPAMP aims to allocate a set of non-rotatable rectangular items on an area with an infinite width W and an infinite height H in order to minimize the resulting area. For this purpose the available area is divided into an arbitrary number of widths for each of which an SPP is applied. By a final comparison of the resulting areas the minimal area is identified [13].

One of the best heuristics for the SPP, regarding its result and time complexity, is the SC-Algorithm [12]. It is based on the idea of dividing an area in levels, whose heights are each defined by its first item's height. At first, all items are listed in the descending order of their height. By going through the list the baseline is filled until no more item can be placed (Fig. 1, 1). Following, the next items are stacked below the upper level border from right to left (Fig. 1, 2). After an item is placed, the gap between the item and the item on the baseline is determined and an attempt is made to fill the gap. After every item placed in the gap, it is tested whether another item can be placed around it. After that the next item is placed below the upper level border and a new stacking process is initiated (Fig. 1, 3) until no further item fits a level and a new one is opened (Fig. 1, 4). This procedure continues until all items on the list are allocated [12].

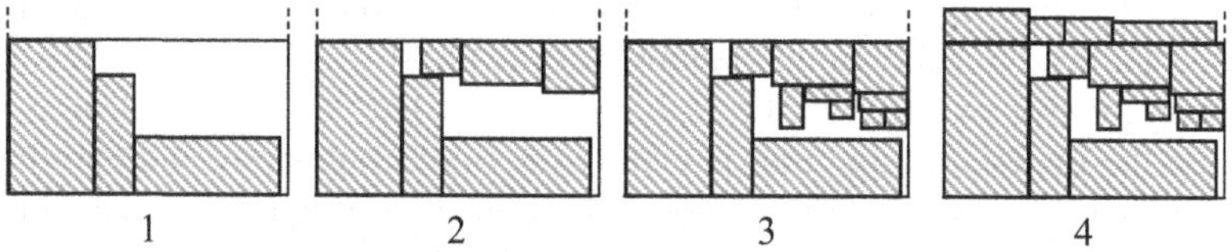

Fig. 1. Procedure of the SC-algorithm

2.3 Related Research

Even though many companies assemble in project shops, almost no approaches to standardize the material supply exist. Most approaches aim to overcome disturbances and interventions in the assembly by identifying alternative target-oriented assembly routes [14–16]. An approach focussing on project shops and the space is presented by [5]. The authors maximize the use of the assembly space by applying a mathematical algorithm that continuously checks the progress of all project shops and reallocates the area edges dynamically. Approaches using packing problems to solve logistical problems have been presented in an innumerable amount of papers, but mainly aim to maximize the packing density and to minimise the number of packing carriers. Comprehensive overviews are given by [17, 18].

Considering the state of the art and the related research, heuristics for packing problems are a promising approach to achieve a high degree of standardisation during the material supply in order to decrease ergonomic risks and delivery delays.

3 Approach to Optimize Throughput Time and Human Ergonomics in Project Shops

To adapt the SC-algorithm for its application in project shops certain requirements have been defined. To reduce and stabilise the throughput time as well as to avoid ergonomic risks the share of waste has to be minimized and walking and transport distances as well as search times and handlings have to be reduced to a minimum. Therefore, the components that are assembled in early stages of the assembly have to be allocated nearer to the point of fit than components assembled in later stages (R1). Furthermore, components have to be allocated in clearly defined locations (R2), components belonging to one module have to be allocated together (R3) and the required space for the material supply has to be minimized (R4). To remain flexible and to be able to apply existing approaches for interventions a permanent accessibility of all components/modules during the assembly process has to be ensured (R5). Since the SPP and RPAMP are NP-hard, the MBS-algorithm has to be solvable in a realistic time (R6).

While SPP fulfil *R2, R4* and *R6,* allocating items by their height violates *R1* and *R3* while the stacking fails *R5.* Since RPAMP find the minimal area, they fulfil *R3.* Therefore, the MBS-Approach consists of two consecutive steps. At first, the minimal area of a module is calculated by an RPAMP, before the modules are allocated on the scheduled provisioning area by using an adapted SC-algorithm called "MBS-algorithm".

Instead of the items' heights, the "MBS-algorithm" allocates the items considering the assembly sequence (*R1*) and ensures their permanent accessibility (*R5*). To fulfil *R1* the MBS-algorithm does consider a list of items sorted by the decreasing height, but a list sorted by a module's position in the assembly process. This approach may increase the height of the required provisioning area and lead to a lower packing density, but it ensures, that modules being assembled earlier have the shortest transport route. To increase the packing density, after each completed baseline the allocated modules are resorted by their heights. By this the modules' heights on the level's baseline decrease from left to right, while the modules' heights below the upper level border decrease from right to left. Furthermore, the accessibility of modules being assembled earlier is bettered, since they are allocated in front of the modules that have to be assembled later, in order to ensure a permanent direct transport route and to lower the transport effort by avoiding lifting operations to transport modules over others.

To resolve the violation of *R5*, the stacking process is limited to the allocation on the baseline and below the upper level border, while after a level is finished and before the new level is started a transport distance is allocated in between the two levels. By this, every item is permanently accessible during the assembly. The dimension of the transport distance depends on the required dimensions of the lifting aid like forklifts, lift trucks or cranes. In this paper a transport distance of 1.8m is recommended and used. The MBS-algorithm for project shops is shown exemplary in Fig. 2.

The starting point of the MBS-Approach is the definition of the dimensions of the module-areas for the RPAMP. Since the SC-algorithm has been identified as the best SPP-Algorithm, the SC-RPAMP is the preferred algorithm to be applied. The selection of the reasonable width values is done with the *WV1* approach by Bortfeldt [13]. After the promising widths for all module-areas have been determined, the modules are sorted by their position in the assembly sequence.

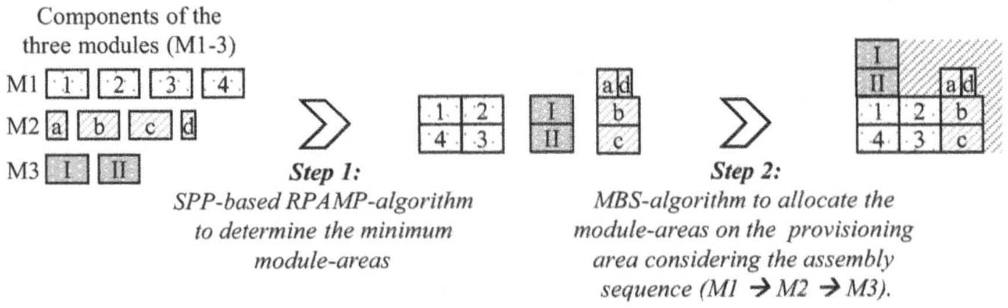

Fig. 2. Example for the procedure of the MBS-Approach

At the start of the MBS-algorithm it is checked whether an unallocated module exists. In this case a new level S_j is opened and filled with modules from the list. Once the allocation on the baseline is complete, the resorting procedure starts. Hereby the modules on S_j are resorted by their height, so that the height decreases from left to right. Afterwards the next items are allocated below the upper level border S_c. Each allocation below S_c is followed by the resort procedure until the allocation below the upper level border is complete. Once a level is completed a transport distance of 1.8m is allocated above S_c and a new run of the MBS-algorithm for the next level is initiated. When all modules have been allocated the MBS-algorithm is finished. To clarify and validate its application, the MBS-Approach is applied to an industrial use case in the next chapter.

4 Case Study: Application to an Industrial Project Shop

The use case is based on an industrial assembly process of a SMM. Scenario 1 is the status quo, while in scenario 2 the use of the MBS-Approach and the results will be presented. Both scenarios have been carried out, while the duration of the tasks has been measured by time studies. A layout of the assembly and provisioning area is shown in Fig. 3.

The case study consists of the three submodules *Preassembly of bearings, Preassembly of flanges* and the *Preassembly of shafts.* At the beginning of the production the material is supplied on the provisioning area (4m x 14m), allocated on the left. Except of the shafts (Fig. 3, Component 22 & 23) all components are supplied on pallets. While the preassembled bearings are buffered next to their preassembly, the preassembled flanges are buffered on the provisioning area (Fig. 3, top left). Both, the bearings and the flanges are preassembled into the shafts (Fig. 3, top right). Afterwards the shaft is mounted into the housing, before the motors and gears are mounted on their lower sides (Fig. 3, below). The entire assembly process takes 2509 min respectively 41.8 h. While 60.7% (1524 min) of the time is value adding, 39.3% (985 min) is non-value adding. The non-value adding time can be divided into 317 min for search times, 318 min for transporting the components to their point of fit and 350 min for additional tasks like unpacking, (dis)mounting transport aids or getting tools and equipment.

According to the MBS-Approach the first step is the definition of the module-areas by using the SC-RPAMP-algorithm. Afterwards the modules are allocated by the MBS-algorithm regarding their positions in the assembly sequence. The width of the strip is 14m. The resulting layout is shown in Fig. 3. To prevent distortions due to a different execution of the assembly process and to focus on the reduction of the non-value adding time, the value adding time from scenario 1 (1524 min.) will adopted for the second scenario.

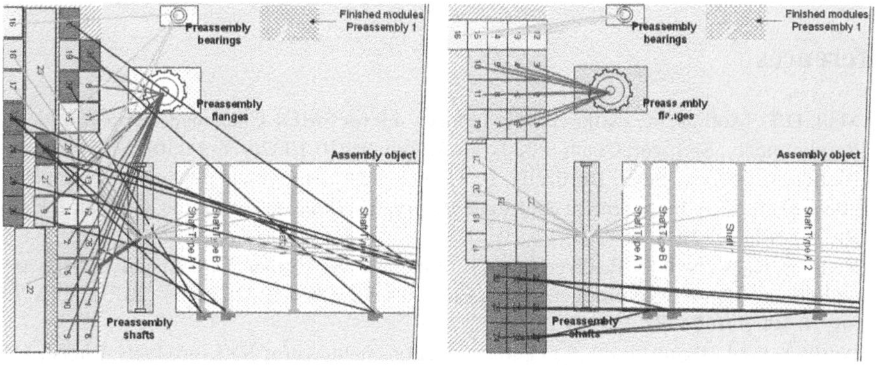

Fig. 3. Material flow before (left) and after (right) using the MBS-Approach

As shown in Fig. 3 the MBS-Approach leads to a clearly straightened material flow. Also, the non-value adding time is reduced to 761 min which equals a decrease of 6% compared to the first scenario. According to this, productivity has increased up to 66.7%. The non-value adding time is composed of 126 min for search times, 291 min of transporting times and 344 min for additional tasks.

The throughput time to complete the task in scenario 2 was decreased by 8.9% to 2285 min. This optimisation is solely achieved by optimising the allocation of the components during material supply and is completely traceable to the reduction of non-value adding time. The main improvement is traced to the increased degree of standardisation that leads to a reduction of the search time of 60%.

The main improvement regarding the ergonomic factors is the reduction of the walking and transporting routes that have been shortened by almost 1 km and the prevention of unnecessary handlings. These improvements not only lower the ergonomic risks but also improve the employee satisfaction and productivity.

5 Conclusion

The presented approach reduces and stabilises the throughput time and prevents ergonomic risks in project shops. Based on the preliminary work presented in literature and papers regarding project shops and packing problems, the applicability of heuristics for packing problems to solve logistical problems has been shown.

On the basis of these findings, the MBS-Approach using heuristics for packing problems was developed to standardize the material supply in order to increase the delivery performance and to lower ergonomic risks. The MBS-Approach consists of two consecutive steps. At first an SC-RPAMP-algorithm is used to calculate the necessary area for each module. In the second step the developed MBS-algorithm is applied to allocate the modules on the provisioning area regarding the assembly sequence.

Finally, the potential of the MBS-Approach was applied to an industrial use case. It was proven that the reduction of the non-value adding time reduced the throughput time by 8.9%, while the ergonomic risk for the workers (walking and transporting times: -10%) was minimized and unnecessary handlings were avoided completely.

References

1. Matt, D.T., Modrák, V., Zsifkovits, H.: Industry 4.0 for SMEs. Challenges, Opportunities and Requirements. Springer, Cham (2020). https://doi.org/10.1007/978-3-030-25425-4
2. Bieling, G., Stock, R.M., Dorozalla, F.: Coping with demographic change in job markets: how age diversity management contributes to organisational performance. German J. Human Resour. Manag. **29**, 5–30 (2015)
3. Behrens, B.-A., Nyhuis, P., Overmeyer, L., Bentlage, A., Rüther, T., Ullmann, G.: Towards a definition of large scale products. Prod. Eng. Res. Dev. **8**(1–2), 153–164 (2013). https://doi.org/10.1007/s11740-013-0503-1
4. Potthast, J.-M., Baumgarten, S.: Lean-production-methods for XXL-products. Logist. J. 1–4 (2011)
5. Zenker, M., Prinzhorn, H., Böning, C., Strating, T.: An optimization model for the arrangement of assembly areas considering time dynamic area requirements. World Acad. Sci. Eng. Technol. **11**, 241–248 (2017)
6. Szwejczewski, M., Jones, M.: Learning from World Class Manufacturers. Palgrave Macmillan, Basingstoke (2013)
7. Noha, H. and Abderrazak, B.: Toward a global approach for value chain optimization, based on Lean management. In: International Colloquium on Logistics, pp. 1–6 (2019)
8. Huang, G.Q., Zhang, Y.F., Jiang, P.Y.: RFID-based wireless manufacturing for walking-worker assembly islands with fixed-position layouts. Robot. Comput.-Integr. Manuf. **23**, 469–477 (2007)
9. Lodi, A., Martello, S., Vigo, D.: Recent advances on two-dimensional bin packing problems. Discret. Appl. Math. **123**, 379–396 (2002)
10. Chen, M., Tang, X., Zeng, Z., Liu, S.: An efficient heuristic algorithm for two-dimensional rectangular packing problem with central rectangle. J. Ind. Manag. Optim. **13**, 1–16 (2017)
11. Christensen, H.I., Khan, A., Pokutta, S., Tetali, P.: Approximation and online algorithms for multidimensional bin packing: a survey. Comput. Sci. Rev. **24**, 63–79 (2017)
12. Ortmann, F.G.: Heuristics for Offline Rectangular Packing Problems. Stellenbosch University, Stellenbosch (2010)

13. Bortfeldt, A.: A reduction approach for solving the ractangle packing area minimization problem. Eur. J. Oper. Res. **224**, 486–496 (2013)
14. Qian, C., Zhang, Y., Jiang, C., Pan, S., Rong, Y.: A real-time data-driven collaborative mechanism in fixed-position assembly systems for smart manufacturing. Robot. Comput.-Integr. Manuf. **61**, 101841 (2020)
15. Rochow, P., Burggräf, P., Reuter, C., Prinzhorn, H., Wagner, J., Schmitz, T.: Identification of alternative assembly sequences for large-scale products (2014)
16. Guo, D., et al.: Towards Assembly 4.0: graduation intelligent manufacturing system for fixed-position assembly islands. IFAC-PapersOnLine **52**, 1513–1518 (2019)
17. Perboli, G., Gobbato, L., Perfetti, F.: Packing problems in transportation and supply chain: new problems and trends. Soc. Behav. Sci. **111**, 672–681 (2014)
18. Eliiyi, U., Eliiyi, D.T.: Applications of bin packing models through the supply chain. Int. J. Bus. Manag. **1**, 11–19 (2009)

Organizational Risk in Custom Manufacturing of Complex Products

Anna K. Stasiuk-Piekarska(✉) and Magdalena K. Wyrwicka

Institute of Safety and Quality Engineering, Faculty of Management Engineering, Poznań University of Technology, Rychlewskiego 2 Street, 60-965 Poznań, Poland
{anna.stasiuk-piekarska,magdalena.wyrwicka}@put.poznan.pl

Abstract. Every organization needs to use the experience and knowledge of its people to identify threats with a potential to cause adverse events, i.e. disruptions, and prepare appropriate contingency plans to address them. Doing so forms a basis of risk management. Any identifiable threats should be assigned to relevant risk categories. A response to emerging threats will be a test not only of an organization's preparedness but also of its vigilance, speed of response and flexibility. The article discusses the essence of organizational risk, which it examines in relation to the custom production of highly complex products with particular focus on unscheduled times. All performance times outside of such schedules are referred to as "unscheduled" and are reported with a view to identifying their causes. The authors analyze historical data (from the year 2018) on reported unscheduled times and assign them to separate categories to identify their causes.

Keywords: Organizational risk · Manufacturing systems · Risk factors · Risk categories · Customization

1 Introduction

Any manufacturing company aspiring to achieve sustained competitive advantage in a global market needs to improve its processes. To that end, it is imperative that it identifies weaknesses in its manufacturing system. Ever since the advent of Industry 4.0 and the outbreak of COVID-19, companies have been building their global position by delivering personalized products, cutting production costs and automating their processes. Significant support in the development of auto-controlled production systems has come from the Internet of Things[1], which ensures efficient data exchange to support smart manipulation of matter with a view to creating an environment in which the product will "organize itself into existence" [2, 8, 12] [18]. Another aim is to create systems that more rapidly respond to changing customer preferences and reduce failure rates while

[1] Future communication systems "include vehicles that drive, maneuver and park autonomously, systems for shipment tracing and shopper behavior analytics, automatic picking for shipping points, smart power grids, remote medical diagnostics (e.g. to assess mental and physical condition of drivers), air conditioning systems notified in advance to adjust indoor conditions ahead of user arrival and a host of other smart solutions" [16].

S. Trzcielinski et al. (Eds.): AHFE 2021, LNNS 274, pp. 100–107, 2021.
https://doi.org/10.1007/978-3-030-80462-6_13

helping conserve resources and energy. For the approach to work, it is vital to set up a management system in which humans no longer directly engage in executing processes but rather initiate, design, integrate and oversee them. Such a new approach to organizing manufacturing processes is therefore flexibility-oriented. The resulting systems are not free of disruptions and breakdowns, which require additional unplanned work that ultimately causes delays. The time wasted to perform such work is a possible measure of organizational risk. In this paper, organizational risk relates to organizing production (designing manufacturing systems) and setting up process execution. The causes of such risk vary depending on the level of management. At the strategic level, threats may result from a mismatch between the components of a given system and its designated purpose. At the tactical level, the problem may lie in task assignment and in defining task order and/or task overlaps in time and space. In the immediate time horizon at the operational level, organizational risk has to do with disruptions caused by managing the demand for the resources or their combinations that are needed to produce the final product.

This paper explains the concept of organizational risk and invokes a real-life case of highly-customized manufacturing of complex products to make a case for gathering historical data as a frame of reference for organizational risk management.

2 Organizational Management in Industry 4.0

2.1 Organizational Risk in Customized Manufacturing Systems

A manufacturing system should operate as continuously and in as stable a manner as possible. There has been growing recognition of late that proper risk management can be very useful for achieving such a state (e.g. [14, 15]). The job of production managers is to strive to meet the objectives put to their system. Each part of the system should be fine-tuned to achieve the prescribed higher-level outcomes and make up a well-arranged work environment. Hence, focus is placed on organizing and thereby creating the desired environment by defining the basic parameters[2] in space and time. Viewed in this manner, the activity of organizing is superior to that of preparing, designing, implementing and verifying [11].

The aim in arranging resource allocation and assigning workloads to individual workstations is to achieve the objectives put to the system. However, doing so may cause disruptions that need to be deliberately anticipated, prevented and prepared for to avert their impacts. Organizational risk is associated with [12, 13]:

- The design of structures (the assignment of Tasks and Responsibilities and the granting of Authorizations) although, as individual orders are fulfilled, such structures may impede operation,
- The setting up of processes (process design over time: defining labor intensity, duration and delay margins for process delivery, designing the spatial layout of process execution, identifying the information needs of the management and selecting work methods, i.e. production technologies), and with such threats and disruptions as may result from poor coordination of the above factors.

[2] Organizational relationships are described in terms of tasks, authorizations and responsibilities [9].

Organizational risk is a function of the rate of occurrence of undesirable states multiplied by the extent of the consequences of negative events that require the re-assessment of the demand for the resources and combinations thereof that are critical for the manufacture of final goods and/or for their allocation, multiplied by the likelihood of occurrence of such negative events. Such a risk coincides with operating risk and is tied to the basic current (short-term) operating parameters of the manufacturing system in space and time.

Most impacts of organizational risk can be seen at the operational level in the form of disruptions at workstations. However, their root causes lie at the tactical and strategic levels where organizational decisions (such as task assignment, the choice of production model, the adoption of concurrent operation or the choice of operational priorities) are made. While the primary impact of organizational risk is on planning and organizing, worker incentivization and control are also significantly affected.

In the course of deploying Industry 4.0, previously known manufacturing solutions that are already in place are redesigned. During this process, digital solutions are adopted and links are established among means of production, means of labor, goods, value chains and business models. The aim is to deploy automation with real-time data transfers and new technologies and methods for production oversight [1, 7, 10].

In Industry 4.0, management is geared towards delivering custom-made products at the price of mass-produced goods. Therefore, the very process of organizing production centers on anticipating disruptions not only on the basis of historical orders (as is done in the case of repetitive production), but also by considering new factors and existing but changing variables that require the business organization to approach each successive order as if it were developing a prototype that is also the final product (as in the case of customized production).

A comparison of the setup of serial production of new products and of customized production to the order of individual customers is presented in Table 1.

As customized production entails risks akin to those seen in project-based systems, it is crucial to identify threats that are unique for a given contract.

A study of a sample of Industry 4.0 deployment has guided the authors to focus their research on highly-customized production, which was found to be partially representative of the concept in question. The biggest advantage has been seen in the ability to simulate the operation of a production line in a way that, in a digitized system, may not only help prevent disruptions but also significantly facilitate the process of setting up a manufacturing system [3, 6]. However, to run such a simulation, data and information on boundary conditions are necessary.

2.2 Significance of Historical Data - Case Study

To create smart factories, manufacturers need not only to define the successive steps to be taken in running production processes, install proper machinery and equipment and embrace new technologies but also acquire data analysis experience on the basis of prior production events [17]. Information gleaned from historical data is critical for reducing response time. Such information is available provided that the manufacturer has analyzed and learned from past data and used it to anticipate future events and develop future scenarios (by way of system analysis).

Table 1. Comparison of serial and highly-customized production of new products

No	Standard design for the production of new goods (technical production setup)	Example of highly-customized manufacturing[a]	Remarks
1	Preliminary design of new product submitted to R&D organization for research aimed at defining technical requirements for a future product	The ability to meet tender specifications, as enumerated in RFP, is assessed	An assessment of the risk of entering into a tender procedure followed by an assessment of the ability of production to meet tender requirements
2	The product, its parts and/or components are designed as part of organizing production processes	The design of individual product and its components	Problems may arise in connection with the specification and the related documents
3	A prototype is made to test engineering solutions and materials	The prototype is to enable approval testing. Commonly, the prototype is also the final product (the sole unit made or the first unit made to a given order)	Production may be delayed by solution design flaws
4	If the results of the above tests are positive, the technological/organizational phase of production setup begins. During this phase, production/assembly technologies are designed and organizational arrangements are made	Production relies on an existing manufacturing system. In certain cases, additional time is allowed and new suppliers are brought in	Additional risks arise due to having to handle multiple production orders simultaneously
5	Before serial production is launched, a trial run is executed. That trial run is vital for testing technological and organizational solutions and testing the product in sales and operation	The production is customized. No trial run is performed	With no trial run to draw on, the technical solution may need to be altered necessitating modifications to an entire seemingly completed product/batch

(*continued*)

Table 1. (*continued*)

No	Standard design for the production of new goods (technical production setup)	Example of highly-customized manufacturing[a]	Remarks
6	The operation of production line and work center is simulated. Production options are assessed for cost, product quality, product sophistication and suitability of manufacturing process	Options are not analyzed, production line operation is not simulated. Line operation is designed for existing time settings and standard time margins Product quality is not checked until the product is made	Problems may arise in connection with production planning and the way the stakeholders (top executives, contractors, customers) understand quality
7	Serial production begins, sales is launched	The product is sold before it is made. To encourage buying, it is common to display the prototype at a trade fair or provide it to a customer for testing either on returnable basis or on other terms	
8	A distribution system is set up	Not applicable. Each customer presents their own preferences for collecting the product	

Source: [Own work] based on [4].
[a]Based on the case of the automotive industry.

Industry 4.0 presumes an integration of all supply chain participants (suppliers, customers, consumers, and employees) to facilitate information processing, learning and decision-making [8]. Organizations seeking to improve their operations should focus on managing organizational risk. To this end, they need to collect historical data and break it up into categories to help decision makers select scenarios for which threats have been clearly identified. In customized production, any unique risks associated with individual orders will additionally need to be assessed. An organization's response to emerging threats will demonstrate not only how well it is prepared for contingencies but also how vigilant it is in detecting early signs of disruption and how flexibly it can make adjustments as events unfold.

The authors studied highly-customized production systems in 2016–2018 with a focus on the so-called unplanned times and on the identification of the responsible business units and threat categories. To ensure a synthesis of categories, a separation of factors and a comprehensive approach, disruptions in the flow of process inputs were identified. Both input flow disruptions (missing technical solutions during production, parts missing during assembly, missing workstation manuals, supplier errors, errors in the production of frame, new customer requests, and item preparation for assembly), particularly those concerning material and information, and facility-related disruptions (such as damage to goods in process, deviations from process) were reported.

In 2015, a study of 20,667 reports of the so-called unplanned times[3] coming from production floor workers using the company's electronic workflow system, showed that such reports were ill-suited for research purposes. Therefore, the study was limited to data from 2016 and subsequent years. The results are shown in Table 2.

Table 2. Unplanned times in manhours in 2016 and 2018

2016	2016 Manhours	2018 Manhours
Damage to goods in process	189855.5	54622.30
Technical solution missing during production	68621	26705.50
Part missing during assembly	61286	19821.05
Production process	165521	12152.50
Missing workstation manual	18272	4104.60
Other	4260	3162.10
Supplier error	2889	3027.25
New customer request	3724	2019
Frame production error	1620	551.2
Item preparation for assembly	0	61
TOTAL	516048.5	126226.5

In 2016, the total average unplanned time was 258 FTEs. By 2018, that number was reduced to about 63 FTEs. Some such manhours resulted from the assignment of overtime to workers and from the need to hire additional workers.

Damage rates were assessed based on the total number of instances of damage to goods in process. In 2016, the most reported category of damage was process-related. This shows that organizational arrangements required excessive retooling, additional questions to establish customer preferences and task reassignment. While the top unplanned-time category in 2018 was damage, the severity of each individual instance was fairly low. As a result, the repair time was shorter than it was in "Other". The ratio of working time per task not included in the production plan (Table 1) to the number of reports in a given category (Table 3) offers an indirect indication of the severity of disruption impacts.

An analysis of unplanned times shows that the factors of key importance in custom production are not only standard operations (such as damage-related disruptions and frame production errors) but also design considerations (such as missing technical solutions during production and missing parts). The study shows that the biggest

[3] In the organization in question, unplanned times are used to identify differences between the planned and actual lead times.

Table 3. Number of unplanned-time reports in 2016 and 2018

YEAR/category	Total contracts	Damage to goods in process	Technical solution missing during production	Part missing during assembly	Production-process related	Missing workstation manual	Other	Supplier error	New customer request	Frame production error	Item preparation for assembly	Total reports
2016	39	793	391	348	1151	161	15	19	37	9	0	2963
2018		7656	3891	3730	3217	735	247	744	312	124	11	20667

share of unplanned times resulted from damage sustained in production and missing technical solutions during production. This demonstrates the existence of organizational risks in highly-customized production. Their existence is related not only to operational activities, but also to decisions made at the strategic and tactical levels.

3 Summary

A comparative analysis of 2016 and 2018 data by disruption category shows a marked reduction in the number of unplanned manhours despite a comparable, if not greater, number of reports concerning a given type of disruption. This proves that the production system "learned" and improved its responses to irregularities, as they were detected. This feature is central to Industry 4.0, which, although itself being an innovative approach to company operation, must focus on setting up cyber-physical systems that meet market requirements.

Acknowledgments. This study has been financed by funds received to support the statutory activities of "DS Młoda Kadra" of the Faculty of Management Engineering of the Poznań University of Technology (No. 503/227/11/141/DSMK/0578). The article has been financed by the Research Projects fund of the Faculty of Management Engineering (No. 0811/SBAD/1016).

References

1. Avunduk, Z.B., Kazan, H.: Üretim Teknolojisinde Yeni Eğilimler ve Endüstriyel Değişimler, Sosyal Bilimlerde Yeni Araştırmalar In: İslamoğlu, E., YpP, S. (eds.) Ankara (2019)
2. Ceruti, A., Marzoca, P., Liverani, A., Bil, C.: Maintenance in aeronautics in an Industry 4.0 context: The Role OF AR and AM. https://books.google.pl/books?id=bo1xDwAAQBAJ&pg=PA70&dq=industry+4.0&hl=pl&sa=X&ved=2ahUKEwiPy8bczdzuAhXSDewKHXYnALAQ6AEwAHoECAAQAg#v=onepage&q=industry%204.0&f=false
3. Hozdić, E.: Smart factory for Industry 4.0: a review. Int. J. Mod. Manuf. Technol. ISSN 2067–3604, **VII**(1) (2015)

4. Kawecka-Endler, A.: Organizacja technicznego przygotowania produkcji- prac rozwojowych, Wyd.Politechniki Poznańskiej, Poznań (2004)
5. Lis, S. (red.): Rytmiczność procesu produkcyjnego. Zakłócenia i ich kompensacja, Państwowe Wydawnictwo Ekonomiczne, Warszawa (1982)
6. Mrugalska, B., Stasiuk-Piekarska, A.: Readiness and maturity of manufacturing enterprises for Industry 4.0. In: Mrugalska, B., Trzcielinski, S., Karwowski, W., Di Nicolantonio, M., Rossi, E. (eds.) AHFE 2020. AISC, vol. 1216, pp. 263–270. Springer, Cham (2020). https://doi.org/10.1007/978-3-030-51981-0_33
7. Oks, S.J., Fritzsche, A., Lehmann, C.: The digitalisation of industry from a strategic perspective. In: R&D Management Conference 2016 "From Science to Society: Innovation and Value Creation", 3–6 July 2016, Cambridge, UK (2016)
8. Oláh, J., Aburumman, N., Popp, J., Asif Khan, M, Hassad, H., Kitukutha, N.: Impact of Industry 4.0 on environmental sustainability. Sustainability, **12**(11), 4674 (2020). https://doi.org/10.3390/su12114674. https://www.mdpi.com/2071-1050/12/11/4674/htm
9. REFA, Moduł 3210364: Ustalanie wymagań (1997)
10. Report PwC: Oswoić ryzyko. Jak odporna jest Twoja firma? grudzień (2016). https://www.pwc.pl/pl/pdf/oswoic-ryzyko-raport-2016-pwc.pdf
11. Rzeszotarska-Wyrwicka, M.: Organizowanie systemów pracy. Materiały pomocnicze, Wydawnictwo Politechniki Poznańskiej, Poznań (1998)
12. Stasiuk-Piekarska, A.: Innowacyjne wykorzystanie założeń metody FMEA dla potrzeb zarządzania ryzykiem organizacyjnym w systemach produkcyjnych, Problemy Jakości, 6, Wyd. Sigma-NOT, pp. 26–31 (2017). https://doi.org/10.15199/48.2017.6.4
13. Stasiuk-Piekarska, A., Wyrwicka, M., Hadaś, Ł.: Ryzyko organizacyjne w produkcji, Wyd. Politechniki Poznańskiej, Poznań (2020)
14. Urbaniak, M.: Wykorzystanie koncepcji zarządzania ryzykiem w doskonaleniu systemów zarządzania jakości. Problemy Jakości **48**(6), 11–16 (2016)
15. Wallis, A.: Rola informacji i wiedzy w zarządzaniu ryzykiem ekonomicznym w przedsiębiorstwie. Zeszyty Naukowe Politechniki Częstochowskiej Zarządzanie, **24**(2), 27–37 (2016)
16. Wyrwicka, M.: Rewolucja czy ewolucja w logistyce? Logistyka **3**, 9–11 (2014)
17. Zawadzki, P., Żywicki, K.: Smart product design and production control for effective mass customization in the Industry 4.0 concept. Manag. Prod. Eng. Rev. **7**(3), 105–112 (2016). https://doi.org/10.1515/mper-2016-0030. https://journals.pan.pl/dlibra/publication/121910/edition/106257/content/

Reduction of Human Effort in Technical Cleanliness Inspection Through Advanced Image Processing Approaches

Thorben Panusch[1(✉)], Roman Möhle[1], Ronny Zwinkau[1], and Jochen Deuse[1,2]

[1] Institute of Production Systems, TU Dortmund University, Leonhard-Euler-Straße 5, 44227 Dortmund, Germany
{Thorben.Panusch,Roman.Moehle,Ronny.Zwinkau, Jochen.Deuse}@ips.tu-dortmund.de

[2] Centre for Advanced Manufacturing, University of Technology Sydney, Ultimo, Australia

Abstract. The inspection of component surfaces for size, number and type of particulate contamination is carried out using the standard cleanliness analysis described in ISO 16232 and VDA 19.1. Currently, the detection, measurement, counting and classification of particles is performed by an analysis-system comprised of an incident light microscope and corresponding software evaluation followed by manual results review. We propose advanced image processing algorithms to enhance the particle recognition for the technical cleanliness inspection. This includes extended-depth-of-field (EDoF), shape-from-focus (SFF) and periodic noise removing. Furthermore, a robust image stitching algorithm is introduced. Combined, these algorithms enable a higher degree of automation and enhance the information received by the human examiner. This leads to a significant reduction of human interaction and intervention during the technical cleanliness inspection.

Keywords: Human factors · Human-systems Integration · Systems engineering · Technical cleanliness · Stitching · Image fusion · EDoF · SFF · Periodic noise removing

1 Introduction

The object of investigation of *technical cleanliness* are contaminants in technical systems and their different damage potential. Our work is focused on particle contamination as an important quality requirement on functionally relevant components especially in automotive, medicine, electronic and food industry. Cleanliness limit values of components are mostly specified by the property, number and size. Furthermore, because of varying damage potential and electrical conductivity, a distinction is made between metallic and non-metallic particles as well as fibers [1].

The aim of the technical cleanliness inspection according to ISO 16232 and VDA 19.1 is to be able to perform an assessment of size and number as well as classification of the detected particles. For the standard cleanliness test accompanying series production,

S. Trzcielinski et al. (Eds.): AHFE 2021, LNNS 274, pp. 108–115, 2021.
https://doi.org/10.1007/978-3-030-80462-6_14

particles are extracted manually from the component surface and then examined with an automatic analysis system (incident light microscope and corresponding software evaluation). Today, this has to be done in a laboratory environment by trained personnel, resulting in a lead time that permits only a low inspection frequency.

Due to the inadequacies of the industrially established automatic analysis systems, misclassification and unrecognized particles can be detected, which require extensive manual post-examination and post-processing by the laboratory staff. The process of the cleanliness analysis accompanying series production is thus strongly influenced by the examiner and is associated with high expenditure of time and personnel.

We demonstrate that advanced image processing algorithms can significantly increase the performance of an analysis system. This reduces the manual effort in post-evaluation of particle images.

2 Background

The standards ISO 16232 and VDA 19.1 specify the necessary parameters for the optical system regarding specific optical properties. There is a multitude of devices that can be used for an analysis, such as various microscopy devices and scanners. More specifically, there are no explicit requirements on the type of device that can be utilized in this context. The standardization specifies a minimum size criterion for the particle size that can be measured with the system. A minimum size of 10 pixels in its longest dimension is given as a minimum. We found that a minimum resolution of 5 μm/pixel for 50 μm and bigger particles is necessary for the successful application of most image processing algorithms [1].

The sample to be analyzed is characterized by a round surface with a diameter of 47 mm. In addition, there is an unknown number of different particles with an undefined orientation in all dimensions on a white background (cellulose or PET mesh).

In order to improve the results of the analysis and at the same time minimize manual effort and influence we had to assess the entire workflow containing focus, image stitching, segmentation, measurement and classification. The metallic/non-metallic classification based on particles images using deep learning algorithms have been discussed by Zwinkau et al. [2]. Zwinkau also showed that particle images can be classified by their formation mechanisms [3]. The basis for these approaches, which aim to improve the use of information from particle images, is the efficient acquisition of high-quality images and low inspection cycles. On the one hand, this requires the minimization of manual effort and influence and, on the other hand, the use of advanced algorithms also in SMEs, as laboratories usually are.

To automate the focus process and to obtain more information about the surface of the particles, we introduce the extended-depth-of-field and shape-from-focus algorithms. Information from blurred images can only be processed to a very limited extent. In addition, only a limited area of the particle is automatically focused today. We also investigate image stitching algorithms, which are already used in state-of-the-art technology to stitch together the individual images of the sample. Our goal is to make this more robust against disturbances caused by vibrations or mechanical insufficiencies. Using periodic noise removing we improve the segmentation of the particles, which is currently improved by a manual post-control.

3 Methodology and Results

3.1 Extended-Depth-of-Field

Optical systems only have a limited depth of field. This causes areas of a specimen that are not within the focal range of the objective to be defocused and thus not sharply imaged [4]. To overcome this limitation, we employ an extended-depth-of-field (EDoF) algorithm for image fusion of a focus sequence. A focus sequence refers to multiple images acquired successively at different focus positions, resulting in a z-stack of images containing all available focused information of the entire sample [5]. From each slice, the focused pixels are selected to reconstruct a single fusion image that has all areas in focus. Therefore, a focus measurement algorithm is applied pixel-wise to each image in the sequence [6].

Figure 1 depicts the results of the image fusion. The images were taken at a magnification of 65x, with a depth of field of 0.6 mm. The offset between the images is 0.3 mm. The algorithm was able to construct a completely sharp and accurate image.

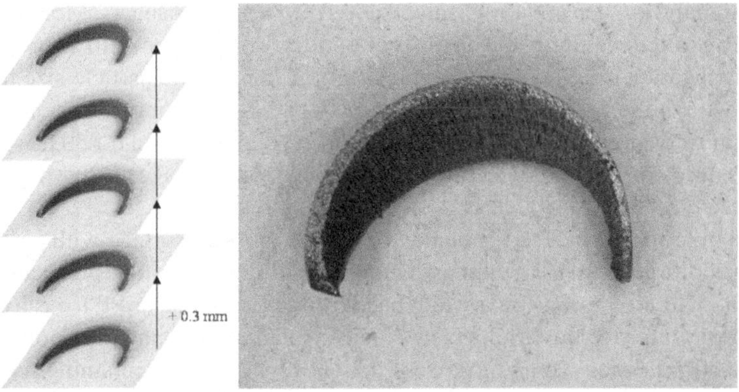

Fig. 1. Focus sequence (left) with an increasing z of 0.3 mm and the resulting all-in-focus image (right) from EDoF

The Tenengrad algorithm has been found to be the most accurate and robust focus algorithm for this purpose. A window size of 17 × 17 proved to be the most accurate. In addition, a threshold of 10 was applied to mask out variations in blurred pixels and noise.

However, it must be noted that due to the different imaging conditions, different scaling as well as small rotations or shifts occur in the images of the focus stack. These misalignments lead to significant problems during image fusion. It is therefore essential for image fusion that the individual images and image information are first adequately geometrically aligned using an image registration process [7, 8].

For feature detection and description, we use the AKAZE algorithm [9], which for our use case finds the highest number of features, has the fewest relative errors and works fast. The nearest-neighbor distance ratio method with two neighbors was used as the feature matching strategy together with the brute force search algorithm, as it was also utilized by Lowe [10]. Based on the feature matches, the homography matrix is created, which provides the perspective transformation of a second image with respect to the reference image. As a robust probabilistic method for outlier rejection, we implemented random sample consensus (RANSAC) [11]. The result is an image, which has all visible parts in focus providing complete image information.

3.2 Shape-From-Focus

Shape-from-focus (SFF) is a passive optical method for 3D shape recovery [12]. Zwinkau [3] already explored the possibility of classifying particles based on their surface topography. Consequently, SFF can provide even more surface information in this context.

SFF is accomplished by determining the depth of each point of the object from the camera lens according to the focus position of the highest focus value (Tenengrad algorithm) in a focus sequence, similar to the EDoF-algorithm process [13]. The result is a depth map that is subsequently refined using special approximation techniques [14]. Therefore, we implement a widely used method proposed by Nayar and Nakagawa [15]. It is assumed that the focus value of a pixel behaves like a Gaussian distribution depending on the focus position. The peak of the distribution corresponds to the depth at which the pixel is actually located. The respective depth is then estimated by interpolating a Gaussian function for each pixel based on the measurements of the focus sequence using the maximum value F_m and the neighbors F_{m-1} and F_{m+1}. Figure 2 displays the corresponding depth map of an all-in-focus image of an examined particle obtained from a focus sequence using the proposed EDoF algorithm.

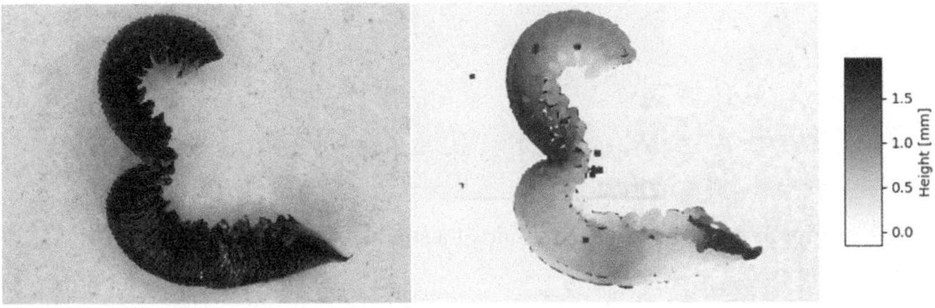

Fig. 2. Resulting depth map of an examined particle (formation mechanism: turning)

Figure 2 provides an impression of the topography and the shape of the particle. It is possible to determine the maximum height and the height profile of a particle. Relevant shape characteristics and distinctive surface topographies can also be identified and used as features for future automated classification approaches. Further, it is conceivable that this approach can be used to detect particle overlays in the segmentation process, which, as of now, leads to measurement errors and demands manual rework.

3.3 Image Stitching

Stitching is the term used to describe image processing methods that allow images of individual sub-areas to be combined to form a mosaic of the overall image [16].

This is not an explicitly necessary process step in light microscopic analysis. It arises implicitly in this context, because a high resolution is to be achieved for the analysis of the entire sample, as displayed in Fig. 3.

The creation of a mosaic of subimages has proven to be a useful technique for expanding the field of view and maximizing the resolution. This allows for the maximum magnification of the optical component to be used and to capture every part of the sample with the image acquisition device [17].

To get the most accurate translation between images, we use phase correlation, an idea introduced by Kuglin and Hines [18]. We use a meta-algorithm for the process of image registration and alignment-improvement [19]. This includes the selection of the optimal maxima with the Quickselect-Algorithm [20]. The selected number of maxima is selected according to the current state of the art [21].

The optimization and final stitching of the image mosaic is executed with the correlation coefficient and the minimum spanning tree (MST)-algorithm [22]. The results of stitched images of a sample are displayed in Fig. 3.

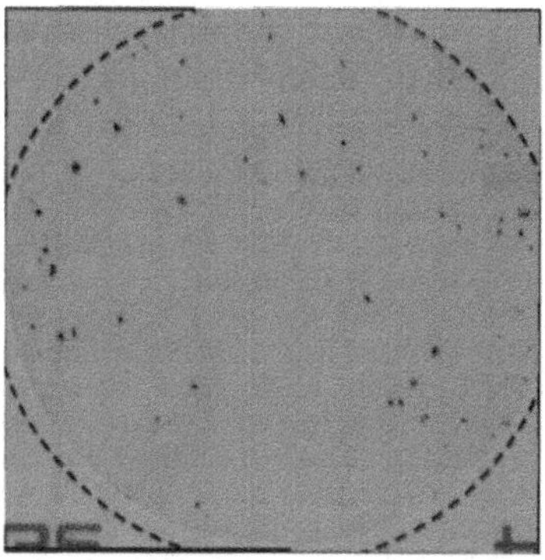

Fig. 3. Resulting image mosaic of a stitched sample (192 images)

The optimal results are achieved with polarized light. This enables a better stitching because the metallic glance of particles is not overexposed. This effect is also seen in other areas of image processing, like segmentation. Improvements to the current setup are the selection of an optimal blending algorithm and the reduction of memory usage of the computer. A significant advantage of the developed approach is the robustness against disturbances caused by vibrations or mechanical insufficiencies in the capturing process. The ability of the algorithm to compensate for these inaccuracies enables the use of a smaller and lighter mechanical system. In our example, 192 single images are used to assemble the image mosaic of the particles on the sample.

3.4 Periodic Noise Removing Filter

The distinct texture of the PET mesh in the particle background can pose a major problem in the segmentation task, especially for low-contrast particles. Thereby, the PET threads create differing brightness in the image because of their superimpositions in the mesh pattern. The result is an intensity pattern that provides the background with a varying intensity range, while also pronouncing the edges of the mesh pattern clearly. Since the pattern consists of recurring, overlapping lines at fixed intervals, it can be understood as a moiré pattern corrupting the image as periodic noise. Accordingly, it is composed of individual defined frequencies, which can be identified as distinct peaks in the frequency domain and thus removed by applying a notch reject filter [23]. We implemented an automatic notch reject filter that detects and removes the defined frequency peaks of the mesh pattern as local maxima. Because the amplitude values decrease with increasing frequency, the image is divided into nine adjacent ring regions. In these, local maxima are identified as the 1.2% largest amplitudes. A small region around the zero frequency is omitted. The zero frequency represents the average intensity value of the image and the low frequencies correspond to the slowly varying intensity components that reflect important intensities of the particles [23].

The application of the proposed periodic noise removing filter is shown in Fig. 4. The impact on a segmentation task can also be identified. The degradation of the raw image by the mesh pattern is clearly visible. However, the results in the filtered image illustrate that the mesh pattern could be almost completely removed, so that the image is free of critical interference effects. This is a significant improvement: the relevant particle information is displayed more distinctly and is easier to extract. This facilitates processing in subsequent applications, like image segmentation, as shown below.

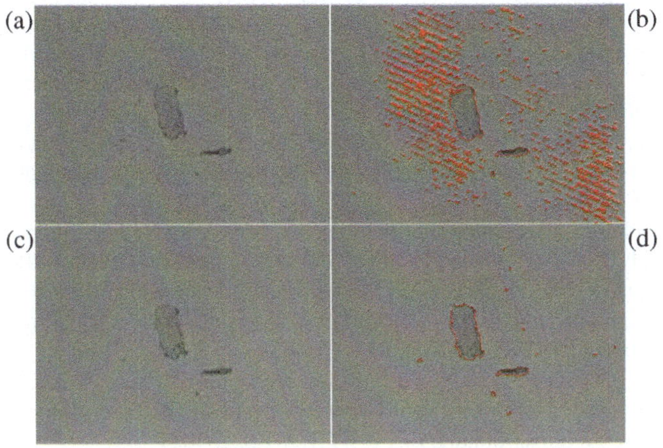

Fig. 4. Raw (a) and filtered image (c) with the respective segmentations (b, d)

4 Discussion and Conclusion

The combined approaches permit a reduced human effort in post-processing and provide a wider range of particle information in an automated way. Furthermore, the stitching algorithm allows the entire sample area to be captured automatically. Another positive aspect of this implementation is the ability to compensate for mechanical distortions during the capture process. This leads to a possible reduction of maintenance, manual calibration and human intervention.

Additionally, the implemented EDoF and periodic noise removing algorithms enhance the particle images, thereby reducing the human effort. This is supplemented by the additional geometric information provided by the SFF algorithm. Combined this allows the examiner to receive detailed information about the particles found in a technical cleanliness inspection.

This opens up possibilities of more use cases and reduces the necessity of specially trained examiners. In our view, one area of interest is bringing cleanliness inspections closer to the production processes reducing inspection lead time. This enables a higher throughput that, coupled with faster improvement cycles, allows faster intervention in cases of limit exceeding. Further areas of research are the possibilities of automated integration of the cleanliness inspection into the production system and the role of image processing algorithms in reducing human effort and interaction.

References

1. ISO 16232 Road vehicles—Cleanliness of components and systems. Véhicules routiers — Propreté des composants et des systèmes, 1. Aufl. International Organization for Standardization, Vernier, Genf, Schweiz (2018)
2. Zwinkau, R., Frentrup, S., Möhle, R., Deuse, J.: Automatic particle classification through deep learning approaches for increasing productivity in the technical cleanliness laboratory. In: Nunes, I.L. (ed.) AHFE 2019. AISC, vol. 959, pp. 34–44. Springer, Cham (2020). https://doi.org/10.1007/978-3-030-20040-4_4
3. Zwinkau, R.: Adaption von Verfahren der Bilderkennung auf die Regelung partikelsensibler Produktionssysteme, 1. Aufl. Schriftenreihe Industrial Engineering, Band 26. Shaker Verlag, Düren (2019)
4. Foster, J., Richards, F.B.: The gibbs phenomenon for piecewise-linear approximation. Am. Math. Monthly **98**(1), 47–49 (1991). https://doi.org/10.1080/00029890.1991.11995703
5. Aguet, F., Van De Ville, D., Unser, M.: Model-based 2.5-d deconvolution for extended depth of field in brightfield microscopy. IEEE Trans. Image Process. **17**(7), 1144–1153 (2008). https://doi.org/10.1109/TIP.2008.924393
6. Pertuz, S., Puig, D., Garcia, M.A., Fusiello, A.: Generation of all-in-focus images by noise-robust selective fusion of limited depth-of-field images. IEEE Trans. Image Process. **22**(3), 1242–1251 (2013). https://doi.org/10.1109/TIP.2012.2231087
7. Zitová, B., Flusser, J.: Image registration methods: a survey. Image Vis. Comput. **21**(11), 977–1000 (2003). https://doi.org/10.1016/S0262-8856(03)00137-9
8. Billiot, B., Cointault, F., Journaux, L., Simon, J.-C., Gouton, P.: 3D image acquisition system based on shape from focus technique. Sensors **13**(4), 5040–5053 (2013). https://doi.org/10.3390/s130405040

9. Alcantarilla, P., Nuevo, J., Bartoli, A.: Fast explicit diffusion for accelerated features in nonlinear scale spaces. In: Burghardt, T., Damen, D., Mayol-Cuevas, W., Mirmehdi, M. (eds.) Proceedings of the British Machine Vision Conference 2013, pp. 13.1–13.11. British Machine Vision Association (2013)
10. Lowe, D.G.: Distinctive image features from scale-invariant keypoints. Int. J. Comput. Vis. **60**(2), 91–110 (2004). https://doi.org/10.1023/B:VISI.0000029664.99615.94
11. Tareen, S.A.K., Saleem, Z.: A comparative analysis of SIFT, SURF, KAZE, AKAZE, ORB, and BRISK Invent, innovate and integrate socioeconomics development. In: 2018 International Conference on Computing, Mathematics and Engineering Technologies - iCoMET 2018, 3–4 March 2018, pp. 1–10. IEEE, Piscataway (2018)
12. Pertuz, S., Puig, D., Garcia, M.A.: Analysis of focus measure operators for shape-from-focus. Pattern Recogn. **46**(5), 1415–1432 (2013). https://doi.org/10.1016/j.patcog.2012.11.011
13. Malik, A.S., Choi, T.-S.: Consideration of illumination effects and optimization of window size for accurate calculation of depth map for 3D shape recovery. Pattern Recogn. **40**(1), 154–170 (2007). https://doi.org/10.1016/j.patcog.2006.05.032
14. Mahmood, M.T., Majid, A., Choi, T.-S.: Optimal depth estimation by combining focus measures using genetic programming. Inf. Sci. **181**(7), 1249–1263 (2011). https://doi.org/10.1016/j.ins.2010.11.039
15. Nayar, S.K., Nakagawa, Y.: Shape from focus. IEEE Trans. Pattern Anal. Mach. Intell. **16**(8), 824–831 (1994). https://doi.org/10.1109/34.308479
16. Bajcsy, P., Chalfoun, J., Simon, M.: Web Microanalysis of Big Image Data. Springer, Cham (2018). https://doi.org/10.1007/978-3-319-63360-2
17. Kumar, R., Anandan, P., Irani, M., Bergen, J., Hanna, K.: Representation of scenes from collections of images Representation of visual scenes. In: Workshop: Papers. IEEE Computer Society Press, Los Alamitos (1995)
18. Kuglin, C.D., Hines, D.C.: The phase correlation image alignment method. In: Proceedings of the IEEE International Conference on Cybernetics and Society. IEEE Press, Piscataway (1975)
19. Chalfoun, J., et al.: MIST: accurate and scalable microscopy image stitching tool with stage modeling and error minimization. Sci Rep **7**(1), 4988 (2017). https://doi.org/10.1038/s41598-017-04567-y
20. Hoare, C.A.R.: Algorithm 65: find. Commun. ACM **4**(7), 321–322 (1961). https://doi.org/10.1145/366622.366647
21. Yu, Y., Peng, H.: Automated high speed stitching of large 3D microscopic images. In: 2011 IEEE International Symposium 3/30/2011, pp. 238–241 (2011)
22. Kruskal, J.B.: On the shortest spanning subtree of a graph and the traveling salesman problem. Proc. Am. Math. Soc. **7**(1), 48–48 (1956). https://doi.org/10.1090/S0002-9939-1956-0078686-7
23. Gonzalez, R.C., Woods, R.E.: Digital Image Processing, 4th edn. Pearson, New York (2018)

Method for the Targeted Selection and Installation of an IoT - Platform

Stefan Kugler(✉), Cordula Czwick, Hoang Phuong Linh Nguyen, and Reiner Anderl

Technische Universität Darmstadt, Otto-Berndt-Str. 2, 64287 Darmstadt, Germany
kugler@dik.tu-darmstadt

Abstract. Technologies and development in the field of Internet of Things (IoT) are growing rapidly. With the digitalization of industries, the importance of IoT-platforms is increasing steadily. The large number of IoT-platform providers on the market is representing a vast selection of IoT-platforms with all sorts of functionalities. Due to the vast selection of IoT-platforms, companies and users have difficulties to select a suitable IoT-platform for their application.

This paper presents the results of a method for the targeted selection and installation of an IoT-platform depending on user's requirements. For the evaluation of IoT-platforms evaluation criteria are derived for the selection of a suitable IoT-platform. Furthermore, the method includes a guideline, which presents a feasible way for the installation. For the validation of the method a use case has been defined and a demonstrator has been developed.

Keyword: Platform · Internet of Things (IoT) · Industry 4.0

1 Introduction

In recent years, the Internet of Things (IoT) has developed rapidly. The survey results of the study by IDG Research Services [1] regarding the current and future relevance of IoT in companies clearly show that there is already a strong awareness of the added value of IoT and that the trend for the future is increasing. While corporations and large enterprises have already recognised the economic potential of IoT in the early initial phases, small and medium-sized enterprises (SMEs) have only in recent years recognised the economic potential of IoT for themselves and are now attempting to bring IoT added value into the company [2]. With the increasing digitalization of industries, the importance of IoT-platforms as a basic framework for IoT solutions is steadily increasing. According to the results of the IDG study, 55 percent (2018) of companies regard the IoT-platform as the most important technology for the IoT. Suppliers have already recognized this relevance, so that more than 450 different providers already exist for IoT-platforms [3]. The number of IoT-platform providers is constantly changing as new platform providers are founded and existing platform providers merge [4]. This multitude makes it difficult for companies to select a suitable IoT-platform in a targeted manner. It is of particular importance to select a suitable IoT-platform in order to enable the success of the application and to extend the progress of IoT within the company.

S. Trzcielinski et al. (Eds.): AHFE 2021, LNNS 274, pp. 116–123, 2021.
https://doi.org/10.1007/978-3-030-80462-6_15

Choosing an unsuitable IoT-platform carries the risk that the goal of the IoT solution will not be achieved and that it will inhibit the progress of digitization in the enterprise. This means that, among other things, unnecessary investment costs can arise in the redesign of the infrastructure, which represents a high investment risk for companies, especially for SMEs. Therefore, the choice of the IoT-platform should be carefully considered in order to minimise this risk.

2 IoT Ecosystem

For the selection and implementation of a best-fitting IoT-platform, it is necessary to understand the role of IoT-platforms in an IoT environment. Regarding the architecture of IoT there are no standardized models, which is why different papers propose or refer to different reference architectures [5]. Depending on the point of view, IoT reference architectures differ in number, definition and detailing of its layers [6, 7]. In general, the different proposals of IoT reference architectures can be best classified by the number of layers regarding functions and components. The models therefore illustrate the relationship between different key technologies, their functions and components [8]. This paper will refer to the five-layer model by Wu et al. [9] to explain the role of IoT-platforms in an IoT environment. It consists of the perception layer, network layer, middleware layer, application layer and business layer. The perception layer is characterized by different edge technologies and embedded systems with sensors and actuators. In this layer, the perception and manipulation of the environment can be achieved with the help of physical objects, which are referred to as devices. For this reason, this layer is often referred to as the edge-, device- or perception layer [9, 10]. Perception is the acquisition of data in which physical parameters of the objects and environment, for example by sensors, are acquired and converted into an electrical signal for further processing with tools provided by an IoT-platform [6, 8, 11]. Manipulation is the control of physical objects via actuators which act on the environment and thus change physical parameters according to instructions resulting from the data processing on IoT-platforms [6, 11, 12]. The network layer deals with the various network technologies and gateways in IoT to integrate IoT- and non-IoT devices into a network [11]. In order to process the data acquired by sensors, the data needs to be transported to the IoT-platform via different network types with different protocols [13]. Typical technologies of the network layer are for example Ethernet and Wireless Local Area Network (WLAN) in Local Area Network (LAN), ZigBee and Bluetooth in Personal Area Network (PAN) and mobile technologies in Wide Area Network (WAN) [5, 6, 8]. If the necessary communication protocols are supported by the device, these network technologies are used to enable direct communication with the middleware layer. If required protocols are not supported by the device, gateways are needed to translate between the protocol and the device [11]. For further information and a detailed overview of individual technologies, standards and protocols in IoT, please refer to the book by Cirani et al. [14]. The middleware layer makes use of various technologies, ranging from databases, ubiquitous computing and cloud computing to IoT-platforms [9, 13]. From a functional perspective, the middleware layer often acts as an integration layer for all physical objects, embedded systems, applications and other sources [11]. The data from devices and other sources are collected, stored, analysed and processed with the help of tools and services of an IoT-platform [9, 10, 13]. This layer is not only responsible for receiving data, but also for sending data and instructions to applications and devices. Therefore, one of the goals of this layer is to enable an autonomous decision-making process based on collected data, existing tools and analysis methods [10]. In addition, the middleware layer enables the integration of different applications through various interfaces, which allows those applications to have access to the data captured by devices and processed with tools and services of an IoT-platform [11]. The functions in the middleware layer are not limited to the previously mentioned functions but range from device management over to user role management to visualization options of processed data. The application layer represents the applications and users of an IoT system [9–11]. In this context users can be either a human or a digital system. The digital system can be a software and application that interacts with the IoT system via an application programming interface (API) [12]. This allows digital users to access data processed in the middleware layer and further process it for specific use cases or to create new application areas [5, 9, 10]. The business layer is introduced with the five-layer model by Wu et al. [9]. The focus of this layer is primarily on further development possibilities and the potential of the IoT system. Therefore, not only the administration of applications takes places in this layer, but also the strategic control of business affairs as well as the development of new business models and user security within the framework of IoT is part of this layer [5, 7, 9]. As shown and explained with the five-layer model, IoT-platforms have a central role within the IoT system. They provide functions, tools and services to connect and manage devices as well as to process and analyse data to optimize and create value for business processes. Thus, an IoT-platform acts as an integration basis and intermediary for various technologies and applications. In order to successfully deploy an IoT project and generate possibilities for future growth, a best-fitting IoT-platform needs to be selected. An IoT-platform is best-fitting, if it fulfils the requirements of a given application and enables its successful deployment, while providing opportunities for future growth and deployment of more complex applications.

3 Concept for the Selection and Implementation of a Best-Fitting IoT-Platform

After a brief explanation of the IoT system and the role of IoT-platforms in Sect. 2, the concept for the selection and implementation of a best-fitting IoT-platform will be described in this chapter.

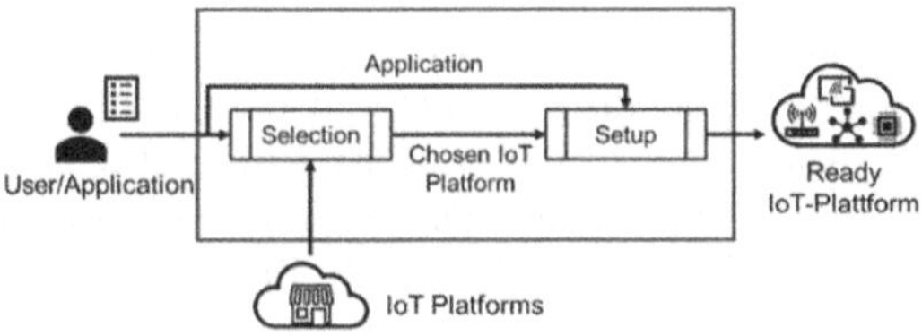

Fig. 1. Two-part concept for the selection and implementation of a best-fitting IoT-platform

The proposed concept, shown in Fig. 1 is a two-part concept consisting of a selection and an implementation method. Due to the large and dynamic landscape of IoT-platforms, their differences in range of functions as well as the different needs of applications, a systematical approach is needed to select a best-fitting IoT-platform and implement the chosen IoT-platform. The concept is ensuring the selection of a best-fitting IoT-platform and supporting its successful implementation by considering the requirements of the application as well as mandatory prerequisites for the setup of an IoT-platform as early as possible. After deriving requirements for a given application and deciding on an initial selection of IoT-platforms, the selection method will be applied and provide a best-fitting IoT-platform. This platform will be setup with help of a proposed method considering the application and derived requirements. The result is an operational IoT-platform with a successful deployed IoT project.

3.1 Selection Method

For the selection of a best-fitting IoT-platform a modified benefit analysis is used. The standard benefit analysis allows the user to prioritize important requirements for the evaluation of IoT-platforms by assigning weights to the criteria. Since the evaluation of IoT-platforms usually considers many criteria, the modified benefit analysis, shown in Table 1, is working with both groups of criteria and each criterion individually for the evaluation.

Table 1. Modified benefit analysis for the evaluation of a best-fitting IoT-platform

		IoT Platform 1		IoT Platform 2	
Criteria	Weight	Value	W. Value	Value	W. Value
Criteria Group A	60%	4,5	2,7	3	1,8
Criterion A1		6		3	
Criterion A2		3		3	
Criteria Group B	...	...	...	...	...
Sum	100%	Total Value		Total Value	

The first step for the modified benefit analysis is to develop criteria and groups of criteria. For the evaluation of a best-fitting IoT-platform following seven groups of criteria are developed based on previous scientific work [15]: *basic functions and platform services*, *interoperability and connectivity*, *data infrastructure and processing*, *security and reliability*, *pricing and license model*, *documentation and support* as well as *usability*.

- *Basic functions and platform services*: This group of criteria includes criteria for the evaluation of various functions of the IoT-platform. The aim is to investigate which functions and to what extent those functions are provided by the platform.
- *Interoperability and connectivity*: This group of criteria contains criteria for the interoperability, connectivity and communication of IoT-platforms. Thus, it describes the ability of an IoT-platform to interact with other systems.
- *Data infrastructure and processing*: For this group of criteria, criteria on infrastructure and processing of data are considered for the evaluation of the IoT-platform. The focus will be on how the IoT-platform deals with data.

- *Security and reliability*: With this group different criteria related to security and reliability are considered for the evaluation of the IoT-platform.
- *Pricing and license model*: Criteria in this group relate to the pricing and license models offered by IoT-platform providers.
- *Documentation and support*: This group of criteria contains both criteria for the evaluation of the IoT-platform in terms of documentations as well as support levels provided by the IoT-platform provider.
- *Usability*: In this group of criteria, the focus is on the usability of the IoT-platform. With criteria of this group, the user-friendliness of the functions of an IoT-platform and the transparency of its user interface are evaluated based on the principles of dialogue design according to DIN EN ISO 9241-110 [16].

A collection of criteria can be taken from previous work [15] and can be extended if necessary.

The second step is to assign weights to the developed groups of criteria according to the priorities of the given application. Therefore, the modified benefit analysis assigns weights to the groups of criteria (A), which will contain the average value of underlying criteria (A1, A2). To rate the underlying criteria, a general rating scale must be defined. Following rating scale is defined for the proposed concept:

- 0 – Criterion is not fulfilled or no information available
- 3 – Criterion can be fulfilled to a limited extent with the help of additional parties
- 6 – Criterion can be fulfilled with minimum requirements
- 9 – Criterion can be excellently fulfilled and has the potential for future growth and scalability

With the above defined rating scale, the criteria will be rated and provide values for the calculation of the average value of the criteria group. Then the average value of the criteria group will be weighted so that the user will receive a weighted value for the group of criteria. The sum of all weighted values of groups of criteria will be the total value, which will be used to evaluate and compare the IoT-platforms with each other. For this concept and benefit analysis, the IoT-platform with the highest score is the best-fitting IoT-platform for the given application. This method is open in terms of numbers of criteria, groups of criteria as well as IoT-platforms. Thus, the user can add requirements and IoT-platforms as needed and is not limited by a certain amount.

3.2 Setup Method

After selecting a best-fitting IoT-platform with the proposed method, the following setup method will introduce steps for the implementation of a given application with a selected IoT-platform. The setup method consists of six steps: *Preparation*, *Setup IoT-platform*, *Setup Devices*, *Establishing Connection*, *Create Applications* and *Test of operational IoT-platform*.

- *Preparation*: The first steps towards setting up an IoT-platform are already taken with the requirements of an application, the subsequent benefit analysis and thus the selection of a best-fitting IoT-platform. The application simultaneously defines application goals as well as framework conditions, such as hardware compatibility, interfaces and protocols, so that aspects of the technical integration are already considered during the selection process of a best-fitting IoT-platform. Furthermore, the user should specify the supported network technology as well as interfaces and protocols that will be used for the communication between the hardware and the selected IoT-platform. In this step, the user should also define aspects regarding the data sources, data points and intervals of sending data.

- *Setup IoT-platform*: After finishing preparation, the user needs to access the selected IoT-platform. Depending on the IoT-platform provider, the user can access the IoT-platform in the web browser either directly after registration or after installing required software packages. In the console the user can select the requested service to create an instance of the service and start an IoT project. Initially, the user should set up various user roles and access rights with provided functions for identity and access management (IAM). Then the devices must be registered in the IoT-platform before they can be connected to the IoT-platform. In the IoT-platform, a virtual object, containing properties and status of a connected device, is then created and provides authentication options for the device.
- *Setup Devices*: After the setup of the selected IoT-platform and the registration of devices, the devices and sensors need to be connected to power and the network. Afterwards the development environment needs to be setup on the devices. This includes, for example, the download of device drivers, necessary programming libraries, software applications, software development kits and certificates provided by the selected IoT-platform.
- *Establishing Connection*: With the provided interfaces a connection between devices and the selected IoT-platform can be established according to the defined specifications in *Preparation*. For this purpose, the user should familiarize himself with the interfaces provided by both the device and the selected IoT-platform. Together with the provided certificate or authentication options, the user can initiate a message client for message exchange between the IoT-platform and device. Thus, a connection between the selected IoT-platform and device is established.
- *Create Applications*: Once the message client has been initiated and messages can be exchanged between the devices and the selected IoT-platform, the user can start creating applications according to the application goals. Depending on the application objective and the selected IoT-platform, different IoT tools and cloud services are available for the creation and deployment of IoT applications.
- *Test of operational IoT-platform*: Once the selected IoT-platform has been set up, the system can be tested for functionality. At the beginning, the applications should be selected as simple as possible so that the basic functions can be checked, and problems can be solved at an early stage before they become more complex.

This setup method is formulated in a general way so that it can be used as a guideline for setting up an IoT-platform for a chosen application.

4 Application: Processing Real-Time Visualization of Temperature and Humidity Data

For the selection of a best-fitting IoT-platform and its setup, a simple application is used. The objective of the chosen IoT application is to acquire the room temperature and humidity with the help of sensors. The acquired data is then sent to an IoT-platform, which will process the data and visualize it in real-time. Tinkerforge modules (Master Brick 2.1, Humidity 2.0 Bricklet and LCD 20x4 Bricklet) and a Raspberry Pi 3 are used for the chosen application. The proposed concept and its setup are applied for the chosen IoT application. Criteria for the selection of a best-fitting IoT-platform for the chosen application are taken from previous work [15].The first steps after defining the application goal is to specify the communication concept and other aspects for data processing. In this case, the programming language will be set to Python and the communication protocol Message Queuing Telemetry Transport (MQTT) is used to exchange data from the Tinkerforge sensor between the Raspberry Pi and the IoT-platform. In addition, the interval of sending data to the IoT Hub is set to 1 s. Now, the user must set up the device and the development environment. Necessary python libraries for the connection with the IoT-platform must be installed prior for establishing the connection between the Raspberry Pi 3 and the IoT Hub instance. For this case, the necessary python libraries are installed on the Raspberry Pi 3. With the help of the documentation, the user can create a python script, which use the primary key and authentication token to initiate the MQTT client and establish a secure connection between the Raspberry Pi 3 and its endpoint in the IoT Hub instance. After establishing the bidirectional communication between Raspberry Pi 3 and the IoT Hub instance, data can be sent to the corresponding endpoint with an interval of 1 s. For the chosen application a service of the provider for analytics is used to analyse and stream the data from the receiving endpoint to another defined endpoint. Since the application goal is to visualize the data in real-time, an application is defined as the final endpoint for the analytics service. As the last step of the proposed concept, the user needs to check whether the operated IoT-platform achieves the application goals and whether the IoT application functions as desired.

5 Conclusion and Outlook

To conclude, the aim of this paper is to provide a systematical approach for the selection of a best-fitting IoT-platform and its setup according to a chosen application. This is necessary due to the large and dynamic landscape of IoT-platforms, which offer differences in range of functions for different application scenarios and requirements. In this paper, a best-fitting selection can be understood as a selection of an IoT-platform that meets the requirements and goals of the application. At the same time, it takes into consideration the prerequisites for the setup of a selected IoT-platform. By incorporating the prerequisites for setting up an IoT-platform into the selection process, it is ensured that the selected IoT-platform can be setup purposefully according the chosen application. For the evaluation and selection of a best-fitting IoT-platform with a modified benefit analysis, the user can adapt criteria from previous work [15] or add own criteria to carry out the evaluation as described in this paper. Depending on the requirements of the chosen application, the user can prioritize certain groups of criteria, which supports the selection of a best-fitting IoT-platform. For the selected IoT-platform with the best fit, a concept with general steps for the setup of an IoT-platform is proposed. The user can use the setup concept as a guideline to build a selected IoT-platform and deploy a chosen application. This paper is using criteria from [15] to apply the proposed concept for a simple application and evaluate several IoT-platforms to define the best-fitting IoT-platform. The proposed concept for the selection of a best-fitting IoT-platform and its setup can be used as a basis for the evaluation of further IoT-platforms and applications. Moreover, this concept can be used to evaluate individual adapted IoT-platforms in order to check the extent of the fulfilment of given requirements.

References

1. Maurer, J.: Studie Internet of things (2019). https://www.q-loud.de/hubfs/Kundendownloads/IDG-Studie_IoT_2018_2019.pdf
2. Wischmann, S., Wangler, L., Botthof, A.: Industrie 4.0 - Volks- und betriebswirtschaftliche Faktoren für den Standort Deutschland. https://vdivde-it.de/system/files/pdfs/industrie-4.0-volks-und-betriebswirtschaftliche-faktoren-fuer-den-standort-deutschland.pdf
3. IoT Analytics, IoT-platforms Market Report 2018-2023. https://iot-analytics.com/product/iot-platforms-market-report-2018-23/
4. Krause, T., et al.: IT-Plattformen für das Internet der Dinge (IOT): Basis intelligenter Produkte und Services. Fraunhofer Verlag, Stuttgart (2017)
5. Pethuru, R., Raman, A.C.: Internet of Things: Enabling Technologies, Platforms, and Use Cases. CRC Press - Taylor & Francis Group LLC, Boca Raton (2017)
6. Bloom, G., Alsulami, B., Nwafor, E., Bertolotti, I.C.: design patterns for the industrial internet of things, imperia, pp. 1–10 (2018)
7. Sethi, P., Sarangi, S.R.: Internet of things: architectures, protocols, and applications. J. Electr. Comput. Eng. **1**, 1–25 (2017)
8. Patel, K.K., Patel, S.M.: Internet of things: definition, characteristics, architecture, enabling technologies, application & future challenges, pp. 6122–6131
9. Wu, M., Lu, T.J., Ling, F.Y., Sun, J., Du, H.Y.: Research on the architecture of Internet of Things. In: 3rd International Conference on Advanced Computer Theory and Engineering, Chengdu, pp. V5-484–V5-487 (2010)
10. Antao, L., Pinto, R., Reis, J., Goncalves, G.: Requirements for testing and validating the industrial internet of things. In: IEEE International Conference on Software Testing, Verification and Validation Workshops, Vasteras, pp. 110–115 (2018)
11. Guth, J., et al.: Comparison of IoT-platform architectures: a field study based on a reference architecture. In: Cloudification of the Internet of Things, Paris, pp. 1–6 (2016)
12. Internet of Things (IoT) - Reference architecture, ISO/IEC 30141:2018 (2018)
13. Happ, D., Karowski, N., Menzel, T., Handziski, V., Wolisz, A.: Meeting IoT platform requirements with open pub/sub solutions. Ann. Telecommun. **72**(1–2), 41–52 (2016). https://doi.org/10.1007/s12243-016-0537-4
14. Cirani, S., Ferrari, G., Picone, M., Veltri, L.: Internet of Things: Architectures, Protocols and Standards. Wiley, Hoboken (2019)
15. Kugler, S., Czwick, C., Anderl, R.: Development of a valuation method for IoT-platforms. In: Fortin, C., Rivest, L., Bernard, A., Bouras, A. (eds.) PLM 2019. IAICT, vol. 565, pp. 293–301. Springer, Cham (2019). https://doi.org/10.1007/978-3-030-42250-9_28
16. DIN EN ISO 9241-110: Ergonomie der Mensch-System-Interaktion - Teil 110: Grundsätze der Dialoggestaltung (2020)

Predictive Model of Rock Fragmentation Using the Neuro-Fuzzy Inference System (ANFIS) and Particle Swarm Optimization (PSO) to Estimate Fragmentation Size in Open Pit Mining

Betty Vergara[1], Maria Torres[1], Vidal Aramburu[1], and Carlos Raymundo[2(✉)]

[1] Ingeniería de Gestión Minera, Universidad Peruana de Ciencias Aplicadas, Prolongación Primavera 2390, Lima 15023, Peru
{U201611305,U201611203,pcgmvara}@upc.edu.pe

[2] Direccion de Investigacion, Universidad Peruana de Ciencias Aplicadas, Prolongación Primavera 2390, Lima 15023, Peru
carlos.raymundo@upc.edu.pe

Abstract. The objective of this research is to generate a predictive model to estimate rock fragmentation size using the Neuro-Diffuse Inference System (ANFIS) in combination with Particle Swarm Optimization (PSO). To build the predictive model, 92 blasting events were investigated and the rock fragmentation values were chosen, as well as three effective parameters on rock fragmentation, that is, burden, burden / spacing ratio, overdrilling and power factor. Likewise, they were separated into training and test data (70%–30%) for the generation of the fuzzy rules of the model. Based on statistical functions, correlation coefficient (R2) and mean square error (RMSE), it was found that the ANFIS-PSO model (with R2 = 0.85 and RMSE = 0.78) can be used as a reliable and acceptable model in the field. prediction of rock fragmentation.

Keywords: ANFIS-PSO · Prediction · Rock fragmentation · Open pit mining

1 Introduction

In the mining industry, one of the most influential parameters in productivity performance is the quality of rock fragmentation. The rock fragmentation size is a general term used to define the granulometric distribution of the blasted material, which is why it is the main result of the blasting process, in which the pressures generated by explosives confined within holes drilled in the rock originate a high zone of energy concentration that produces the fragmentation and displacement of the rock [1]. However, only 20–30% of the energy produced is actually destined for rock fragmentation [2], which hinders the success of the blasting process and, therefore, different sizes of fragmentation are generated. Likewise, there are a series of factors that directly or indirectly intervene in the fragmentation size, that are mutually dependent or that are related to each other [1,

S. Trzcielinski et al. (Eds.): AHFE 2021, LNNS 274, pp. 124–131, 2021.
https://doi.org/10.1007/978-3-030-80462-6_16

3]. These factors are divided into blast design factors and rock mass properties, also called controllable and uncontrollable parameters.

The objective of this research is to generate a predictive model of rock fragmentation using the Neuro-Diffuse Inference System (ANFIS) and Particle Swarm Optimization (PSO) techniques to estimate the size of fragmentation that may impact operational processes. and overall costs in open pit mines. The ANFIS and PSO techniques are related creating a hybrid model that works very well for the prediction of fragmentation, since the PSO is an optimizing algorithm that is easy to understand and implement [2].

2 State of the Art

2.1 Predictive Model of Rock Fragmentation in Open Pit Mining

Through the blasting process, an optimal rock fragment size that facilitates subsequent unit processes, for which, in the last decade, various models or experimental equations have been proposed with the aim of estimating the fragmentation size of the rock product of blasting, however, these still do not meet the expected level of precision [2, 3]. The empirical models are based on data obtained in several blasts, collecting the parameters that are considered relevant but with the risk of presenting problems in relation to the variation of geological conditions, blast design configurations and the amount of effective parameters that are analyzed., since it is not possible to collect the parameters with the greatest influence on fragmentation when information is sometimes not available or is difficult to quantify; For this reason, predictive models have taken a greater position in the engineering field, being able to accept qualitative or quantitative variables, in addition to accepting changes in the model without compromising predictive performance [2, 4, 5].

2.2 Predictive Model of Rock Fragmentation Applying the Neuro-Diffuse Inference System (ANFIS) in Open Pit Mining

The neuro-fuzzy inference system (ANFIS) is a hybrid method that combines the benefits of fuzzy inference systems (FIS) and neural networks (ANN) to solve complex non-linear systems, this is evidenced in the interpretability of human language which uses fuzzy rules and the ability to understand the structure of a database with a high capacity for adaptation and rapid learning [6–8]. The data selection stage must be meticulous, since knowledge about the geological conditions of the rock mass will be used, which are not modifiable, but which facilitate the selection of the explosive characteristics and the blasting parameters, avoiding experimental work that they are more hectic and expensive [6, 7]. In addition, modeling by ANFIS implies different adjustments of the input parameters, seeking to find the appropriate type and amount of membership functions and fuzzy rules, all this is achieved through multiple simulations for the data training stage, that is, vital, since at that point the type of function belonging to each variable entered into the system is defined, which will later be applied in the test stage [1, 9].

2.3 Predictive Model of Rock Fragmentation Applying the Particle Swarm Optimizer (PSO) Algorithm in Open Pit Mining

Among the variety of algorithms developed for optimizing predictive models of artificial intelligence is the PSO algorithm, which is a powerful computational method that optimizes the input and output parameters of a predictive model by adjusting the membership functions of each parameter; by means of the PSO algorithm it is possible to significantly improve the predictive performance in the data training stage [10, 11]. In the field of geotechnics, various techniques have been tested that can handle the complexity of the geotechnical behavior analysis, which is subject to multiple rock variables that generate problems to estimate a non-convex and discontinuous objective function such as the fragmentation size of stone. Therefore, the PSO algorithm, being an easy to understand and implement tool, provides great capabilities to optimize input parameters adaptable to any predictive model without affecting the characteristics and variables that compose it, in addition to maintaining a complete balance of precision and more interpretability. adaptable for the problematic [10, 12].

3 Input

3.1 General Contribution

The present study proposes the creation of a predictive model combining the ANFIS and PSO techniques in order to obtain an efficiency advantage to estimate the size of fragmentation since it is a model capable of handling complex problems and possessing high-level learning capacities [3], making it more systematic and less dependent on expert knowledge [8]. Therefore, the ANFIS-PSO predictive model will be created in the MATLAB system from the data load. Where the study of rock fragmentation will be carried out through the historical blasting processes of the open pit mine and the selection of the most significant parameters that will be input to the model (Fig. 1).

3.2 Detailed Contribution

Database Preparation. The input database consisted of 3 blast design parameters Burden (B), Spacing (S) and Load Factor (PF) different from those used by Hasanipanah, Amnieh, Arab & Zamzam, 2018 since parameters of drilling and explosives. Likewise, the database was separated into 2 stages: training and testing, which had a relationship of 70% -30% of a total of 92 data from blasting processes. On the other hand, frequency diagrams were made for the input and output parameters in order to visualize the values within each one and, therefore, if there is a possibility of data dispersion.

Estimation of ANFIS Fragmentation. In the estimation with the use of the MATLAB R2020a software, the ANFIS model used the membership functions that were Gaussian for all the input parameters. In it, a rule of the numbers mfs 5 was used, which generated a difference than that proposed by Hasanipanah, Amnieh, Arab & Zamzam, 2018 in the number of nodes, linear parameters, non-linear parameters, training parameters and rule parameters diffuse.

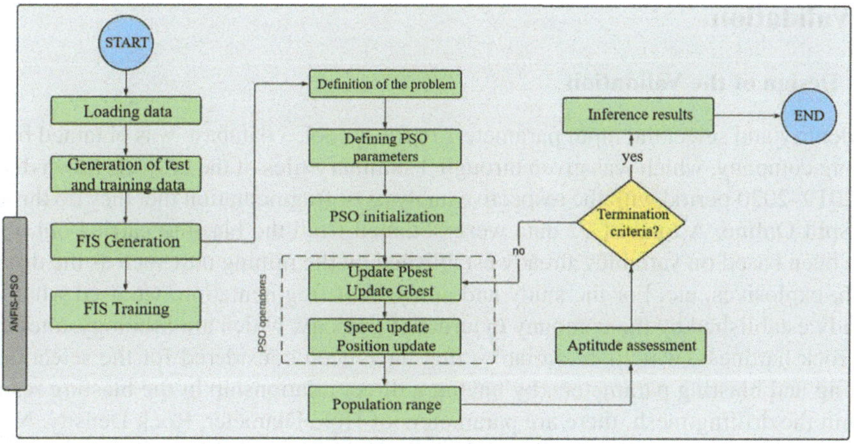

Fig. 1. Proposed model

Analysis of Data and Results ANFIS-PSO. In the present investigation, the analysis of data and results generated by MATLAB was used two statistical criteria for the comparison of results and the acceptance of the model consisted of the correlation coefficient (R2) and the mean square error (RMSE), unlike Zhou, Li, Arslan, Hasanipanah & Amnieh, 2019 that make use of other statistical values to value the predictive capacity of the model.

3.3 Indicators

If the values theoretically give 1, 0 for R2 and RMSE respectively, they show that there is a perfect agreement between all the measured and predicted values. Therefore, if the results of the statistical values of the present study are far from the results, other alternatives would have to be evaluated since the closer they are, it means that the model has reached a high performance capacity.

Root Mean Square Error (RMSE). Allows you to measure the error between actual and predicted fragmentation.

$$RMSE = \sqrt{\frac{1}{n} \times \sum_{i=|}^{n} [(x_i - x_p)^2]} \tag{1}$$

Determination Coefficient (R2). Lets you measure the relationship between actual and predicted fragmentation.

$$R^2 = \frac{\left[\sum_{i=1}^{n} (x_i - x_{mean})^2\right] - \left[\sum_{i=1}^{n} (x_i - x_p)^2\right]}{\left[\sum_{i=1}^{n} (x_i - x_{mean})^2\right]} \tag{2}$$

4 Validation

4.1 Design of the Validation

To identify and select the input parameters to the model, a database was obtained by the mining company, which was given through 3 summary files of the projects flown during the 2019–2020 period with the respective analysis of fragmentation that they do through the Split Online. A total of 92 data were obtained from the blasting carried out, these have been based on variables already established by the mining unit such as the drilling mesh, explosives, etc. For the study and analysis of fragmentation, we used what was already established by the company in terms of geology, which are lithology, alteration and rock hardness, which are variables that have been considered for the selection of drilling and blasting parameters. by having a direct relationship in the blasting results. Within the drilling mesh, there are parameters of Hole Diameter, Rock Density, No. of holes, Burden (B), Spacing (E), Bench Height (H) and Overdrilling (SD).

The identification of the input parameters to the model was based on a choice of the controllable parameters determining the rock fragmentation. Based on the previous studies analyzed, it was determined that the parameters of the drilling mesh and explosives usually have an important role in the general results of the blasting, they must also be related to the characteristics of the rock mass, in this case they are related to lithology and alteration. The selected input parameters covered both determining variables, therefore, Burden (B), Relationship between Burden and Spacing (E/B), Overperforation (SD) and Power Factor (PF) were those established for the present investigation.

On the other hand, frequency diagrams were made for the different selected parameters in order to facilitate the information contained in the data. In this you can see the number of times that the input and output parameters take a certain value and, therefore, shows the distribution of each of them (Figs. 2 and 3).

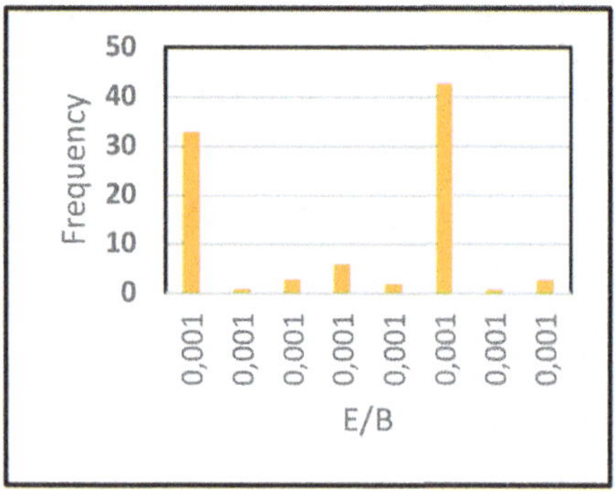

Fig. 2. One frequency of the burden/spacing ratio parameter

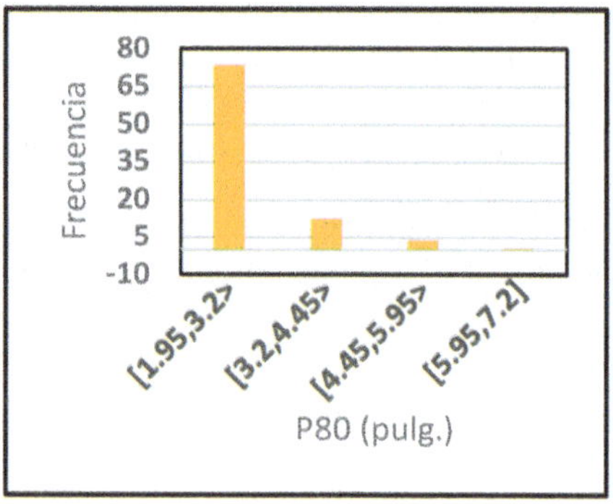

Fig. 3. Frequency of parameter P80

In order to create the rock fragmentation model, the most influential parameters in the rock fragmentation that entered the system as input parameters were identified: burden, burden-spacing relationship, over-drilling and power factor. The mentioned parameters were selected and used for the prediction of rock fragmentation in the present research

Table 1. Parameters for Genfis 3

Parameters	Value
Number of clusters	10
Number of rules	10
Exponent U	2
Maximum number of iterations	100
Minimum number of upgrades	1e−5

Table 2. Test results

Iterations				
	100	300	500	1000
RMSE	0.9102	0.8304	0.7814	0.8506
Mean Error	0.11	−0.05	−0.08	−0.27
Error St. D.	0.9104	0.8304	0.7814	0.8506

work. The database considers 92 blasting events that include the values measured in the field of the parameters mentioned above with the fragmentation results (P80).

For the creation of the fragmentation model, the Matlab R2015a software was used, it has the ANFIS editor that consists of 4 steps for its implementation.

In order not to compromise the merging process in the following section, the data must be separated randomly, fulfilling a ratio of 70–30. The data that makes up 70% of the records are applied for the training phase, where the ANFIS system will create the rules of the predictive structure, and 30% for the test phase that will serve to verify the efficiency of the structure already created.. Therefore, 64 records are used in the training stage and 28 records are used for the test.

Next, a fuzzy inference system is generated. For the nature of the data, which is related to the geological and geotechnical characterization of the study area, it was decided to apply a fuzzy clustering that determines the relationship that exists between parameters by means of a Genfis3 function, which the program offers Matlab. The type of input membership function is' gaussmf, which obtained the highest precision result compared to the other alternatives. By default, the output membership function type is 'linear'. From this, it is possible to generate the inference rules of the model (Table 1).

The model chosen from the generation of the FIS begins its training process applying the PSO algorithm, however, it is important to determine the number of iterations that suits the model. Given this, it is recommended to test with different values, so that the optimum can be determined (Table 2).

The model chosen from the generation of the FIS begins its training process applying the PSO algorithm, however, it is important to determine the number of iterations that suits the model. Given this, it is recommended to test with different values, so that the optimum can be determined. Variations were made with iteration values of 100,300,500 and 1000, of which it was determined that with the value of 500 iterations the RMSE value drops considerably and its mean error, despite being negative, is close to 0. By identifying the iterations that the model requires, the rest of the parameters can be calculated.

Finally, the test data is applied to evaluate the training performance, obtaining the training and test results. This data can be extracted from the program for its corresponding evaluation.

With the results given by the ANFIS-PSO model, the value of R2 is analyzed, which the closer its value is to 1, the greater the fit of the model to the variable that we are trying to explain. Conversely, the closer to zero, the less tight the model will be and therefore the less reliable it will be. As can be seen in the scatter graphs, the training data obtained a value of 0.9315, which indicates that, for a reevaluation of the training data, the ANFIS model responds adequately according to what has been learned. On the other hand, the test data obtained a result of 0.8511, which indicates that the model is responding to expectations, complying with giving a level of precision in accordance with previous research.

In addition, it was also considered at what level the PSO algorithm contributed in the training stage, which can be seen below (Table 3).

Table 3. Summary of final results

Model	Statistical indices			
	R^2		RMSE	
	Train	Test	Train	Test
ANFIS	0.7519	0.6129	1.03	0.91
ANFIS-PSO	0.9315	0.8511	0.67	0.78

4.2 Analysis of Results

A correlation index of 0.85 and a mean square error of 0.78 were obtained, being within the range of the various scenarios analyzed. Regarding the influence of the entered parameters, the power factor was the most effective on the fragmentation of the rock in the present case, this suggests that making changes in the parameter with the greatest influence in the course of further studies to improve fragmentation.

5 Conclusions

The results generated by the application of the ANFIS-PSO hybrid model to predict the fragmentation of rock product of blasting in open pit mining, confirm the effectiveness and reliability of the methodology used, since a precision level of 85% was obtained in the ANFIS-PSO model, which compared to previous research is within the acceptance range of the model, which determines that if the model does not reach at least a value of R2 = 0.70, then it is not reliable, the present study shows the high performance of the predictive model as long as it is created properly, this considers the correct selection of the input parameters, as well as the standardization of the database. Likewise, it is important to consider the parameters widely used in previous research, which prioritize the burden, spacing, specific load and block in the selection of parameters, since they have given the best results, of which burden and spacing were taken for the present study.

In the research process, it was identified that for the generation of the FIS the most appropriate type to the fragmentation problem is the sugeno, with which the results of R2 = 0.85 were obtained, on the other hand, the option of the Mamdani type FIS gave As a result, R2 = 0.56, for which the first option was chosen.

Finally, it was determined that the methodology applied in the study can be replicated in different open pit mining scenarios in Peru, for this it is essential to have a solid database and that simulations are carried out varying the parameters of both the ANFIS and the PSO until you find the right model for the stage.

References

1. Asl, P.F., Monjezi, M., Hamidi, J.K., Armaghani, D.J.: Optimization of flyrock and rock fragmentation in the Tajareh limestone mine using metaheuristics method of firefly algorithm. Eng. Comput. **34**(2), 241–251 (2017). https://doi.org/10.1007/s00366-017-0535-9
2. Adebola, J.M., Ajayi, O.D., Elijah, P.: Rock fragmentation prediction using Kuz-Ram model. J Environ Earth Sci **6**(5), 110–115 (2016)
3. Basser, H., et al.: Hybrid ANFIS-PSO approach for predicting optimum parameters of a protective spur dike. Appl. Soft Comput. J. **30**, 642–649 (2015). https://doi.org/10.1016/j.asoc.2015.02.011
4. Lyana, K.N., Hareyani, Z., Shah, A.K., Hazizan, M.M.: Effect of geological condition on degree of fragmentation in a Simpang Pulai marble quarry. Procedia Chem. **19**, 694–701 (2016)
5. Farid, S., Mojtahedi, F., Ebtehaj, I., Hasanipanah, M., Bonakdari, H., Amnieh, H.: Proposing a novel hybrid intelligent model for the simulation of particle size distribution resulting from blasting. Eng. Comput. **35**(1), 47–56 (2019). https://doi.org/10.1007/s00366-018-0582-x
6. Ebrahimi, E., Monjezi, M., Khalesi, M.R., Armaghani, D.J.: Prediction and optimization of back-break and rock fragmentation using an artificial neural network and a bee colony algorithm. Bull. Eng. Geol. Env. **75**(1), 27–36 (2015). https://doi.org/10.1007/s10064-015-0720-2
7. Gao, W., Karbasi, M., Hasanipanah, M., Zhang, X., Guo, J.: Developing GPR model for forecasting the rock fragmentation in surface mines. Eng. Comput. **34**(2), 339–345 (2018). https://doi.org/10.1007/s00366-017-0544-8
8. Hajihassani, M., Jahed Armaghani, D., Kalatehjari, R.: Applications of particle swarm optimization in geotechnical engineering: a comprehensive review. Geotech. Geol. Eng. **36**(2), 705–722 (2017). https://doi.org/10.1007/s10706-017-0356-z
9. Hasanipanah, M., Jahed Armaghani, D., Monjezi, M., Shams, S.: Risk assessment and prediction of rock fragmentation produced by blasting operation: a rock engineering system. Environ. Earth Sci. **75**(9), 1–12 (2016). https://doi.org/10.1007/s12665-016-5503-y
10. Akbari, M., Lashkaripour, G., Bafghi, A.Y., Ghafoori, M.: Blastability evaluation for rock mass fragmentation in Iran central iron ore mines. Int. J. Min. Sci. Technol. **25**(1), 59–66 (2015)
11. Sharma, A., Mishra, A.K., Choudhary, B.S., Meena, R.: Impact of blast design parameters on rock fragmentation in building stone quarries. Curr. Sci. **116**(11), 1861 (2019)
12. Jug, J., Strelec, S., Gazdek, M., Kavur, B.: Fragment size distribution of blasted rock mass. In: IOP Conference Series: Earth and Environmental Science, vol. 95, no. 4, p. 042013. IOP Publishing (2017)

Managing the Team by Individualization. How Can Artificial Intelligence Help to Manage a Team Effectively and Increase Its Efficiency?

Ewa Jochheim[1(✉)], Maciej Orzechowski[1], and Pawel Nowodzinski[2]

[1] Aia Concept Sp. z o.o., Slaska, Czestochowa, Poland
ewa@aiaconcept.com
[2] Faculty of Management, Czestochowa University of Technology, Armii Krajowej 19B, 42-200 Częstochowa, Poland

Abstract. This article aims to present how manage the team using Individualised Management to support efficient achievement of specific business objectives. An individualised approach to employees and the whole concept of Individualised Management works for individual employees, hence the approach has been expanded to incorporate entire teams. An IT tool i.e., an application that uses AI has been developed to enable individualised management and support managers in analysing the internal needs of employees and teams, and in using this knowledge to effectively manage their teams. Knowledge from various fields was used to develop the tool: Artificial Intelligence (AI) as well as psychology of management and needs.

The tool presented here provides managers with recommendations on how to manage teams, thus supporting organisations in fast and effective achievement of business objectives through an individualised approach and utilising state-of-the-art information technology. The IT solution operates on the basis of proprietary algorithms and models developed during research.

Keywords: Team management · Setting goals for the team · Effective team · Needs assessment · Individualised management · Artificial

1 Introduction

The goal of the research was to figure out how could be possible manage the team using Individualised Management to support efficient achievement of specific business objectives. It was assumed that the above research objective could only be achieved by answering the following research question: why do some teams work effectively and achieve the set objectives, while some do not? Too often team management overlooks the role of individual employees and the conceptualisation of the team as a multi-level structure consisting of individuals (multilevel entities) [1]. More and more research are focusing on group processes, looking at them not only from the level of the whole team, but paying more attention to individual employees [2, 3]. In our article we intend to present

S. Trzcielinski et al. (Eds.): AHFE 2021, LNNS 274, pp. 132–139, 2021.
https://doi.org/10.1007/978-3-030-80462-6_17

a method for managing teams which focuses on individual resources of employees, profiling their needs and applying individualised management using Artificial Intelligence. The diagnostic app we are developing is a response to the needs of many companies and the problems addressed in research on the impact of individual differences on team effectiveness and the decline in commitment.

2 Individual Differences vs Team Efficiency and Commitment

Research into the development of interpersonal relations, which includes among other studies of the individual fit between the worker and the team [4] and impact of personality on interactions with fellow team members, suggests that individual character traits have a significant impact on relations and effectiveness of the team [5]. Bell and Outland (2017) [6] demonstrated in their study that individual character traits are an important factor affecting team cohesion.

In teams composed of different professionals, there are three levels of factors that support or inhibit group processes: organisational, group and individual. The way individual members understand cooperation and the way they communicate with fellow team members influences the quality and effectiveness of relations [7]. The lack of consensus in how different aspects of group work are interpreted has a significant impact on the understanding of group roles, decisions on compromises in communication and thus on group dynamics and the learning process.

Differences in perception of group work have an adverse effect on efficiency. A shared vision, good communication and understanding of group roles all result in improved efficiency [7]. Research also suggests a significant role of different perceptions of group work and the matching of values of individual employees. On the one hand, different values differentiate employees into, for example, those who prefer hierarchy or open communication. On the other, depending on the perception of group work, it is important for employees to share a common vision, as a team [8].

Studies of Tasa, Taggar, and Seijts [9] demonstrated that individual task-related expertise and the feeling of efficacy are key predictors of group behaviour at the individual level. Research also points to other individual characteristics, such as empathy, self-confidence, affect management, identification with the company or expression of emotions [10]. Particularly in areas that require specialised knowledge (e.g., product development, software and hardware design; [11, 12]), the requirements related to interactions with other team members or effective information sharing are high [13], since mastering them is crucial to ensuring team effectiveness [3]. In order to develop an efficient team working model it is essential to focus on the behaviour of not just the group as a whole, but also individuals in the context of group work, to identify predictors of individual behaviour and their impact on the dynamics of group processes [14].

Many managers and team members share the same views on problems in the company or their daily working experience [15] One of the more visible challenges to teamwork is the lack of commitment of individual members [16]. Research shows that individual members of the group differ in their levels of commitment even at the level of specific group tasks [17]. Another difficulty with teamwork is the dominance of overly aggressive individuals [18], who can have a negative influence on less assertive persons. Considering the importance of an individualised approach and seeing the team through the lens of the individual, we have decided to develop an application which would, while referring to numerous studies of group processes, focus not only on the team as a group but on individual persons and their individual needs.

3 Diversity Management. Opportunities and Threats

Diversity in teams is an increasingly popular topic in the field of organisational culture. Individual resources of the employee contribute various experiences, skills and problem-solving capacity to the team. In the age of globalisation having culturally diverse teams is becoming increasingly common, however diversity should also be understood more locally – as individual differences between team members. An individual skill set significantly influences the team's ability to innovate. At the same time, a high level of diversity may result in communication problems and conflicts caused by differences of opinion, ultimately resulting in lower motivation to get involved in teamwork [19].

Many studies have shown a link between cohesion and positive group performance [20]. Team cohesion is identified as a priority value for effectiveness. It is defined as perception of group experiences shared by the team members [21]. There are two aspects of cohesion: social and task-related [22]. While social cohesion represents the interpersonal nature of the construct, task cohesion represents activities focused on the task [23].

Diversity in the team may lead to conflicts, which in turn result in reduced effectiveness of the whole group [24]. On the other hand, research points to positive impact of task-related diversity on the achievement of group goals [25]. The analysis of 41 teams, performed at the University of Haifa, made it possible to establish which personality traits a group should have in order to be innovative. The data shows that the team should include people who are flexible, easy to work with and highly self-confident [26].

The interviews with team leaders in international projects mentioned the lack of knowledge of the employee [27], insufficient communication skills and lack of face-to-face contact [28] as some of the most important reasons for unsatisfactory cooperation and conflicts. Vakola and Wilson [27] warn against underestimating the importance of the human factor and the way employees communicate for team effectiveness. Important elements of satisfying cooperation included communicating change and keeping employees informed, a pleasant working environment, a positive atmosphere as well as highlighting positive results and rewarding employees for a job well done [27]. Moreover, accepting the employee's weaknesses and matching their tasks to their strengths also played a key role in an effective and fruitful collaboration. The conflicts emerging in the team were attributed to personality differences. Group members who, according to the employees, had a negative impact on group cooperation were described as "over-extroverts", "difficult people" "aggressive people who are always right", and "introverts who never share their knowledge" [29].

In summarising the above research, the significance of an individualised approach for effectiveness and commitment in the team should not be overlooked. On the other hand, it is easy to notice the inconsistent results related to managing diversity in the team. Social psychology research shows that we like people who are similar to ourselves [30]. However, the diversity of individuals in a team positively influences the creativity and variety of proposed solutions to problems.

The application we developed enables the user to analyse the needs of individual team members, identifying an ideal set of characteristics to ensure a positive influence on work effectiveness and the achievement of goals. What is more, the app also supports the manager, providing them with the necessary set of recommendations to manage the diverse needs of individual team members.

4 Research Method

Social research is based on defining and finding a method of measurement for the phenomenon being measured. The psychometric tools available on the market are often expensive to use and difficult to interpret. Moreover, the use of tools in social sciences has recently raised many questions among researchers. There is a lack of transparent and reliable methodologies that would make it possible to collect the right content, interpret it quickly and draw conclusions [31]. Barry and colleagues [32] showed in their research that 40–93% of measurement methods used in studies published in scientific journals lacked confirmation of the accuracy of the tool.

The selection of psychometric tools is one of the most important steps in the measurement of variables. The social sciences offer at least 280 scales to measure depression [33], 65 different scales to measure emotions, 19 of which are dedicated to the measurement of anger [34]. The app used in our study exploits the potential of Artificial Intelligence and machine learning. Artificial Intelligence provides many innovative techniques in various areas of business and science. Intelligent agents [35] are used to assist in the search for information and manage systems in organisations. Expert systems used in companies take over tedious and repetitive tasks such as protocol analysis, simulations or document sorting [36].

The starting point for our tool (AIA), which aims to manage employees in an individualised way to ensure quick and effective achievement of business objectives, is the analysis of the individual needs of the employee. On the basis of the research into needs analysis, the number of needs was limited to 32.

Every human being has all the needs mentioned above, but with different saturation. In order to determine the level of saturation on a scale of 0 to 100 percent, you first need to fill in the questionnaire in the AIA application. Once the value of the need is defined, the system develops recommendations for the manager. The application suggests which actions should be taken towards the employee to boost their effectiveness. The role of the supervisor is limited to following the suggestions from the system and providing feedback on whether a given recommendation was useful or not. Through this system of active cooperation between the manager and the application, the AIA app continues to learn about every person, and its suggestions become more and more relevant. The development of this system has only now become possible with the use of Artificial Intelligence and deep learning. Until now, the manager had to not only watch the employee to be able to analyse their individual needs but also to draw conclusions on how to manage the specific employee and the entire team. The development of an individualised management strategy was not only tiring and time-consuming but also fraught with many potential errors.

The AIA app used in this study is easy to interpret and delivers a high level of relevance and reliability. A set of questions and needs was defined during system design.

The number of needs was then reduced to 32 during the development, validation and tests. The mobile app performs an online survey, and a profile is developed individually for the tested person with the use of Artificial Intelligence. The fact that AI is used ensures that the profile is calculated from the responses taking all parameters embedded in the app into account. AI looks for correlations in the calculations obtained. The application improves the model as new surveys are delivered along with new data, including user confirmations/denials at later stages.

The figure below shows the process of deploying the application to a company. The organisation identifies all existing teams. The system analyses the individual needs of all employees, and then aggregates them. The identified teams are analysed for strengths and weaknesses and potential conflicts. The system develops team matrixes based on the individual needs of employees along with recommendations for the manager formulated in such a way so as to reach as many individual team members as possible. The recommendations include: guidelines for setting goals for the team, how to communicate, how to build a positive work environment and how to effectively motivate the team. In addition, every team member (employee of the organisation) receives guidance on communication within the team.

In situations where the team is ineffective, a "rotation" is suggested; the system currently only examines needs related to personality, it does not consider competence. In the future we plan to develop the system further by adding competence, in order to strengthen the results of the team survey and expand the value of the IT system.

A sales team can serve as an example of individualised analysis of team members. The research compiled suggests that the most effective sales staff have a set of three needs:

- High need of social contacts,
- High need of status,
- Low need of acceptance.

In order to check whether the team under examination has the necessary set of characteristics to sell effectively, you need to analyse the needs of the team's members and check the strength of these needs e.g. for social contacts the median should be above 50 percent. At the same time, the standard deviation of the needs mentioned above should be as low as possible, while other needs should remain diverse (Fig. 1).

5 Conclusion

Effective teams are characterised by the fact that they consist of people with diverse personality types. At the same time, the more diverse the team, the more difficult it is to manage. While individual employees are relatively easy to manage, managing entire teams represents a challenge, as you are dealing with very different people, with varied value systems and different needs. The market lacks tools to support the manager in team management, tools capable of providing clear, precise and reliable support.

The app, developed on the basis of empirical research, analyses the individual needs of team members and presents them in aggregate form on the matrix of team needs. The

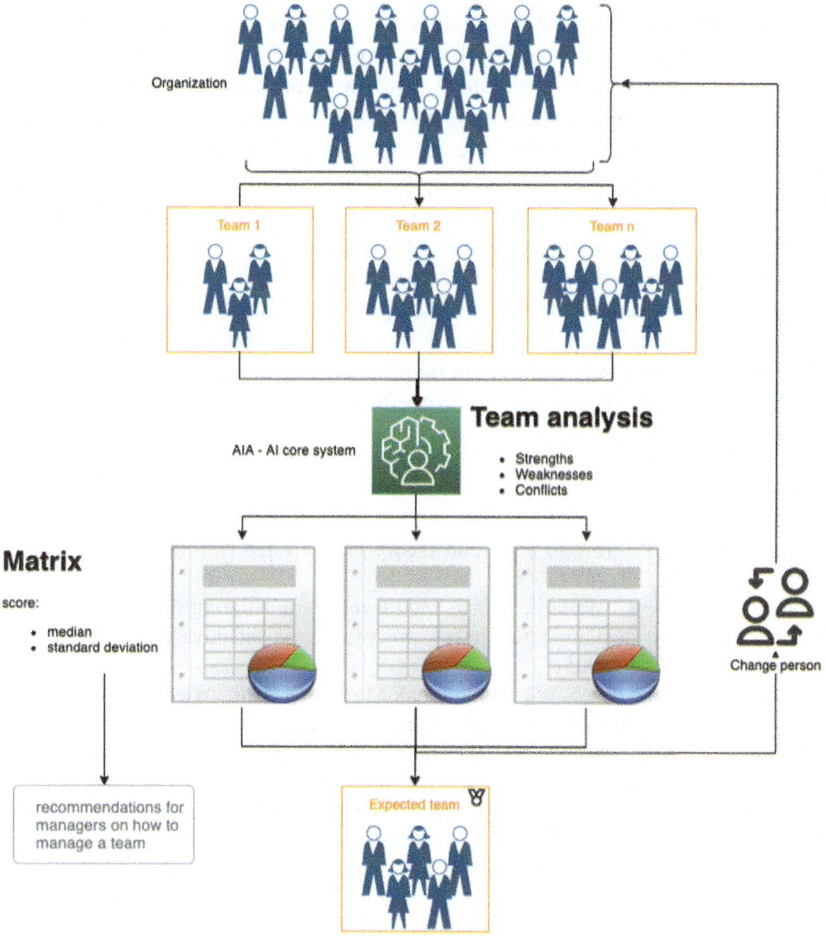

Fig. 1. Description of the AIA system based on artificial intelligence Source: Own work.

software uses neural networks. Collected samples were used during the deep learning process to develop the correct model. The software works on a case-by-case basis. The app analyses the team's results to determine those individual needs of team members, which are consistently shared by all of them. The results are analysed further and ultimately the manager is presented with a system of recommendations and guidelines on how to manage the entire team, set goals and motivate it in a way that is consistent with the needs of team members. The software learns about the team and supplies optimum suggestions based on the user's assessment of its recommendations.

Apart from the most consistent needs of the teams, the system also analyses those that differentiate the teams the most, as they are a likely reason of conflicts. Awareness of these areas coupled with the right suggestions for the manager, also developed by the app, are an indispensable element of the tool that has been created to support the manager in their daily work with people.

The paper describes a new approach to team management through individualisation, at the same time addressing the needs identified in research on group processes and team effectiveness. Using Artificial Intelligence as well as expertise and experience in studying motivation, we offer support in managing and analysing internal needs of employees and teams, and in using this knowledge in effective team management. By analysing needs at individual level, companies will not only be able to achieve their business goals more effectively but also to create specialised and highly motivated employee teams. The key advantage of the proposed solution is not only the fact that it automates the process of needs analysis but also that it offers support to managers with specific recommendations. The needs assessment allows you to check the potential of the team, identify its strengths and weaknesses and analyse potential conflicts. Through a multiple needs analysis and by including and removing individual team members it is possible to create a homogenous team with skills and needs matched to a specific task.

References

1. Chen, G., Kanfer, R.: Toward a systems theory of motivated behavior in work teams. Res. Organ. Behav. **27**, 223–267 (2006)
2. Chen, G., Thomas, B., Wallace, J.C.: A multilevel examination of the relationships among training outcomes, mediating regulatory processes, and adaptive performance. J. Appl. Psychol. **90**, 827–841 (2005)
3. DeShon, R.P., Kozlowski, S.W.J., Schmidt, A.M., Milner, K.R., Wiechman, D.: A multiple-goal, multilevel model of feedback effects on the regulation of individual and team performance. J. Appl. Psychol. **89**, 1035–1056 (2004)
4. Kristof, A.L.: Person-organization fit: an integrative review of its conceptualizations, measurement, and implications. Pers. Psychol. **49**, 1–49 (1996)
5. Asendorpf, J.B., Wilpers, S.: Personality effects on social relationships. J. Pers. Soc. Psychol. **74**, 1531–1544 (1998)
6. Overtveit, J.: Making the team work. Prof. Nurse **5**, 284–288 (1990)
7. Freeman, C.M., Nick Ross, M.: The impact of individual philosophies of teamwork on multi-professional practice and the implications for education. J. Interprof. Care **14**(3), 237–247 (2000). https://doi.org/10.1080/jic.14.3.237.247
8. Kellett, J., Humphrey, R., Sleeth, R.: Empathy and the emergence of task and relations leaders. Leadersh. Q. **17**, 146–162 (2006). https://doi.org/10.1016/j.leaqua.2005.12.003
9. Tasa, K., Taggar, S., Seijts, G.H.: The development of collective efficacy in teams: a multilevel and longitudinal perspective. J. Appl. Psychol. **92**, 17–27 (2007)
10. Sawyer, S.: Software development teams. Commun. ACM **47**(12), 95–99 (2004)
11. Walz, D.B., Elam, J.J., Curtis, B.: Inside a software design team: knowledge acquisition, sharing, and integration. Commun. ACM **36**(10), 63–77 (1993)
12. Reus, T.H., Liu, Y.: Rhyme and reason: emotional capability and the performance of knowledge-intensive work groups. Hum. Perform. **17**, 245–266 (2004)
13. Brodbeck, F.C.: Communication and performance in software development projects. Eur. J. Work Organ. Psy. **10**, 73–94 (2001)
14. Franz, T.M., Larson, J.R.J.: The impact of experts on information sharing during group discussion. Small Group Res. **33**, 383–411 (2002)
15. Brown, R.W.: Autorating: getting individual marks from team marks and enhancing teamwork. In: Proceedings Frontiers in Education 1995 25th Annual Conference. Engineering Education for the 21st Century (n.d). https://doi.org/10.1109/fie.1995.483140

16. Sonnentag, S.: High performance and meeting participation. An observational study in software design teams. Group Dyn. Theory Res. Pract. **5**, 3–18 (2001)
17. Christianson, D.: New Cumcula, SpectralLines. In: IEEE Spectrum, vol. 29, no 7 (1992)
18. Agrawal, V.: Managing the diversified team: challenges and strategies for improving performance. Team Perform. Manag.: Int. J. **18**(7/8), 384–400 (2012). https://doi.org/10.1108/13527591211281129
19. Evans, C.R., Dion, K.L.: Group cohesion and performance: a meta-analysis. Small Group Res. **43**, 690–701 (2012)
20. Carron, A.V., Bray, S.R., Eys, M.A.: Team cohesion and team success in sport. J. Sports Sci. **20**, 119–126 (2002)
21. Dion, K.L.: Group cohesion: from "field of forces" to multidimensional construct. Group Dyn. Theory Res. Pract. **4**, 7–26 (2000)
22. Severt, J.B., Estrada, A.X.: On the function and structure of group cohesion. In: Salas, E., Vessey, W.B., Estrada, A.X. (eds.) Team Cohesion: Advances in Psychological Theory, Methods and Practice, pp. 3–24. Emerald Group publishing limited, West Yorkshire (2015)
23. Wickramasinghe, V., Nandula, S.: Diversity in team composition, relationship conflict and team leader support on globally distributed virtual software development team performance. Strateg. Outsourcing: Int. J. **8**(2/3), 138–155 (2015)
24. Horwitz, S.K., Horwitz, I.B.: The effects of team diversity on team outcomes: a meta-analytic review of team demography. J. Manag. **33**(6), 987–1015 (2007). https://doi.org/10.1177/0149206307308587
25. Miron-Spector, E., Erez, M., Naveh, E.: To drive creativity add some conformity. Harvard Bus. Rev. **90**, 30 (2012)
26. Hinds, P.J., Bailey, D.E.: Out of sight, out of sync: understanding conflict in distributed teams. Organ. Sci. **14**(6), 615–632 (2003)
27. Oertig, M., Buergi, T.: The challenges of managing cross-cultural virtual project teams. Team Perform. Manag.: Int. J. **12**(1/2), 23–30 (2006). https://doi.org/10.1108/13527590610652774
28. Vakola, M., Wilson, I.E.: The challenge of virtual organisations: critical success factors in dealing with constant change. Team Perform. Manag. **10**(5/6), 112–120 (2004)
29. Byrne, D.: An overview (and underview) of research and theory within the attraction paradigm. J. Soc. Pers. Relat. **14**, 417–431 (1997)
30. Flake, J., Fried, E.: Measurement Schmeasurement: Questionable Measurement Practices and How to Avoid Them (2019). https://doi.org/10.31234/osf.io/hs7wm
31. Barry, A.E., Chaney, B., Piazza-Gardner, A.K., Chavarria, E.A.: Validity and reliability reporting practices in the field of health education and behavior. Health Educ. Behav. **41**(1), 12–18 (2014). https://doi.org/10.1177/1090198113483139
32. Santor, D.A., Gregus, M., Welch, A.: Eight decades of measurement in depression. Measurement **4**, 135–155 (2006). https://doi.org/10.1207/s15366359mea0403_1
33. Weidman, A.C., Steckler, C.M., Tracy, J.L.: The jingle and jangle of emotion assessment: imprecise measurement, casual scale usage, and conceptual fuzziness in emotion research. Emotion **17**, 267–295 (2017). https://doi.org/10.1037/emo0000226
34. Bradshaw, J.D., et al.: Roles for agent technology in knowledge management: examples from applications in aerospace and medicine, White Paper, Boeing Information and Support Services, Seattle, WA (1998)
35. Hendriks, P., Vriens, D.: Knowledge-based systems and knowledge management: friends or foes? Inf. Manag. **35**, 113–125 (1999)
36. Bell, S.T., Outland, N.: Team composition over time. In: Team Dynamics Over Time, vol. 18, pp. 3–27. Emerald Publishing Limited, UK (2017)

Sill Panel Corrosion in Automotive Industry

Dina Diga(✉) and Irina Severin

Faculty of Industrial Engineering and Robotics, University POLITEHNICA of Bucharest, Bucharest, Romania
irina.severin@upb.ro

Abstract. Corrosion resistance of automotive components has become a sensitive topic over time, becoming an important improvement requirement for materials used in the manufacture of automobiles, especially for the functionality of vehicles on roads treated with de-icing salt. By preventing corrosion, an annual expenses of about 25% is estimated to be eliminated.

This study details a failure analysis on doorsill panel returned from Russia after 41 months in service. Customers reported five cases via a Quality Report, of which there are three different cases. Firstly, corrosion forming underneath the door opening seal. Second rework from a previous repair to top of rear doorsill. Third corrosion forming from paint becoming damaged to base of B pillar area. The analysis includes a visual inspection of the corroded surface at high magnification and chemical analysis of the substrate material, corrosion product and surface contaminants. The analysis results consist as input elements for the next step of simulation including corrosion tests, humidity and hot test, dry cycle so as to improve corrosion protection.

Keywords: Sill panel · Automotive industry · Quality inspection · Market · Customer satisfaction

1 Introduction

Corrosion performance can be a key factor in determining the life of the vehicle and is the basis of a corrosion guarantee offered by manufacturers [1]. In line with corporate goals of increasing customer satisfaction and long-term improvement, this paper will present an analysis of corrosion around the surface of the doorsill panel and identify the possible cause of degradation. It is necessary to take into account the aspects related to corrosion control when designing and producing automotive components and subassemblies, respectively for major modifications or their repairs [2]. The weight of the coating is the major factor that determines the corrosion resistance to perforation of existing coatings in automotive environments [3].

A comprehensive review of the process and divert outputs has indicated that the current process requires accelerated focus. The cost of poor components quality may result in increased work-in-progress or unnecessary additional financial expenditure (overtime) and potential customer delivery commitment delays. Failure to comply with key performance indicators will have an impact on the ability to make successful products [4].

S. Trzcielinski et al. (Eds.): AHFE 2021, LNNS 274, pp. 140–147, 2021.
https://doi.org/10.1007/978-3-030-80462-6_18

2 Problem Statement

The study is part of a project carried out within an automotive company, the company manufacturing sill panels used for assembling Side Door Systems. The objective of the paper is to identify and analyze the statement of the corrosion problem in the sill area that caused the perforation of the panel. Table 1 presents problem statement analysis on the issue reported by customers: corrosion around the sill area.

The data presented in the table indicate the need for a detailed analysis of the corrosion problem.

2.1 Customer Concern

In Russia, customers reported five cases of corrosion of the sill panel surface. The still is not sealed correctly to stop the elements entering via the sill seams, causing over time corrosion from the inside out [5]. Corrosion perforation proceeds from the inner side of automotive outer panels to the outer side. Thus, when corrosion is discovered by visual inspection, repair is extremely difficult. Corrosion is considered the most important problem in vehicle resistance [6]. Customer required having no corrosion on their vehicles.

Customers are concerned over Corrosion around the sill area, as presented in Fig. 1.

Customer usage: This vehicle was used on city roads. Sometimes the customer visits the nearest suburb no more 20 km away from the city. This car is not being used in off-road, or swampy areas. The car has been keeping in an indoor parking for last 3 months. Before that, the car had kept outside in an outdoor parking.

The Fig. 2 shows the Sill panel Issue: Body paint above the right sill is damaged.

Investigation: No signs of mechanical impact or body repairs were found.

Presumably, paintwork was damaged due to corrosion from the other side of the sill.

2.2 Supporting Graphical Evidence

The Data collected will support the graphical evidence, as shown in Fig. 3 and 4.

3 Inspection

Many automotive components, especially those metal manufactured, are exposed to the risk of salt corrosion [7]. Corrosion problems can lead to costly repairs and can cause considerable damage. Degradation will occur from either electrochemical or chemical processes with different materials or protective coatings reacting at varying rates. Providing results in a relatively short time, chemical analysis can help engineers select types of materials and review projects to minimize susceptibility to the effects of corrosion [8].

The task now is to find out, through a chemical analysis, performed in the test laboratory, if the product will continue to function as it was designed during the useful life of the vehicle [9, 10].

Table 1. Problem statement: Corrosion around the sill area

IS/IS NOT	Problem statement: corrosion around the sill area			
	Problem description	IS	IS NOT	Get information
WHAT	What object	Sill corrosion		
	What Defect	Corrosion on the sill panel from inside out		What the Russians use on the roads in winter conditions
WHERE	Where on object	Sill door shut panel face and where the door aperture seal fits		
	Where first observed	First seen in Russia high time in service	No reports from any other market	Temperature/climatic condition at time of year
	Where seen since	Belarus		
WHEN	When first observed	Russia, 21st of Apr		
	What pattern since	All different dates and Different Km's but all in Russian markets		
HOW BIG	How many affected	Each Object could have several sill sealing issues letting in the elements		
	What size	5 cases to date, i per 1000 = 0.22	On every vehicle built in the market	
	Defects per object	Several	One	
	Trend	Expecting to Increase due to cold weather temperature and the use of the salt and grit used on the roads	Decreasing	

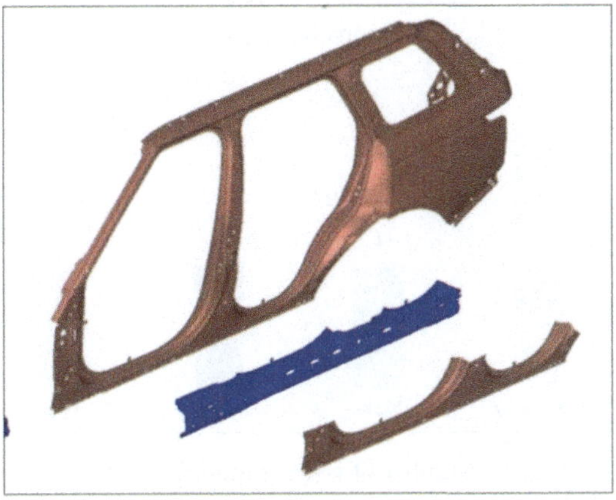

Fig. 1. Sill panel design

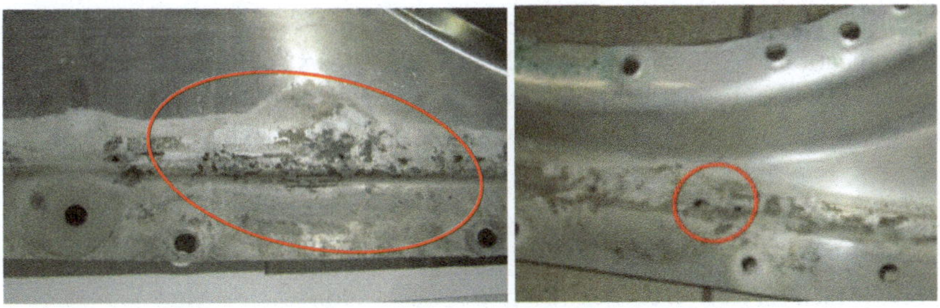

Fig. 2. Sill panel internal corrosion

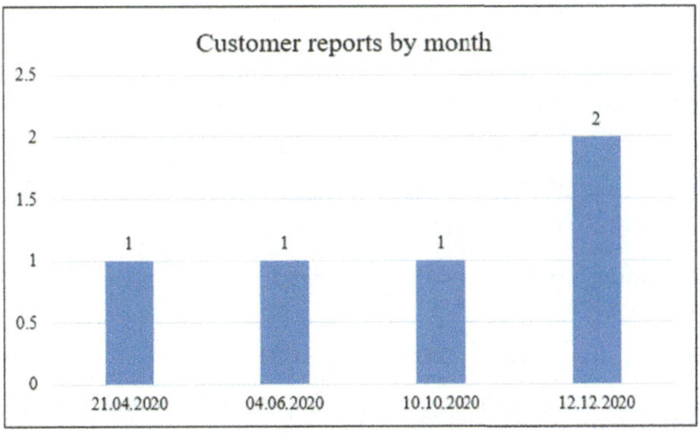

Fig. 3. Number of complaints reported by customers monthly

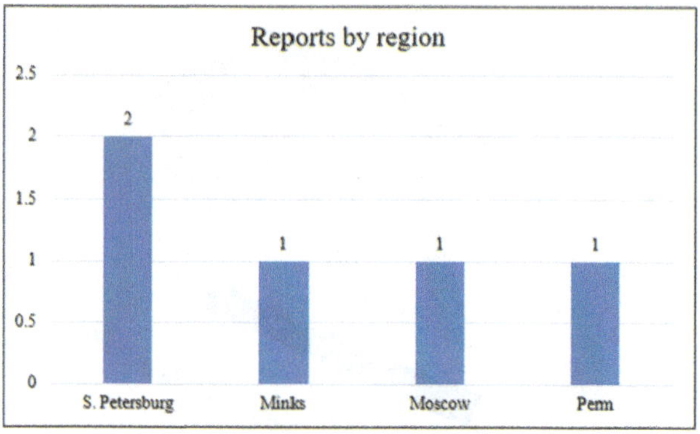

Fig. 4. Number of issues reported by region

3.1 Laboratory Quality Report

Chemical analysis of body side material confirms the alloy composition of the sill section is to specifications as shown in Table 2. Figure 5 shows the inner surface had no e-coat coverage and a distinct water line, where corrosive media filled the lower section, corroding the bare aluminum until complete perforation through to the outer surface.

Table 2. Problem statement: Corrosion around the sill area

Element (%)	Fe	Si	Cu	Mn	Mg
Specification [max]	0.35	0.3–0.95	0.25	0.15	0.4–0.85
EOS	0.19	0.604	0.147	0.10	0.73
Element (%)	Cr	Zn	Ti	V	Ni
Specification [max]	0.05	0.15	0.15	0.05 -0.25	0.05
EOS	0.012	0.003	0.024	0.074	0.006
Element (%)	Ca	Co	Pb	Sn	Sr
Specification [max]	0.05	0.05	0.05	0.05	0.05
EOS	0.0001	0.001	0.001	0.001	0.0001

*Equation of State (EOS) is a semi-empirical functional relationship between pressure, volume and temperature of a pure substance.

Figure 6 shows magnified images of a paint blister highlighted in Fig. 4. The paint layer and underlying corrosion product was removed to reveal a hole in the panel that was associated with a wider area of pitting corrosion on the inner surface. This clearly shows that corrosion initiated from the inside surface and perforated outwards.

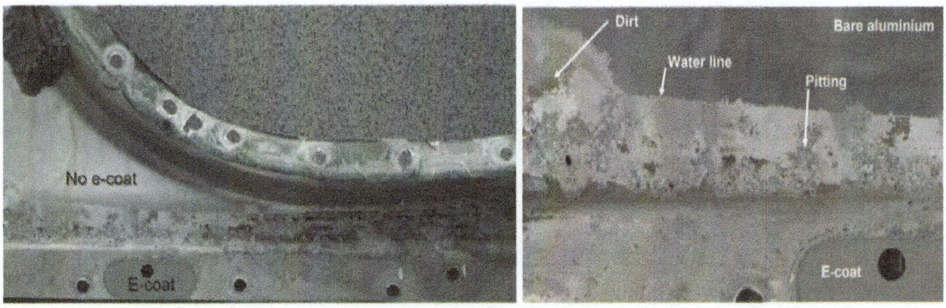

Fig. 5. Left hand doorsill outer surface (top) and inner surface (bottom). Inside surface showing 'water' line, dirt and pitting corrosion

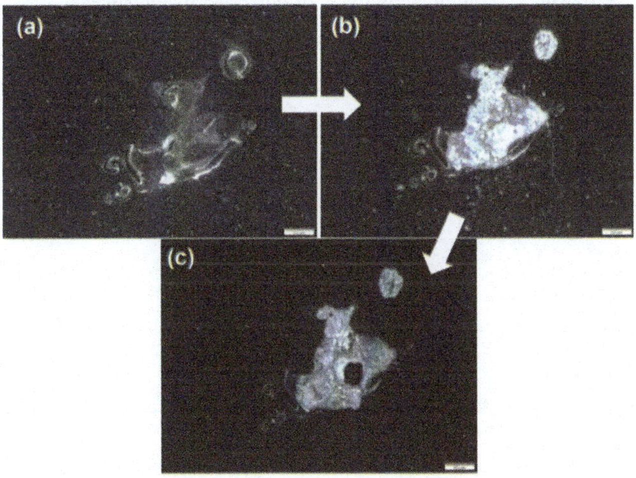

Fig. 6. Outer surface paint blister (a), paint layer removed (b), corrosion product removed (c)

Chemical analysis of the corroded inner surface (Fig. 7) shows major elements associated with aluminum oxide corrosion product as expected and minor elements associated with corrosive chloride-containing road de-icing salts of sodium, magnesium and possibly calcium. Calcium and Silicon would be expected as a constituent of mud (SiO_2 and $CaCO_3$).

Chemical analysis of mud collected from the inner surface (Fig. 8) shows major elements associated with SiO_2 as expected in mud, minor elements associated with corrosive road de-icing salts as before, but also includes a potassium peak, which suggests the use of potassium chloride as a road de-icer, which is also a common ingredient in fertilizers.

From the chemical analysis, we obtained adequate results that prove once again that corrosion has perforated the panel and we can say that the production of panels should be improved in the future.

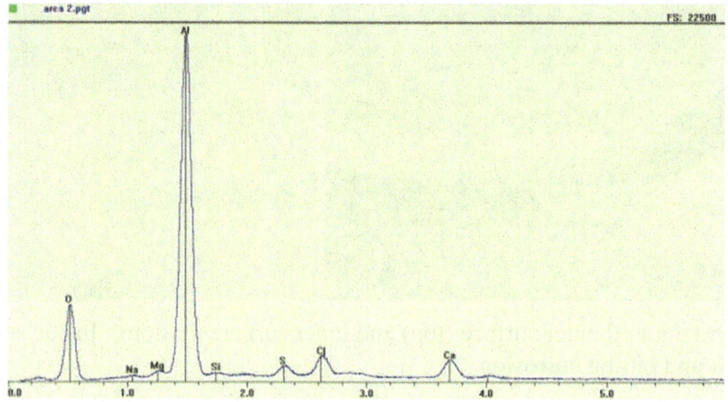

Fig. 7. Chemical analysis of corroded inside surface

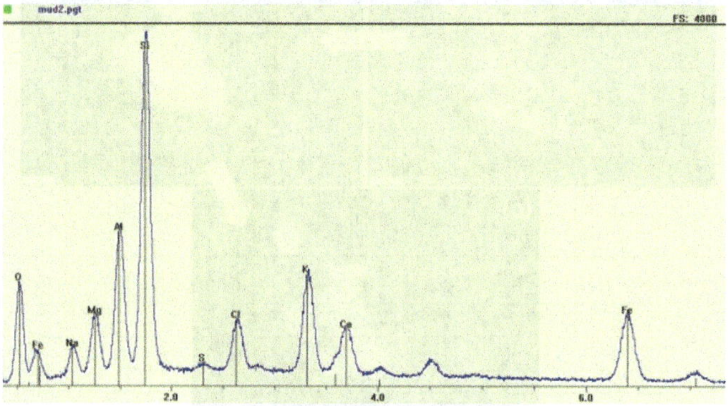

Fig. 8. Chemical analysis of mud collected from inner surface

4 Conclusions

The results of the chemical analysis of the corroded inner surface shows that corrosion has perforated the panel.

The inspection investigates the corrosive inner surface, inside surface and mud collected from the inner surface. The inspection shows without doubt that corrosion initiated from the inner surface resulting in complete panel perforation. A corrosive chloride-containing poultice on bare aluminum with no e-coat protection caused corrosion. The composition of the poultice would retain moisture at low humidity levels allowing continuous corrosion attack of the passive film by aggressive chloride ions.

Following the analysis performed, corrosion tests, humidity and hot test, dry cycle was proposed for simulation. The test will be performed in laboratory conditions and will be investigated from a qualitative and quantitative point of view. Improvements in corrosion protection resulting from significant advances in design, material selection, protective coating systems and application methods, as well as sealants and application methods will help to improve corrosion protection on new vehicle components.

References

1. Hayashi, K., Miyoshi, Y., Ito, Y.: Corrosion protection of galvanized steel sheet-corrosion investigation of field vehicle. SAE Trans. **100**, 1183–1195 (1991). Section 5: Journal of materials & manufacturing
2. Winston Revie, R.: Corrosion and Corrosion Control, An Introduction to Corrosion Science and Engineering, p. 512. Wiley (2008). E-book
3. Daisuke, M.: Automotive corrosion and accelerated corrosion tests for zinc coated steels. In: Steel Research Laboratory, vol. 58, pp. 1562-1568. JFE Steel Corporation (2018). https://doi.org/10.2355/isijinternational.ISIJINT-2018-159
4. Elayaperumal, K., Raja, V.S.: Corrosion Failures. Theory, Case Studies, and Solution, p. 556. Wiley (2015). E-book
5. Uchida, M., Mochizuki, K.: Corrosion behavior of vehicles operated in north america for 5 years. Automotive materials and instrumentation and process control, Kawasaki Steel Technical Report, No. 43 (2000)
6. Murchie, R.J: An Accelerated Track Test for Corrosion of Vehicles, Ford Motor Co. Ltd. (1967). https://doi.org/10.1243/PIME_CONF_1967_182_289_02
7. Anuar, L., Amrin, A., Mohammad, R., Ourdjin, A.: Vehicle accelerated corrosion test procedures for automotive in Malaysia, Department of Mechanical Engineering, in UTM Razak School of Engineering and Advanced Technology Universiti Teknologi Malaysia, MATEC Web of Conference, vol. 90, p. 01040 (2017)
8. Cajani, M.: Ageing airframes and corrosion maintenance. ART, Engineering Department. **25**(3–4) (2007). https://doi.org/10.1515/CORRREV.2007.25.3-4.263
9. ISO/IEC 17025:2017 General requirements for the competence of testing and calibration laboratories
10. IATF 16949:2016 Quality management system for organizations in the automotive industry, ISO (2016)

Improved Sulfuric Acid Discharge Through Combined Task and Risk Analysis

Jefferson dos Santos Mascarenhas(✉), Igor Santos Araujo, Salvador Ávila Filho, Danillo Ramos Camargo, Leila Maria Aguilera Campos, and Luiz Antônio Magalhães Pontes

UFBA - Universidade Federal da Bahia, Salvador, BA, Brazil
leila.campos@unifacs.br, uolpontes@uol.com.br

Abstract. Industries face challenges in achieving and maintaining economic and environmental sustainability indicators. The risks must be reduced in terms of time and the possibility of an accident. This work intends to apply tools to minimize operating costs with the reduction of the unloading cycle and with the risk analysis of this activity. The applied concepts and tools the study of times and movements, as well as the application of the LOPA tool for the success of this activity. The main intention is to prevent human error. This activity was applied in the case of a chemical industry in Brazil. It is including: (1) Operational context; (2) Scenario and Task Analysis; (3) Time Study; (4) Application of the LOPA; (5) Recommendations in the areas of checklist, automation, training, and task review. This job authorizes new activities and reduces the risk. An annual savings of 1.5 million in the local currency.

Keyword: LOPA · Times and movements · Risk analysis · Chemical process industry

1 Introduction

The relationship that employees have with the company and its organizational culture will reflect on how the worker validates and understands the company's mission, vision and values, affecting their performance and satisfaction [11]. The good work performance of employees is fundamental to the survival of companies within the current global economic scenario. Competitiveness in the economy demands continuous improvement of processes [1] to reduce production costs without losing product quality and maintaining the safety standard. This set of initiatives strengthens the economic, environmental and social survival of businesses. Within this context, the concept of lean manufacturing is discussed, which aims to reduce waste in companies through the analysis of hazards and risks for the promotion of reengineering of facilities, respecting the HSE pillars.

The principles of process safety and lean manufacturing when combined bring about an improvement in productive activities as in this particular case, where there is an improvement in the sulfuric acid discharge system in the Camaçari Pole Industry, located in Brazil [11]. The work has as one of the main objectives to reduce human error, so the

S. Trzcielinski et al. (Eds.): AHFE 2021, LNNS 274, pp. 148–158, 2021.
https://doi.org/10.1007/978-3-030-80462-6_19

human reliability will act in order to guarantee the adequate performance of the operator in the execution of the task, aiming at the non-occurrence of failures for a certain period. A study of time and movement applied using the chrono analysis technique indicates which activities add value and how to minimize or eliminate unproductive activities due to their high cost of production and possible environmental impact. The idleness of the acid truck unloading operator was considered as one of the main activities that reduces the value of manufacturing in the production system [2].

Following the principles for Process Safety Projects, we seek to design safeguards such as instrumented systems, procedures and training to partially or completely mitigate the process risks identified in the risk assessments. The LOPA method was applied for risk assessment [3] in the activity. Independent Protection barriers (IPL) incorporated or revised include administrative controls, communication devices, and improved human responses to ensure production and safety. logistics for unloading sulfuric acid in the case of the truck handling station. The LOPA is a semi-quantitative method and it involves the identification and measurement of failures within the independent protection layers to achieve an expected standard indicated by the literature and standards. The LOPA [5] when applied in the Chemical Industry shows through scenarios what are the risk levels associated with dangerous activities and provides necessary information for decision making avoiding events of environmental accident or process safety, whenever possible through installation of fail-safe devices, called poka-yoke.

2 The Process

This activity studied in the logistics area in order to discharge sulfuric acid, is composed of steps listed in Table 1 and due to the risks associated with handling sulfuric acid, they must be accompanied by occupational safety instructions. For the discharge of the acid there are preparations and final activities. Of the activities listed in the table, the acid unloading step from the tanker takes the longest, at a duration of about 1 h, as it requires the responsible operator to remain. Most of this period (40 min) involves the centrifugal pump, present in topic 6 of Fig. 1. This pump operates to discharge the acid; however, the list does not include the action of the performer. Under normal conditions, the operator only watches the pump re-press the acid, and will only intervene if there is a leak or to stop the pump at the end of the discharge. This assessment was the trigger for change in the process in order to generate greater operational productivity.

Figure 1 illustrates the block diagram of the acid discharge preparation and the sulfuric acid discharge task. Including the time to put on and take off the special clothes.

Table 1. Unloading steps of sulfuric acid

Steps	Description
1st	Truck entrance at the industrial unit
2nd	Parking at the unloading dock
3rd	Connecting the discharge hoses
4th	Opening of the pressure relief system
5th	Opening of the relief valves (manhole)
6th	Connecting the sulfuric acid discharges pumps
7th	Transfer the analysis of acid discharge to the control room
8th	Close the valves after the discharge is complete
9th	Disconnect the hoses after unloading
10th	Release the truck to exit the unloading area
11th	Release the truck to leave the industrial unit

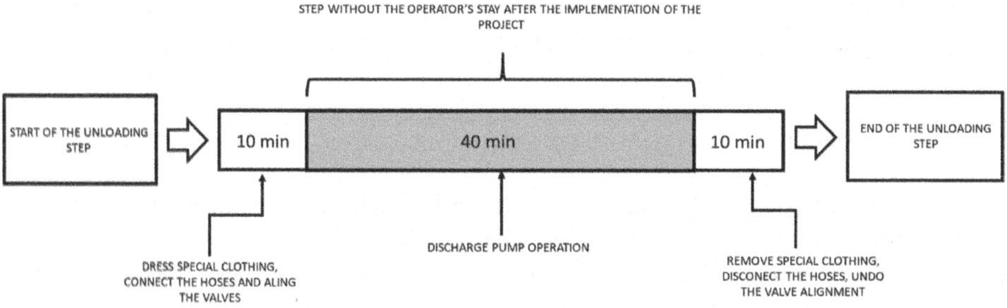

Fig. 1. Flowchart for unloading the acid

3 Human Factors with Study of Times and Movements

The stimuli to weakening the performer's mental map are enhanced both in the shift change and in embarkation and disembarkation situations, in the case of an oil platform for example. The details of these motivations that detract from the operator's focus should be studied for further treatment [14] [15]. According to SPARH [12] and API770 [10], worldwide reference documents for the analysis of the human factor. A study of times and movements would be necessary to detail the operations, identifying important factors, namely: Stress level, availability of time to perform critical steps, audit the quality of procedures, assess the interfaces between the operator and the machines, understand the complexity of the task, assessing the seniority of the team, as well as the stages of the process and the conditioning of the performers to perform them. Thus, it is also necessary to diagnose the team with characteristics of individuals such as: Consolidated competences, communication, cooperation and commitment [13]. Thus, allowing the analysis of activities that do not add value or are considered waste, they would possibly be minimized or eliminated. Based on this study, it is possible to calculate and measure the efficiency and productivity of the manufacturing process, making the organization more competitive and being an economical differential in the production chain. The TOLL TIME study tool was applied to increase the productivity of the discharge of sulfuric acid seeking to contribute to achieving lean manufacturing. The work enabled the new performers of the task the possibility of online visual monitoring and flow interlock, unifying the control location in the main operating room (Same location) and also allowing the control room operators to have all the same information as before. viewed only by the field (Same crew).

3.1 Evaluation of Shift E&I

The studied company operates on a 24-h basis in an uninterrupted shift, with six classes following a relay table. The demands for electrical maintenance and instrumentation are met by an electrical and instrumentation technician per shift. The latter is responsible for allocating the unit in the event of power outages and carrying out minor maintenance that puts the plant's operational continuity at risk.

The study of times and movements indicated the average times for using the maintenance resource in the shift, where 39% of the time of the person performing the shift maintenance was waiting for the equipment to break, as on duty. About 3 h and 12 min, on average, were not transformed into useful work, excluding the time of physiological needs, meal, commuting, and rest performed by the performer. Only 13% of the time was used for maintenance work, as shown in Fig. 2.

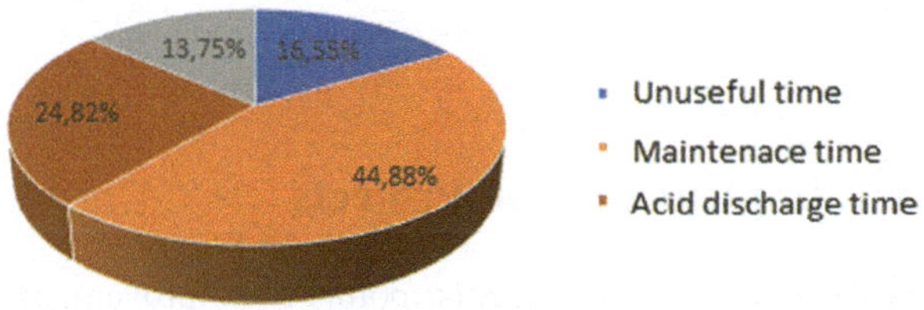

Fig. 2. Study of shift maintenance times and movements – before changes.

3.2 Merger Proposal Between Functions

By compiling the data on the time spent by the operator for the discharge of acid together with the available time of the shift performers, a great opportunity was identified to use the shift E&I in the discharge and mitigate the idle time, increasing operational productivity. Table 2 shows the unloading demand with the availability of the maintenance worker in the shift.

Table 2. Comparison of the maintenance performer's availability times with the time required for acid discharge in a day.

Evaluation items	Results
Time required by trailer	20 min
Time available per player	192 min
Unloading capacity	9,6 tanker

Seeking to ensure that the shift executor will have full availability to perform the unloading, after the implementation of the project, the activities of inspection, carrying out routes and pending issues arising from the administrative shift have been removed from the obligations of this function. The useful time was broken down into maintenance time, together with unloading time. Figure 3 illustrates the change in the distribution of times and movements.

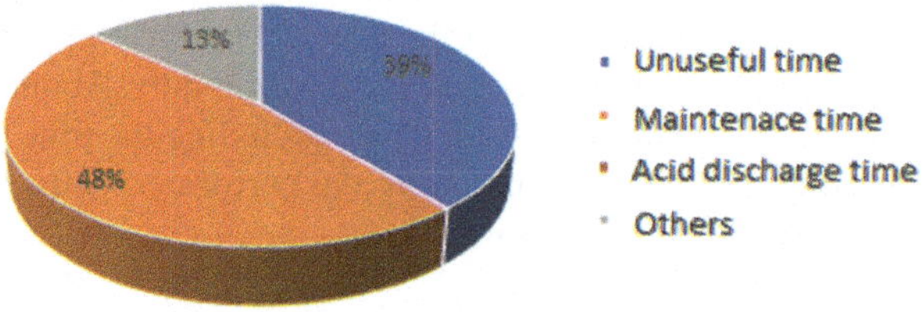

Fig. 3. Study of shift maintenance times and movements – after changes.

4 Improvement Implementation Projects

Through the study of times and movements, points of improvement were identified, which will be presented in the Fig. 4.

4.1 Collapse Protection

The step of opening the relief system valves is important to avoid the collapse of the tanker when unloading the input, the truck is equipped with a relief valve, following the requirements of NR13 [16], but this opening is an unwanted occurrence. In order to prevent this, it is necessary for the tank opening at the top of the tanker to be open. Human factors engineering, or ergonomics, is the design of equipment, operations, procedures, and work environments that are compatible with the capabilities, limitations and needs of the workers [10]. According to the incompatibility of the design of the truck manhole and operator access, there are two risks in this activity: the operator falling on the tanker to open the BV, indicated in Fig. 4 in topic 8 and exposure of the operator to gases emitted from sulfuric acid while conducting this same task. The time spent to carry out the activity totalled about 20 min, counting the time of connection and disconnection of

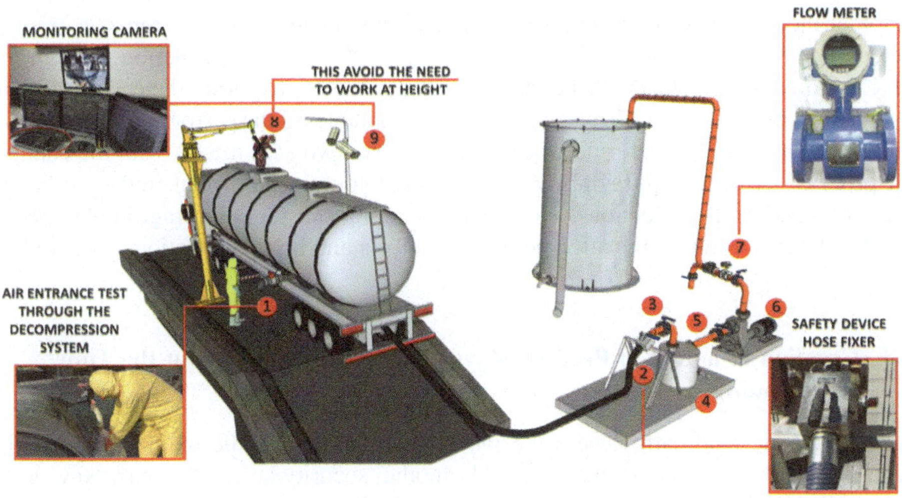

Fig. 4. Scheme for improving acid discharge.

the tankers. After the time studies, it was recommended to install a pneumatic valve to relieve the tanker as indicated in Fig. 4 in topic 1, without the need for access at height, as it is at ground level.

Using the new system, the operator activates the manual air supply switch for the pneumatic actuator of the tank relief valve, promoting its opening. This solution removes the performer from the risk of working above 2 m and removes the performer from gases derived from sulfuric acid at the moment of opening the manhole. The actuation to close the valve will be through the air supply of the truck's air balloon, produced by its compressor. In the event of a shortage of air in the pneumatic actuator of the valve during the discharge, by means of mechanical systems, the latter must ensure that the relief valve remains in the last actuated state, being closed or even open. For this reason, the valve has no spring return as is commonly used in pneumatically operated valves, but it is necessary to activate the air both to open the relief valve and to close it. Within the piping circuit, between the trailer tank and the acid discharge pump, there is an intermediate tank called the vacuum break tank, in Fig. 4, topic 4. This tank, with opening at atmospheric pressure, was assembled with purpose safety having the role of receiving acid from the carts through gravity. This equipment adds yet another safety barrier because if the duct responsible for the intake of air to the trailer tank is blocked, all the acid in the vacuum breaking tank will be sucked and the pump will cavitate. This sequence of events occurs to prevent the collapse of the cart, ensuring that all the acid inventory in the tanker is safe.

4.2 Deployment of Online Camera Monitoring

The monitoring of the unloading site was installed using high precision cameras in the field and imaging in the control room. As it is an inhabited place 24 h a day, as indicated in Fig. 4, topic 9, it frees the unloading operator to carry out other productive activities.

Another measure was the installation of an acid flow meter (in Fig. 4, topic 7), used to detect the end of the acid level in the tanker, generating a process action through the SDCD, commanding the discharge pump to stop automatically after the end of the acid flow in this instrument. In this way, 40 min were available for the performer for every two unloaded trailers. The performer spent 1 h to unload two acid trucks, with the change, the process started to require only 20 min, 10 min to put on the special clothes and align the valves and hose connections, and after the acid transfer for the storage tank to finish, another 10 min to return the alignment of the valves and hose connections. Totaling 20 min.

4.3 Creation of a Tool to Prevent Movement of the Acid Cart at the Time of Unloading

As the sulfuric acid carts access the industrial unit through the main entrance. The security guard receives the tanker and adds another security key to the tanker key, using a special keyring. This security key is used in a fail-safe device that will prevent the driver from leaving with the tanker inappropriate moments that pose risks to the process during the sulfuric acid unloading step. Figure 5A illustrates the key ring before adding together with tanker key; Fig. 5B illustrates the key ring add together with tanker key both are fixed at Interlocking panel; Fig. 5C illustrates; Fig. 5C illustrates the point to insert the key released from interlock panel to release the hose used to discharge sulfuric acid.

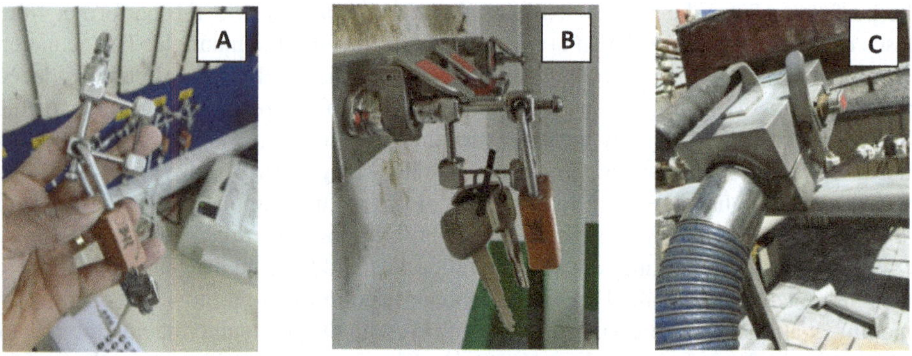

Fig. 5. A, B, C Devices systems to avoid tanker movement at inappropriate time.

This fail-proof device is used to prevent potential tanker driver despairing behaviour, as this professional, being unfamiliar with the reality of the industrial unit, may be frightened by ordinary events, such as the opening of a trap or an emergency alarm test. In the event of an event that leads this professional to remove the tanker from the site, even if connected, it will not be able to, due to this set of keys developed to protect the system from this action. Figure 6 indicates the flow that the safety switch travels from the truck's entrance to the main entrance, through the acid discharge area to the exit.

5 Risks Involved in Change

Human errors have either directly caused or significantly contributed to many major accidents in the process industries [10]. The LOPA methodology ware use in this case to detect human errors and mitigate their effects. The Protection Analysis Layer [4–6] is a method that assesses the risks of possible incidents and provides guidance on the adequacy of the independent protection layers (IPLs) to mitigate the risk. During an LOPA, the team is responsible for assessing the risk of the process from various deviations and determining the consequence of possible incidents, usually using order of magnitude categories to initiate frequency of events and for the probabilities of failure of IPLs [5]. The difference between the level of real risk and the tolerable risk is the necessary level of risk reduction, also called the risk reduction factor (RRF).

RRF is the relationship between the real risk presented by the installation and the risk that must be achieved as a goal based on the acceptance criteria [7]:

RRF = Risk Real risk/tolerable target risk

Therefore, this methodology was implemented to evaluate the new process proposed for the discharge of sulfuric acid in a TiO2 industry, highlighting the 3 main scenarios highlighted in Fig. 2 and topics 1, 2 and 4. Figure 7 shows the LOPA table for the scenarios.

The function of the columns in the table is described below [7]:

- Accident event and the potential severity level (columns 1 and 2). These come from an earlier hazard identification study, e.g., HAZOP, WHAT IF, etc.
- Description of the cause and its probability (columns 3 and 4) The single cause of the dangerous event. The probability is usually expressed and the frequency per year.
- Independent protection layer, IPL (columns 5 and 8) These columns include protection layers to mitigate the accident event. An IPL is a device, system or action capable of preventing a scenario from proceeding to its unwanted consequence, regardless of the initial event and any other layer of protection. An IPL must meet these conditions: 1) it must be effective in preventing the consequence; 2) independent of the initial event; and 3) auditable.
- Conditional modification (columns 9 and 10) One of several probabilities included in the scenario's risk calculations that modify the frequency of the dangerous event.

Risk analysis using the LOPA methodology was used in the scenarios defined below in order to identify the safety status of activities in the current scenario and to implement layers of protection to mitigate risks. The company has a risk criterion [8], which indicates two categories:

- High severity level: serious injuries or a fatality; tolerable risk: 1 x 10^{-4} per year
- Exceedingly high severity level: more than 1 fatality; tolerable risk: 1 x10^{-5} per year

5.1 Scenario 1 - Moving the Trailer at the Time of Unloading

The initial event corresponds to human error during a routine task that is performed more than once a day, with a failure rate of 0.01 / year. These protection layer failure probability values were obtained from CCPS [9]. The severity level of the event is conceived as low, with the potential for 1 fatality. As IPL, human action was identified in 10 min due to the presence of the operator at the unloading, culminating in an RRF factor equal to 10. The addition of the locking device for the truck keys increased safety, adding an IPL, bringing the RRF to 0, as evidenced in the LOPA in Fig. 6.

1	2	3	4	5	6	7	8	9	10	11	12	13
				Protection Layer (probability of failure)				Conditional modify				
Impact event description	severity level	initiating cause description	initiation likelihood (freq per year)	Basic control system	Alarms & operator action	IPL	Other mitigation measures	occupancy factor	Probability of ignation	Intermediate event likelihood	tolerable risk likelihood	Risk reduction factor
Truck movement at the time of acid unloading at the industrial unit	low (potencial for 1 fatality)	Human error	0.01 / per year			Human action in 10 minutes (0.1) Truck keys locking device (0.1)				$1x10^{-4}$ / year	$1x10^{-4}$ / year	0

Fig. 6. LOPA of scenario 1

5.2 Scenario 2 - Collapse of the Cart Due to Acid Exit in Atmospheric Air Entry to Compensate the Acid Mass Leaving the Cart

The initial event for the cart collapse corresponds to human error during the task of opening the cart manhole to enter atmospheric air in order to compensate the mass of acid that leaves the cart, with a failure rate of 0.01 / year, is due to human error. These protection layer failure probability values were obtained from CCPS [9]. The severity level of the event is conceived as high, with the potential for more than 1 fatality. As IPL, human action was identified in 10 min, culminating in an RRF factor equal to 100. The addition of the Air Inlet Test Valve through the relief system increased safety, adding an IPL, bringing the RRF to 10, as evidenced in the LOPA in Fig. 7.

1	2	3	4	5	6	7	8	9	10	11	12	13
				Protection Layer (probability of failure)				Conditional modify				
Impact event description	severity level	initiating cause description	initiation likelihood (freq per year)	Basic control system	Alarms & operator action	IPL	Other mitigation measures	occupancy factor	Probability of ignation	Intermediate event likelihood	tolerable risk likelihood	Risk reduction factor
Colapse of the truck caused by acid leaving the air inlet to compesate for acid mass leaving the truck	low (potencial for 1 fatality)	Human error	0.01 / per year			Human action in 10 minutes (0.1) Truck keys locking device (0.1)				$1x10^{-4}$ / year	$1x10^{-4}$ / year	0

Fig. 7. LOPA of scenario 2

5.3 Scenario 3 - Drop and Inhalation of Acid Gases During the Opening of the Manhole on Top of the Cart by the Operator

The initial event for the drop and inhalation of acid gases during the opening of the manhole on top of the cart by the operator, which is performed every time the unloading of a cart starts with a failure rate of 0.01 / year, it is human error and the breaking of the supporting cable. These protection layer failure probability values were obtained from CCPS [9]. The severity level of the event is conceived as high, with the potential for more than 1 fatality. As IPL human action was identified in 10 min, culminating in an RRF factor equal to 10. The addition of the Air Inlet Test Valve through the relief system increased safety, adding an IPL, bringing the RRF to 9, as evidenced in the LOPA in Fig. 8. The test valve made it unnecessary to climb into the tank to open the manhole in order to avoid the collapse of the cart in the discharge of acid.

1	2	3	4	5	6	7	8	9	10	11	12	13
impact event description	severity level	initiating cause description	initiation likelihood (freq per year)	Protection Layer (probability of failure)				Conditional modify		Intermediate event likelihood	tolerable risk likelihood	Risk reduction factor
				Basic control system	Alarms & operator action	IPL	Other mitigation measures	occupancy factor	Probability of ignation			
Fall and inhalation of acid gases during the opening of the manhole on the truck by the operator	low (potencial for 1 fatality)	Human error Rupture of the sustaining cable	0.01 / per year			Human action in 10 minutes (0.1) Air entrance test valve through descompression system (0.1)				$1x10^{-4}$ / year	$1x10^{-4}$ / year	0

Fig. 8. LOPA of scenario 3

6 Conclusion

Using methodology of studies of times and movements and applying new technologies, this work presented gains both improving the safety level of the task, reducing the chance of human error, and bringing better financial results to the business. With a process less dependent on the operator that was directed to perform tasks of greater technical requirement and less manual. All risks studied were properly controlled. The risks of working at heights, inhalation of gases from sulfuric acid and the event of a trailer collapse were eliminated, after the development of the safe fail valve the operation no longer required access at heights greater than 2 m. It also removes the possibility of inhaling sulfuric acid vapors and allowing air to enter to compensate for the release of sulfuric acid during unloading. The risk of removing the truck with the hoses connected was also eliminated, as the creation of the interlocking method starting from keys between the activation of the trailer, central panel and easels of the hoses, made possible a fail-safe solution and/or human error. The reduction in the execution cycle was made possible by the automatic stop of the discharge by the flow monitoring through the unit's DCS. The remote detection system, monitored by the control room, complemented the productivity gains, making the performer available to work with more noble tasks even during unloading.

References

1. Cury, P.H.A., Saraiva, J.: Time and motion study applied to a production line of organic lenses in Manaus Industrial Hub. Gestão & Produção **25**(4), 901–915 (2018)

2. Houts, L.: Minimizing the seven wastes at Betts Spring Manufacturing. Int. J. Bus. Appl. Soc. Sci. **2**(4), 23–28 (2016)
3. Tixier, J., Dusserre, G., Salvi, O., Gaston, D.: Review of 62 risk analysis methodologies of industrial plants. J. Loss Prev. Process Ind. **15**, 291–303 (2002)
4. Summers, A.E.: Introduction to layers of protection analysis. J. Hazard. Mater. **104**, 163–168 (2003)
5. Layers of protection analysis: simplified process risk assessment, Center for Chemical Process Safety, American Institute of Chemical Engineers, New York (2001)
6. Dowell III, A.: Layer of protection analysis for determining safety integrity level. ISA Trans. **37**(3), 155–165 (1998)
7. Torres-Echeverria, A.: On the use of LOPA and risk graphs for SIL determination. J. Loss Prev. Process Ind. **41**, 333–343 (2016)
8. HSE: Reducing Risks, Protecting People. HSE's Decision-making Process. Health and Safety Executive. Her Majesty's Stationery Office, Norwich, UK (2001)
9. CCPS: Guidelines of initiating events and independent protection layers in layer of protection analysis. american institute of chemical engineers, Center for Chemical Process Safety. Wiley-AIChE, New York (2014)
10. API – AMERICAN PETROLEUM INSTITUTE. API 770: A Manager's Guide to Reducing Human Errors: Washington, D.C., March 2001
11. Ávila Filho, S., Souza, M.L., Ramalho, G., Passos, I.C., Oliveira, A.R.: Análise Do Modelo De Confiabilidade Para Modo De Falha Interdependente , Uma Discussão Sobre O Erro Humano, Fatores Organizacionais e Ambientais. In: CONEM – Congresso Nacional de Engenharia Mecânica. 5., 2018, Salvador. Anais [...] Salvador: [S.I.]: [s.n.], pp. 1–9 (2018)
12. Boring, R., Blackman, H.: The origins of SPAR-H method's performance shaping factors multipliers. Idaho – EUA, August 2007
13. Ávila Filho, S., Drigo, E.S.: Ferramenta C4t para avaliação de Fatores Humanos. Congresso Associação Brasileira de Análise de Risco, Segurança de Processos e Confiabilidade – ABRISCO, Rio de janeiro (2017)
14. Ávila Filho, S.: Risks in the platform: social and human guidelines avoid accidents and alter standards in oil production. Revista Proteção (2013)
15. Abreu, M.N.G., Ávila Filho, S.: Assessment of risk in the exploration and production of oil at pre-salt layer – Brazil, a discussion about social-technical factors. In: Proceedings of Rio Pipeline Conference & exposition, Rio de Janeiro (2015)
16. NR 13: NORMA REGULAMENTADORA 13. ENIT – Escola Nacional da Inspeção do Trabalho: Portaria SEPRT, Brasília (2019)

Tacit Knowledge Awareness and Sharing as a Focal Part of Knowledge Production. Polish-US View on IT, Healthcare, and Construction Industry

Wioleta Kucharska(✉)

Gdańsk University of Technology, Gdańsk, Poland
wioleta.kucharska@pg.edu.pl

Abstract. In the knowledge economy era, knowledge production and dissemination are of key interest to individuals, organizations, and economies. Tacit knowledge results from experience, leading to innovation. The learning culture can facilitate the transformation of errors into experiences. This study explores whether mistake acceptance facilitates tacit knowledge awareness and sharing in the information technology, healthcare, and construction industries in Poland and the United States. The findings show the influence of mistake acceptance on knowledge production and the differences between countries and industries. The US showed a higher level of mistake acceptance, which was similar across the three industries, than did Poland, which showed differences between sectors. In general, the higher the acceptance of mistakes, the greater the effect of tacit knowledge awareness on sharing. This study shows that there is no knowledge production without learning and no learning without mistake acceptance.

Keywords: Knowledge production · Tacit knowledge awareness · Tacit knowledge sharing · Learning culture · Mistake acceptance · Error management

1 Introduction

In the knowledge economy era [1], the production, capture, and dissemination of knowledge is of key interest to individuals, organizations, and economies. How is knowledge produced? According to the SECI (socialization–externalization–combination–internalization) model [2], tacit knowledge is converted to explicit knowledge through socialization and externalization, while explicit knowledge contributes to tacit knowledge through combination and internalization. Thus, given that tacit knowledge is characterized as novel, personal, produced and stored in human mind, and it is the root of one's entire knowledge [3], the human factor is key in knowledge production. Therefore, any new knowledge is based on tacit knowledge. Unlike the production of tangible assets, the production of knowledge, especially tacit knowledge, is usually not formalized. While explicit knowledge results from structured and sequential analysis, the production of tacit knowledge is unstructured and nonsequential. New ideas are generated in

S. Trzcielinski et al. (Eds.): AHFE 2021, LNNS 274, pp. 159–167, 2021.
https://doi.org/10.1007/978-3-030-80462-6_20

the human mind when it is ready. Therefore, typical error management based on mean values and deviations is not suitable. Since tacit knowledge production is not formalized, then management of mistakes during this production also can't be formalized. Therefore, managing of it can only rely on company culture that supports critical and creative thinking. The culture of learning, collaboration, and safety is the best what organizations can do to support somehow tacit knowledge production.

The capture and sharing of tacit knowledge are crucial challenges in knowledge management [4]. Mueller [5, 6] notes that while both knowledge and learning cultures are vital, the latter is essential for continuous development. According to Kucharska and Bedford [7], a continuous learning culture includes factors such as a learning climate and mistake acceptance. Zappa and Robins [8] argue that the essence of organizational learning is to identify and modify errors. Therefore, this study explores how tacit knowledge awareness affects tacit knowledge sharing, facilitated by the learning cultural factor of mistake acceptance. Therefore, the research question is: *Does the acceptance of mistakes facilitate tacit knowledge awareness and sharing?* Studies by Farnese et al. [9, 10] have stressed the importance of cultural orientation in learning from mistakes. Vanderheiden and Mayer [11] highlighted the cultural factor of mistake perception. Given that we cannot learn from mistakes if we do not accept them as part of the learning process, mistake acceptance should be included in the organizational culture of learning [7]. Therefore, this study aims to explore cultural perceptions of mistake acceptance in the production of knowledge in the information technology (IT), healthcare, and construction industries in Poland and the United States (US).

2 Theoretical Framework

The theoretical framework for this study is based on Olaisen and Revang's [12] study of tacit knowledge. Tacit knowledge is gained from personal experience, repeated on-the-job activities, and those who are willing to share their knowledge. Asher and Popper's [13] "onion" model posits different layers of tacit knowledge and describes knowledge as a matter of degree along a single axis, being more explicit at one end and more tacit at the other, with elements of both comprising the middle. Knowledge ranges from being explicit and explainable at one end to being virtually impossible to explain and only demonstrable (perhaps verging on insight/intelligence) at the furthest reaches of the tacit end. Kucharska and Erickson [14] show that the awareness of tacit knowledge affects the sharing of such knowledge. Moreover, recent studies by Farnese et al. [9, 10] have demonstrated the importance of cultural orientation in learning from errors. Kucharska and Bedford [7] have proposed a scale that measures the mistake acceptance factor in the learning culture. Love and Smith [15, 16] and Love et al. [17, 18] assume that the learning culture can transform error events into experiences, leading to tacit knowledge [12]. Therefore, based on all of the above, the following hypotheses are proposed:

H1: Tacit knowledge awareness fosters sharing.
H2: Mistake acceptance as a learning culture factor enhances the relationship between tacit knowledge awareness and sharing.

Figure 1 presents the conceptual framework of the current research.

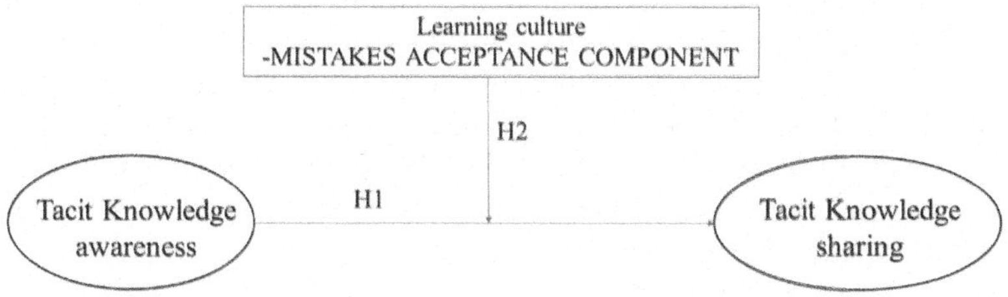

Fig. 1. Theoretical framework

3 Method

To verify the hypotheses and answer the research question, the collected data were analyzed using structural equation modeling and visualized using the PROCESS macro in SPSS version 3.4 [19]. Data were collected from January to February 2020 using Qualtrics (US) and ASM (Poland). Knowledge workers, employees whose main input and output of work is knowledge; in other words: they produce new knowledge using existing knowledge; the mind is their key tool, and knowledge is their key resource when working, were targeted. Specifically, knowledge workers from the IT, healthcare, and construction industries in both countries were invited to participate in the survey. Final sample sizes exceeded 1,000 cases for each country. The questionnaire began with a short introduction providing an overview of the study, including a definition of tacit knowledge to ensure respondents understood what they were being asked. All constructs were measured on a 7-point Likert scale based on the items presented in the appendix. Assessment of sample quality began with invariance, followed by the Kaiser–Meyer–Olkin test of sampling adequacy (Poland = .789, US = .871) [20, 21] and Harman's one-factor test [22]. Both achieved an acceptable result, with 31% and 38% for Poland and the US, respectively, and a common method bias of 19% and 30% for Poland and the US, respectively. Total variance extracted exceeded 68% for both samples, confirming that sample quality was good and enabling further analysis.

Based on recommendations for large groups [23, 24], the measurement tool was verified to be nationally invariant: Δ comparative fit index (CFI) = 0.02; Δ Tucker-Lewis index (TLI) = 0.03; Δ root-mean-square error of approximation (RMSEA) = 0.01. The validity and reliability of applied measurement scales and quality of the model were confirmed: CMIN/df = 3.7 and 4.61, CFI = .971 and .971, TLI = .948 and .959, and RMSEA = 0.51 and 0.57 for Poland and the US, respectively [25].

4 Results

Tacit knowledge awareness was strongly related to tacit knowledge sharing in the US ($\beta = .74, p < .001$) and weakly related to tacit knowledge sharing in Poland ($\beta = .27, p$

< .001). However, in Poland, this relationship was more strongly moderated by mistake acceptance (β = .27, $p = .001$) compared with the US (β = .10, $p < .001$). This was because the acceptance of mistakes as part of the learning culture is lower in Poland than in the US. The mean value of this variable for Poland was 5.14 (*SD* = 1.4) compared with 5.66 (*SD* = 1.34) for the US. Mean tacit knowledge awareness was 5.83 (*SD* = 1) for Poland and 5.93 (*SD* = 1.06) for the US, while mean tacit knowledge sharing was 5.90 (*SD* = 1.09) for Poland and 6.18 (*SD* = 0.94) for the US. Figure 2 presents the general results, while Fig. 3 details how mistake acceptance influenced this relationship for each industry studied (Table 1).

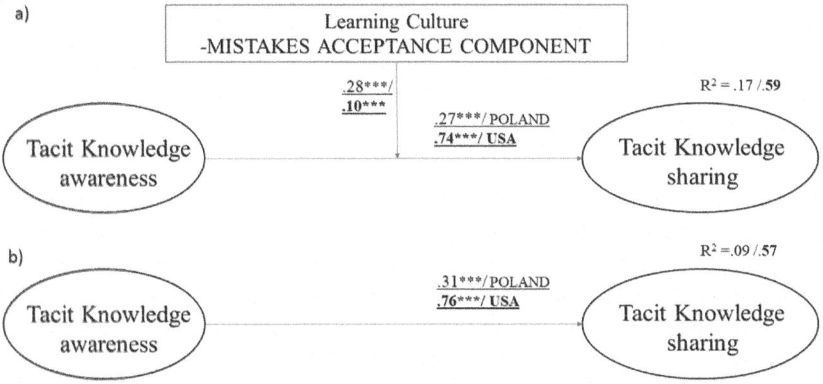

Fig. 2. Results for Poland and United States: a) CMIN/df = 3.7/4.610; CFI = .971/.971; TLI = .948/.959; RMSEA = 0.51/0.57; b) CMIN/df = 4.45/5.07; CFI = .971/.971; TLI = .944/.962; RMSEA = 0.57/0.60. Note: Poland (n = 1,050); United States (n = 1,118), ML–standardized results. CMIN/df; CFI: comparative fit index; TLI: Tucker–Lewis index; RMSEA: root-mean-square error of approximation. $^{***}p < .001$, $^{**}p < .01$, $^{*}p < .05$.

Table 1. Basic statistics, correlations, and root square of average value explained.

Variable	Mean	*SD*	Cronbach's α	CR	AVE	MA	TKA	TKS
MA	5.14/5.66	1.4/1.34	.86/.84	.87/.82	.66/.53	**.825/.729**		
TKA	5.83/5.93	1.0/1.06	.70/.76	.74/.77	.50/.53	.143/.327	**.708/.728**	
TKS	5.90/6.18	1.1/0.94	.71/.81	.74/.80	.55/.56	.319/.324	.310/.723	**.742/.743**

Note: MA: mistake acceptance; TKA: tacit knowledge awareness; TKS: tacit knowledge sharing

The results confirm the influence of mistake acceptance in knowledge awareness and sharing in Poland and the US. All hypotheses were supported. The model using "mistake acceptance" (see Fig. 2a) as a control variable fit the data better compared with the model without the control variable (Fig. 2b), justifying the control variable imputation [26]. Moreover, R-squared obtained for the US (59%) was much higher than that obtained for Poland (17%). The difference in R-squared between the models with and without the mistake acceptance factor was larger for Poland (9%/17%) than for the US (57%/59%). This means that mistake acceptance in the learning culture in Poland is more problematic (imputation of this factor makes a difference) than in the US. Therefore, the acceptance of mistakes contributes more significantly to tacit knowledge awareness and sharing in Poland than in the US.

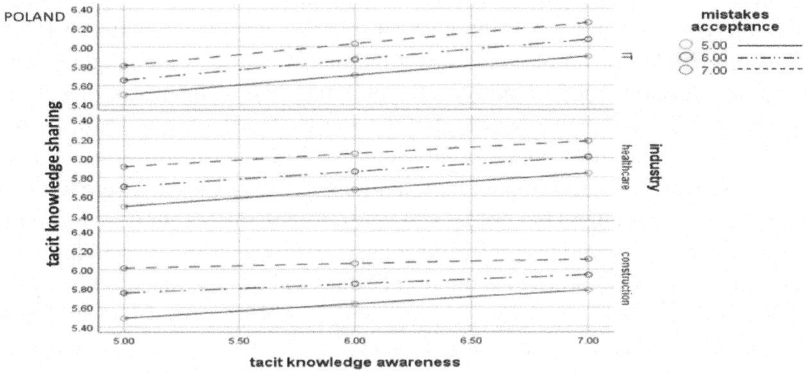

a)

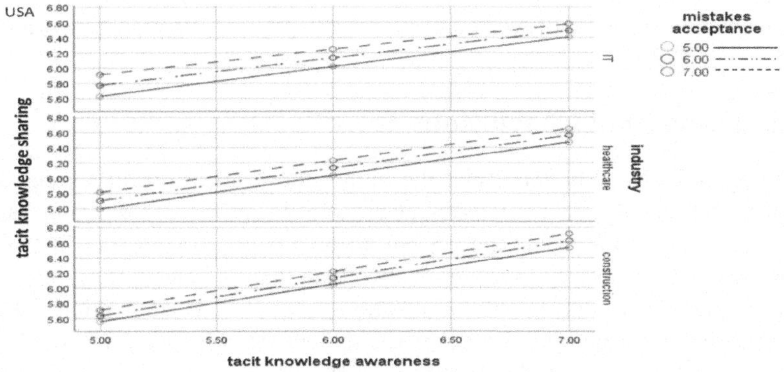

b)

Fig. 3. Results for each industry: a) Poland; b) United States. Note: Level of confidence for all confidence intervals in output: 95.0000.

5 Discussion

The findings confirm that the acceptance of mistakes in a learning culture is vital in the relationship between tacit knowledge awareness and sharing. So, it matters for the learning organizations [27]. This study has been empirically shown that there is no learning without mistakes [28]; thus, by accepting mistakes, we can learn faster, which was confirmed by the presented results. Moreover, tacit knowledge is a source of innovation [29, 30]; therefore, individuals in industries that shift their learning cultures and include mistake acceptance in their policies will learn faster and develop and share tacit knowledge more effectively, thus will be more innovative. The findings show that mistake acceptance enhances tacit knowledge awareness and sharing in the IT, healthcare, and construction industries in Poland and the US. In the US, where the general level of mistake acceptance is higher, this transformation is smooth for all industries. In contrast, in Poland, there are differences in the level of mistake acceptance between industries. Specifically, the highest level of tacit knowledge sharing via tacit knowledge awareness was observed in the construction industry, but only for the lowest level of mistake acceptance. When mistake acceptance was high, tacit knowledge sharing was similar for each awareness stage (as measured on a 7-point Likert scale). Thus, mistakes are perceived in mostly negative way in Poland's construction industry, and employees were likely to collaborate on how to avoid them but not how to learn from them. So, they are more focused on error prevention than on learning from their occurrence at the organizational level. For the healthcare and IT industries, the higher the level of tacit knowledge awareness, the higher the sharing of knowledge, a process that was supported by the acceptance of mistakes. The acceptance of mistakes was slightly higher in the IT industry than in healthcare, while tacit knowledge awareness and sharing was generally greater in healthcare than in IT. In conclusion, mistake acceptance influences tacit knowledge production and, as a potential source of learning for novelty, competitive advantage, and the overall innovativeness, is a source of worthy of more in-depth investigation across nations and industries.

6 Limitations and Conclusions

The findings of this study show that the higher the level of mistake acceptance, the stronger the effect of tacit knowledge awareness on knowledge sharing. This effect was similar for all industries in the US but differed between industries in Poland. The US economy is undoubtedly more innovative than that of Poland; thus, accepting and learning from mistakes may support innovation because of the authenticity of learning. Learning cultures that ignore mistakes waste the opportunity to learn through mistakes and it is critically important. As Kucharska and Bedford [7] note, if an organization's learning culture is inauthentic and ignores mistakes, employees may learn from their mistakes to avoid problems but will learn poorly or not at all. Moreover, in the Polish model R-sq, of the relationship between tacit knowledge awareness and sharing moderated by the mistake acceptance factor was only 17% compared with 59% in the US. This shows the existence of cultural differences in the perception of mistake acceptance and exposes that other factors in Poland may contribute to the unexplained 83% by the presented model factors influencing tacit knowledge production in Poland (compared with only 41% unexplained factors for the US). Thus, further studies are encouraged. The main limitation of the study was that it involved a comparison of only two nations and three sectors. The presented findings empirically show that accepting and learning from mistakes is complex and depends on national culture. Thus, more in-depth investigations into the phenomenon are warranted. This study has shown that there is no knowledge production without learning and no learning without mistake acceptance.

Funding Acknowledgement. The presented research is a result of the project Tacit Knowledge Sharing Influence on Innovativeness: The Sector Analysis (No. UMO-2018/31/D/HS4/02623), which was funded by the National Science Centre in Poland.

Appendix: Measurement of Constructs

Construct	Source	Items (authors' compilation based on source)
Tacit knowledge awareness	Kucharska and Erickson (2021)	I can create and explain new ideas or insights Even if my idea is hard to explain, I am able express it or demonstrate it Sometimes I am sure about a new idea but find it difficult to express As I have accumulated experience, I find it is easier to express
Tacit knowledge sharing	Kucharska and Erickson (2021)	I share knowledge learned from my own experience I have the opportunity to learn from others' experiences Colleagues share new ideas with me Colleagues include me in discussions about best practices
Learning Culture -mistake acceptance factor	Kucharska and Bedford (2020)	People know that mistakes are learning consequences and tolerate it up to a certain limit Most people freely declare mistakes We discuss problems openly without blaming Mistakes are tolerated and treated as learning opportunities

References

1. Powell, W.W., Snellman, K.: The knowledge economy. Ann. Rev. Sociol. **30**, 199–220 (2004)
2. Nonaka, I., Takeuchi, H.: The Knowledge-Creating Company: How Japanese Companies Create the Dynamics of Innovation. Oxford University Press, New York (1995)
3. Polanyi, M.: The Tacit Dimension. University of Chicago Press, Chicago (1966)
4. Koriat, N., Gelbard, R.: knowledge sharing motivation among IT personnel: integrated model and implications of employment contracts. Int. J. Inf. Manag. **34**, 577–591 (2014)

5. Mueller, J.: A specific knowledge culture: cultural antecedents for knowledge sharing between project teams. Eur. Manag. J. **32**, 190–202 (2014)
6. Mueller, J.C.Y.: Effective knowledge management and organizational learning in the context of sustainable development. Electron. J. Knowl. Manag. **16**, 56–69 (2018)
7. Kucharska, W., Bedford, D.A.D.: Love your mistakes! they help you adapt to change: how do knowledge, collaboration and learning cultures foster organizational intelligence? J. Org. Change Manag. **33**, 1329–1354 (2020)
8. Zappa, P., Robins, G.: Organizational learning across multi-level networks. Soc. Netw. **44**, 295–306 (2016)
9. Farnese, M., Zaghini, F., Caruso, R., Fida, R., Romagnoli, M., Sili, A.: Managing care errors in the wards: the contribution of authentic leadership and error management culture. Leadersh. Org. Develop. J. **40**, 17–30 (2019)
10. Farnese, M.L., Fida, R., Picoco, M.: Error orientation at work: dimensionality and relationships with errors and organizational cultural factors. Curr. Psychol. (2020)
11. Vanderheiden, E., Mayer, C.-H.: Mistakes, Errors and Failures Across Cultures: Navigating Potentials. Springer, Cham (2020). https://doi.org/10.1007/978-3-030-35574-6
12. Olaisen, J., Revang, O.: Exploring the performance of tacit knowledge: how to make ordinary people deliver extraordinary results in teams. Int. J. Inf. Manag. **43**, 295–304 (2018)
13. Asher, D., Popper, M.: Tacit knowledge as a multilayer phenomenon: the "onion" model. Learn. Org. **26**, 264–275 (2019)
14. Kucharska, W., Erickson, G.S.: Tacit knowledge awareness, sharing, and influence on innovation: a Polish/US cross-country study. Int. J. Inf. Manag. (2021)
15. Love, P.E.D., Smith, J.: Error management: implications for construction. Construct. Innov. **16**, 418–424 (2016)
16. Love, P.E.D., Smith, J.: Toward error management in construction: moving beyond a zero vision. J. Construct. Eng. Manage. **142**, 04016058 (2016)
17. Love, P.E.D., Smith, J., Ackermann, F., Irani, Z.: The praxis of stupidity: an explanation to understand the barriers mitigating rework in construction. Prod. Plan. Control. **29**, 1112–1125 (2018)
18. Love, P.E.D., Smith, J., Teo, P.: Putting into practice error management theory: unlearning and learning to manage action errors in construction. Appl. Ergon. **69**, 104–111 (2018)
19. Hayes, A.: Introduction to Mediation, Moderation and Conditional Process Analysis: A Regression-based Approach, 2nd edn. Guilford Press, New York (2018)
20. Podsakoff, P.M., Organ, D.: Self-reports in organizational research: problems and prospects. J. Manag. **12**(4), 531–544 (1986)
21. Hair, J.F., Anderson, R.E., Babin, B.J., Black, W.C.: Multivariate Data Analysis: A Global Perspective. Pearson Education, Upper Saddle River (2010)
22. Kaiser, H.: An index of factor simplicity. Psychometrika **39**, 31–36 (1974)
23. Putnick, D.L., Bornstein, M.H.: Measurement invariance conventions and reporting: the state of the art and future directions for psychological research. Dev. Rev. **41**, 71–90 (2014)
24. Rutkowski, L., Svetina, D.: Assessing the hypothesis of measurement invariance in the context of large-scale international surveys. Educ. Psychol. Meas. **74**, 31–57 (2014)
25. Byrne, B.M.: Structural Equation Modeling with Amos. Routledge, Abingdon (2016)
26. Aguinis, H., Vandenberg, R.J.: An ounce of prevention is worth a pound of cure: improving research quality before data collection. Annu. Rev. Org. Psychol. Org. Behav. **1**, 569–595 (2014)
27. Frese, M., Keith, N.: Action errors, error management and learning in organizations. Annu. Rev. Psychol. **66**, 661–687 (2015)
28. Van Dyck, C., Frese, M., Baer, M., Sonnentag, S.: Organizational error management culture and its impact on performance: a two-study replication. J. Appl. Psychol. **90**, 1228–1240 (2005)

29. Pérez-Luño, A., Alegre, J., Valle-Cabrera, R.: The role of tacit knowledge in connecting knowledge exchange and combination with innovation. Technol. Anal. Strateg. Manag. **31**, 186–198 (2019)
30. Ganguly, A., Talukdar, A., Chatterjee, D.: Evaluating the role of social capital, tacit knowledge sharing, knowledge quality, and reciprocity in determining innovation capability of an organization. J. Knowl. Manag. **23**, 1105–1135 (2019)

Evaluation Method in Multiple Scenarios by Means of Datamine Applying the Law of Variable Cutting in Optimization of the NPV in Open Pit

Elmer Flores-Zuniga[1], Diego Rivera-Ascencios[1], Edgar Alayo-Leon[1], Vidal Aramburu-Rojas[1], and Carlos Raymundo[2](✉)

[1] Ingenieria de Gestion Minera, Universidad Peruana de Ciencias Aplicadas, Prolongación Primavera 2390, Lima 15023, Peru
{U201411122,U201221621,pcgmeala,pcgmvara}@upc.edu.pe
[2] Direccion de Investigacion, Universidad Peruana de Ciencias Aplicadas, Prolongación Primavera 2390, Lima 15023, Peru
carlos.raymundo@upc.edu.pe

Abstract. This research is proposed to better detail the economic information that is handled and is necessary for the evaluation of a mining project, and in turn to reduce the uncertainty that exists regarding the economic results that the mining operation will obtain. In the same way, it is expected to optimize the extraction process and consequently increase the income generated in the operation, this will be evaluated with the economic measurement tool: net present value (NPV). The research will be developed through the use of mining software such as Datamine OP and NPV Scheduler, in which simulations of the various scenarios will be carried out where the cutoff grade will be the main variable to modify in order to verify the increase in the VPN in the mining operation.

Keywords: Cutoff grade · VPN optimization · Mining costs · Economic evaluation · Datamine

1 Introduction

The mining industry is currently, and as it has been for many years, one of the most important industries for the economic and social development of the country, it is also an activity where there is a high risk mainly due to large investments that must be carried out to start a mining project, as well as the different social conflicts that exist in our country and that can, as has been seen in recent years, paralyze an entire mining operation for many years.

It is for these reasons, and others, that the studies carried out to define the viability of a mining project, based mainly on the economic value of the project, are of great importance and require a broad analysis of all the variables that influence, directly and indirectly, in said process.

S. Trzcielinski et al. (Eds.): AHFE 2021, LNNS 274, pp. 168–175, 2021.
https://doi.org/10.1007/978-3-030-80462-6_21

The importance of this study lies in the fact that by being able to correctly handle said economic variable, it will allow us to maximize the economic value that the exploitation of the deposit will generate, thus generating a more optimal and efficient sequence of exploitation. This is expected to improve the economic income generated by the project, while at the same time reducing the time and costs used in all the processes of the operation. The economic tool: NPV will be the main factor that will indicate the changes that arise at the time of the simulations applying the variable cut-off grade methodology. Within the calculation of the variable cut-off grade, some factors will be taken as constants, in the multiple scenarios, and only a few will change, this will define the scope of the investigation.

2 State of the Art

2.1 VPN Optimization in Open Pit Mining Project

Obtaining the cutoff grade, appropriately, in a mining project will define the future production and planning of the mine, in turn this is an important indicator of the amount of economic income that will be obtained period by period throughout the life of the mine and will also influence the correct implementation of the concentrator plant [1]. The choice of an optimal cut-off grade directly influences the net present value (NPV) that the mining project will generate throughout the life of the mine, this results in the determination of a correct cut-off grade is of vital importance in In the planning and evaluation stage of any mining project, a poor choice of the same can lead to an over or under estimate of the project, generating, in the worst case, considerable economic losses for the company [1–3].

2.2 VPN Optimization in Open Pit Mining Project Through Computer Analysis

For the simultaneous optimization of production sequences and dynamic cut-off grades in an open pit mining operation. It consists of simulating the support of the points of a dense grid that covers the region of interest and then taking the average within the mining unit (s) to be analyzed or in the relevant blocks in order to perform a correct simulation. This simulation process has two main disadvantages, specifically in the computational area [4]. First, this process consumes a significant amount of time to perform all the necessary operation. Secondly, a computer or Workstation with a high-level processing capacity is required, for these reasons instead of performing a grid simulation that discreet the blocks, it is recommended to perform a direct simulation on the support blocks [5, 6].

2.3 Optimization of NPV in Open Pit Mining Project Through the Use of Variable Cut-Off Grades

In relation to the aforementioned, an investigation has quantifiably demonstrated that the use of a fixed cut-off grade strictly applied to all mineralized bodies that generally exist within a mining operation can generate a considerable economic loss in the income generated by the mine. This, added to the large dimensions of mining operations in general, can be deduced that the optimization of variable cut-off grades can generate great economic benefits in most mining operations [7–9].

2.4 Optimization of the NPV in an Open Pit Mining Project Applying Variable Cut-Off Grades Using a Computer

When using stochastic algorithm optimization software such as NPV Mining, higher NPV values were generated than those obtained with other simulations, this is explained by the fact that in this type of models different parameters are included that are not incorporated in the other models. most used [10, 11]. This shows that superior results are obtained when incorporating parameters such as uncertainty in classical models such as Lane's cut-off method [12].

3 Contributions

3.1 General Contribution

The research proposal will contribute to the search for the optimization of the Net Present Value of a mining project, this through the analysis and simulations of various scenarios with the variations of some variables in the calculation of the cutoff grade, thus generating that it may fall modified within the same scenario, in this way you can get to more detail about a better extraction. Through the use of software such as Datamine, for modeling and NPV Scheduler, to carry out scenario runs, the different scenarios will be analyzed, thus obtaining the result that maximizes the NPV of the project.

To be able to do this, the data of a mining project that has certain variables within it must first be collected, with this data the scenario simulations are carried out, changing some variables to designate the laws of variable cuts and therefore the costs. Finally, the economic calculation is carried out and the NPV factor is obtained as a result and when comparing them, the scenario that provides the highest NPV will be found (Fig. 1).

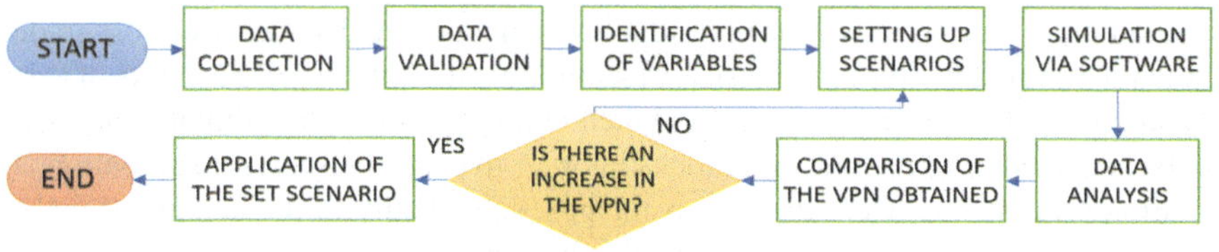

Fig. 1. Proposed model

3.2 Contribution to Detail

3.2.1 Identification of the Parameters for the Cutoff Grade

With the information already filtered and validated, data will be sought for the variables that manage to define the cut-off laws to be followed, for this an analysis must be made of both the equipment that will be used up to the prices to be taken. It is known that with greater knowledge of the project it will be possible to have more accurate results, for this research the information is limited so we focus on covering the most necessary variables.

In general, a single cut-off grade is calculated and through this, the calculations for the modeling of a mining project are generated, either by underground or surface methods. By finding different cutting laws with this, the design itself can be optimized and thus, for example, reduce waste rock extraction or otherwise, extract profitable mineral that with the first design would not be counted as such.

3.2.2 Preparation of the Base Scenario

The elaboration of the base scenario will be given precisely with the initial information that has been filtered and validated, once this is done, the economic calculation is carried out to see how much your NPV amounts to, which will later be compared with the scenarios that arise.

With the base scenario already carried out, it can be tested through the use of software, changes in production, costs, thus generating other scenarios with which we will end up comparing your final VPNs. More factors can be established for the creation of scenarios such as a distribution of the deposit by zones, identifying the mineral grades and cut-off grades for each zone.

3.2.3 Software Application

Here, both modeling and evaluation softwares will be used, using these softwares they will obtain the results of the alternative scenarios that were proposed, in the same way the use of these provides a certain degree of reliability because they are handled under very advanced estimation algorithms and data processing (Fig. 2).

For this case, 4 scenarios will be proposed apart from the base scenario, in order to achieve a notable increase in the economic value. By applying the variable cut-off grade method, it will be possible to identify costs by sectors and mineral grades by sectors.

4 Validation

4.1 Initial Study

Within the study carried out using Datamine, the Net Present Value (NPV) is calculated, under this metric the performance of the simulations of the different scenarios that are proposed is evaluated, as well as the same base or initial scenario.

The NPV or NPV is calculated using the following formula (Fig. 3):

$$VAN = \sum_{t=1}^{n} \frac{F_t}{(1-f)^t} - I_0 \quad (1)$$

Where:

Ft: Money flows by period
I0: Initial investment
n: Number of periods
k: Discount rate

CAPEX		
Plant	US$/tpd	**10,000**
Unforeseen		**0%**
Production cost		
mine cost		7
EAST		5
NORTH		7
WEST		6
SOUTH		4.5
freight	0.38	1.52
EAST	0.25	1
NORTH	0.5	2
WEST	0.55	2.2
SOUTH	0.2	0.8
Plant cost 300 tpd		18
Other costs		1
Mine + Silver US$/TM		**27.52**
EAST		**6**
NORTH		**9**
WEST		**8.2**
SOUTH		**5.3**
Other costs		**5%**
Selected plant		**300**

Fig. 2. Cost

Assumption		
Reserves	TM	**15,000,000**
Content by TM		
EAST		
Cu	%	**0.50%**
Mo	%	**0.17%**
Ag	oz/t	**0.250**
Au	oz/t	**0.002**
NORTH		
Cu	%	**0.42%**
Mo	%	**0.25%**
Ag	oz/t	**0.200**
Au	oz/t	**0.003**
WEST		
Cu	%	**0.38%**
Mo	%	**0.31%**
Ag	oz/t	**0.400**
Au	oz/t	**0.000**
SOUTH		
Cu	%	**0.46%**
Mo	%	**0.39%**
Ag	oz/t	**0.430**
Au	oz/t	**0.001**
Sale price		
Cu	US$/TM	**6,311**
Mo	US$/TM	**14,570**
Ag	US$/oz	**16.39**
Au	US$/oz	**1229.52**
Recovery Level		
Cu	%	**93%**
Mo	%	**75%**
Ag	%	**75%**
Au	%	**70%**

Fig. 3. Assumption

The NPV can be calculated using the Datamine itself and using Microsoft Excel, where economic parameters such as costs, prices, production, mineral grades must be established. There are limitations since it is not very detailed information, it will not be possible to make calculations related to the different geologies of the project, as is known, the geology will change and therefore costs will change, either due to drilling or locations. With the objective of this research, we will seek to increase this value within the same time period or without having to affect other variables that, on the other hand, may harm the project.

The evaluated project has an initial economic value of US $ 22 million, based on this factor, comparisons are made with the results obtained from the simulated scenarios where the variable cutoff law will be applied.

4.2 Validation Design

Initially, the necessary information for the calculations is collected, this data must be validated, that is, it must be filtered to avoid data that distorts the values such as 0 values or erroneous or atypical data. The variables to be taken into account to calculate the cut-off grades are identified, mostly that refer to production costs.

When these data have been established, the scenarios and the parameters that will follow are planned, the exploitation sequence and the calculation of tonnage that will be estimated per extraction period. Once the scenarios have been established, the scenarios are run using optimization software and economic calculations.

Once the results of these are obtained, they are compared against the base scenario in order to define the variation of the NPV where a significant increase is expected to be found.

5 Discussion

5.1 Results

Four scenarios were executed where variations in costs and daily production were chosen, basically a sectorization of the deposit was made in order to identify the difference in costs. These results were compared against the study or base scenario.

For scenario one, the EAST, NORTH, WEST and SOUTH fronts were identified. The following mineral grades were defined per front, in addition, the costs for each production front were estimated.

The prices of metals and concentrates will be constant for the different scenarios, in addition to mineral recovery. There will be a plant of 300 tons per day. Operating costs were calculated for each zone, taking into account geology, rock quality and location. Transportation costs were determined by the distances of the different fronts to the mineral stockpile. After the economic analysis and simulation of the scenario, a NPV of 25 million dollars was obtained.

For scenario 2, a NPV of US $ 24 million was obtained with an IRR of 35.3%, this reduction in the IRR is mainly due to the implementation of a higher capacity process plant compared to scenario 1. For this case, managed a 500 ton plant, this generates a higher initial investment and therefore the IRR will be lower.

In scenario 3, a NPV of almost 24 million dollars was obtained with an IRR of 43.1%. Although the plant conditions were similar to those of scenario 1, the distribution of the mineral grades by phases did not allow to maximize revenues as expected, otherwise it generated higher costs.

In the last scenario, a NPV of 22 million was obtained with an IRR of 34%, this low IRR is mainly due to the increase in production. This caused the investment cost to be much higher compared to the base scenario.

5.2 Analysis of Results

Under the results obtained, we see that scenario 1 is where there is a greater increase in the NPV.

This increase turns out to be 13% more or almost 3 million dollars in terms of the valuation of a mining project (Table 1).

In the results table shown, it can be seen that the variation of the values obtained in the investigations carried out by other authors applying a similar methodology are in the same range as those of the present investigation, except for only one of the cases [2] in which the methodology was used in a long-term planning evaluation, in which the variations of a series of operational factors were not taken into account, thus the results obtained in said investigation would have been greatly diminished.

Table 1. Results

Cases	Base VPN (US$)	Final VPN (US$)	Difference (US$)	Difference %
1	22,188,132	25,095,360	2,907,228	13.10
2	22,188,132	24,019,365	1,831,233	8.25
3	22,188,132	23,969,838	1,781,706	8.03
4	22,188,132	22,893,842	705,711	3.18

6 Conclusions

Following the structured and proposed design, it was possible to obtain a significant increase in the NPV of 13% for an open pit mining project. It is a result that, based on other research that sought to increase the NPV, is within the average.

These results could be improved if the scenarios were worked with a broader amount of information, as well as having knowledge of the terrain or area where the project will be developed, since the locations of the facilities will influence the calculations.

References

1. Zhang, K., Olawoyin, R., Nieto, A., Kleit, A.N.: Risk of commodity price, production cost and time to build in resource economics. Environ. Dev. Sustain. **20**(6), 2521–2544 (2017). https://doi.org/10.1007/s10668-017-0003-0
2. Kržanović, D., Kolonja, B., Stevanović, D.: Maximizing the net present value by applying an optimal cut-off grade for long-term planning of the copper open pits. Acta Montanistica Slovaca **20**(1), 49–61 (2015). https://doi.org/10.3390/ams20010049
3. Eggert, R.G.: Staking a claim for cambodia. mineral exploration and development: risk and reward, May 2010, pp. 1–16 (2010)
4. Esmaeil, R., Ehsan, M., Reza, S., Mehran, G.: Optimized algorithm in mine production planning, mined material destination, and ultimate pit limit. J. Cent. South Univ. **25**(6), 1475–1488 (2018). https://doi.org/10.1007/s11771-018-3841-5
5. Navia-vásquez, C., Monsalve-hinestroza, M.C., Franco-sepúlveda, G.: NPV analysis as a function of the discount rate and cost of re- handling implementing SIMSCHED DBS to open pit mining. Análisis del VPN en función de la tasa de descuento y el costo de remanejo implementando SIMSCHED DBS para una explotación minera a cie **41**, 81–86 (2017)
6. Hosseini, S.A., Asghari, O., Emery, X.: Direct block-support simulation of grades in multi-element deposits: application to recoverable mineral resource estimation at Sungun porphyry copper-molybdenum deposit. J. Southern Afr. Inst. Min. Metall. **117**(6), 577–586 (2017). https://doi.org/10.17159/2411-9717/2017/v117n6a8
7. Asad, M.W.A., Qureshi, M.A., Jang, H.: A review of cut-off grade policy models for open pit mining operations. Resour. Policy **49**, 142–152 (2016). https://doi.org/10.1016/j.resourpol.2016.05.005
8. Martinelli, R., Collard, J., Gamache, M.: Strategic planning of an underground mine with variable cut-off grades. Optim. Eng. **21**(3), 803–849 (2019). https://doi.org/10.1007/s11081-019-09479-6

9. Jewbali, A., Dimitrakopoulos, R.: Stochastic mine planning—example and value from integrating long- and short-term mine planning through simulated grade control, Sunrise Dam, Western Australia. In: Dimitrakopoulos, R. (ed.) Advances in Applied Strategic Mine Planning, pp. 173–189. Springer, Cham (2018). https://doi.org/10.1007/978-3-319-69320-0_13
10. Paithankar, A., Chatterjee, S., Goodfellow, R., Asad, M.W.A.: Simultaneous stochastic optimization of production sequence and dynamic cut-off grades in an open pit mining operation. Resour. Policy **66**(March), 101634 (2020). https://doi.org/10.1016/j.resourpol.2020.101634
11. Ahmadi, M.R., Bazzazi, A.A.: Application of meta-heuristic optimization algorithm to determine the optimal cutoff grade of open pit mines. Arab. J. Geosci. **13**(5), 1–12 (2020). https://doi.org/10.1007/s12517-020-5104-7
12. Githiria, J., Musingwini, C.: A stochastic cut-off grade optimization model to incorporate uncertainty for improved project value. J. Southern Afr. Inst. Min. Metall. **119**(3), 217–228 (2019). https://doi.org/10.17159/2411-9717/2019/v119n3a1

Sustainment Method in Gold-Bearing Workings Using Woodpacks to Mitigate the Induced Stresses that Cause Instability in an Underground Mine that Uses Longwall Exploitation

Carlos Saldaña-Paredes[1], Humberto Pehovaz-Alvarez[1], and Carlos Raymundo[2(✉)]

[1] Ingenieria de Gestion Minera, Universidad Peruana de Ciencias Aplicadas, Prolongación Primavera 2390, Lima 15023, Peru
{u201315868,pcgmhpeh}@upc.edu.pe

[2] Direccion de Investigacion, Universidad Peruana de Ciencias Aplicadas, Prolongación Primavera 2390, Lima 15023, Peru
carlos.raymundo@upc.edu.pe

Abstract. The underground mining is an activity that occurs in the depths of the rocky massif, which makes it possible to enter scenarios that may or may not be safe; This will depend on the quality of the rock and the engineering team. Until now, there is no support method that has been defined only considering the safety factor. For this reason, this research has been developed to use a type of support in gold mantles, in such a way that the spacings can be determined considering the maximum deformation lengths and therefore a safety factor. To determine the spacing between support elements, progressive results are taken as they contribute to the final stage of prediction of support spacing. Through the tests using the Rocscience RS2, spacings with probabilities of 90% compliance and with safety factors of 1.7 were determined; which allows to give more security and confidence in the work of auriferous mantle.

Keywords: Underground mining · Stress concentration · Auriferous mantles · Reduction factor · Woodpack and longwall

1 Introduction

Underground mining is developed in mountain range type deposits, which are presented as veins for which they are also known as filonianos, these deposits are located in the eastern mountain range and in the western mountain range of Peru, some mines that are included in these types of deposits are Marsa, Poderosa and Consorcio Minero Horizonte. As the mining activity has evolved, different conditions created by the tillage activities have appeared, so much so that, until today, in underground mining with gold-bearing mantles, the main problem is the concentration of induced efforts that give rise to the instability of the work, that instability, leads to accidents; The same ones that, according

S. Trzcielinski et al. (Eds.): AHFE 2021, LNNS 274, pp. 176–183, 2021.
https://doi.org/10.1007/978-3-030-80462-6_22

to the Ministry of Energy and Mines, between 2000 and 2019, 29% of accidents were due to rockfall.

2 State of the Art

2.1 Support Method in Underground Mine

There are different types of support in underground mining, in that sense it is very important to take into account the concentration of efforts and the variables that are strictly related to each other such as type of support, rock quality and stability of the work; This will make it possible to avoid generating negative events that can compromise the integrity of the people and the stability of the work, for that the choice of support elements is crucial to support the mining components [1–3]. The conditions of adequate interaction within the "rock support" system are not very effective for any of the supports that are used. This is due to the fact that many times the support capacity of a certain type of support is not taken into account, in other cases it may be that due to operational issues it ends up being oversized in terms of response as support.

2.2 Woodpack as a Support Method in Underground Mining

The support in an underground mine is very complex, since there are different types of support such as shotcrete, friction and anchor bolts, falsework, RS-Bolt, woodpack, among others. One of the many types of support is the Woodpack; the same that is applicable in the construction of gold-bearing works; In this type of reservoir, sometimes it is chosen to leave pillars as support, it is necessary to mention that the behavior of the efforts will be different in the support pillars [4, 5].

In this context, it can be concluded that the horizontal columns suffer a single compression load, while the inclined columns are exposed to combined loads; that is, a compressive load and a shear load [2, 6, 7]. Faced with this behavior of forces in inclined columns, it is necessary to look for more effective mechanisms to sustain the compressive and shear loads. In this sense, the woodpack, due to its characteristics, is a potential element of analysis to determine its behavior in low-angle work.

2.3 Sustainment in Longwall Exploitation in Underground Mining

The longwall or long wall exploitation is used in veins smaller than 30°; that is to say, in mantles, and for this it is convenient to use a type of support that interacts effectively in this type of work with powers and inclinations less than 2 m and 30°. This type of reservoir is usually exploited by the longwall or long-wall exploitation method [8, 9].

Therefore, the working conditions with slopes will give rise to the formation of pillars as the exploitation of the vein progresses, however this is a practice that contributes to leaving buried mineral and at the same time these pillars tend to weaken by the effect of the action of the efforts [2, 10, 11], since the reduction factors depend on the inclination, being that for greater dip the factor is lower and while for smaller dips, the reduction factors they are older.

3 Input

3.1 Proposed Method

The support analysis method is based on three stages, the first being the study of geology at a structural level, where the type of rock that is included in the surroundings of the mining work is taken into account and also the mineralization it contains. the area of exploitation. The second is the evaluation of the geotechnical parameters where the quality of the rocks is defined and for this, field tasks are practiced making on-site analysis of the main characteristics of the rocky massif; Here too, an evaluation of the mining work is made, which consists of identifying where it is located and what existing works may be close to it. Finally, in the third phase with the information collected in the field, that is, the geomechanical mapping, the geomechanical classification of the rock is made and based on it, it is located in the RMR ranges using the classification of the rock suggested by Bieniawski of the year. 89; with this, the application of the support is made and therefore the exploitation of the mineralized zone. According to research, a proposed method is presented (Fig. 1).

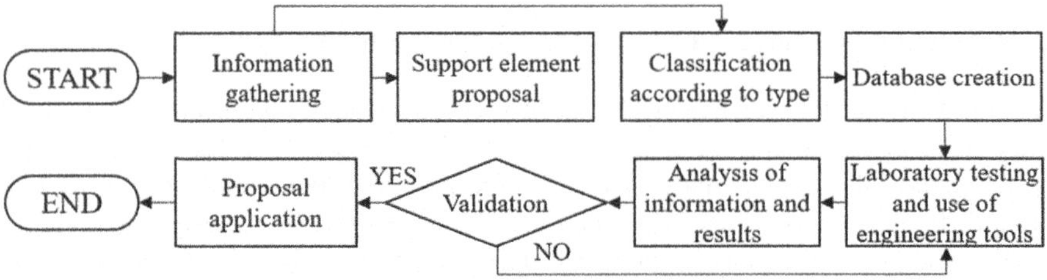

Fig. 1. Mechanism for method proposed

3.2 Components of the Method

The proposed method is based on three stages. The first is to determine the RMR (Rock Mass Rating) in the exploitation area. The second is to identify the area of the zone of deformation of the rock, according to dimensions of work. The third, evaluate the woodpack spacing according to the RMR and the support capacity of the support system.

Phase 1: Evaluation of the Rock Massif to Determine the RMR
First, it was necessary to identify the study area where the problem is located, after that, information was collected that consists of obtaining references of the structural geology, this allowed to identify the main characteristics of the rock both of the boxes and of the mineralized zone. After identifying the type of rock, we proceeded with the geomechanical mapping to obtain input data, after which simulations were carried out using the Monte Carlo method, with which ten thousand iterations were made; keep in

mind that, with more iterations, the results are better compared to when they are done with few iterations.

Phase 2: Identification of the Deformation Zone

The configuration of the project was made in the RS2 engineering software, that is, in the general tab, stages, stress analysis, groundwater, strength Reduction, and assign the project name. Subsequently, the geometry was assigned and the discretization (triangulation) was made, the initial stress field to which the excavation is subjected was entered; since before making the excavation the field is subjected to stresses and deformations. Likewise, it is also necessary to enter the constants of the rocky massif; To find these constants of the boxes (roof and floor) and vein, the RocLab was used, since through it the calculations of the rock mass constants (a, mb and s) are simplified using the generalized Hoek Brown criterion. When processing the information entered through the RS2, the isovaloric curves were obtained from which, based on their scope, the maximum deformation distance was deduced.

Phase 3: Determination of the Type of Support and Its Spacing

To achieve phase 3, correlative information was taken that was obtained in phases 1 and 2, of which, a fundamental factor to find the spacing between woodpacks was the deformation length and the iterations of different spaces between the proposed support element.. Thus, a spacing in meters was achieved, a distance that would be the radius of influence of each woodpack, taking into account that according to studies carried out on eucalyptus (wood to be used), as a whole, it reaches a support capacity of 120 tons; and on the side of the unstable load comprised in the area of the plastic zone, it also generates a tonnage that acts as a function of its specific weight and gravity. With this information, safety factors are obtained, which to be taken into account must be greater than one and based on this, the spacing between each support element can be established.

3.3 View of Indicators

The indicators used in the development of the phases are the geomechanical tables where it indicates and can determine the quality of the rock mass, the zone of deformation of the rock included in the surroundings of the work that is opening and the factors of security to determine spacings.

- To determine rock quality - RMR, its indicator is a table geomechanics.
- To determine deformations, its indicator is isovaloric curves.
- To determine spacing, if indicator is a safety factor.

4 Validation

4.1 Description of the Scenario

The area where the sustaining method was applied is in underground mining in which mining activities are carried out through tillage that consists of generating tunnels in the rock mass. To obtain the mineral by means of underground mining, it is necessary to make a series of excavations or holes in such a way that it can reach the mineralization zone. The tillage system is done from the beginning of excavation until reaching the mineralized zone; In the process of exploitation of the mineral, unstable areas appear, where it is necessary to adopt certain mechanisms to create conditions of safety in the work area.

4.2 Initial Diagnosis

In the first place, there is the Mercedes vein, which has orientation and variable dips East - West and South, Northwest-South East; It has a dip between 10° and 30°, vein width (power) ranging from 0.02 m to 2 m and is located on granodiorite rocks; the mineralization is together with quartz, pyrite, arsenopyrite and sphalerite; the quartz fracture and micro-fracture system are filled with sulfides; this vein is influenced by main faults (La Brava, Corihuarmi and María; North, Center and South).

Secondly, the Glorita 2 vein, the orientation of this vein is to the Northwest with slopes to the Northeast ranging from 0° to 30°, the power of the vein is from 0.02 m to 2 m and it is located on granodiorite rocks; the mineralization is together with quartz, pyrite, arsenopyrite and sphalerite; the quartz fracture and microfracture system are filled with sulfides; This vein is affected by the vein called "La Lima", which is acting as an inverse fault and is found in the ceiling box, the Glorita 2 vein and in the floor box, the Esperanza vein is located.

Finally in third place, there is the El Carmen vein, here there is a system of parallel veils and it is located at the ceiling of the vein called "Padre de Dios"; El Carmen has a power of 0.1 m to 0.8 m, it is shown in the form of quartz lenses with various veil systems, it contains galena, pyrite, sphalerite, arsenopyrite and contains high oxide contents, which are from sulfides; in contact with the boxes, it contains high concentrations of oxides and with free gold content whose grades are on average 2 Onz-Au/TM.

4.3 Application of the Contribution on Stage

Determination of the RMR in the Exploitation Area

To determine the RMR, the Montecarlo method was used, for this, the information obtained on-site, was processed and used as input parameters; From these inputs, RMR values were selected differentiating by box (floor box and roof box), in the floor boxes all the values belong to the Regular A rock type (RMR between 51 to 60), while in the roof boxes, the values have RMR between 41 to 60; that is, some have RMR that belong to Regular B and others Regular A. Therefore, for the purposes of the investigation, two tasks were considered, one with RMR in the ceiling and floor box from 51 to 60 and 41 to 50; that is, Regular A and Regular B, while the other was considered with RMR in the ceiling and floor box from 51 to 60 and 51 to 60; that is, Regular A for both boxes. The following table shows the iterations for the ceiling box and for the floor box.

To define the scenarios, a frequency table was built for each rock quality (Regular A and Regular B), then ten thousand iterations were carried out; As can be seen in the tables below, for this purpose, probabilistic tools were used; This process was repeated

Table 1. Summary of statistical data from probabilistic RMR

Statistical data roof box RMR IIIB	
Average	38,12167792
Standard deviation	0,508948345
Statistical data roof box RMR IIIA	
Average	46,03912267
Standard deviation	1,225639019
Statistical data floor box RMR IIIB	
Average	45,78399738
Standard deviation	0,785325526
Statistical data roof box RMR IIIA	
Average	52,73809873
Standard deviation	0,593037747

Table 2. RMR calculation from probabilistic data

Probabilistic calculation_montecarlo	
P90 (RMR) roof box with RMR IIIB	
90%	39
P90 (RMR) roof box with RMR IIIA	
90%	48
P90 (RMR) floor box with RMR IIIB	
90%	47
P90 (RMR) floor box with RMR IIIA	
90%	53

four times considering the deterministic results, that is, one job with a box and ceiling of RMR 41 to 50 and another job with a different RMR for each box; that is to say, from 41 to 50 and 51 to 60. The following tables show the summary of statistical data that was obtained after doing the 10 thousand iterations with the Monte Carlo method and with it the results of the RMR with a probability of 90% (Tables 1 and 2).

Determination of the Deformation Area
To finish the plastic area deformation area, calculations of the rock constants (a, mb and s) were first performed, for this the Rocscience software was used since with this tool the calculation of the mentioned constants is simplified.

For the process of determining the plastic zone, fundamentally the use of some software, for this case Phase 2D or RS2 was used (Fig. 2, Table 3).

Spacing According to the Support Capacity of the Support
To achieve the third objective, correlative information was taken that was obtained in objectives 1 and 2, of which, a fundamental factor to find the spacing between woodpacks was the deformation length and the iterations of different spaces between the proposed support element.. Thus, a spacing of 3m was obtained, a distance that would be the radius of influence of each woodpack, taking into account that according to studies

Table 3. Rock constants

Constant	a	Mb	s
Roof box	0.511	0.815	0.0001
Vein	0.544	0.180	0.0000544
Floor box	0.503	2.2846	0.0027

carried out on eucalyptus (wood to be used), as a whole, it reaches a support capacity of 120 tons; while on the side of the unstable load comprised in the area of the plastic zone, it generates a load of 70.2 tons; the interaction of these two loads gives a factor of safety of 1.7 (Fig. 3).

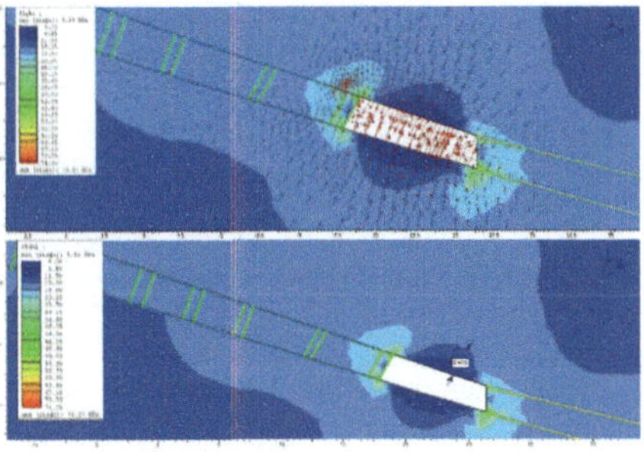

Fig. 2. Length of deformation area

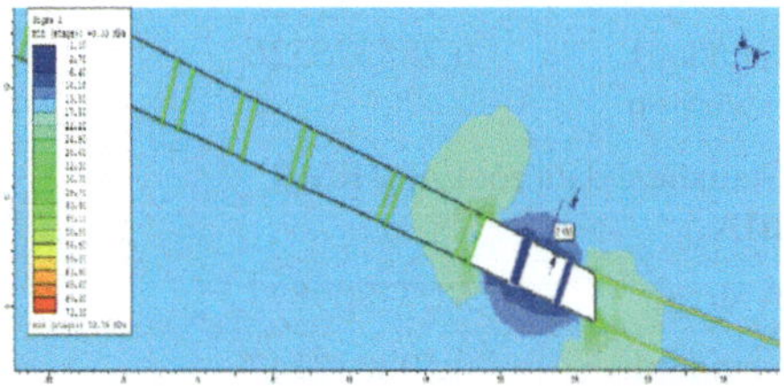

Fig. 3. Determination of woodpack spacing

5 Conclusions

Using the Monte method Carlo, it was possible to improve the analysis of the results by 7.8% on average since, by complementing with the data obtained from the field, they allowed to generate more robust scenarios because it considers iterations of thousands and thousands of scenarios; This is equivalent to making multiple mappings in the study area and while you have more data, you can work with greater accuracy and improve the results.

In underground mining, the workings are exposed to stress concentration and consequently deformations of the rock that is included in the surroundings of the workings. For this, it is necessary to carry out an engineering analysis in such a way as to be able to determine the parameters that serve as input elements for the software such as mb, s and a, which allow the analysis of rock deformations.

The results obtained were compared with those obtained in on-site mappings, it was found that 90% have concordance; In other words, by taking into account statistical tools for geomechanical mapping, I help determine the quality of the rock in the study area with quantitative results with RMR from 39 to 47 and RMR from 48 to 53; that is, a rock Bad A/Regular B and Regular B/Regular A.

After the analysis after generating the opening, a plastic zone is created around the excavation with an approximate deformation length of 2,493 m, which, when supporting the work, it was observed that the distance or reach of the deformation curve tend to be reduced.

Based on the tests, considering different spacings between woodpacks, it was determined that the spacing of the supporting element is 3.0 m, thus achieving a safety factor of 1.7, while, if there are spacings greater than three meters, said factor was less than 1. In operational practice, not much is taken into account; However, doing this test is better since it allows us to identify safety factors greater or less than 1 and thus make the decision.

References

1. Radu, S.M., Vereş, I.S., Pleşea, V., Cucu, I.: Metallic support with reinforced elements for the execution of underground excavations found in difficult geo-mechanical conditions (4) (2017)
2. Xing, Y., Kulatilake, P.H.S.W., Sandbak, L.A.: Investigation of rock mass stability around tunnels in an underground mine in USA by 3-D numerical modeling. In: 35th International Conference on Ground Control in Mining, pp. 213–221 (2016)
3. Weng, L., Li, X., Taheri, A., Wu, Q., Xie, X.: Fracture evolution around a cavity in brittle rock under uniaxial compression and coupled static–dynamic loads. Rock Mech. Rock Eng. **51**(2), 531–545 (2017). https://doi.org/10.1007/s00603-017-1343-7
4. Jessu, K.V., Spearing, A.J.S.: Effect of dip on pillar strength. J. Southern Afr. Inst. Min. Metall. **118**(7), 765–776 (2018). https://doi.org/10.17159/2411-9717/2018/v118n7a10
5. Jessu, K.V., Spearing, A.J.S., Sharifzadeh, M.: Laboratory and numerical investigation on strength performance of inclined pillars. Energies **11**(11) (2018). https://doi.org/10.3390/en11113229
6. Rezaei, M., Hossaini, M.F., Majdi, A.: Determination of longwall mining-induced stress using the strain energy method. Rock Mech. Rock Eng. **48**(6), 2421–2433 (2015). https://doi.org/10.1007/s00603-014-0704-8
7. Chung, K.L., Zhang, C., Li, Y., Sun, L., Ghannam, M.: Microwave non-destructive inspection and prediction of modulusof rupture and modulus of elasticity of engineered cementitious composites (ECCs) using dual-frequency correlation. Sensors **17**(12), article no. 2831 (2017)
8. Zhu, W., Chen, L., Zhou, Z., Shen, B., Xu, Y.: Failure propagation of pillars and roof in a room and pillar mine induced by longwall mining in the lower seam. Rock Mech. Rock Eng. **52**(4), 1193–1209 (2018). https://doi.org/10.1007/s00603-018-1630-y
9. Zhang, G.C., et al.: Ground response of a gob-side entry in a longwall panel extracting 17 m-thick coal seam: a case study. Rock Mech. Rock Eng. **53**(2), 497–516 (2019). https://doi.org/10.1007/s00603-019-01922-5
10. Yu, B., Zhang, Z., Kuang, T., Liu, J.: Stress changes and deformation monitoring of longwall coal pillars located in weak ground. Rock Mech. Rock Eng. **49**(8), 3293–3305 (2016). https://doi.org/10.1007/s00603-016-0970-8
11. Ozdogan, M.V., Gonen, A.: Determination of distance required to ensure stope and footwall-drift non-interaction zone based on geological strength index. Earth Sci. Res. J. **23**(1), 17–25 (2019). https://doi.org/10.15446/esrj.v23n1.71879

Operational Reliability Investigation of Turbogenerator Based on Technological Aspects and Human Factors

Igor Santos Araujo(✉), Salvador Ávila Filho, Jefferson dos Santos Mascarenhas, and Danillo Ramos Camargo

UFBA, Salvador, Bahia, Brazil

Abstract. In the search for operational excellence of equipment and systems in the industrial sector, it is essential to analyze their failures and seek solutions to increase the reliability of the process. This article aims to investigate the causes for the failure of a turbogenerator in a refining unit based on reliability tools, analysis of critical variables and elaboration of mental map root cause. The failure occurred after 5 months of operation due to high vibration, with erosion of the vanes and severe unbalance of the assembly. The studied turbogenerator generates electrical energy using the energy from the combustion gases of an FCC (fluid catalytic cracking) unit. The investigation showed that human factors, such as operational discipline, decision-making added to technological factors, such as low efficiency of the combustion gas cyclones of the FCC catalyst, in addition to the design failure of the pipeline contributed significantly to the occurrence.

Keywords: Human factors · Oil and gas · Turbogenerator · Reliability

1 Introduction

In the current global situation, where environmental resources are scarce, it is necessary to seek solutions in the energy field, to make these processes more efficient and more reliable. This implies a scenario with energy savings and reduction of CO2 and greenhouse gas emissions, which impacts society. According to [1], the carbon emissions in the world from energy use grew by 0.5% in 2019, less than half 10-year average growth of 1.1% per year. This means that the world is changing, and large energy companies need to adapt and improve efficiency in their processes to be competitive in this sector.

The petrochemical industry has great relevance in the global energy scenario. In Brazil, the oil production grew by 7.8% in 2019 and the national refineries processed 1.8 million barrels per day [2]. Due to its high complexity, it is important to monitor its parameters and study reliability solutions including human and technological factors with real cases.

This article aims to investigate the causes for the failure of a turbogenerator located in a fluidized catalytic cracking (FCC) unit of an oil refinery in Brazil. When this equipment is operating continuously, it produces electrical energy from the heat generated in the process itself. This work is based on reliability tools, human factors, and analysis of critical variables, to identify the causes of failure, mainly the root cause of the event.

S. Trzcielinski et al. (Eds.): AHFE 2021, LNNS 274, pp. 184–191, 2021.
https://doi.org/10.1007/978-3-030-80462-6_23

2 Literature Review

Oil is basically composed of a mixture of hydrocarbons. Its chemical composition may vary widely. The combustion of petroleum products generates energy to power automobiles, trains, ships and airplanes. There are numerous possible applications using petroleum products and refineries are responsible for transforming oil into its products that are consumed by the society [3]. According to [4], the different refining processes present in a refinery can be classified according to the transformation that aggregate the input stream, which can be separation, conversion and treatment processes. One of the most profitable conversion processes is catalytic cracking. The main purpose of the unit is to convert high-boiling petroleum fractions called gas oil to high-value, high-octane gasoline and heating oil [5].

A catalytic cracking unit operates under severe temperature conditions and uses a catalyst based on silica and alumina in the process. In these units, the energy contained in a flue gas stream in the regenerator can be used to generate steam in boilers and electrical energy in turbogenerators. The role of this equipment integrates the demand for power in this industry with the process of expansion of flue gases. This machine is complex, large and has auxiliary systems that are essential for its continuous operation. Common causes of failure of turbogenerators are discussed in [6], where we highlight: failures due to expander support, unit vibration, rotor rubbing and catalyst deposition and plugging.

The search for operational excellence and operational continuity of the turbogenerator involves the study of reliability to achieve good results. Complex systems involve several factors that can contribute to the failure of a component or equipment. According to [7], a study of accidents in Petrochemical and refining units identified the following causes: equipment and design failures (41%), operator and maintenance (41%), inadequate or improper procedures (11%), inadequate or improper inspection (5%) and miscellaneous causes (2%). This guide aims to understand the causes of human errors and suggest ways to reduce them. The PSF (Performance shaping factor) is anything that affect a worker's performance. So, managers can improve the PSF's and reduce the frequency of human errors.

Human error is discussed in [8] by the traditional view and by the systemic view. According to the traditional view, errors occur due to negligence, lack of commitment and failure to observe rules or procedures. Meanwhile, the systemic view of human error says that they occur due to the complex relationship between factors such as: environment, culture, inadequate guidelines and systems. It is a challenge to create an environment of antecedents appropriate to the behavior that one wishes to achieve in the organization [9]. A model of human reliability analysis is SPAR-H, which is a model that combines elements of the stimulus-response and the information processing approaches and acknowledges the role of environmental factors upon diagnosis and action [10]. Thus, it is possible to identify human factors that contribute to the occurrence of a failure.

3 Methodology

The methodology for investigating this failure was based on technological and human factors. The development process involved: reading and analyzing manuals, books and procedures; interviewing specialists, operators and engineers; application of reliability tools and analysis of the causes of equipment failures. For this, the methodology is described in Fig. 1. In the first step, the operational context of the process in which the turbogenerator is presented and of the equipment itself was defined. In the second step, critical variables and tools were defined and we investigated operational reliability with the appropriate tools applied in the process. In the third step, we discussed the human factors that influenced the failure of the turbogenerator. In the fourth step, a mental map with human and technological factors that led to the failure of the equipment was drawn up and a discussion of the results was made.

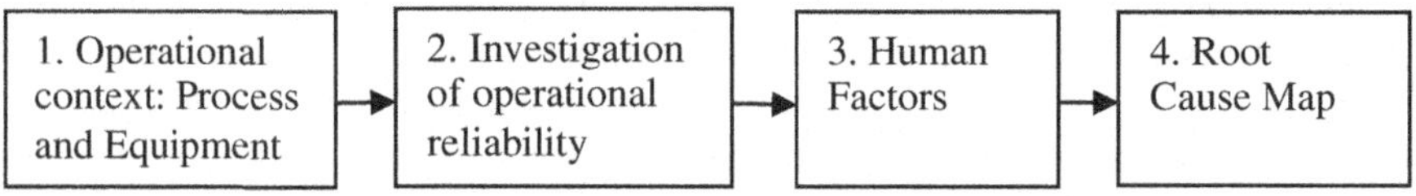

Fig. 1. Block diagram of methodology.

4 Case

This work will be applied to a turbogenerator in a fluidized catalytic cracking unit in an oil refinery in Brazil. The operational context, equipment and analyzes performed will be discussed below.

4.1 Operational Context

The FCC unit uses a fine solid catalyst based on silica and alumina, which behaves like a fluid when it is crossed by a gas stream, such as air or steam. In the studied unit, there is generation of combustion gases capable of generating electrical energy and high pressure steam. This current is named flue gas and goes to a vessel where there is a third phase of cyclones to remove entrained catalyst fines and then it will be used to generate electricity with a turbogenerator and high pressure steam with a heat recovery boiler. The studied turbogenerator is inserted in this flue gas energy recovery system of an FCC in Brazil. Its objective in the process is to generate electrical energy from the heat of the flue gas. For this, the equipment consists of an expander that drives a generator through a gearbox. There is a lubricating oil system that supplies the expander, generator and gearbox bearings. In addition, the expander has a steam sealing and cooling system. To monitor and control the machine at startup, normal operation and shutdown, there are instruments that measure vibration, temperature and speed.

An interlock system is used to protect the machine in situations that are unsafe for it. In summary, the situations that lead to the turbogenerator shutdown are low lubricating oil pressure, high temperature in the expander, low cooling vapor pressure, low cooling vapor temperature, over speed, rotation signal failure, axial displacement, vibration, rotation signal failure, lubricating oil supply and cooling vapor pressure. An operation group is responsible for doing the field and control panel routines. There is also a dynamic equipment engineering team that monitors the operation of the turbogenerator. The equipment manual highlights that a regular operation routine of the equipment depends a lot on the continuous operation or not of the FCC unit. It is worth mentioning that the flue gas that goes to the equipment, despite going through 3 cyclone stages, has catalyst particles in its composition. The piping design that surrounds the system has expansion joints and supports that have the purpose of absorbing expansion and vibrations.

4.2 Investigation of Operational Reliability

The campaign in which this study took place between September 2016 and February 2017. The turbogenerator operated for 5 months until a high vibration shutdown occurred. This investigation consisted of reading the reports of the shift operational group, choosing the appropriate variables to analyze this campaign, analyzing photos of the blades of the rotating set of the turbogenerator and post-parade maintenance reports, in the conversation with operators, engineers, mechanics of the plant and elaboration of the methodology and application of reliability tools.

The figure below shows a PI (plant information system) screen with green and blue vibration instruments, yellow speed, pink power generation, and FCC unit feed in Red. A timeline was drawn up with significant events at the bottom, which are: (A) the startup of the equipment, (B) operation of the turbogenerator with an average generation of 15 MW, (C) shutdown of the turbogenerator for thermal shock, (D) emergency shutdown of the FCC unit, (E) significant increase in vibration and reduction of power generated, (F) shutdown of the turbogenerator due to the need for the process, (G) shutdown of the turbogenerator due to high vibration (Fig. 2).

From this screen, at each shutdown of the turbogenerator, the rate of increase in equipment vibration is higher. This occurs until the limit of vibration supported by the machine is reached. The operational instability of the FCC contributed to the reduction of the turbogenerator's campaign time.

The figure below shows the relationship between the vibration of the turbogenerator and the temperature of the day. There is a significant variation in vibration. This graph shows that the vibration variations with the temperature were not absorbed by the expansion joints, as well as by the pipe support design (Fig. 3).

The photos below were taken after the turbogenerator interlocked by vibration. The image on the right shows the blades of the turbogenerator. This picture shows the process of erosion that these blades have suffered. The low efficiency in separating the flue gas from the catalyst caused excess catalyst particles in the flue gas to go to the turbogenerator, and this erosive jet caused severe unbalancing of the assembly and continuous increase in vibration. The image on the left shows that there was an opening in the closing of the high-pressure box. This indicates that the machine has been over tensioned (Fig. 4).

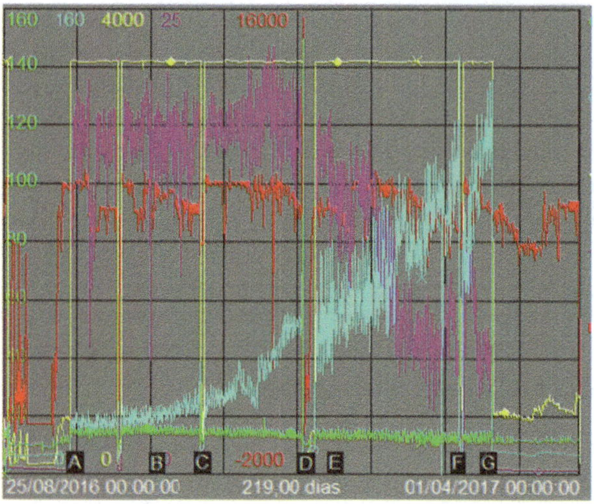

Fig. 2. Critical Variable in Turbogenerator process September 2016 to February 2017. Green and blue vibration instruments, yellow speed, pink power generation, and FCC unit feed in Red.

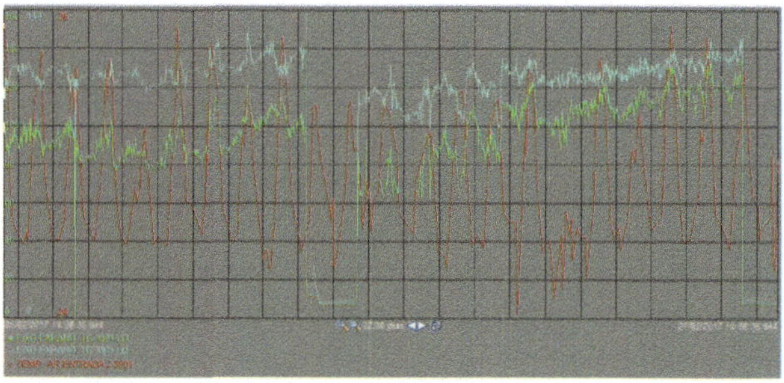

Fig. 3. Variation of the vibration (blue and green) and the temperature (red) of the day during the campaign.

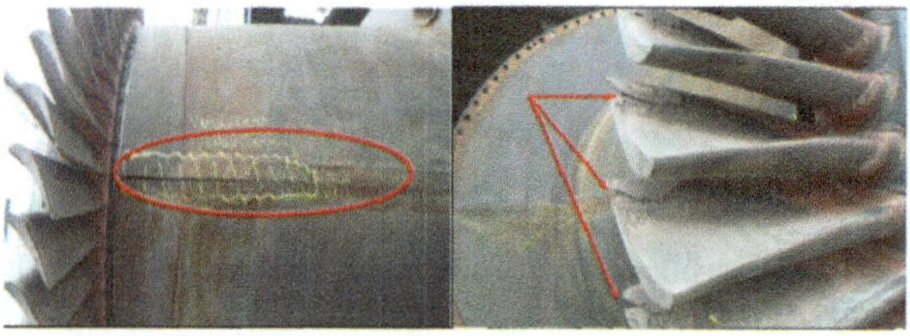

Fig. 4. Photos of the turbogenerator after the interlock by vibration.

4.3 Human Factors

The human being has a fundamental role to guarantee the operational continuity of the studied turbogenerator. Some error likely situations described in API 770 were found here. Regarding the operational group, it was noted after talking with experienced operators in the area that their training in relation to the turbogenerator is insufficient. As it is a complex machine, a specific and continuous training program is necessary, which was not noticed. There are procedures for operational routines and equipment startup and shutdown. However, in this period studied there was no adequate system to ensure that routines were performed. The emotional structure of the operators affects their work and routine. Ergonomic factors such as high noise from the operating area in addition to external personal factors increase the stress of operators and decrease their performance. The engineering team responsible for monitoring the equipment has adequate knowledge about the machine. However, points of improvement in the sense of communication with the operation can be upgraded. In addition, a human analysis is up to managers, as their decision making influences the machine, as they can maintain or modify the third stage of cyclones and the design and piping items of the turbogenerator. Some human factors are discussed below.

Procedure: Starting the plant involves remarkably high flow rates, time and quality targets that are difficult due to the complexity of the process and the possibility of mechanical impact. The checklist for this procedure is completed quickly and there may be slips. In addition, since the FCC unit is large and complex, there may be mistakes in completing routine checklists.

Insufficient knowledge: Training a large turbogenerator requires hours of dedicated study and proper instruction, as well as refresher courses. For not having full knowledge of the equipment, problems can occur and do not receive the proper treatment.

Monitoring of engineering and communication: The dynamic equipment engineering sector serves the entire refinery. There is no specific engineer for the FCC unit alone. As a result, there may be slips in certain analyzes as well as communication failures between this important team and the operational group.

Decision of the Manager and Supervisor: The leaders of the plant have the power to maintain or change a specific project or component of equipment. Keeping the turbogenerator operating with a third stage of cyclones with low efficiency can result in a gas for the equipment with a high number of solid particles, creating an erosive jet in the vanes.

4.4 Root Cause Map

Based on previous discussions on the investigation of operational reliability, this mental map was elaborated with factors that influenced the operational continuity of the turbogenerator, causing it to shutdown due to high vibration with damage to the blades and the set as a whole. The factors were divided into technological (TECH - Technological factors) and human (HUMAN - Human factors). There is also the operational instability factor at FCC unit, which was classified as human and technological (Fig. 5).

5 Conclusion

Failures resulting from the design of the turbogenerator are mixed with human errors due to deficiencies indicated in human factors. The mechanical stress of solids on the

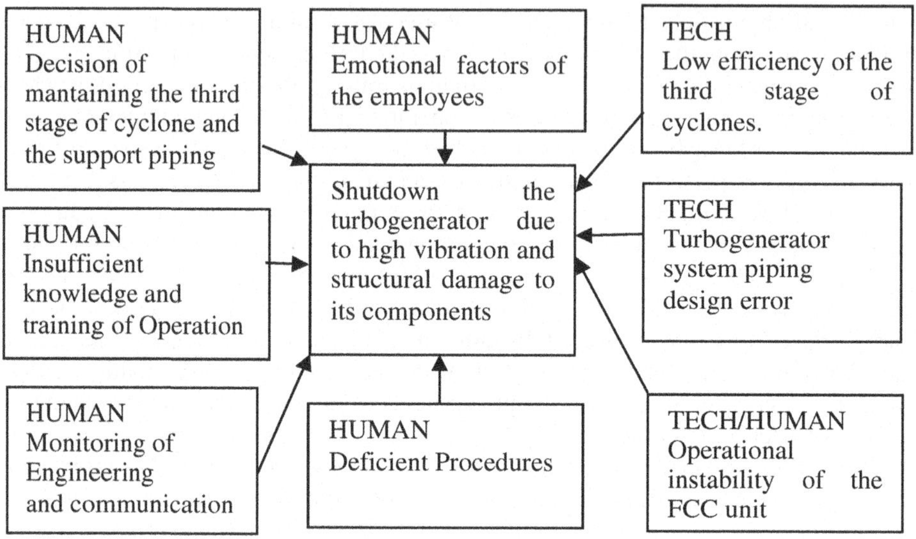

Fig. 5. Root cause map.

blades of the turbogenerator is a technological issue accompanied by undue human factors. The root cause map presented in Sect. 4.4 shows the causes contributing to the shutdown of the turbogenerator due to high vibration. This work showed that the failure of the turbogenerator comes from human and technological factors. In human factors, insufficient training, deficient procedures, automatic check list and decision making to maintain the projects of the third stage of the cyclone and the pipes may have contributed to the event. In technological factors, the main contributions were the low efficiency of the third stage of cyclones and error of piping design, in addition to the operational instability of the FCC unit. Because it is a complex machine, it is difficult to determine the root cause. This will be for a future study.

References

1. Looney, B.: BP statistical review of world energy 2020. Technical report, London (2020)
2. Brazilian National Agency for Oil and Gas and Biofuels – ANP. The Oil, Natural Gas and Biofuels Statistical Yearbook 2020. Technical report, Rio de Janeiro (2020)
3. Farah, M.A.: Petroleum and its derivatives: definition, constitution, applications, specifications, quality, characteristics. LTC, Rio de Janeiro (2013)
4. Brasil, N.I., Araújo, M.A.S., Sousa, E.C.M., Quelhas, A.D.: Oil and Gas processing: oil and its derivatives, primary processing, refining processes, petrochemicals, environment. LTC, Rio de Janeiro (2014)
5. Sadeghbeigi, R.: Fluid Catalytic Cracking Handbook, 2nd edn. Gulf Publishing Company, Houston (2000)
6. Singh, M.P., Drosjack, M.J., Linden, D.H.: Expanders for Oil and Gas Operations: Design Applications and Troubleshooting. McGraw-Hill Education, New York (2014)
7. American Petroleum Institute: A Manager's Guide to Reducing Human Errors. Improving Human Performance in the Process Industries. API Publication 770, Washington, D.C. (2001)

8. Lafraia, J.R.B.: Reliability, maintainability and availability manual. Qualitymark, Rio de Janeiro (2014)
9. Lafraia, J.R.B.: Leadership for SMS: understanding the influence of the mind on risk perception. Qualitymark, Rio de Janeiro (2011)
10. Gertman, D., Blackman, H., Marble, J., Byers, J., Smith, C.: The SPAR-H human reliability analysis method. Idaho National Laboratory, Washington, DC (2005)

Stability Method for Pit Dimensioning Obtained Using the Gradient Boosting Machine Algorithm in Underground Mining

Hernan Camacho[1], Humberto Pehovaz-Alvarez[1], and Carlos Raymundo[2](✉)

[1] Ingenieria de Gestion Minera, Universidad Peruana de Ciencias Aplicadas, Prolongación Primavera 2390, Lima 15023, Peru
{u201612472,pcgmhpeh}@upc.edu.pe

[2] Direccion de Investigacion, Universidad Peruana de Ciencias Aplicadas, Prolongación Primavera 2390, Lima 15023, Peru
carlos.raymundo@upc.edu.pe

Abstract. The diversification of mining in different geological contexts and the need to work at higher depths has shown that the stability graph method has disregarded scenarios with the presence of water and different confinement regimes. It is for this reason that the present investigation sought to incorporate these scenarios through the Gradient Boosting Machine algorithm. For this purpose, scenarios with different levels of water pressure were simulated and the degree of confinement around the excavations was considered. The generated model was based on the binary classification criterion, I feel the predicted classes, "stable" and "unstable"; with which an AUC value of 0.88 was obtained, which demonstrated an excellent predictive capacity of the GBM model. Likewise, the advantages over the traditional method were demonstrated, since a component of rigor and generalization is added.

Keywords: Gradient boosting machine · Active stresses · Chopping · Graphical stability method

1 Introduction

The Gradient Boosting Machine algorithm is a prediction tool for "objective variables" from a set of predictors (independent variables). Its importance lies in the ability to control aspects such as variance and bias; while its success is due to its composition, which consists of decision trees such as "weak learners", where the results of the model are weighed based on the results of the previous tree. Its application in the field of civil engineering and in geomechanical research has been positive and has allowed the development of various studies, which is why the creation of a classification model based on geological and geometric parameters of the pit will be sought to be able to analyze the degree of influence of active stresses on the stability of underground pits.

The pit design phase of a mining project depends fundamentally on the geomechanical characterization, lithology, and disposition of the mineralized body. In the graphical

S. Trzcielinski et al. (Eds.): AHFE 2021, LNNS 274, pp. 192–199, 2021.
https://doi.org/10.1007/978-3-030-80462-6_24

stability method, a large part of these parameters are considered, since the Stability Number (N') is made up of adjustment factors that address the stress relationship (Factor A), the geometric properties (Factor B and Factor C) and the characteristics of the rock mass such as the alteration and roughness of the discontinuities, the RQD index and the number of main families. However, aspects such as the presence of water in the discontinuities (Jw) and the stress reduction factor (SRF) are evaluated with a score equal to unity; so that the quality index of the modified rock mass (Q') is obtained. In such consideration, a dry excavation and with a stress tension state is assumed.

2 State of the Art

2.1 Pitting Stability Method

The graphical stability method has been a design tool with an application history of four decades. In that course, various researchers identified opportunities for improvement to extend the scope and precision of its use. In this way, a probabilistic approach was developed based on the Mathews [1, 2] stability graph method, in order to optimize the structural parameters. Similarly, the parameters of cohesion, stress, deformation modulus and friction angle were considered as stochastic variables to study the deformation of the pits, so that the optimal geometry was selected to optimize the design process [2, 3]. Likewise, it was sought to study the influence of the efforts around the excavation in order to propose new modeling criteria and propose improvements to the existing representations (Effort factor A) [3, 4].

2.2 Grading Method Gradient Boosting Machine

There has been a growing acceptance in the application of artificial intelligence to perform various tasks in the field of geomechanics. Under this approach, one of the inherent challenges is the variability of the rock mass and raises the need to study its properties, for which artificial intelligence algorithms such as the Gradient Boosting Machine and the Support Vector Regression were used [5, 6]. Likewise, the capacity of the algorithms to relate non-linear variables with the stability of the slices was used [7, 8]. There is a plethora of algorithms, which is why the question arises as to which one has the greatest predictive ability. Then a comparative study of 5 algorithms was proposed: [5, 9].

2.3 Cutting Dimensioning Method

Pit dimensioning involves understanding the intrinsic parameters of the rock massif, the stress regime in the ground and the geometric aspects of the pit. The relaxation and creep zones were studied, which are related to the deterioration of the ceiling box and the result of such, the unplanned dilution [10–12]. On the other hand, the geometry and dip of the pit will be characteristics that influenced the stability of a pit by subjecting its dimensions and geometries to a sensitivity evaluation [11, 13].

3 Input

3.1 General Contribution

As observed in Fig. 1, the work criteria and the main components are presented. With the execution of the activities corresponding to the workflow, it is possible to achieve the quantitative effects obtained when comparing the MGE with the model obtained by the GBM algorithm. These results will be able to corroborate the theory around the loss of normal efforts due to the effect of water, as well as to analyze the degree of importance that active forces are having in the stability of the pit. Models based on artificial intelligence were previously carried out that considered the geological and geometric characteristics that influence the behavior of the plowing. However, a simulation phase is added for the incorporation of active efforts, which, according to the research criteria, will allow a better performance of the model and will allow to extend the scope of use of the graphical method to other scenarios.

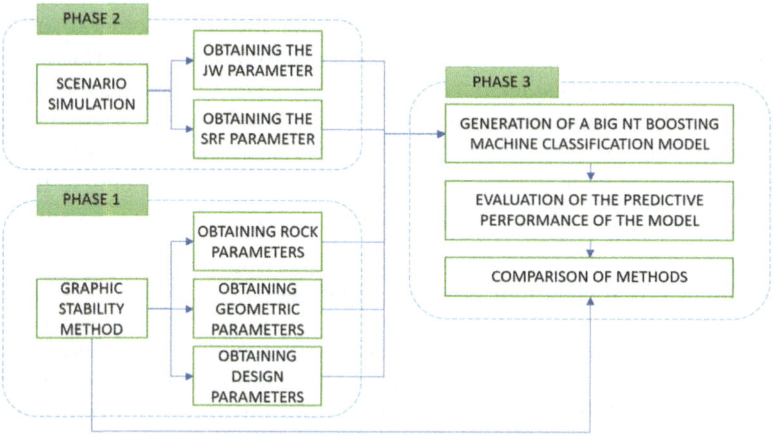

Fig. 1. Main figure of the investigation

3.2 Detail Contribution

3.2.1 Incorporation of Water Pressure (Jw)

The parameter "Jw" describes the water pressure in the geological environment of the pit, which is why it is a data that can hardly be deduced from diamond drilling. The methodology for the present investigation proposes a stochastic simulation of water pressure. The behavior assumption will be obtained by means of a uniform distribution function. The values of this distribution will allow describing various measurements in kilograms per square centimeter (kg/cm^2). Later, these values will be categorized so that they acquire a qualitative meaning, so that three scenarios will be obtained. (1) For a Jw = 1 it will be considered as a dry excavation. (2) In the event that Jw = 0.66, it will refer to a scenario with average presence of flows or water pressure and there will be an occasional backfill of the discontinuities. (3) Finally, for a Jw = 0..33, a large influx and pressure of water and notable washing of the filling of the discontinuities will be recognized. The table that will allow recording the procedure described is set out below.

3.2.2 Incorporation of the Confinement Regime (SRF)

Like the previous parameter, it was necessary to simulate scenarios for the "SRF". Therefore, it was established that, for the present investigation, the parameter represents only the state of confinement around the pit, namely: low confinement, medium confinement and high confinement. The criterion for categorizing the parameter "SRF" will be based on the result of the quotient of the uniaxial compressive stress and the main induced stress ($\sigma c/\sigma 1$); therefore, a range was established that will allow each quotient to be assigned its confinement category.

3.2.3 Grading Model Gradient Boosting Machine

Once scenarios are randomly generated that describe the presence of active efforts; these characteristics will be incorporated into the historical database of pits. This database compiles 169 structured samples that contain geometric and geological features of pits. You can identify the fields that represent the stress factor (A), the adjustment factor by orientation (B), gravitational factor (C), the hydraulic radius (HR), the RQD index, the set of discontinuities (Jn), the alteration of discontinuities (Ja) and roughness (Jr). The consigned characteristics were used for the design and construction of the previously analyzed stability methods. Precisely, if you want to externder and check the scope of the tool in question, you should consider the reduction and adjustment offered by the factor (Jw/SRF).

3.3 Contribution Process

First instance, a data pre-processing will be carried out that consists of cleaning and balancing the samples; followed by dividing the same into two groups: in the test data and the training data. Later, in the learning stage, we will seek to make a hyperparameter adjustment of the model based on the Gradient Boosting Machine algorithm. For the evaluation of the performance of the model, measurement criteria such as the confusion matrix, the ROC curve and the area under the ROC curve (AUC) will be used. Once the internal evaluation of the classification model has been finalized, an external comparison with the original stability method (MGE) will be carried out. In order to make a distinction between the aforementioned methods, the. Finally, an analysis of the importance of the characteristics will be carried out, in order to identify the influence power of the active efforts (Jw and SRF).

3.4 Indicators

3.4.1 Area Under the ROC Curve (AUC)

The objective of the area under the curve is the characterization of the performance of a classification model: 0.7–0.8 Acceptable; 0.8–0.9 Excellent; 0.9–1.0 Outstanding.

3.5 Model Accuracy (ACC)

The goal: to provide general information about misclassification of test samples. The data are obtained from the construction of a confusion matrix.

Calculation formula: the formula that represents the described index is listed below

$$ACC = \frac{TP + TN}{FP + FN + TP + TN} \quad (1)$$

Reference Level: it is understood that on a scale established between 0 and 1, as the result is made to the unit, the precision of the model is greater.

4 Validation

4.1 Initial Validation Scenario

The validation scenario is based, fundamentally, on the compilation of historical plowing data, on which the characteristics corresponding to the active efforts were attached. For this purpose, the python programming language is used; with which the necessary codes for the preprocessing, training and validation of the stability classification model were made. The codes of the environment and the use of libraries allowed the deployment of all the functions for the application of the Gradient Boosting Machine algorithm. The interpreted language of python was complemented with libraries such as Numpy, Scipy, scikit-learn, among others.

4.2 Validation Design

The consequent procedures were carried out in the Python Notebook environment, where libraries were used for data management. The data was partitioned in the test set and the training set. Subsequently, the grid search and cross-validation criteria were used in order to identify the hyperparameters that improve the predictive performance of the GBM classification algorithm.

Once the optimal hyperparameters were obtained, the adjustment of the Gradient Boosting Machine classification model was carried out. Therefore, it was necessary to apply metrics to elucidate the accuracy and predictive performance of the fitted model. The confusion matrix, ROC curve and AUC curve metrics were used to evaluate the model. In Fig. 2 and Table 1.

The previous results denote the achievement of the first objective of the research, which resides in the generation of a classification model of pit stability and which was carried out based on the first group of data. The consequence was to compare the resulting model with the graphical stability method. For that purpose, the second set of data was used.

The data were loaded into the Python environment and the stability class was preceded in the GBM classification model.

Then, of the totality of samples of the second group of data, that is, of the classes predicted according to the Gradient Boosting Machine model, 77 were classified as unstable, while 8 were stable after using the GBM classification model.

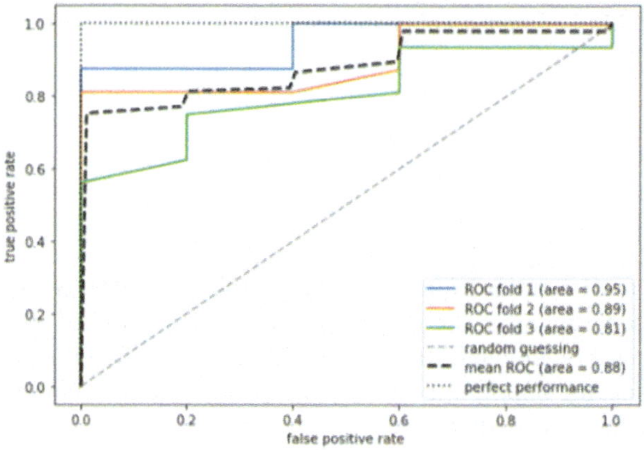

Fig. 2. ROC curve and average AUC value of the trained GBM model

Table 1. GBM model performance metrics

Performance metrics	
AUC value	0.88
Accuracy (ACC)	0.885
Error (ERR)	0.115
Completeness (REC)	0.833

The same samples were classified according to the criterion of Potvin's modified graphical stability method, 47 were classified as unstable, while 38 were stable.

Finally, the analysis of the relative importance of the variables was carried out, in which the following hierarchy was observed: the variable with the greatest influence was the size and shape of the pits, followed by the efforts around the excavations. With a relative importance of 0.175, the orientation of the surface of the pits was placed in third order; followed by the gravitational effect and quality of the rock, the variables corresponding to the presence of water and resistance to shear stresses showed little significance in the prediction model.

5 Results

As shown in Table 2, a considerable difference is observed in terms of class classification. The proposed method was characterized by predicting 90.5% of the pitting as unstable. This would be warning of a possible overestimation of the method compared to the proposed GBM stability method.

A disagreement was found between the variables of greatest importance in both models. However, this difference could be due to the samples used to generate the model. In other words, the research [9] focused on the roof box of the pits, while the

Table 2. Summary of predicted classes in method comparison

Method	Stable	Unstable
MGEM	38	47
GBM	8	77

present research used samples from the different surfaces of the pits. Given this diversity of surface, it is consistent that the shape and dimensions have had a great influence on their stability. On the other hand, since there is little significant influence of the presence of water in the pits, the opportunity was opened to be able to extend the scope and use of the GBM pits stability method.

Regarding the similarity of the predictive performance of both compared models, it can be confirmed that the development of the proposed methodology was carried out appropriately.

6 Conclusions

A stability classification method was generated using the Gradient Boosting Machine algorithm that incorporates active efforts, whose performance was measured based on the criterion of the area under the AUC curve, from which an average value of 0.88 was obtained, demonstrating an excellent predictive capacity..

The most important variable was the shape and size of the pits, this is due to the fact that different plowing surfaces were used. However, this represents an advantage for the generalization of the method.

The stability method obtained by means of the GBM algorithm proved to have an advantage over the graphical stability method, this is due to more rigorous results that allow visualizing the probability of failure of the pit that is being evaluated.

The generated method is susceptible to generalization, allowing the incorporation of different scenarios that describe different geological conditions around the pits. This leverages industry expertise for pit sizing and enables the method to continue to learn from future new experiences to achieve higher predictive performance and range in use.

References

1. Zhang, L., Hu, J.H., Wang, X.L., Zhao, L.: Optimization of stope structural parameters based on mathews stability graph probability model. Adv. Civ. Eng. **2018** (2018)
2. Idris, M.A., Nordlund, E.: Probabilistic-based stope design methodology for complex ore body with rock mass property variability. J. Min. Sci. **55**(5), 743–750 (2019). https://doi.org/10.1134/S1062739119056112
3. Vallejos, J.A., Díaz, L.: A new criterion for numerical modelling of hangingwall overbreak in open stopes. Rock Mech. Rock Eng. **53**(10), 4559–4581 (2020). https://doi.org/10.1007/s00603-020-02179-z
4. Madenova, Y., Suorineni, F.T.: On the question of original versus modified stability graph factors–a critical evaluation. Min. Technol. Trans. Inst. Min. Metall. **129**(1), 40–52 (2020)

5. Qi, C., Fourie, A., Zhao, X.: Back-analysis method for stope displacements using gradient-boosted regression tree and firefly algorithm. J. Comput. Civ. Eng. **32**(5) (2018)
6. Xu, C., Nait Amar, M., Ghriga, M.A., Ouaer, H., Zhang, X., Hasanipanah, M.: Evolving support vector regression using Grey Wolf optimization; forecasting the geomechanical properties of rock. Eng. Comput. **36**, 1–15 (2020)
7. Qi, C., Chen, Q.: Evolutionary random forest algorithms for predicting the maximum failure depth of open stope hangingwalls. IEEE Access **6**, 72808–72813 (2018)
8. Santos, A.E.M., Amaral, T.K.M., Mendonça, G.A., da Silva, D.D.F.S.: Open stope stability assessment through artificial intelligencehttp. Rev. Esc. Minas **73**(3), 395–401 (2020)
9. Qi, C., Fourie, A., Ma, G., Tang, X., Du, X.: Comparative study of hybrid artificial intelligence approaches for predicting hangingwall stability. J. Comput. Civ. Eng. **32**(2) (2018)
10. Sepehri, M., Apel, D., Liu, W.: Stope stability assessment and effect of horizontal to vertical stress ratio on the yielding and relaxation zones around underground open stopes using empirical and finite element methods. Arch. Min. Sci. **62**(3), 653–669 (2017)
11. Kasanda, T.B., Bowa, V.M.: Causes and impact of hangingwall instability in open stopes: a case of Chifupu project in Zambia. Geotech. Geol. Eng. **38**, 475–495 (2019)
12. Heidarzadeh, S., Saeidi, A., Rouleau, A.: Evaluation of the effect of geometrical parameters on stope probability of failure in the open stoping method using numerical modeling. Int. J. Min. Sci. Technol. **29**(3), 399–408 (2019)
13. Abdellah, W.R.E., Hefni, M.A., Ahmed, H.M.: Factors influencing stope hanging wall stability and ore dilution in narrow-vein deposits: Part 1. Geotech. Geol. Eng. **38**(2), 1451–1470 (2019). https://doi.org/10.1007/s10706-019-01102-w

Expanded Polystyrene Wall and Conventional Concrete Wall Sustainability Issues in Housing

Lateef Adewole Adegboyega[1] and Rita Yi Man Li[2(✉)]

[1] The Federal University of Technology, Akure, Nigeria
adegboyegaqsv120225@futa.edu.ng
[2] Hong Kong Shue Yan University, Hong Kong, China
ymli@hksyu.edu

Abstract. The provision of sustainable houses for the multitudes in developing nations has been a priority over the years. Numerous ecological issues, such as the use of unsustainable construction materials and non-economical building materials, are necessitating the search for improved and economical alternatives in housing delivery. As a result of research and technological improvements, modern eco-friendly building materials have been developed to fulfil the requirements of sustainability issues and alleviate environmental problems. One such material is expanded polystyrene (EPS). This paper focuses on EPS wall and conventional concrete wall sustainability issues in housing delivery. A comprehensive literature review was conducted to analyse the sustainable potential, environmental impact, timely delivery and economic implications of EPS and conventional concrete walls. EPS walls provide higher sustainability potential, and are environmentally friendly and cost-effective, with faster construction time than conventional concrete. The conceptual framework developed from this study could help government agencies, real estate developers and other stakeholders to solve the sustainability issues of housing delivery.

Keywords: Concrete wall · Expanded Polystyrene (EPS) wall · Housing · Sustainability

1 Introduction

Urbanisation leads to an exponential population growth, as a result of an extraordinary leap in the quantitative housing needs of the public [1]. These housing needs do not correspond to effective demand, since most of the populace do not have the means to afford satisfactory housing. According to Adejumo [2], the rate of providing sustainable housing stock has lagged severely behind the rate of population growth in Nigeria, resulting in a huge housing deficit; a yearly production of more than 70,000 housing units is required to cope with the population movement. This deficit is more visible in urban centres, where the rapid rise in the population has caused an increase in the standard of living because of the sophisticated demand for urban goods and services. In urban centres, there is scarcity and high cost of land and housing, which are often in short

S. Trzcielinski et al. (Eds.): AHFE 2021, LNNS 274, pp. 200–207, 2021.
https://doi.org/10.1007/978-3-030-80462-6_25

supply and beyond the economic reach of most urban households [3]. The urban centres are occupied by a broad range of people with low income and seasonal unemployment, who are often poor, and are hence forced to occupy limited, insufficient, jam-packed, unfriendly and muddy shelters, and a generally dishonoured environment. These urban poor are exposed to a life marred by hazardous conditions of nutrition and well-being, with few or poor-quality material belongings [4].

As stated by Olotuah and Ajenifujah [5], most urban centres in Nigeria are categorised by high masses of buildings with a dense population. There is no adequate space for natural ventilation between houses, poor well-being, inferior housing, and severe environmental and hygienic difficulties. The scarcity of affordable and decent accommodation for the urban poor is thus a significant housing problem in Nigeria. In large urban centres such as Abuja, meagre housing conditions are often evident in the high numbers of people living in one room and paying excessive rent. This physical congestion is the primary determinant of health risk and destructive social behaviour; as sleeping accommodation is poorly ventilated, transmittable (air and water-borne) diseases spread rapidly under such conditions [6]. More coherent construction methods could be employed, with new production technologies that allow exceptional modular plans, high-strength and load-bearing capacity materials, as well as minimal labour, materials, mass, time and cost. For example, the Expanded Polystyrene (EPS) initiative Adedeji et al. [7] allows the reduction of mass in construction through the use of lightweight and heavily insulated envelope wall panels. Thus, EPS can be classified as a sustainable building material in construction. Consequently, this research work will focus on the sustainable construction issues of EPS walls and concrete walls in housing delivery.

2 Conventional Concrete Walls

Concrete is the most useful material in construction, but leads to the depletion of natural resources and increases the shortage of the ingredients, such as steel, cement and aggregates. Consequently, there is a large demand for these materials in the commercial sector [8]. The global consumption of natural sand is exceptionally high because of the intensive use of concrete. The aggregate properties affect the sturdiness and performance of concrete; hence, fine aggregate – typically natural watercourse or pit sand – is an essential component of concrete. Fine and coarse aggregate constitute 75% of total volume; thus it is essential to obtain the right type and good-quality aggregate at the site, because this forms the main matrix of concrete or mortar [9, 10].

2.1 Concrete Wall Components

According to Fridman [11], the components of concrete and the method of preparation are as follows:

2.1.1 Cement

Concrete without admixtures and of high cement content (over about 460 kg/m^3) will be difficult to pump because of high friction between the concrete and the pipeline. Cement content below 270 to 320 kg/m^3, depending upon the proportion of the aggregates, may also prove difficult to pump because of segregation within the pipeline [12].

2.1.2 Aggregate

The determined size of crushed aggregate is limited to one-third of the smallest inside diameter of the hose or pipe, based on simple geometry of cubical shape aggregates. For uncrushed (rounded) aggregates, the determined size should be limited to 40% of the pipe or hose diameter. The shape of the coarse aggregate, whether crushed or uncrushed, has an impact on the mix proportions, although both shapes can be pumped satisfactorily. The crushed pieces have a greater surface area per unit volume, compared to uncrushed parts, and thus necessitate relatively more mortar to coat the surface. Coarse aggregate of a very bad particle shape should be avoided. Difficulties with pump mixed have often been experienced when too large a quantity of coarse aggregate is used, in an attempt to economise by reducing the amount of cement; such mixes are also more difficult and expensive to finish. The grading of coarse aggregate ought to be as per IS: 383–1970. If aggregates are nominally single sized, then 10 mm and 20 mm are specified in the ratio of 1:2 for a graded coarse aggregate. In the same way, 10 mm, 20 mm and 40 mm aggregates should be used in the ratio of 1:1.5:3 for a graded coarse aggregate. A fine mixture of Zone II as per IS: 383–1970 is mostly appropriate for pumped concrete, provided that 15 to 30% of sand passes through the 300 micron sieve, and 5 to 10% through the 150 micron sieve [12].

2.1.3 Water

Water actively participates in the chemical action with cement. Potable fresh water with pH value of 7 should be used, which is available from local sources and is free from deleterious materials.

2.2 Preparation of Conventional Concrete

Volume batching is not a suitable method for proportioning the material, because of the exertion required to measure granular material in terms of volume. The volume of moist sand in a loose condition weighs much less than the same volume of dry compacted sand. Thus, the effect of bulking should be considered for moist fine aggregate. For low quality concrete or for any small job, concrete may be batched by volume. Weight batching is the correct method for determining materials, as the weight system of batching facilitates accuracy, flexibility and ease. Large weight-batching plants have accurate weighing equipment. On large sites, the weight bucket type of weighing equipment is used.

2.3 Environmental Impact of Conventional Concrete Walls

The production and use of concrete have a wide range of environmental and social consequences. A major component of concrete is cement, which equally exerts environmental and societal effects [13]. The cement manufacturing industry is one of the three prime producers of CO_2, which is the main greenhouse gas (the other two sectors being the energy production and transportation industries). The production of Portland cement has contributed 7% to worldwide anthropogenic CO_2 emissions, mainly due to the sintering of limestone and clay at 1,500 °C (2,730 °F). (Turner et al. [36]). Concrete is used to form hard surfaces that contribute to surface overflow, which can cause heavy soil erosion, water contamination, and submerging; but equally, it can be used to divert, dam, and control submerging. Concrete is also the one of major contributors to the urban heat island effect, though less so than asphalt [14].

3 Construction of Three-Dimensional EPS Walls and Its Merits

Traditional earth materials from soils and rocks are heavy and can cause settlement, instability or lateral pressures; while other lightweight materials, such as foamed concrete, wood chips, clay and wood fibre, have higher densities, limitations in handling, and can be weather-sensitive, thus requiring staged construction and/or preloading, surcharging and draining [15]. The ultra-lightweight property of EPS is evident, as it is 98% air captured within a 2% cellular matrix. The advantages of on-site production and conveyance bring unique economic benefits, while considerably reducing the health and safety risks associated with the use of heavier materials. It is an excellent supernumerary for infill materials and ballast, where it also brings load and fill times down in time-critical building projects [16].

Despite the lightweight nature of EPS, its structure brings the benefits of exceptional compressive strength, rendering it suitable for use in areas of construction and civil engineering applications [16]. In order to prove the strength of EPS, several research institutes across the globe have conducted studies showing that EPS panels can serve as a structural base infill for roads [17], railway and bridge infrastructure [16].

4 Environmental Impact of EPS Compared to Conventional Concrete Walls

Currently, procedures are in place globally to reduce the impact of activities on the environment [18]. For the building and construction industries worldwide, this involves the selection of building materials; specifically, the choice of insulation [19]. Therefore, the use of environmentally friendly materials such as EPS, as an innovation, has improved housing delivery.

Concrete technology is progressing, with many advances and inventions in the construction context, including the production of lightweight concrete, and use of lightweight aggregates and artificial aggregates such as EPS beads, fly ash and slag [20–23]. EPS beads can be added to mixes, either partially or wholly replacing aggregates, subject to the desired strength and properties. Lightweight concrete is beneficial in many applications and is attractive; it is frequently used in the form of cement-foam composites. EPS is commonly accepted as a permanent formwork, as merged construction materials with a sandwiched core are becoming a more conventional construction material.

4.1 Economic Implications of EPS Walls

On the financial side, EPS material technology appears to be very attractive for the key players in the construction industry. Frequently, the clients, project managers, contractors and end users are at loggerheads, in terms of reaching a satisfactory balance on the foremost building industry anxieties of cost, quality and time [24, 25], while end users are concerned about standard houses with excellent facilities at an affordable price. The most appropriate technique to accomplish this is through proper and careful selection of building materials. EPS can help attain high quality, low cost and record-time completion of the building. At all stages of its life cycle, from manufacture to an application, and to recycling or disposal, EPS has shown excellent performance. Indeed, the use of EPS offers considerable cost and environmental advantages, since it contributes positively to a better environment at minimal cost.

4.2 Environmental/Sustainability Potentials of Conventional Concrete over EPS

The report of the World Commission on Environment and Development stated that in "Our common future, in a quest for a possible solution to environmental challenges through the concept of sustainable development. Fundamentally, sustainable development signifies meeting the needs of the present without compromising the ability of the future generations to meet their own needs" [26]. It also means the consumption of resources available to the current generation without depriving the future generation of resources for their effective living. It strives to achieve the goal of social, environmental and economic co-development [27].

Global discussions regarding sustainability emphasise the comparative analysis of building materials, in terms of quality of CO_2 released by the material into the atmosphere; absorption and storage of CO_2; amount of fossil fuel required in the materials production; amount of fossil fuel released into atmosphere; and thermal conductivity. The studies by Ferguson et al. [28] and Lienhard [29], regarding the environmental impact of common building material, revealed that the amount of fossil fuel required to manufacture concrete is 4,800 MJ/m^3; whereas steel uses 266,000 MJ/m^3, and aluminium employs 1,100,000 MJ/m^3. Similarly, of the materials mentioned above, concrete, steel and aluminium cannot store any amount of CO_2.

Regarding the thermal conductivity of building materials, EPS possesses better heat resistance than concrete. The conductivity rate of EPS is 0.05–0.0.15 J/m-Ik-I, concrete is 1.4–2.9 J/mIK-I, and steel is 19.0–21.0 J/m-I k-I; whereas fired clay is 1.0 J/m-IK-I, limestone gravel 0.6 J/m-I k-I, cement board 0.6 J/m-Ik-I, and stone 1.5–3.0 J/m-IK-I [29]. Apparently, EPS buildings do not require artificial insulation inside to maintain the interior temperature. Chen [30], and Ede and Okundaye [31] also stated that EPS is environmentally friendly, since it contains the lowest embodied energy in comparison to other building materials, such as concrete. Despite the great efforts in the development of building materials, little attention has been directed towards the sustainability potential of building materials. Often, information on new innovations, as well as the opportunities offered by new building products, takes a long time to reach building consultants and prospective clients. According to Ljungberg [32], sustainability goals should not only consider the environmental aspects, economic impact and customer needs, but also the market demands. In the quest for sustainability, efforts should be directed towards raising public awareness of the opportunities and environmental demerits of building materials, so that prospective clients and consultants can make best choices for the collective benefit of our environment.

4.3 Economics, Implications and Timely Delivery of EPS Compared to Conventional Concrete Walls

Agabi [33] states that the construction cost of buildings is established by estimating the cost after the building is designed, or by using a cost-control guide during construction. A huge advantage is gained through the reduction of construction cost. To achieve a complete superstructure (less fittings, fixtures and finishes), the polystyrene panel system will cost approximately 10% more in price of materials; however, it saves up to 30% in formwork for suspended slabs and walls, 100% in formwork and reinforced concrete in openings, and up to 30% in floor reinforcements, depending on the design. Finally, a 70% saving on labour is achievable in the erection of walls [34]. Moreover, the additional overhead cost incurred from long construction duration is eliminated on superstructural development by 50% or more, depending on project management. Setting up three (lightweight) polystyrene panels per 45 m^2 wall will take two people one day to set up all four walls, brace them and cut out windows and doors. They can then be sprayed the following day; therefore, four days using a conventional construction technique can be cut down to half the time by using the EPS system [33].

5 Recommendations and Conclusion

Resources are finite and we have to reserve our future on this planet by adopting sustainability approach [35]. This paper has presented the issues regarding EPS and conventional concrete walls in the construction industry, detailing construction processes and methods of construction using EPS and conventional concrete walls. It concludes that EPS walls have higher sustainability potential over conventional concrete walls, as well as being more eco-friendly and economical. Even though EPS is a sustainable building material, it also saves time in housing delivery. Therefore, it is recommended that government agencies, real estate developers and all other stakeholders in the building industry accept the concept of contemporary building materials, to ensure that buildings satisfy the energy requirements for the actualisation of the period's development goals, and significantly reduce the cost of construction. The application of EPS in housing projects should be encouraged; not only to provide cost savings, but also for the delivery of sustainable residential and commercial housing.

References

1. Diogu, J.O.: Housing the poor in Nigeria: the integrated project approach. Assoc. Archit. Educ. Nigeria (AARCHES) J. **2**(1), 1–6 (2002)
2. Adejumo, A.A.: Some thoughts on affordable and social housing in Nigeria (2008). www.nigeriavillagesquare.com/articles/akintokunbo-a-adejumo/some-thoughts-onaffordable-and-social-housing-in-nigeria.html. Accessed 10 Dec 2008

3. Oladapo, R.A., Olotuah, A.O.: Appropriate real estate laws and policies for sustainable development in Nigeria. Struct. Surv. (Spec. Issue), **25**(3/4) 330–338 (2007). www.emeraldinsight.com/10.1108/02630800710772890
4. Olotuah, A.O.: Housing development and environmental degeneration in Nigeria. Built Hum. Environ. Rev. **3**, 45–51 (2010)
5. Olotuah, A.O., Ajenifujah, A.O.: Architectural education and housing provision in Nigeria. CEBE Trans. **6**(1), 86–102 (2009). ISSN: 1745-0322 (Online) 86
6. Amado, M.P., Pinto, A.J., Santos, C.V.: The sustainable building process. In: Proceedings of XXXV IAHS World Congress on Housing Science, Melbourne, Australia, 4–7 September 2007. CD ROM (2007)
7. Adedeji, Y.M.D., Arum, C., Ajayi, B: Affordable housing initiative in Nigeria: use of composite panels. In: Proceedings of West Africa Built Environment Research, 19–21 July Accra, Ghana, pp. 79–86 (2011)
8. Porwal, V., Dwivedi, R.: Studies on tyre aggregate concrete with fly ash. Int. J. Sci. Res. (IJSR) **5**, 708–711 (2016)
9. Hudson, B.P.: Manufactured Sand for concrete. Indian Concr. J. 237–240 (1997)
10. Sahu, A.K., Sunil, K., Sachan, A.K.: Crushed stone waste as fine aggregate for concrete. Indian Concr. J. **77**, 845–847 (2003)
11. Fridman, V.: Method for producing construction mixture for concrete, issued 22 August 1995 (1995)
12. Shukla, P., Manish, K., Sahu, L.S.: International seminar on non-conventional energy sources for sustainable development of rural areas. IJAERD-Int. J. Adv. Eng. Res. Dev. (2017). ISSN: 2348-4470. p-ISSN:2348-6406. Impact Factor: 4.72. 10.21090/ijaerd
13. Worrell, E., Price, L., Martin, N., Hendriks, C., Ozawa Meida, L.: Carbon dioxide emissions from the global cement industry. Ann. Rev. Energy Environ. **26**, 303–329 (2001). https://doi.org/10.1146/annurev.energy.26.1.303
14. Shepherd, S., Woskie, S.: Controlling dust from concrete saw cutting (PDF). J. Occup. Environ. Hygiene **10**, 64–70 (2013)
15. Aabøe, R., et al.: Geofoam blocks in civil engineering applications. In: Arellano, D., Özer, A.T., Bartlett, S.F., Vaslestad, J. (eds.) 5th International Conference on Geofoam Blocks in Construction Applications, pp. 3–38. Springer, Cham (2019). https://doi.org/10.1007/978-3-319-78981-1_1
16. British Plastics Federation (BPF): EPS Briefing for Architects, Architectural Technicians, Building and Construction Specifiers, P.R. Principles Ltd., Nottingham England (2008). www.epsconstructiongroup.com
17. Arellano, D., Stark, T.D.: Load Bearing Analysis of EPS-Block Geofoam Embankments. Taylor and Francis Group, London (2009)
18. Ede, A.N., Okundaye, J.O.: Appraisal of timber as structural members for residential buildings in Nigeria. Int. J. Eng. Technol. **14**(1), 108–112 (2014)
19. EPSASA:Selection Guide Introducing Expanded Polystyrene (EPS), The Expanded Polystyrene Association of Southern Africa Administered by Association of Architectural Aluminium Manufacturers of South Africa (2006)
20. Bonacina, C., Campanale, M., Moro, L.: Analytical and experimental investigations on the heat transfer properties of light concrete. Int. J. Thermophys. **24**(5), 1407–1414 (2003)
21. Babu, K.G., Babu, D.S.: Behaviour of lightweight expanded polystyrene concrete containing silica fume. Cem. Concr. Res. **33**, 755–762 (2002)
22. Concrete Homes; Technology Brief: Plastic Foams for Concrete Homes (2009). http://www.toolbase.org/PDF/DesignGuides/Concrete_Homes_Brief_PlasticFoams.pdf. Accessed 4 Apr 2012
23. Chen, B., Liu, J.: Properties of lightweight expanded polystyrene concrete reinforced with steel fiber. Cem. Concr. Res. **34**, 1259–1263 (2004)

24. Aina, O.O., Wahab, A.B.: Assessment of build ability problems in The Nigerian construction industry. Glob. J. Res. Eng. **11**(2), 2011 (2011)
25. Li, R.Y.M., Poon, S.W.: Future motivation in construction safety knowledge sharing by means of Information Technology in Hong Kong. J. Appl. Econ. Sci. **4**(3(9)), 457–472 (2009)
26. Gustavsson, L., Pingoud, K., Sathre, R.: Carbon dioxide balance of wood substitution: comparing concrete- and wood-framed buildings. J. Mitigat. Adapt. Strategies Glob. Change **11**(3), 667–691 (2005)
27. Li, R.Y.M.: The usage of automation system in smart home to provide a sustainable indoor environment: a content analysis in web 1.0. Int. J. Smart Home **7**(4), 47–59 (2013)
28. Ferguson, I., LaFontarie, B., Vinden, P., Bren, L., Hateley, R., Hermesec, B.: Environmental properties of timber. Forest and Wood Products Research and Development Corporation (FWPRDC), Austria (1996)
29. Lienhard, J.H.: A Heat Transfer Textbook. Phlogiston Press Cambridge, Massachusetts (2003)
30. Chen, Y.J.: Comparison of environmental performance of a five storey building with crosses laminated timber and concrete. Department of Wood Science, University of British Columbia. Vancouver, B.C., Canada (2003)
31. Ede, A.N., Ogundiran, A.: Thermal behaviour and admissible compressive strength of expanded polystyrene wall panels of varying thickness. Curr. Trends Technol. Sci. **3**(2), 110–117, **5**(6), 320–337 (2014). ISSN: 2279–0535
32. Ljungberg, L.Y.: Materials selection and design for development of sustainable products. Mater. Des. **28**(2), 466–479 (2007)
33. Agabi, C.: Delivering affordable housing using polystyrene technology (2016). https://www.dailytrust.com.ng/news/property/deliveringaffordable-housing-using-polystyrene-technology/142818.html. Accessed 18 Sept 2017
34. Haron, N.A., Hassim, I.S., Kadir, M.R.A., Jaafar, M.S: Building cost comparison between conventional and formwork system: a case study of four-storey school buildings in Malaysia. Am. J. Appl. Sci. **2**(4), 819–823 (2005)
35. Li, R.Y.M.: Resistance and motivation to share sustainable development knowledge by web 2.0. J. Inf. Knowl. Manage. **9**(3), 251–262 (2010)
36. Turner, L.K., Collins, F.G.: Carbon dioxide equivalent (CO2-e) emissions: a comparison between geopolymer and OPC cement concrete. Constr. Build. Mater. **43**, 125–130 (2013)

Impact of Globalization on Current Practices in the Supply Chain Management of SMEs

Ayodeji Dennis Adeitan[1](✉), Clinton Aigbavboa[2], and Oyindamola Cynthia Olubiyo[3]

[1] SARChl in Sustainable Construction Management and Leadership in the Built Environment, Postgraduate School of Operations Management, University of Johannesburg, Johannesburg 2028, South Africa

[2] SARChl in Sustainable Construction Management and Leadership in the Built Environment, Faculty of Engineering and the Built Environment, University of Johannesburg, Johannesburg 2028, South Africa

[3] Postgraduate School of Operations Management, University of Johannesburg, Johannesburg, South Africa

Abstract. In recent times, the market strategy in which most nations that are already developed SMEs operation has witnessed a tremendous change in the past, due to the implementation of globalization, highly advanced technologies and supply chain management to its highly competitive market landscape. Nigeria SMEs have not fully taken advantage of these management tools because of lacking the quality or quantity required knowledge of globalization, research and supply chain management. The over dependence on crude oil as the major financial of its economy has caused a major neglect of other aspects of the economy such as garment industry, shoe making industry and textiles industries through inadequate or no funding. These reasons had prevented Nigeria SMEs to enhance its profitability and competitive advantage with the developed nation's global market. The aim of this paper is to present a literature review on the impact of globalization on current practices in the supply chain management of Nigerian SMEs.

Keywords: Globalization · Supply chain · Market strategy · Management tools

1 Introduction

Nigeria is a country with diverse natural and human re-sources that could make it a leading economy in the world within two decades of its independence. It is a country with the most population growth in Africa and currently places fifth in the world [18]. The over dependence on crude oil as the major financial of its economy has caused a major neglect of other aspects of the economy such as garment industry, shoemaking industry and textiles industries through inadequate or no funding [14]. Owing to the recent fall in the oil prices which is the driving force of Nigeria economy, there is an urgent need to actualize and stabilize the economy which implies channeling all re-sources and factors that can reshape its economy to achieving a purported growth rate. In this regard, Small and medium enterprises (SME) have been identified as a reliable

S. Trzcielinski et al. (Eds.): AHFE 2021, LNNS 274, pp. 208–214, 2021.
https://doi.org/10.1007/978-3-030-80462-6_26

employer of labor and a source of financial security for most Nigerian youths. Supply chain management is a technique that involves the coordination of the business activities within a particular firm and across businesses within the supply chain with the sole aim of achieving the long-term performance of the separate companies and the supply chain [20]. As a result of these developments, small and medium enterprises (SMEs) in Nigeria realized that they must partake in the logistics management of suppliers responsible for goods delivery, aftermarket service of the goods to the final/end customers and efficient & effective monitoring of day-to-day activity in the manufacturing of finished goods and services. The incessant increase in customers' demands for speed of goods delivery or service, higher quality and more flexibility all at lower cost have necessitated a new trend in the global SMEs industries [1]. As a result of these, the recent global competition in manufacturing is increasingly difficult for SMEs industries while trying to remain safe and competitive.

Globalization is defined as the technique used to access rise in social and cultural association, political interdependence, and financial integrations [12]. To compete in a global environment, organizations have had to be flexible in their strategies to sustain growth and break new boundaries. As a result, most industries have trans-formed completely from a manual way of manufacturing to automated technologies and strategies. Entrepreneurs in various sectors of manufacturing/enterprises in nations across the world have seen the benefit of supply chain management, Logistics management, globalization and lean manufacturing practices and have invested in its implementation [11]. For example, industries such as Amazon, Toyota, Dell, DHL have se-cured a strong competitive advantage due to the use of these practices. Most Small and Medium Enterprises (SMEs) in Nigeria are yet to embrace fully the globalization practices to their operations thus preventing them to compete in transforming to world-class organizations. The aim of this work is to present a literature review on the impact of globalization on current practices in the supply chain management of Nigerian SMEs.

2 Supply Chain Management

Supply Chain Management can be defined as the constant and coordinated flow of goods and services from origins through suppliers into and through the organization and onto the consumers in a way to maximize profit and minimize cost [5]. It is all about the connections between links of the immediate seller/buyer relationship into a longer series of occurrence. This means that industry suppliers have their own suppliers, which are direct customers other than the ultimate consumers. [21] reported that a competitive advantage can be achieved in supply chain management if only the activities are integrated through improved supply chain relationships, associated flows of funds and information. Supply chain management highlights sales increment, expenses reduction while making full potential of organizational resources by direct participation in teamwork and information sharing along the supply chain. Over the years, manufacturing companies are more involved in customers and supplier's ways of conducting business due to globalization. It is a very challenging rule because to be successful in today's competitive business environment, entrepreneurs and business owners should be able to effectively integrate the functions within a firm and also link them efficiently with external operations of suppliers and supply chain members. They need to focus on supply chain management practices that have an impact on enhancing supply chain activities and performances [3]. The functions of supply chain management are not restricted not only on the supplier and the manufacturer; it also involves customers, transporter, and retailers. These functions include new product development, marketing, operations, distribution, finance, and customer service [6]. The planning and execution of upstream and downstream activities with suppliers and customers to deliver effective, efficient and superior customer value at minimizing cost is the function of supply chain management [7]. It summarizes the activities involved in converting the product, processing it, creating a market and finally fulfilling all logistics management activities. Supply chain management is a process that goes into improving the way a company finds the raw components, transforming it needs to make a product and finally deliver it to customers. The five basic components of supply chain management as recommended by [26] are planning, source, make, delivery and return. The benefits of Supply chain management include; the supply of quality, quantity number of parts for production, cost-effective and safe transportation of material for delivery, smooth manufacturing without hindrance due to transportation. The flexibility in producing goods at lower prices and distributing to consumers quicker and maintaining the cost of purchased products and parts at acceptable levels. The simplistic view of supply chain management is presented in Fig. 1.

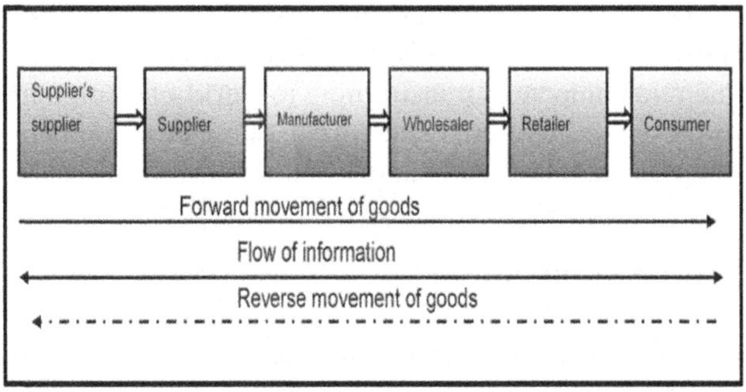

Fig. 1. The simplistic view of a supply chain (Adapted from [17])

Figure 1 above, highlights the roles of a supply chain in moving of physical goods as raw materials to the manufacturers and then as a finished commodity from same manufacturers to the final consumers. It is evident that supply chain management incorporates reverse movement of goods to indicate the level of economic value to organization and related information flows. According to [15], supply chain management needs to be assessed in businesses for its performance in order to evolve an efficient and effective supply chain. SMEs and manufacturing firms need to measure the performance of their supply chain management practices by applying the five dimensions of measurement, namely Supply Chain Flexibility, Supply Chain Integration, and Responsiveness to Customers, Supplier Performance, and Partnership Quality [22].

3 Globalization

Globalization can be referred to as the intent of expanding capital, world trade, capital markets and social formations [16]. Globalization is the interaction that brings closer contact within cultures by identifying their cultural differences, resource management and government roles in the national economy while focusing on project changes and its effect. The effect of globalization can be felt in all cities economic activity by increasing capital mobility, advanced manufacturing, information & communications technology and logistics technology. In recent time, globalization is seen as extension of forces that determine expanded market through the integration of economies around the world. It is a term that is characterized over time because of its promises and consequences, it is happening, and it is difficult to stop.

As reported by [25], Globalization consists of five progressive stages which are; exporting, translation of sales and marketing, localization of production etc. This is best explained through population migration; trade in goods & services, movements of capital and financial markets. The primary role of globalization is the continuous expansion of the business through a capturing share of the expanded market. The continuous expansion is achieved by further improvement in the economies of scope, control of supply chains and production control decisions rather than involved in the supply networks. As a part of the operation, supply chain management with Globalization holds an important position in maintaining the flow of materials from the production units to the supply of finished products to the end users [6]. As a result, the drivers of these operation processes include (a) improvement in the technology of logistics to reduce transportation and production cost (b) ease in government policies to facilitate speed of globalization (c) preferential change in a product to avoid taking advantages in the price reduction of transportation and communication. Even though the importance of globalization cannot be overemphasized, underdeveloped and developing nations like Nigeria continue to lose out on the benefits [9]. Therefore, there is a need to identify various challenges facing globalization in these nations and thereby proffers solution to reap its benefit if they want to be relevant in this present-day competitive market.

4 SMEs in the Nigerian Industry

According to [13], SMEs can be defined as an enterprise whose annual turnover not exceeding EUR 50 million and employs less than 50 people. SMEs have more flexible procedures, simple coordination, less standardization and control systems which allow it to accomplish its daily activities. However, there are many limitations SMEs experience in order to achieve products quality such as human resource, capital, information system, material resources, and manufacturing. The limitations in information acquisition and human resource experience by SMEs will lead to higher capital, higher transaction cost [8]. With an estimated 170 million inhabitants, Nigeria is the most populous nation in Africa, and it is now acknowledged as the continent's largest economy [19]. However, the budget and government spending were characterized by rising or fall in the price of oil and the effects were felt across the sectors. Nigeria remains highly reliant on oil, about 75% of Federal revenue comes from taxes on the oil and gas industry, and oil & gas make up more than 90% of exports, providing the critical source of foreign exchange to aid Nigeria's consumption of imports [19]. In recent years, oil production has dropped, and this has caused massive rate in unemployment and low earnings in federal revenue. In an effort to revive the economy, the Nigerian government has relatively diversified its economy and focus on SMEs industry. The Nigeria SMEs industries includes the garment industry, shoe industry, food/beverage industry, gaming industry and logistic companies etc. They are part of the supply chain management which activities include manufacturing goods, services within the packaging line and the distribution of goods through logistic services to the end consumers. The Nigeria SMEs industries have harness re-sources and relevant innovations to stabilize and reshape the nation's economy to purposeful growth. According to Global Competitiveness Report on ease of doing business (2016), Nigeria compares absolutely well with the big emerging markets such as China and India in some key metrics used by World Bank to compare general business environment.

For example, the garment industry has been identified to be an employer of labor and a source of living for Nigerian youths [4]. [24] also reported that the Nigerian garment/textile industry had been identified to be a major employer of labor (about 60%) in the manufacturing sector. It implies that SMEs has contributed immensely to the social and economic development of the nation. Unfortunately, the factors limiting the garment industry include inadequacy of locally manufactured raw materials, manpower development, taste for foreign textiles, and dependency on foreign technology, electricity, and manufacturing of low-quality materials, among others. If these factors are addressed, it will result in the ability of Nigerian garment industry to reduce unemployment among the youth, compete against large enterprises and less vulnerable to changes of the market. Nigeria SMEs needs to integrate globalization in its operations to boost technical and social knowledge of the market. It is also important to apply to [22] supply chain management five dimensions of measurement to assess their performance instead of a one-way process.

5 Analysis of Impact of Globalization in the Supply Chain of Nigeria SMEs

Globalization impact on the supply chain of many SMEs is very evident in human migration, the transaction in goods and final market (stocks and bonds). In recent years, supply chain and logistics sector had been facilitated by the spread of globalization. This implies that most of the production is been moved to less developed nations due to an average length of transport distances. Globalization will position organizations in a highly competitive framework if the five basic components of supply chain management such as planning, source, make, delivery and return implemented. Impact of globalization will enable Nigeria SMEs to integrate information systems and processes through the supply chain to transportation, inventory control, and market forecasts. The supply chain management will become limitless and flow of resources will not be limited to a certain place. At this point, Nigerian SMEs supply chain management will see a drastic reduction in the cost of materials, cost of labor and ease of access to raw materials. Presently, SMEs in Nigeria operate in a very competitive environment and the only way to survive in these business environments is to invest in effective logistics arrangements, cost-effectiveness,

Product innovation, expertise and timely execution of contracts. However, the level to which these expectations can be met is to have functional, effective supply chain management that can translate the strengths and vast opportunities into a competitive advantage. The aim of all SMEs is to maximize profit through the reduction in inventories, delivery time and cost of production while still maintaining quality needs to customers. It has been reported in various articles how globalization and Supply chain management could determine the survival in the SMEs industry of countries like USA, China, and South Korea etc. However, there has not been any empirical model to measure the performance or justification for implementing these programs to Nigerian SMEs industry. Therefore, there is an urgent need for an empirical study to provide insight into the performance of Nigerian SMEs industry.

6 Conclusion

The impact of globalization on current practices in the supply chain management of Nigeria SMEs, have been successfully discussed. This paper examines the importance of globalization and Supply chain management to Nigeria SMEs, and how it has improved the GDP of Nigeria economy. The literature has shown that SMEs is a major employer of labor among Nigerian youth, it has also contributed to the socio-economic development of the nation. Also, it has highlighted the problems facing Nigeria SMEs such as inadequacy of locally made raw materials, human resource, electricity, capital, information system and high logistic problem. It was discovered that much has been said in supply chain management to Nigeria economies, but no literature was found on the impact of globalization on the nation SMEs. This inadequacy has shown that there is a need for more empirical research on globalization impact to Nigeria industries.

References

1. Abioye, T., Bello, E., Olabanji, O.: A roadmap to lean production implementation for small scale Nigerian wire and cable companies. J. Mech. Eng. Autom. **1**, 119–125 (2011)
2. Adebayo, I.T.: Supply chain practices in nigeria today: impact on supply chain management performance. Eur. J. Bus. Soc. Sci. **1**(6), 107–115 (2012)
3. Arawati, A.: Supply chain management, product quality and business performance. In: International Conference on Sociality and Economic Development, vol. 10, pp. 98–102 (2011)
4. Aremu: Tackling Unemployment in Nigeria. Leadership Newspaper, Lagos, Nigeria (2014)
5. Baily, P., et al.: Purchasing Principle and Management. 9th edn. Hampshire, pp. 88-92. Prentice Hall (2005)
6. Chopra, S., Meindl, P.: Supply Chain Management: Strategy, Planning & Operation. 3rd edn. Pearson Prentice Hall (2007). SBN 0 – 13 – 173042 – 8
7. Christopher, M.: Logistics and Supply Chain Management. Great Britain: Financial Time. 2nd edn. Prentice Hall (2005)
8. Coviello, N.E., McAuley, A.: Internationalization and the smaller firm: a review of contemporary empirical research. Manage. Int. Rev. **39**(3), 223–256 (1999)
9. Daramola, I., Oyinade, B.: Impact of cultural globalization on African: the role of western media. Int. J. Educ. Res. **3**(3), 2201–6740 (2015)
10. DHL global connectedness index. The State of Globalization in an Age of Ambiguity (2016). www.dhl.com/gci

11. Doolen, T.I., Hacker, M.E.: A review of lean assessment in organizations: an exploratory study of lean practices by electronics manufacturers. J. Manuf. Syst. **24**(1), 55–67 (2005)
12. Eden, L., Lenway, S.: Introduction to the symposium multinational: the janus face of globalization. J. Int. Bus. Stud. **32**(3), 383–400 (2001)
13. European union: The European Commission User Guide to SME Definiton (2015). www.europa.eu
14. Forrest, T.: Politics and economic development in Nigeria. Updated Edition. Routledge (2019)
15. Gunasekaran, A., Patel, C., McGaughey, R.E.: A framework for supply chain performance measurement. Int. J. Prod. Econ. **87**(3), 333–347 (2004)
16. Henry, C., Springborg, R.: Globalization and the Politics of Development in the Middle East. Cambridge University Press, Cambridge (2001)
17. Lummus, R.R., Krumwiede, W.D., Vokurka, J.R.: The relationship of logistics to supply chain management: developing a common industry definition. Ind. Manage. Data Syst. **101**(8), 426–432 (2001)
18. Makinde, D.O., Fajuyigbe, M.O., Ajiboye, O.J.: Nigerian textile industry: a tool for actualizing economic stability and national development. Eur. J. Bus. Soc. Sci. **4**(8), 331–344 (2015)
19. Mckinsey Global Institute, Nigeria's Renewal: Delivering Inclusive Growth in Africa's Largest Economy (2014). www.mckinsey.com
20. Mentzer, J.T., et al.: Defining supply chain management. J. Bus. Logist. **22**(2), 60–70 (2001)
21. Monezka, R.: Purchasing and Supply Chain Management. 2nd edn., pp. 82–87, 572–579. R.R. Donnelley and Sons Company, Crawfords Ville (2002)
22. Muhammad, B.H.: Supply Chain Management: Practices, Performance and Its Impact on Business Performance. M.Sc thesis of the University of Utara, Malaysia, pp. 1-67 (2004)
23. Ohmae, K.: The Borderless World: Power and Strategy in the Interlinked Economy. Harper Business, New York (1999)
24. Oloyede, Z.O.: An assessment of factors militating against fabric production in contemporary nigerian textile industry. Paper Presented at a Workshop Organized by The Entrepreneurial Unit of Obafemi Awolowo University. Ile-Ife (2014)
25. Pieterse, J.N.: Globalization and Culture: Global mélange. Rowman & Littlefield (2019)
26. Vickery, S., Jayaram, J., Droge, C., Calantome, R.: The effects of an integrative supply chain strategy on customer service and financial performance: an analysis of direct versus indirect relationship. J. Oper. Manage. **21**, 523–525 (2003)

Achieving the Success of Sustainability Systemic Design Through Data Visualization Approach

Ying Chuai[1] and Haiwei Yan[2](✉)

[1] Liquid Studio of Accenture, 1687 ChangYang Road, 5F, Building C4, Shanghai, China
ying.chuai@accenture.com

[2] Beijing Institute of Technology, 5 South Zhongguancun Street, Haidian District, Beijing, China

Abstract. Under the SDG goals put forward by the United Nations, all things, races, universities, governments, institutions and organizations on the planet are working together for sustainable development. The blue economy, which goes beyond the Globalized and the Green Economy. This economic philosophy was first introduced in 1994 by Prof. Gunter Pauli when asked by the United Nations to reflect on the business models of the future in preparation for COP3 in Japan where the Kyoto Protocol was decided. Starting from economy, ecology and culture, it is increasingly clear that it is possible to generate more revenue, while generating more jobs and still compete on the global market. But in this way of sustainable design, we need a very professional economics, ecological background, and long-term research and observation, for designers, it is difficult to quickly capture an effective solution, so benefiting from data system is an inevitable necessity for company to direct the sustainability business operation.

This article introduces new ideas about sustainability system design with data and elaborates on how to collect data, which can be collected from computer modeling or real-world measurements; analyzes all inputs and outputs in the system; calculates carbon emissions and other impact; and presents all the data in a visualized way, so designers can choose the most appropriate solution.

Keywords: Sustainability · Systemic design · Data visualization · Supply chain

1 Introduction

1.1 Systems Thinking

"Systems Thinking is a context for seeing wholes. It is a framework for seeing interrelationships rather than things, for seeing patterns of change rather than static snapshots" [1].

Looking through a systems lens can be overwhelming. As a system theorist Donella Meadows said System Thinking needs to see all dynamic changes and potential changes brought about by it [2]. The major characteristics of this systemic thinking are as shown in Fig. 1.

Therefore, systematic thinking can help us see the problem in a macroscopic view. In the current complex environment, it is easier to find the root cause of the problem than linear thinking.

S. Trzcielinski et al. (Eds.): AHFE 2021, LNNS 274, pp. 215–222, 2021.
https://doi.org/10.1007/978-3-030-80462-6_27

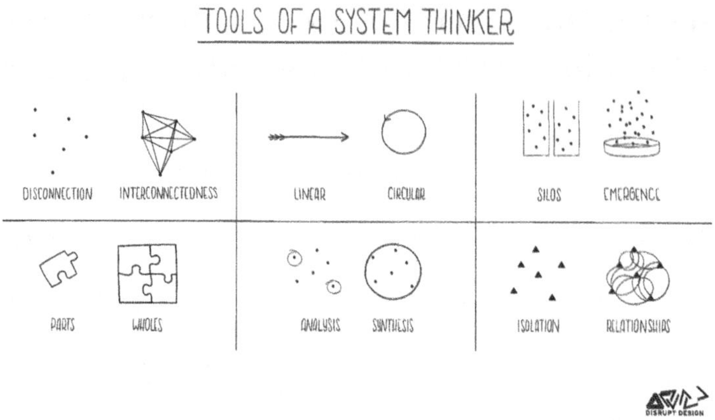

Fig. 1. Tools for systems thinkers: the 6 fundamental concepts of systems thinking (resource: Disrupt Design) [3].

1.2 Systemic Design

Systemic Design integrates systems thinking and human-centered design, with the intention of helping designers cope with complex design projects (also called Wicked Problems [4]).

Traditional design methods are inadequate to face the recent global challenges stemming from increased complexity as globalization, migration, and sustainability.

Systemic Designers need improved tools and methods to design responsibly while avoiding uninterested consequences/side-effects.

2 Systemic Design Methods and Tools

2.1 Systemic Design Methods

Since Karl Ludwig von Bertalanffy (1968) put forward the "general system theory" [5], system design has gradually been applied to artificial systems and management organizations. Treating productive organizations as complex adaptive systems allows a new management model to emerge in economical, social and environmental benefits (Pisek and Wilson, 2001 [6]). With the further improvement of society to the environment, systemic design has gradually produced theories such as industrial ecology (Frosh and Galopoulos, 1989 [7]) with multiple design dimensions. Therefore, systemic design requires strong interdisciplinary and cross-disciplinary design in the design stage. Disciplinary competence (Fuller, 1981 [8]). The systemic design initiative is addressing this problem by seeking new connections and relations between systems thinking and designerly ways of working. This is also the systemic design process and method that is more acceptable to designers.

2.2 Systemic Design Tools

One of the well-known tools is the system design toolkit (as shown in Fig. 2). Built by Namah in collaboration with shiftN, MaRS and SDA, the Systemic Design Toolkit is a methodology and a library of tools based on academic research and human-center design expertise [9].

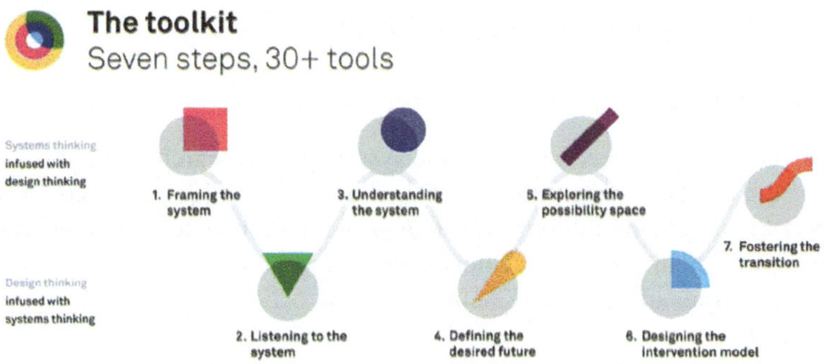

Fig. 2. The structure of the Systemic Design Toolkit (resource: Massimo Curatella. https://curatella.com/systemic-design-toolkit-virtual-design-thinking-barcamp).

It's based on the principle that Systems Change should be co-designed and co-created within the system and with the actor of the system, preferably, with the stakeholders in the same room. And provides tools to foster dialogue between the parts without requiring participants to master its inner working and principles.

The Systemic Design Toolkit is composed by seven steps: Framing the system (Systems Thinking); Listening to the system (Design Thinking); Understanding the system (Systems Thinking); Defining the desired future (Design Thinking); Exploring the possibility space (Systems Thinking); Designing the intervention model (Design Thinking); Fostering the transition.

2.3 Challenges

Among them, the first, third, fifth, and seventh steps require systems thinking. However, for designers and even corporate decision makers, the current systemic design has several difficulties that are difficult to be widely used:

- System thinking is very abstract and the cost of learning is high;
- Current system design theories and tools are biased towards manual drawing and analysis, which is less efficient;
- A research is only for one complex system, it is difficult to examine the relationship between multiple systems from a higher dimension, and it is difficult to produce a multi-system cooperation scheme;

Therefore, this paper proposes a systemic design visualization model. Using data sharing and visualization methods. Firstly, quantify the key factors in each system, digitize abstract thinking, and lower the threshold of use; Secondly, data can be automatically analyzed by the platform to efficiently generate valuable information, and by adjusting the data to achieve rapid comparison; Third, through the sharing of part of the system data by enterprises, a matching network can be built together to achieve a more local, low-cost, and recyclable economic cooperation forum.

3 Systemic Design Visualization Model

3.1 System Unit

In order to visualize the core content of the system, this research first clarifies the core elements of the system. System theorist Donella Meadows [2] proposed "A system is not just any old collection of things, a system is an interconnected set of elements that is coherently organized in a way that achieves something", and he concluded a system must consist of three kinds of things: elements, interconnections, and a function or purpose (as shown in Fig. 3).

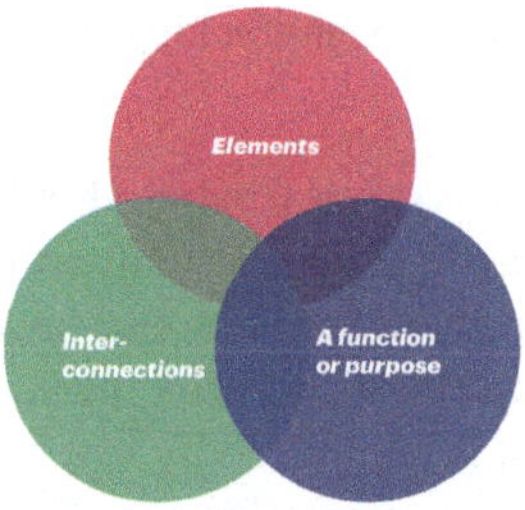

Fig. 3. Three kinds of things in a system.

Therefore, based on this, this paper establishes a systemic design visualization model. From the three necessary components of the system, the system unit in the system is deduced. Through the system unit, the data input dimension is clarified, and the visualization chart is displayed by obtaining the corresponding data (as shown in Fig. 4).

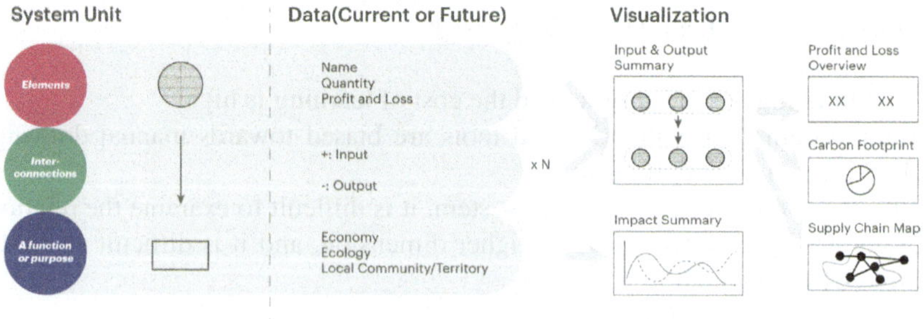

Fig. 4. Systemic design visualization model.

3.2 Data

Elements. In order to be able to calculate the scale of the elements and the corresponding benefits and costs, each element needs to have a clear name, quantity, and profit or loss per unit quantity.

Interconnections. The interconnections developed within the system generate the open system itself, and in the Systemic Design: the outputs (wastes) of a system become the inputs (resources) of another one [10]. Therefore, in order to be able to perform subsequent input and output calculations and supply chain network analysis, interconnections need to be marked as "+" (input) and "–" (output).

A function or Purpose. Since in the systemic design, the design integrates the three perspectives of economy, humanities, and engineering, this model defines the function or purpose of each system unit as whether there are three aspects of influence, namely economy, ecology, and local community/territory is shown in Fig. 5 [11].

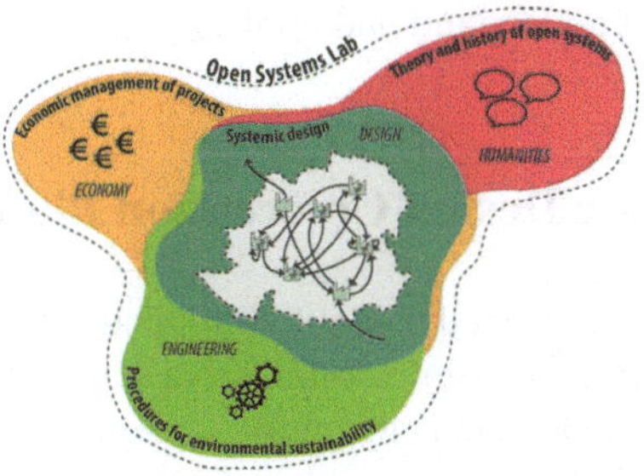

Fig. 5. Relationships between the four courses of the lab [11].

3.3 Visualization

Through data collection and corresponding calculation formulas, the following data visualization charts can be generated:

Input and Output Summary. Through the calculation of all the elements, quantities and interconnections (input/output) in the system unit, the corresponding total chart can be obtained. Through this chart, the designer can clearly distinguish the total input and the total situation of certain companies from the system.

Impact Summary. By judging the purpose or function of each transition in the system unit, the impact of each type of system design in the full life cycle of the production process can be restored, and the designer can adjust the plan according to the specific alternative.

Profit and Loss Overview. In order to monitor the economic impact of the system solution on the enterprise in real time, through all the elements, quantity and direction (input/output) in the system unit plus the calculation of its profit and loss, the most simplified profit and loss overview can be obtained. The designer checks this table in real time to ensure the profitability of the company.

Carbon Footprint. In order to ensure that the design plan can have a sustainable impact on the environment, the carbon footprint of the plan can be viewed in real time by inputting and outputting summary data, plus the local Carbon emission calculation formula.

Supply Chain Diagram. By inputting and outputting summary data, combined with each company's own information and local location, the system can automatically calculate supply chain costs and environmental impact, and finally recommend a suitable supply chain network map.

Therefore, this model can convert steps 1, 3, and 5 of the 7-step process in the above-mentioned System Design Toolkit into a data visualization method to analyze the current system, and directly explore possibilities by adjusting the data to achieve systemic design is easier to understand and apply.

4 Enterprise Systemic Design Co-creation Platform

This platform could have transformed the first, third, and fifth steps of the 7-step process in the aforementioned Systemic Design Toolkit into a data visualization method, making the system design easier to understand and apply. Therefore, the 7-step process for using the platform is:

Step1: Framing the system (Systems Data). Complete the platform registration and enter the basic information of the company, including name, headquarters, number of employees, fixed assets, etc.
Step2: Listening to the system (Design Thinking). Designers or decision makers need to disassemble the system to minimize the production steps (i.e. system unit) according to the industrial process of the enterprise, and put the elements involved in each step according to name, quantity, profit and loss, whether it is input or output, the target or function classification is input into the platform.
Step3: Understanding the system (Systems Thinking). The platform will analyze according to the corresponding data and calculation relationship. The designer can place the auxiliary understanding system according to the following icons. Take a Chinese animal husbandry company as an example is shown in Fig. 6.
Step4: Defining the desired future (Design Thinking). Based on the above-mentioned visualization plan, alternatives can be adopted to make stakeholders consistent in value. By imagining how we can improve the future environment of individuals, organizations and society, together we design an ideal future.

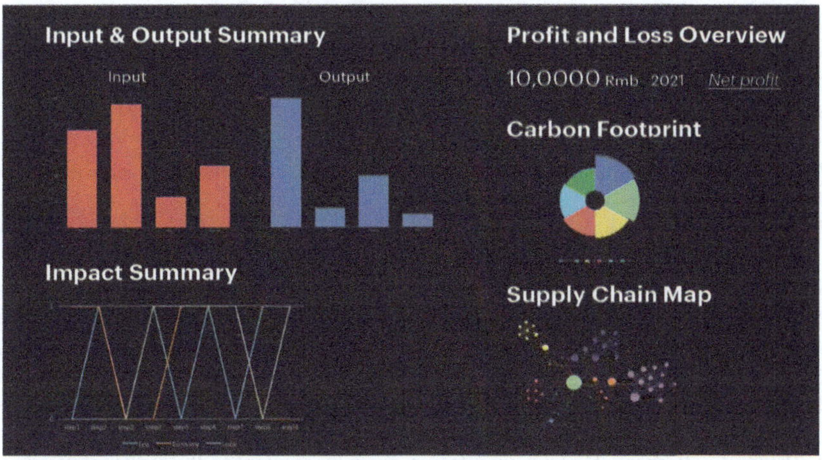

Fig. 6. Systemic data visualization of a Chinese animal husbandry company.

Step5: Exploring the possibility space (Systems Thinking). In order to make the whole process reasonable, designers need to explore various types of possible interventions to ensure that they cover the conditions that occurred in the initial research activities. Through brainstorming, designers can try to adjust the quantity, type, manpower or transportation cost of each raw material on the platform, and see the corresponding visual chart changes in real time.
Step6: Designing the intervention model (Design Thinking). Based on the above exploration, clearly improve the innovative plan, and find a suitable supply chain partner according to the supply chain map.
Step7: Fostering the transition. Plan the transition towards the desired goal by moving from the Minimum Viable Product (maybe the Minimum Viable Solution in this case) to the full implementation of the intervention model.

5 Discussion and Conclusions

In response to the current complex environment and social progress, this article introduces the use of data visualization to improve system design methods and promote new ideas for system design to be widely used. It explains in detail the visualization model of the system design, including how to divide the data dimensions, how to use the data to analyze the input and output of the system, calculate carbon emissions, and visualize better supply chain methods. However, due to the lack of practice in real companies, this theory needs further exploration and iteration in application. The design of the data dimension is only a preliminary stage, and it needs to be verified and supplemented by practice.

References

1. Senge, P.M.: The Fifth Discipline: The Art & Practice of a Learning Organisation. Random House, London (1990)

2. Meadows, D.H.: Thinking in systems: a primer. Earthscan (2010)
3. Acaroglu, L.: Tools for systems thinkers: the 6 fundamental concepts of Systems Thinking. Medium, Disruptive Design (2017)
4. Buchanan, R.: Wicked problems in design thinking. Des. Issues **8**(2), 5–21 (1992)
5. Von Bertalanffy, L.: General System Theory. George Braziller Inc., New York (1968)
6. Plsek, P.E., Wilson, T.: Complexity, leadership, and management in healthcare organisations. BMJ **323**(7315), 746–749 (2001)
7. Frosch, R.A., Gallopoulos, N.E.: Strategies for manufacturing. Sci. Am. **261**(3), 144–153 (1989)
8. Kuromiya, K., Fuller, R.B.: Critical Path, p. 331 (1981)
9. Curatella, M.: Systemic design toolkit by Kristel van Ael of Namahn | Massimo Curatella. Massimo Curatella (2020). https://curatella.com/systemic-design-toolkit-virtual-design-thinking-barcamp
10. Barbero, S., Agnese, P.: Systemic design and policy making: the case of the Retrace project. FormAkademisk-forskningstidsskrift for design og designdidaktikk **11**(4) (2018)
11. Battistoni, C., Barbero, S.: Systemic design, from the content to the structure of education: new educational model. Des. J. **20**(sup1), S1336–S1354 (2017)

An Assessment of Supply Chain Resilience in Catastrophic Events: A Case of South Africa

Ifije Ohiomah[1(✉)], Nita Sukdeo[1], and Clinton Aigbavboa[2]

[1] Department of Quality and Operations Management, University of Johannesburg, Johannesburg, South Africa
nsukdeo@uj.ac.za
[2] CIDB Centre of Excellence, University of Johannesburg, Johannesburg, South Africa
caigbavboa@uj.ac.za

Abstract. The objectives of this study are to explore the effects of disruption events on the manufacturing supply chain and investigate the strategies to minimise the impact of disruption events. A survey of professionals involved in supply chain and have dealt with disruptions in the supply chain was carried out with the aid of a well-structured questionnaire. A mean item score was carried out and the findings from the analysis of the effect of catastrophic events to manufacturing supply chain revealed death and injuries. Furthermore, Creation of risk management that include top leadership and strong integration/teamwork, using information technology are the main strategies to minimise the impacts of catastrophic events on manufacturing supply chain. The study recommended that organisations in the manufacturing industry can use the strategies to minimise the impact of future catastrophic events on Manufacturing supply chain for creating resilient manufacturing industry.

Keywords: Supply chain · Resilience · Supply chain resilience · Manufacturing · South Africa

1 Introduction

In today's rapidly changing environment, supply chains are more easily exposed to risks due to factors such as increased globalisation, higher customer expectations, environment volatility, and the occurrence of internal and external risk events [1–3] (Chen, Sohal, and Prajogo 2013; Ivanov, Sokolov, and Dolgui 2014; Aqlan and Lam 2015). Scholars and practitioners emphasize that managing risk and disruption in the supply chain is a crucial capability for firms, in order to compete in the today's increasingly turbulent and unpredictable marketplace [4–7] (Melnyk et al. 2009; Colicchia and Strozzi 2012; Nooraie and Parast 2016; Yang, Pan, and Ballot 2017). This has led to the concept of supply chain risk management (SCRM), SCRM is the management of supply chain's risk through coordination or collaborations among the supply chain partners to ensure continuity of the supply chain. This was emphasized as the supply chain executives in IBM believe that SCRM is the second most important issue for the company (Butner, 2010) [8].

S. Trzcielinski et al. (Eds.): AHFE 2021, LNNS 274, pp. 223–230, 2021.
https://doi.org/10.1007/978-3-030-80462-6_28

Supply chain risks can be classified into disruption and operational risk (Chen, Sohal, and Prajogo 2013) [1]. Disruption risks can be defined as events unplanned that restricts a supply chain system, this may be resultant from man-made or natural events such as economic downturns, technology changes, hurricanes, terrorist attacks (Parast and Shekarian 2019) [9]. Operational risks are more about the events such as supply-demand coordination, which often is as a result of inadequate processes, control, people or systems (Nooraie and Parast 2016 [9]; Yang, Pan, and Ballot 2017 [7]). Thus, operational risks are relatively more controllable than disruption risks. Moreover, catastrophic events such as 9/11 in 2001, the tsunami in 2004, hurricane Katrina in 2005, hurricane Harvey in Houston in 2017, and hurricane Maria in Puerto Rico in 2017 are motivation for supply chain researchers to consider supply chain risk and its effects on the design of supply chains [10, 11].

Supply chain resilience (SCRES) reduces the impact of disruptions by identifying strategies that allow a supply chain to react to a disruption while recovering to its original functional state or better. SCRES has received more attention in recent years as one of the main characteristics of a firm to improve responsiveness to unexpected dynamics in the business environment [12]. Melnyk et al [13] asserted that SCRES is at the heart of current supply chain management thinking. Based on Tukamuhabwa et al. (2015) [14], more than 80% of companies were concerned about their SCRES. Hence, SCRES is of a great interest for firms and organisations [15, 16].

The study aims to address the following objectives, the effects of catastrophic events to manufacturing supply chain, the strategies to minimize the impacts of catastrophic events on manufacturing supply chain.

2 Literature Review

Organisations worldwide are increasingly vulnerable to high impact/low probable events [17]. Assertions by Stecke and Kumar (2009) [18] showed that there has been a marked increase in the frequency and losses economically from natural and man-made catastrophes. Despite all this effects, Vanny et al 2009 [19] highlighted that there has been less attention to the catastrophic events in the SCM literature. Catastrophic events have been identified in various ways under supply chain risks.

Catastrophic events can be defined as events beyond the ordinary and are characterized by being of low probability but high consequence [20]. Brindley (2004) [21] categorised potential supply chain risk based on probability and severity perspectives. Mitroff and Alpaslan (2003) [22] classified catastrophes into seven categories such as economic crises (recessions, hostile takeovers), physical crises (industrial accidents, product failures), personnel crises (strikes, exodus of key employees, workplace violence or vandalism), criminal crises (product tampering, act of terrorism), information crises (theft of proprietary information, tampering with company records), reputation crises (logo tampering, rumour mongering) and natural disasters (floods, fires). Wagener and Bode (2008) [23] recognised natural hazards, socio-political instability, civil unrest, economic disruptions and terrorist attacks as catastrophic events. Stecke and Kumar (2009) [18] broadly classified catastrophes into two main parts: man-made and natural catastrophes.

2.1 Impact of Catastrophic Events on Supply Chain

From time to time, frequent disruptions as well as rare catastrophes disrupt supply chain operations and every organisations supply chain is susceptible to a diverse set of risks. Research has shown evidence to prove that failure to manage supply chain risks effectively may lead to a significant negative impact on organisations. Furthermore, discovery by Stecke and Kumar (2009) [18] showed that business losses constitute a major percentage of the total losses caused by the catastrophes. Failure to manage risks effectively, has led to wider consequences as identified by Cousins et al 2004 [24], these consequences are not limited to financial losses but also product quality reduction, damage to property and equipment, suppliers and the wider public and delays in delivery days. loss of reputation in the eyes of customers. Knemeyer et al. [25] stated that if, for example, a manufacturing plant is lost due to a catastrophic event, the consequences affect supply chain operations, financial flows and possibly information flows too.

Manufacturing industry is also highly susceptible to catastrophic events. The impacts of catastrophic events on manufacturing industry include manufacturing supply chain breakdowns, information and communication breakdowns, significant damages to property and infrastructure, increased demand for reconstruction, injuries and deaths. The impacts lead to increase cost and time, reduce quality and devote more management time for crisis handing. Catastrophic events may have wider implications such as extensive delay or plant failure, significant business failure, loss of money and loss of reputation for all concerned [25]. Extreme weather events such as floods, hurricanes and storms have significant effects on the manufacturing supply chain, and according to Wedawatta et al. 2010 [26], supply chain disruptions due to extreme weather events can create a substantial impact on a manufacturing company. Therefore, potential consequences of catastrophic events on manufacturing supply chain are wide ranging and long-lasting.Developing strategies to mitigate disruptions has grown to become a necessity, this is so because the supply chain system has grown complex [27]. This has been revealed as there is more research is focused on the on the vulnerability off the supply chain and we need for organization to assets a more systemic systematic analysis off their supply chain for ability. This one vulnerability has led organizations to develop plans to protect against recurrent low impact risk in the supply chain although there are many who have ignored the high impact of low likelihood risk. A requirement of action plan was sanctioned by Norrman and Jansson 2004 [28] in their study as it required to ensure continued operations in case of a catastrophic events in the supply chain.

Stecke and Kumar (2009) [18] introduced the following four strategies to make supply chain components robust. Proactive strategies—Help a company to avoid or decrease the impact of possible types of future disruptions. Advanced warning strategies—Gain benefits from advance information (forecast) of a catastrophe. Coping strategies—Flexibility and redundancy in various supply chain components to mitigate catastrophe. Survival strategies—Aid companies to reduce losses and duration of disruptions.

Stecke and Kumar 2009 [18] stated that proactive strategies can help a company to avoid or decrease the likelihood of certain forms of disruptions. Researchers have further emphasised that a well developed and implemented proactive strategies can lower the need of mitigating strategies. Furthermore, developing strategies that can help in forecasting a catastrophe under the advanced warning strategies was put forward by stecke and Kumar. The researchers highlighted that these advanced warning strategies can provide valuable preparation time to align its capabilities to minimize disruption effects or may avoid complete prevention of a disruption. Flexibility and redundancy in various supply chain components help in defining coping strategies, which help to mitigate disasters [5]. A severe catastrophe and/or lack of proactive and coping strategies may result in the breakdown of the supply chain, which can leave an organisation out of operations. Survival strategies can be used by companies in such situations [5]. Survival strategies can be implemented in two stages: immediate response to a catastrophe (i.e. save life and property) and steps taken to recover (i.e. reorganise resources to restart supply chain operations).

Tang (2006) [29] came up with nine different robust supply chain strategies that aim to improve a firm's ability to achieve supply and/or demand better under normal circumstance and to improve a business's capability to sustain its operation when a major disruption happens. The nine strategies include. postponement, where a strategic stock was developed, by employing a flexible supplier base, make and buy, offer economic supply incentives, flexible transportation, revenue management via dynamic pricing, assortment planning and silent product rollover. Tang (2006) [29] was of the opinion that although robust supply chain strategies enable establishments to utilize the corresponding contingency plans when disruption happen, these businesses would become less susceptible if they could reduce their risk exposure. Hence, several possible ways to reduce the impact of disruption on the supply chain operations such as proactively form strategic alliances with other suppliers in different countries; reduction of the lead time by redesigning the supply chain network and establish a recovery planning system to gain visibility of inventories, sales and shipments.

3 Methodology

To adequately achieve the objective of this study which is to investigate the effects of disruption events on the manufacturing supply chain, identify the strategies to minimise the impact of disruption events. A quantitative research method was adopted. This method is used by researchers to "provide solutions to predetermined research questions and objectives, by analysing the relationship between the different variables through the use of an experimental or non-experimental data collection instrument (survey), in turn yielding results which are numerical, thus can be analysed using statistics" (Creswell, 2014) [30]. The study focused on the three main municipalities that made up Gauteng province of SA which are "City of Ekurhuleni Metropolitan Municipality (EKU), City of Johannesburg Metropolitan Municipality (COJ) and City of Tshwane Metropolitan Municipality (COT) all in the Gauteng province of South Africa. Respondents were chosen from professionals who have knowledge about supply chain resilience, risks. The total number of the target population could not be ascertained as the study requires respondents to have a minimum of 5 years experience in the manufacturing industry coupled with adequate knowledge of risk and resilience of supply chain. This, therefore, led to the adoption of snowball sampling technique to reach out to the respondents, and a total of 51 were identified. To carry out the snowball sampling technique, a primary network of professionals currently in the supply chain domain was identified first and were sent the questionnaires electronically.

They were further obliged to distribute the questionnaire to colleagues who meet up with the study's target population criteria.

A five point Likert scale was used to gauge the opinions of the respondents. Test of reliability and validity was carried out on the retrieved data using Cronbach's alpha, and the result showed 0.978 value for effects of catastrophic events to construction supply chain and 0.981 for strategies to minimise the impacts of catastrophic events on manufacturing supply chain. According to Tavakol and Dennick (2011) [31], with the Cronbach's alpha value ranging between 0.70 and 0.95, the statistical tool is reliable, and values close to 0.95 represents a high level of inter-dependence between the different concepts being measured or investigated in the study. Retrieved data was analysed using Mean Item Score (MIS), which is the main analysis carried out to rank the variables based on respondents' agreement to individual variables.

4 Findings and Discussion

Effect of Catastrophic Events to Manufacturing Supply Chain
Respondents believe that death and injury is a major impact of catastrophic events on the supply chain with a mean of 4.33 respectively. While breakdown of information and communication among the supplier and organisation was ranked the least by respondents with a mean of 3.84. The findings from the survey proved that failure to manage supply chain risks effectively may lead to a significant negative impact on organisations. The findings from the study also is consistent with findings from stecke and Kumar [18] who stated that effects of catastrophic events leads to business losses. Furthermore, findings from cousins et al. [24] that financial losses but also reduction in product quality, damage to property and equipment, loss of reputation in the eyes of customers, suppliers and the wider public and delays in delivery days are the effects of catastrophic events on manufacturing supply chain, which is consistent with our findings (Table 1).

Strategies to Minimise the Impacts of Catastrophic Events on Manufacturing Supply Chain
Respondents believe creation of risk management that include top leadership and strong integration/teamwork is one of the primary strategies to minimise the impacts of catastrophic events on manufacturing supply chain with a mean of 4.39. This finding is consistent with the findings of [28] whereby they were of the opinion where they stated that an adequate risk management framework must be developed. Furthermore, the constructs measured in this study were consistent with variables and possible strategies from Tang 2006 [29] and Stecke and Kumar 2009 [18] (Table 2).

Table 1. Effect of catastrophic events to manufacturing supply chain.

Effects	Mean	Ranking
Deaths	4.33	1
Injuries	4.33	2
Important damage to infrastructure and Property	4.29	3
Machinery harm	4.22	4
Loss of revenue	4.18	5
Manufacturing stoppages	4.16	6
Decreased production quality	4.14	7
Failure of business	4.12	8
Delay in transport	4.10	9
Production delays	4.10	10
Logistics breakdowns	4.08	11
Loss of organisation reputation	4.06	12
Closure of plants	4.00	13
Reconstruction cost	3.96	14
Senior management time devoted to crisis management	3.84	15
Information and communication breakdown among suppliers and organisations	3.84	16

Table 2. Strategies to minimise the impacts of catastrophic events on manufacturing supply chain.

Effect	Mean	Ranking
Creation of risk management that include top leadership and strong integration/teamwork	4.39	1
Using information technology	4.29	
Contingency scheduling	4.27	
Supplier selection	4.25	
Ensuring agility in the supply chain	4.24	
Increasing flexibility in the supplier base	4.22	
Building logistics capabilities	4.20	
Increasing velocity	4.18	
Inventory management	4.16	
Redundancy creation	4.10	
Increased visibility	4.14	

5 Conclusions and Recommendations

Catastrophic events are unique among supply chain risks due to low probability of occurrence, difficulty in prediction and severity of impact. Vulnerability of MSC for various types of catastrophic events has been substantiated in the literature. The study further identified that the impact of such catastrophes is highly diverse and has different effects on MSC performance. According to the findings, the most significant impact of catastrophic events is a business failure and least significant impact is loss of focus to work. The research finally established the need to implement strategies not only after catastrophe occurs but also before the catastrophe in order to avoid severe consequences on construction organisations. The catastrophic event risk minimisation strategy developed in this study would be useful for manufacturing organisations to identify suitable strategic actions according to the risk level that they faced. The organisations in the manufacturing industry can use the strategies to minimise the impact of future catastrophic events on MSC for creating resilient manufacturing industry.

References

1. Chen, J., Sohal, A.S., Prajogo, D.I.: Supply chain operational risk mitigation: a collaborative approach. Int. J. Prod. Res. **51**(7), 2186–2199 (2013)
2. Ivanov, D., Sokolov, B., Dolgui, A.: The Ripple effect in supply chains: trade-off 'efficiency-flexibility-resilience'in disruption management. Int. J. Prod. Res. **52**(7), 2154–2172 (2014)
3. Aqlan, F., Lam, S.S.: Supply chain risk modelling and mitigation. Int. J. Prod. Res. **53**(18), 5640–5656 (2015)
4. Melnyk, S.A., et al.: Mapping the future of supply chain management: a Delphi study. Int. J. Prod. Res. **47**(16), 4629–4653 (2009)
5. Colicchia, C., Strozzi, F.: Supply chain risk management: a new methodology for a systematic literature review. Supply Chain Manage. Int. J. (2012)
6. Nooraie, S.V., Parast, M.M.: Mitigating supply chain disruptions through the assessment of trade-offs among risks, costs and investments in capabilities. Int. J. Prod. Econ. **171**, 8–21 (2016)
7. Yang, Y., Pan, S., Ballot, E.: Mitigating supply chain disruptions through interconnected logistics services in the Physical Internet. Int. J. Prod. Res. **55**(14), 3970–3983 (2017)
8. Butner, K.: The smarter supply chain of the future. Strategy & Leadership (2010)
9. Parast, M.M., Shekarian, M.: The impact of supply chain disruptions on organizational performance: a literature review. Revisit. Supply Chain Risk 367–389 (2019)
10. Ho, W., Zheng, T., Yildiz, H., Ta, S.: Supply chain risk management: a literature review. Int. J. Prod. Res. **53**(16), 5031–5069 (2015)
11. Kamalahmadi, M., Mellat-Parast, M.: Developing a resilient supply chain through supplier flexibility and reliability assessment. Int. J. Prod. Res. **54**(1), 302–321 (2016)
12. Yilmaz Borekci, D., Rofcanin, Y., Gürbüz, H.: Organisational resilience and relational dynamics in triadic networks: a multiple case analysis. Int. J. Prod. Res. **53**(22), 6839–6867 (2015)
13. Melnyk, S.A., et al.: Understanding supply chain resilience. Supply Chain Manage. Rev. **18**(1), 34–41 (2014)
14. Tukamuhabwa, B.R., et al.: Supply chain resilience: definition, review and theoretical foundations for further study. Int. J. Prod. Res. **53**(18), 5592–5623 (2015)

15. Scholten, K., Schilder, S.: The role of collaboration in supply chain resilience. Supply Chain Manage.: Int. J. (2015)
16. Ivanov, D.: Revealing interfaces of supply chain resilience and sustainability: a simulation study. Int. J. Prod. Res. **56**(10), 3507–3523 (2018)
17. Sheffi, Y., Rice, J.B., Jr.: A supply chain view of the resilient enterprise. MIT Sloan Manage. Rev. **47**(1), 40–48 (2005)
18. Stecke, K.E., Kumar, S.: Sources of supply chain disruptions, factors that breed vulnerability, and mitigating strategies. J. Mark. Channels **16**(3), 193–226 (2009)
19. Vanany, I., Zailani, S., Pujawan, N.: Supply chain risk management: literature review and future research. Int. J. Inf. Syst. Supply Chain Manage. **2**(1), 16–33 (2009)
20. Gilbertson, A., Kappia, J., Bosher, L., Gibb, A.: Preventing Catastrophic Events in Construction. HSE, London (2011)
21. Brindley, C.: Supply Chain Risk. Gower Publishing, Ltd. (2004)
22. Mitroff, I.I., Alpaslan, M.C.: Preparing for evil. Harvard Bus. Rev. **81**(4), 109–115 (2003)
23. Wagner, S.M., Bode, C.: An empirical examination of supply chain performance along several dimensions of risk. J. Bus. Logist. **29**(1), 307–325 (2008)
24. Causins, P.D., Lamming, R.C., Bowen, F.: The role of risk in environment-related supplier initiatives. Int. J. Oper. Prod. Manage. **24**(6), 554–565 (2004)
25. Knemeyer, M.A., Zinn, W., Eroglu, S.: Proactive planning for catastrophic events in supply chains. J. Oper. Manage. **27**(2), 141–153 (2009)
26. Wedawatta, G., Ingirige, B., Amaratunga, D.: Developing a conceptual framework for investigating the resilience of construction SMEs and their supply chains against extreme weather events. In: RICS Construction and Building Research Conference, 2–3 September 2010. RICS, Paris, London (2010)
27. Marley, K.A.: Mitigating supply chain disruptions: Essays on lean management, interactive complexity and tight coupling [thesis]. The Ohio State University (2006)
28. Norrman, A., Jansson, U.: Ericsson's proactive supply chain risk management approach after a serious sub-supplier accident. Int. J. Phys. Distrib. Logist. Manage. **34**(5), 434–456 (2004)
29. Tang, C.: Robust strategies for mitigating supply chain disruptions. Int. J. Logist.: Res. Appl. **9**(1), 33–45 (2006)
30. Creswell, J.W., Creswell, J.D.: Research Design: Qualitative, Quantitative, and Mixed Methods Approaches. Sage Publications (2017)
31. Tavakol, M., Dennick, R.: Making sense of Cronbach's alpha. Int. J. Med. Educ. **2**, 53 (2011)

Influence of Effective and New Information Flow on Logistics Management

Ayodeji Dennis Adeitan[1(✉)], Clinton Aigbavboa[2], Adedamola Donald Adeitan[3], and Adebayo Oluwasegun Adewole[4]

[1] SARChl in Sustainable Construction Management and Leadership in the Built Environment, Postgraduate School of Operations Management, University of Johannesburg, Johannesburg 2028, South Africa

[2] SARChl in Sustainable Construction Management and Leadership in the Built Environment, Faculty of Engineering and the Built Environment, University of Johannesburg, Johannesburg 2028, South Africa

[3] Department of Agric Economic and Extension, University of Ibadan, Ibadan, Nigeria

[4] Agricultural Finance and Rural Credit Division, Agricultural Development and Management (ADEM) Department, Agricultural and Rural Management Training Institute (ARMTI), Ilorin, Nigeria

Abstract. The growth of any logistics firm and its ability to compete globally with developed nation's businesses depends on effective information flow. Information flows in the manufacturing sector is vital in enhancing both economic development and material flow in today's competitive economy. The manufacturers had to reexamine their method to ways of getting efficient, quality and up-to-date information within the enterprises and ways of dispatching the information. Sadly, most logistics firms in Nigeria are among companies who are slow to realize the benefit of new information flow on modern logistics management. Therefore, this article is built on a review of literature on the influence of effective and new information flow on logistics management in Nigeria. This study elaborates the benefits of effective and competent information flow to Nigerian firms, types of logistics information, and influence of information flow to the management of logistics firms.

Keywords: Information flow · Logistics · Management · Logistics information

1 Introduction

The growth of any logistics firm and its ability to compete internationally with developed nations businesses depends on effective information flow. According to [6] the usefulness and influence of new information flow in the logistics chain have been identified together with material flow. Even though the aim of every logistics firm is to make a profit through quality services and cost-effective means, they still suffer from inefficiency and insecurity due to an omission of information flow processes in their daily activities. As logistics chain becomes bigger and larger, it will become more difficult to keep up with

S. Trzcielinski et al. (Eds.): AHFE 2021, LNNS 274, pp. 231–236, 2021.
https://doi.org/10.1007/978-3-030-80462-6_29

efficient and effectiveness of information flow. Therefore, there is a need to improve on the implementation of logistics in reducing waste such as inventory and prime time from suppliers to end consumers [8]. The assessment of information flow within a logistics chain can be referred to as the accomplished profits which surpass the charges convoluted [4]. This cost mentioned includes; communication and administration costs, costs by suppliers for providing new information and information systems investment. These costs can be compressed significantly, as reported by [24], new effective information flow in logistics chain can help improve logistics disruptions, optimize inventory levels, achieve higher efficiency, deliver products to customer on time and analytics on business partner performance.

Sadly, some logistics firms in Nigeria are among firms who are slow to realize the benefit of new information flow on modern logistics management. They still apply the manual information processing (paper-based inventory management system) and as a result, they experienced prolonged cycle time and lead time in delivering products to customers. These practices have managed to reduce the profitability, customer satisfaction and increased overhead cost of this logistics firms due to inefficient logistics operations or management. Therefore, this article is built on a review of literature on the influence of effective or new information flow on logistics management in Nigeria.

2 The Concept of Logistics Management

The development of logistics management function started during the 1990s in a bid to control ever-rising costs and to address the problem of logistics being overlooked by more developed functions like purchasing, marketing, and production [1]. Logistics in contrast to supply chain management is the arrangement that its important in moving and positioning inventory throughout a supply chain. Integrated logistics serves to link and accompany the overall supply chain as a continuous process and is vital for a desired result in supply chain connectivity [2]. The Council of Logistics Management (1998) defines Logistics management as that element of the supply chain system that plans, implements, and tests the efficient, effective flow and storage of goods, services, and new information flow from the initial point to the consumption stage in order to meet customers' specifications. This implies that Logistics management deals with the short-term aspect of a supply chain where products are moved into a facility, properly stored, handled and moved out.

The transportation management, distribution management, management of 3PLs are identified as 'logistics activities' and their management are vital to the performance of the supply chain and the organizations as well [20]. Thus, logistics management is involved with the effective management of activities all along the supply chain. Logistics management target is to get the best out of the firms and to have an outlining perspective that seeks to create a single plan for the flow of goods and information through a business [4]. Furthermore, if logistics is mentioned as a combine concept that seeks to develop a system-wide view of the firm, it is still primarily a planning concept and the goals of logistics management are to create a single plan mindset within the business. The combination of physical supply and physical distribution came in handy to the high cost of logistics and management's desire to minimize these costs while also elaborating on customer service [15].

3 Information Flow In Logistics Management

According to [23], Information can be defined as a cluster of facts that is structured in such a way that they possess extra worth beyond the worth of the certitudes themselves. It comprises of raw facts which comprise, employee's information, number of hours, list of inventory and order of sales. The info flows in the manufacturing sector is vital in enhancing both economic development and material flow in today's competitive economy. The manufacturers had to reexamine their method to ways of getting efficient, quality and up-to-date information within the enterprises and ways of dispatching the information [18]. [21] and [17] mentioned that the availability of the information resource such as hardware and software only is not enough. This implies the organizations ought to participate in information flow activities because if they operate alone, information flow and sharing will only be limited to a single organization. Influence of new information flow is the way dissimilar structures of network can be linked to make the organization among logistics chain associates closer. The importance of this coordination includes; rise in effective Information flows, reduction in uncertainties and the clients will collect higher goods quality with a minimal cost in a short period [7]. Customers' needs can be met through the planning and management of the information flows and resources from the source, via the operating environs of the company and to the suppliers [4]. [10] mentioned that the flow of material cannot be set-apart from flow of information; there are association between the physical flow and the info that flows between the sources. In present times, the information flow within the logistics chain has become essential since it permits logistics firms to perform in real time with precise and efficient data. Firms today classify flow of information as an asset since it is unfeasible to have well-organized and accurate flow of material without it [12].

This implies that the information flow ensures the logistics chain's quality, from the place of departure at the supplier's right through to the final consumer. The information flow must not be directed to suppliers alone both also on the customers' ability to source and determine what they need or requires in a shorter ordering time. New effective information flow enables firms to make better planning decision in their operations such as; improved utilization of resources, supply and logistics chain costs minimization and better responsiveness to customer's needs [9, 13].

3.1 Types of Logistics Information Flow

[25] expresses in a research that there are various dissimilar sorts of effective info that can be apportioned within a chain of logistics, such as; Inventory Info, Sales Dataset, Forecasting of Sales, order info, Information of Product Ability, and Exploitation Info of New Merchandises. Two types of logistics information have also been established by [15], they are strategic and operational info. Operational information involves quantitative information about daily sales on inventory levels and Its importance includes; demand sequence times reduction, reduction of inventory levels, and to upgrade customer service. Tactical information involves qualitative issues about the company's business strategies and Its importance includes; enhanced cooperation among associates of logistics chain and future forecast on operations of logistics based on scheme variations. As per [2] who further characterizes logistics flow of information in two major procedures

namely; Planning and Operations. Planning/management is used to facilitate capacity limitations, logistics needs, inventory preparation, manufacturing needs, procurement demand, and forecasting while Operations is used to facilitate order processing, order project, distribution operations, and inventory management.

3.2 Benefits of Logistics Information Flow

Numerous studies have been carried out by many authors to ascertain the benefits of information flows in logistics companies' management. Some of these sources include; [11, 17], and [24]. These authors have emphasized on the information flow benefit within the supply/logistics management chain, it is done by identifying distinct parties within the firm's network and finding a way to improve the business contacts. The benefits of information flow include; 1) increase in safety of stock, 2) inventory reduction, 3) decrease in cost, 4) reduction of uncertainties, 5) enhanced resource utilization, 6) Better tracing and tracking, 7) expanded network, 8) negotiate better contracts, 9) find less expensive transportation modes, 10) efficient and precise reporting 11) better understanding of the various types of cost to serve customers.

3.3 The Implication of Information Flow in Nigerian Organizations

In a review by [19], the introduction of new information flow to logistics management can help in knowledgeable resolutions which can save a firm up to 40% on expenditures of logistics, so one of the most excellent logistics management applications is to make improvements on logistics information flow. Therefore, the implementation of logistics information flow will help to bring ease in firm's decision-making process and increase in timely execution of contracts. [14] mentioned that usage of intensive information resource, logistics firms can gain access to the active flow of information, less transaction costs, and develop a capacity for information gathering. According to [3] presently, logistics firms in Nigeria operate in a very competitive market and the only way of surviving in these business environments is to incorporate information flow into their firms' operations. For example, Nigerian wine industry dealers face various challenges in logistics for an assortment of the wine stocks imported from South Africa, European countries and the rest of the world. Some of these logistics challenges include; distribution limitations, limited storage facilities, and the problem of inventory management. This means that competitive advantage growth is mainly required in the dispersal costs acquired and efficiency in the procedure of transporting cargo to final consumers from distributors. Therefore, the competitive advantages of the suppliers in the Nigerian wine industry can be boosted using logistics management information flow in their daily activities that will help to bring ease in consignments ordering process and increase in timely execution of distributions. The incorporation of effective information to logistics management will bring about innovation, expertise, timely execution of contracts, and cost-effectiveness. However, the only way these expectations can be met is to have operational and strategic information flow that can convert the strengths and opportunities of their organizations into what could boost their competitive advantage, thereby giving a continual survival and relevance in the industry. The urge for competitive advantage in today's global market has brought the whole supply chain environs in a continual developing market, so the functions of logistics must be supple. To determine the influence of effective flow of information on logistics management, analysis of the existing logistics processes must be performed [19]. Therefore, the following questions are to be used for the analysis; 1) Is there a plan that defines when an item should be cataloged and when an item should be sent directly to a consumer? 2) Would your firm prefer to have a third-party logistics firm manage some or all aspects of your logistics functions? What cost effective and customers service applications must you take into detail before making this decision? 3) What part of your delivery network should be improved? 4) Could a change of mode of transportation save cost? 5) Do you channel too much inventory? Too little? 6) What are your definite customer service goals? Smooth item returns? The speed of delivery? Safety of the consignment? 7) What form of future business operations will affect logistics management functions?

4 Conclusion

This study provides an insight on the influence of new information flow in logistic management, how it has increased the growth of developed nation's logistic chain. It was discovered that many logistics firms in Nigeria are slow to realize the benefit of new information flow on modern logistics management. As a result, they experienced a prolonged cycle time and lead time in delivering products to consumers. Therefore, this study has highlighted the benefits of effective and competent information flow to Nigerian firms, types of logistics information, and influence of information flow to Nigerian logistic firms.

References

1. Ballou, R.H.: The evolution and future of logistics of logistics and supply chain management. Eur. Bus. Rev. **19**(4), 332–348 (2007)
2. Bowersox, D., Gloss, D., Cooper, M.: Supply Chain Logistics Management, International McGraw-Hil, New York (2002)
3. Business Monitor International. 'African Markets' (2017). www.businessmonitor.com. Accessed 15 Dec 2017
4. Christopher, M.: Logistics and Supply Chain Management. Financial Times. 2 nd edn. Prentice Hall (1998)
5. Christopher, M.: Logistics and Supply Chain Management: Strategies for Reducing Cost and Improving Service. Prentice-Hill, London (1998)
6. Closs, D.J., Goldsby, T.J., Clinton, S.R.: Information technology influences on world class logistics capability. Int. J. Phys. Distrib. Logist. Manage. **27**(1), 4–17 (1997)
7. Fiala, P.: Information sharing in supply chains. Omega **33**, 419–423 (2005)
8. Lai, K.H., Ngai, E.W.T., Cheng, T.C.E.: Adoption of information technologies in Hong Kong's logistics industry. Transp. J. **44**(4), 1–9 (2006)
9. Lee, H.L., So, K.C., Tang, C.S.: The value of information sharing in a two-level supply chain. Manage. Sci. **46**(5), 626–643 (2000)
10. Lumsden, K.: Logistikens Grunder, Studentlitteratur, Foundations of logistics, Lund (1998)

11. Marshall, C.C., Bly, S.: Sharing encountered information: digital libraries get a social life. In: The 4th ACM/IEEE-CS joint Conference on Digital Libraries, pp. 218-227. IEEE, Tuscon (2004)
12. Mattsson, S.-A.: Logistik i försörjningskedjor, Studentlitteratur, Lund (2002)
13. Mentzer, J.T.: Fundamentals of Supply Chain Management: Twelve Drivers for Competitive Advantage. Sage Publications Inc., Thousand Oaks (2004)
14. Minton, S.: Nordic nations still top information index. The world paper (2003). http://www.worldpaper.com/2003/oct02/isi1.htm. Accessed 13 Aug 2004
15. Moberg, C.R., Whipple, T.W., Cutler, B.D., Speh, T.W.: Do the management components of supply chain management affect logistics performance? Int. J. Logist. Manage. **15**(2), 15–30 (2004)
16. Moberg, C., Cutier, B., Gross, A., Speh, T.: Identifying antecedents of information exchange within supply chains. Int. J. Phy. Dis. Log. Manag. **32**(9), 321–328 (2002)
17. Mourtzis, D.: Internet based collaboration in the manufacturing supply chain. CIRP J. Manuf. Sci. Technol. **4**, 296–304 (2011)
18. Nunes, M.B., Annansingh, F., Eaglestone, B., Wakefield, R.: Knowledge management issues in knowledge-intensive SMEs. J. Doc. **62**, 101–119 (2006)
19. PLS Logistics Services. Logistics Management Best Practices (2015). www.plslogistics.com
20. Rahman, S.: An exploratory study of outsourcing 3PL services: an Australian perspective. Benchmark. Int. J. **18**(3), 342–358 (2011)
21. Rosen, B., Furst, S., Blackburn, R.: Overcoming barriers to knowledge sharing in virtual teams. Organi. Dyn. **36**, 259–273 (2007)
22. Rutner, S.M., Langley, C.J., Jr.: Logistics value: definition, process and measurement. Int. J. Logist. Manage. **11**(2), 73–82 (2000)
23. Stair, R., Reynolds, G.: Principles of Information Systems: A. Managerial Approach, Course Technology. Memphis continental Traffic, Sweeden (2001)
24. Yang, T.M., Maxwell, T.A.: Information-sharing in public organizations: a literature review of interpersonal, intra-organizational and inter- organizational success factors. Govern. Inf. Q. **28**, 164–175 (2011)
25. Zahra, L., Muriati, M., Shahnorbanun, S.: Information sharing in supply chain management. In: The 4th International Conference on Electrical Engineering and Informatics. Procedia Technology, vol. 11, pp. 298–304 (2013)

Structural Equation Modelling of Resource Commitment Constructs as a Predictor of Implementation of Reverse Supply Chain Management

Ohiomah Ifije[1(✉)], Clinton Aigbavboa[2], and Nita Sukdeo[1]

[1] Department of Quality and Operations Management, University of Johannesburg, Johannesburg, South Africa
nsukdeo@uj.ac.za

[2] Centre of Excellence, University of Johannesburg, Johannesburg, South Africa
caigbavboa@uj.ac.za

Abstract. This paper reports the use of Structural Equation Modelling technique to test the influence of resource commitment as a predictor of reverse supply chain management implementation. The study was conducted among professionals involved in supply chain and reverse supply chain using a survey method for data collection. At total of 314 questionnaires were filled completely and retuned back. The data gathered was analysed using structural equation modelling, which was used to assess the factorial structure of the constructs, the structural equation modelling software used was the Amos 26.0. the factorial structure, reliability and validity of resource commitment indicator variables were investigated. The finding revealed a positive influence on the outcome of the implementation of RSC. Further SEM analysis revealed that the Rho and the Cronbach's alpha coefficients of internal consistency were over 0.70 criterions for acceptability, and the constructs shows a good mode fit to the sample data.

Keywords: Manufacturing · Resource commitment · South Africa · Reverse supply chain

1 Introduction

In today's business world, reverse supply chain is growing with immense importance to becoming a competitive necessity. Returns in recent times have become more liberal, an increase in the use consignment inventory, shorter product lifecycles, more demanding customers has translated to more returned products [1]. As a result of more returned products, organisations are seeking ways to find more efficient was to reclaim, redistribute and/or dispose of returns. In this respect, end of life take back laws have been enacted in past periods in the European Union (EU) and United States of America [2], examples of such takeback laws are Waste and Electrical and Electronic Equipment (WEEE) and End-of-Life Vehicles (ELV) [3].

S. Trzcielinski et al. (Eds.): AHFE 2021, LNNS 274, pp. 237–244, 2021.
https://doi.org/10.1007/978-3-030-80462-6_30

These directives have led to the buzzwords of reverse logistics, reverse supply chain spoken about in recent times. Reverse supply chain is the management of a series of activities vital to recover a product from a customer to dispose of the product or recover value of the product [4]. In the same length, reverse logistics as defined by Rogers and Tibben-Lembke (1999) [5] as the process of planning, implementing and controlling the efficient, cost-effective flow of raw materials, in-process inventory, finished goods and related information from the point of consumption to the point of origin for the purpose of recapturing or creating value or for proper disposal.

For this study, the research focuses on the role resource commitment plays in the implementation of reverse supply chain for manufacturing companies.

2 Resource Commitment and Reverse Supply Chain

In order to attain an effective RSC, it is important to incorporate the commitment of resources, whether it is managerial (such as day to day decision making) or financial, such as an investment in training personnel on the new technology, and information at levels on what should be done [6]. An organisations resources include all assets, capabilities, Organisational processes, firm attributes amongst others controlled by a firm that enable it to conceive of and implement strategies that improve its efficiency and effectiveness [7]. In order for organisations to implement reverse supply chain, organisations have to allocate their limited resources to enable the implementation of an effective reverse supply chain [8]. It has been suggested by Sweeny and Szwejczewski (1996) [9], that as managers commit more financial human and physical resources to a program, superior financial performance may be realized [10].

Regarding RSD specifically, the commitment of RSC resources has positive impacts on the achievement of RSC program goals, including environmental regulatory compliance, reduced inventory investment, improved profitability, and increased economic performance [11]. Recent studies have suggested that resource commitment makes RSC programs efficient and more effective [12].

Resource commitment to reverse supply chain should be a priority because of the possibility for enhancing the implementation of reverse supply chain. In addition, it offers a strategic way of developing lasting connections with customers [13]. However, allocation of insufficient financial and personnel resources has been identified as one of the principle obstacles to development of a good RSC program [14].

3 Methodology

A quantitative methodology was selected for this study, as it based on the positivist approach which describes the connecting variables that establish realism [15]. As well considered, in order to carry out a quantitative method, it is dependent on collection of data and analysed by statistical analysis and mathematical formula. The statistical analysis selected was the structural equation modelling software, Amos 26.

Data was collected using an electronic method. A total population of 512 firms was sampled, from which 302. The Survey instrument used on this study was designed to obtain the perceptions of respondents to determine the influence of resource commitment on the implementation of reverse supply chan. The instrument consisted of a demographic's sections. In addition, the construct of resource commitment. Individuals items were valued on a 5-point Likert scale.

Factor analysis and internal consistency tests were conducted on the constructs to determine reliability and validity. Confirmatory factor analysis (CFA) was selected, as it provides a more rigorous testing of theory as compared to exploratory factor analysis (EFA). Furthermore, CFA ensures that the researcher is able to evaluate a priori relationships whereas EFA does not require priori specifications for relationships theorised priori to model testing. CFA allows the indicators to load on certain pre-selected factors, while EFA allows the data and statistical techniques to determine the measurement of the model [15].

Internal consistency of the questionnaire was tested for the Cronbach alpha, with an acceptable cut off criteria of 0.70 and 1.0 is deemed better. But Hair et al (1998) [15] stated that the Cronbach alpha accepted when performing an EFA is 0.6. For this study, the Cronbach alpha is 0.965, for the following constructs to measure RC as shown in Table 1 and for Table 2 the Cronbach alpha is 0.975.

Absolute Fit Indices: These indices determine how good a priori fits thc data samplc [16] and further shows the model which has a superior fit. These include the foremost fundamental sign of how well the anticipated theory fits the data. Calculation of SEM does not rely on comparison with a baseline model but instead measure of how well in comparison the models fit.

This calculation does not rely on comparison with a baseline model but is instead a measure of how well the model fits in comparison to no model at all [17]. Included in this study to be measured are the chi-squared test, RMSEA, GFI, the RMR, SRMR and the reliability.

Root Mean Square Error of Approximation (RSMEA)

RSMEA reports how well the model, with unknown but optimally chosen parameter estimates, would fit the populations covariance matrix [18]. Up to the early nineties the cut-off criteria of RSMEA ranged between 0.5 to 0.10 with values above 0.10 considered a poor fit [18]. It was formerly assumed that a RMSEA of between 0.08 to 0.10 provides a mediocre fit and values below 0.08 shows a good fit [19]. Recently, a cut-off value close to .06 [20] or a strict upper limit of 0.07 [21] is considered a good-fit and it has been the general consensus amongst authorities in this area.

Root Mean Square Residual (RMR) and Standardised Root Mean Square Residual (SRMR)

The RMR and SRMR are the square root of the difference amongst the residuals of the sample covariance and the hypothesised covariance model. The range of RMR is grounded on the type of Likert scale used which could hold varying levels ranging from 1–5 while others range from 1–7. This often makes the interpretation of RMR difficult [23]. Hence SRMR resolves this difficulty and its interpretation is clear. The values of SRMR range from 0 to 1 with models' well-fitting obtaining values less than 0.05 [19, 24]. However, values as high as 0.08 are considered acceptable [20]. For a perfect, the value of SRMR is 0 but it is seen that SRMR will be low when there is a high number of parameters and affected by large sample size.

Normed-Fit Index (NFI)
These indices assess the model by relating the $\chi 2$ value of the model to the $\chi 2$ of the null model. The null/independence model scenario is the worst case as it puts all measured variables as uncorrelated. The statistical range varies between 0 to 1 with the recommendation from Bentler and Bonnet (1980) [26] putting the values to be greater than 0.90. However, in recent times, the range has been put at a cut-off criterion be $\geq$ .95 [20] One of the limitations of this index is its sensitivity to size of sample, thus miscalculating sample fit for sample size fewer than 200 [27, 28]. This led the NFI not to be depended on solely (Kline, 2005). This limitation was corrected by the non-normed fit index, also identified as the Tucker-Lewis index, an index with a preference for simple models. In situations where the sample sizes used are small, the results of the NNFI can show poor fit in spite of all the added indices pointing to a good fit [23, 28, 29]. One more limitation of the NNFI is that owing to its non-normed nature, values can go above 1.0, making it hard to interpret [30]. Values of 0.08 and less have been recommended as preferred; however, Bentler and Hu suggested greater or equal to 0.95 as threshold.

Comparative Fit Index (CFI)
The CFI is a revised form of NFI as this takes account of the sample size [30]. It also performs even when the sample size is small [29]. [28] introduced it and later included it as one of the fit indices in the SEM programme [23]. Just like the NFI, the statistics assumes that the latent variables are uncorrelated and brings into comparison the sample covariance matrix as well as the model. The cut-off criteria range from 0 to 1 with values closer to 1 showing a good fit. However, recent research has shown that a CFI value greater or equal to 0.95 is indicative of a good fit.

Goodness of Fit Statistic (GFI) and the Adjusted Goodness-of-fit Statistic (AGFI)
The goodness of fit was created as an alternative to the chi-square test. GFI computes the measure of variance that is taken into by the population variance [29]. GFI takes account of the covariance accounted for by the model. It also proves how carefully the model comes to repeating the observed covariance matrix [24]. The cut-off criteria are between 0 to 1. Related to the GFI is the AGFI which adjusts the GFI based upon degrees of freedom, with more saturated models reducing fit [29]. Thus, more parsimonious models are preferred while complicated models are penalised. In addition to this, AGFI tends to increase with sample size. As with the GFI, values for the AGFI also range between 0 and 1 and it is accepted that values of 0.90 or greater show well-fitting models. Given the often-detrimental effect of sample size on these two fit indices, they are not relied upon as a standalone index. However, given their historical importance they are often reported in covariance structure analyses.

4 Analysis and Discussion

Structural equation modelling was used alongside CFA to analyse and evaluate the model proposed and hypothesis. SEM is a multivariate statistical technique that defines and estimates the relationships between the endogenous and exogenous variables simultaneously [35]. Firstly, before a CFA is carried out, there will be a descriptive analysis carried out to determine the variables the respondents considered important, as shown below in the Table 1. In addition, the KMO and Bartlett test of sphericity where within acceptable range of 0 to 1 and for this study it is 0.948 [29]. The Bartlett test for sphericity was significant at 0.00. For the test for communalities, the variables should be greater than 0.5 to enable it to be good enough to carry out a CFA (Pallant, 2013).

Table 1. Descriptive analysis for resource commitment

Constructs	Mean	Std Dev	Communalities
Having personnel to run the RSC	3.98	1.073	0.722
Having resources to train personnel regularly	3.98	1.107	0.841
Having resources to store and remanufacture most of the returned goods	3.94	1.089	0.807
Having resources to manage third party suppliers	3.93	1.101	0.825
Having resources to ensure proper selection of suppliers	3.91	1.091	0.827
Having resources for necessary information technology	3.87	1.086	0.747
Having necessary financial resources	3.72	1.082	0.793
Kaiser-Meyer-Olkin Measure of Sampling Adequacy		0.948	
Bartlett's Test of Sphericity	Approx.Chi-square	2512	
	Df	6	
	Sig.	0.000	

Table 1 reveals the convergent validity and internal consistency of the resource commitment construct. The threshold of average variance extracted (AVE) is above 0.5 as recommended by Fornell and Larcker [37] and Hair et al. [39]. The composite reliability (CR), threshold as recommended by Litwin [36] is 0.5, but Fornell and Larcker [37] and Wong [38] recommended 0.7. The AVE finding of this study is above 0.5, meeting the cut-off criteria. The CR for this construct is 0.96 because they indicate that all the indicator variables are measuring the same phenomenon [39].

The X^2 divided by the degree of freedom (Df) revealed a good fit of 3 to 5 as noted by Hu and Bentler [20], while the goodness-of-fit (GFI) must be from 0 to 1 as observed by Doloi et al. [40]. Similarly, the CFI met the minimum threshold of at least 0.80 set by Hu and Bentler [20] and a cut-off of 1 as set by Hu and Bentler [20]. The normed fit index (NFI) can fall within the 0.6 to 1.0 threshold observed in Doloi et al. [40]. The chi-square was 4.6 which makes it a good fit. The GFI, CFI, NFI, RFI and TLI all met the cut criteria of >0.90 and >0.95 respectively. The RSMEA gave a fit 0.08, which makes it acceptable and the SRMR is 0.05. This means that all the variables of resource commitment are fit for acceptance into the final structural equation model.

5 Conclusion, Limitations and Recommendation

This study focused on the role resource commitment play in the implementation of reverse supply chain. The study focused specifically on the impact from a reverse supply chain standpoint of view, of resource commitment and implementation of reverse supply chain. The findings from this study is consistent with the theory of resource-based view which suggests that much of the differences in performance or among organisations results from resource heterogeneity (Christmann 2000). However, a firm leverage on the committed resources is important (Cho, Kim and Rhee 1998). Resources of an organisation include all assets, capabilities, organisational processes, attributes of the firm amongst others which are controlled by the organisation to allow it to comprehend and implement strategies such as RSC that improve its efficiency and effectiveness (Hall 1992). Firms have to allocate resources which are limited resources between two key processes of creating value and appropriating value which includes extracting profits in the marketplace (Mizik and Jacobson 2003). Isobe et al (2000) stated that a high resource commitment has positive impact in overall customer satisfaction. The commitment of resources, specifically to RSC has positive impacts on the achievement of RSC, as seen in the findings including greater profitability (Daugherty et al. 2002).

The limitation of the study is that data was collected from only manufacturing organisations in Gauteng, this may limit the findings. Also, the methodology used is also a limitation. As well, the sample size could also be limitation, as a bigger sample size could lead to a better result.

Recommendations, the geographical region must be spread, which simply means other metros, provinces in South Africa must be explored. More variables could be added to measure the construct of resource commitment and the output of the implementation of RSC. Also, a mixed method is also recommended to get the best of results.

References

1. Meade, L., Sarkis, J.: A conceptual model for selecting and evaluating third-party reverse logistics providers. Supply Chain Manage.: Int. J. (2002)
2. Prahinski, C., Kocabasoglu, C.: Empirical research opportunities in reverse supply chains. Omega **34**(6), 519–532 (2006)
3. Directive, E.C.: Directive 2012/19/EU of the European Parliament and of the Council of 4 July 2012 on waste electrical and electronic equipment, WEEE. Off. J. Eur. Union L **197**, 38 – 71 (2004)
4. Defee, C.C., Randall, W.S., Gibson, B.J.: Roles and capabilities of the retail supply chain organization. J. Transp. Manage. **21**(2), 5 (2009)
5. Rogers, D.S., Tibben-Lembke, R.S.: Going backwards: reverse logistics trends and practices, vol. 2. Reverse Logistics Executive Council, Pittsburgh (1999)
6. Richey, R.G., Genchev, S.E., Daugherty, P.J.: The role of resource commitment and innovation in reverse logistics performance. Int. J. Phys. Distrib. Logist. Manage. (2005)

7. Daft, R.L.: Organization theory and design, Minnesota. West Publishing Company. Delery, JE, and Doty, HD (1996), Modes of theorizing in strategic human resource management: types of universalistic, contingency, and configurational performance predictions. Acad. Manage. J. **39**(4), 802–835 (1983)
8. Mizik, N., Jacobson, R.: Trading off between value creation and value appropriation: the financial implications of shifts in strategic emphasis. J. Mark. **67**(1), 63–76 (2003)
9. Sweeney, M.T., Szwejczewski, M.: Manufacturing strategy and performance. Int. J. Oper. Prod. Manage. (1996)
10. Isobe, T., Makino, S., Montgomery, D.B.: Resource commitment, entry timing, and market performance of foreign direct investments in emerging economies: the case of Japanese international joint ventures in China. Acad. Manage. J. **43**(3), 468–484 (2000)
11. Daugherty, P.J., Myers, M.B., Richey, R.G.: Information support for reverse logistics: the influence of relationship commitment. J. Bus. Logist. **23**(1), 85–106 (2002)
12. Tan, A.W.K., Yu, W.S., Arun, K.: Improving the performance of a computer company in supporting its reverse logistics operations in the Asia-Pacific region. Int. J. Phys. Distrib. Logist. Manage. (2003)
13. Sheet, W.E.R.C.: In through the out door. WERC Sheet **26**, 1–3 (2003)
14. Quattrone, P.: Constructivism and accounting research: towards a trans-disciplinary perspective. Acc. Audit. Acc. J. (2000)
15. Hair, J., et al.: Multivariate Data Analysis. vol. 5, no. 3. Prentice Hall, Upper Saddle River (1998)
16. Jöreskog, K.G., Sörbom, D.: LISREL 8: Structural equation modeling with the SIMPLIS command language. Sci. Softw. Int. (1993)
17. McDonald, R.P., Ho, M.-H.R.: Principles and practice in reporting structural equation analyses. Psychol. Methods **7**(1), 64 (2002)
18. Byrne, B.M.: Structural Equation Modeling with LISREL, PRELIS, and SIMPLIS: Basic Concepts, Applications, and Programming (1998)
19. MacCallum, R., Austin, J.: Applications of structural equation modeling in psychological research. Ann. Rev. Psychol. **51**(1), 201–226 (2000)
20. Hu, L., Bentler, P.M.: Cutoff criteria for fit indexes in covariance structure analysis: Conventional criteria versus new alternatives. Struct. Eqn. Model.: Multidisc. J. **6**(1), 1–55 (1999)
21. Steiger, J.H.: Understanding the limitations of global fit assessment in structural equation modeling. Pers. Ind. Differ. **42**(5), 893–898 (2007)
22. McQuitty, S.: Statistical power and structural equation models in business research. J. Bus. Res. **57**(2), 175–183 (2004)
23. Kline, R.B.: Principles and Practice of Structural Equation Modeling. Guilford Publications (2015)
24. Diamantopoulos, A., Siguaw, J.A., Siguaw, J.A.: Introducing LISREL: A Guide for the Uninitiated. Sage (2000)
25. Miles, J., Shevlin, M.: A time and a place for incremental fit indices. Pers. Ind. Differ. **42**(5), 869–874 (2007)
26. Bentler, P., Bonett, D.G.: Significance tests and goodness of fit in the analysis of covariance structures. Psychol. Bull. **88**(3), 588 (1980)
27. Mulaik, S.A., et al.: Evaluation of goodness-of-fit indices for structural equation models. Psychol. Bull. **105**(3), 430 (1989)
28. Bentler, P.M.: Comparative fit indexes in structural models. Psychol. Bull. **107**(2), 238 (1990)
29. Tabachnick, B.G., Fidell, L.S., Ullman, J.B.: Using Multivariate Statistics, vol. 5. Pearson, Boston (2007)
30. Hair, J.F., et al.: Multivariate Data Analysis. Uppersaddle River (2006)

31. Bollen, K.A., Long, J.S.: Tests for structural equation models: introduction. Sociol. Methods Res. **21**(2), 123–131 (1992)
32. Holmes-Smith, P.: Structural equation modeling (using AMOS): From the fundamentals to advanced topics. Sreams, Melbourne (2011)
33. MacCallum, R.C., Austin, J.T.: Applications of structural equation modeling in psychological research. Ann. Rev. Psychol. **51**(1), 201–226 (2000)
34. Rutter, M.: Proceeding from observed correlation to causal inference: the use of natural experiments. Perspect. Psychol. Sci. **2**(4), 377–395 (2007)
35. Bollen, K.A., Curran, P.J.: Wiley series in probability and mathematical statistics. Structural Equations with Latent Variables, Random Variables (1989)
36. Fink, A., Litwin, M. S.: How to Measure Survey Reliability and Validity, vol. 7. Sage (1995)
37. Fornell, C., Larcker, D.F.: Evaluating structural equation models with unobservable variables and measurement error. J. Mark. Res. **18**(1), 39–50 (1981)
38. Wong, K.K.-K.: Partial least squares structural equation modeling (PLS-SEM) techniques using SmartPLS. Mark. Bull. **24**(1), 1–32 (2013)
39. Hair, J.F., et al.: When to use and how to report the results of PLS-SEM. Eur. Bus. Rev. (2019)
40. Doloi, H., Iyer, K.C., Sawhney, A.: Structural equation model for assessing impacts of contractor's performance on project success. Int. J. Proj. Manage. **29**(6), 687–695 (2011)
41. Christmann, P., Taylor, G.: Globalization and the environment: Determinants of firm self-regulation in China. J. Int. Bus. Stud. **32**(3), 439–458 (2001)
42. Chou, Y.C., Huang, H.Y., Kuo, C.H.: A framework of economic analysis for tapered technology-manufacturing alliances. Int. J. Prod. Econ. **127**(2), 238–248 (2010)
43. Hall, J.: Environmental supply chain innovation. In: Sarkis, J. (ed.) Greening the Supply Chain, pp. 233–249. Springer, London (2006). https://doi.org/10.1007/1-84628-299-3_13

Method of Implementation of Air Chambers Using Recycled Rubber Based on Positioning for the Optimization of Fragmentation in Open-Pit Blasting

Cristhian Chapoñan-Chinchay[1], Jhonatan Diaz-Cueto[1], Vidal Aramburu-Rojas[1], Yaneth Vasquez-Olivera[1], and Carlos Raymundo[2](✉)

[1] Ingenieria de Gestion Minera, Universidad Peruana de Ciencias Aplicadas, Prolongación Primavera 2390, Lima 15023, Peru
{pcgmyvas,u201522876,u201516739}@upc.edu.pe

[2] Direccion de Investigacion, Universidad Peruana de Ciencias Aplicadas, Prolongación Primavera 2390, Lima 15023, Peru
carlos.raymundo@upc.edu.pe

Abstract. This article sought to carry out an experimental study for the optimization of fragmentation through the implementation of a rubber plug to apply the air chambers technique; For this, based on past research, the best position of the accessory and the length of the air chamber that provide the best results in the size of the fragment were also identified. The tests were carried out in an open pit mine, with proven geological and geomechanical characteristics and calculated in geomechanical laboratory tests by compression and tensile tests. There were 3 proposals for the position of the air chamber accessory with the determination of the corresponding length. After the blasting process, the comparative granulometry measurements of the proposal were made versus the granulometry obtained without the application of the accessory, for which the use of an image analysis software was proposed to determine a P80 of fragments passing through said mesh.

Keywords: Drill and blast · Air chamber · Fragmentation · Rubber plug · Open pit

1 Introduction

The most important purpose of the blasting is to reduce the rock mass to smaller particles, called fragments, allowing their practical loading and subsequent transport with the use of machinery to the concentration plant or waste pile. Blasting, thanks to the detonating power of explosives, produces shock waves and, through a reaction, releases gases at high pressure and temperature in a substantially instantaneous way, to start, fracture or remove a quantity of material according to the design parameters of the blasting itself [1].

S. Trzcielinski et al. (Eds.): AHFE 2021, LNNS 274, pp. 245–253, 2021.
https://doi.org/10.1007/978-3-030-80462-6_31

The fragmentation of rocks by blasting plays an essential role in hard rock mining since it is considered the most important effect of blasting, which will depend on certain factors such as: quantity of explosive, accessories, techniques, etc.; the efficiency of these will influence that of subsequent systems such as loading, transportation, crushing and grinding in mining operations [2]. The fragmentation distribution in most of the blasts that are carried out does not reach the optimum size required; even obtaining oversized fragments that affect the following loading, transportation and crushing processes. The distribution of the sizes of the blown material reveals the efficiency of the parameters that were considered and used in the development of the blast to obtain the optimal sizes necessary [3].

2 State of Art

2.1 Implementation of Plugs in the Air Chambers

Plugs or plugs as air chamber generating agents can improve the utilization of explosive energy to a considerable and considerable extent. Research confirms this statement by the fact that an obvious improvement has been achieved in the use of the energy released as a result of the blasting, with the application of the air chambers. The generation of air spaces requires an important accessory that allows the formation of this empty space, which is why the plug that is used or also called Stemming Plug Augmenting Resistance to Stemming in Holes or by its acronym SPARSH, is next to the high-strength polypropylene caps one of the available devices that can be used effectively as an accessory to generate the void space or air chamber [3, 4].

2.2 Air Chambers in the Blasting of Mineralized Banks of an Open Pit

Air chambers are emerging as an alternative in open-cast manganese mines. The air chamber in the fragmentation of the rock mass of a manganese open-pit mine was investigated, revealing that air chamber blasting improves the degree of fragmentation and produces a more uniform fragmentation compared to conventional blasting. The aforementioned regarding the benefits of applying a new blasting technique is reinforced by what has been said by those who expose the study of a new technique; in parallel compared between conventional blasting, the air chamber and plastic tube [7, 8]. A comparative is also presented with the evaluation of rock fragmentation from these techniques. In addition, among the factors that complement the use of the technique, the non-restriction on the type of material used for the plug or the length that it may have are considered [9–11].

2.3 Positioning of the Air Chamber Within the Explosive Column

This technique was evaluated in a scale scenario, but recreated in tests with concrete for experimentation; since, the investigation would test the study to scale in a laboratory. The reason for the use of concrete is due to the property that allows to eliminate the effects of uncertainties and geological irregularities such as: fractures, folds, faults and joints in blasting results. The study phases took place in a base scenario (without technique), then the use of the air chamber was implemented and the chamber was positioned in a superior, middle and inferior way; finally, the appropriate length was sought by testing 4 lengths corresponding to 10%, 20% and 40% of the total length of the hole. The use of the air chamber gave a reduced fragmentation screen in the middle and lower position [12–14].

3 Input

3.1 General View

The present investigation presented a proposal in order to remedy or solve the problem of the fragmentation of the rock mass produced by the blasting of banks in surface mining (Fig. 1).

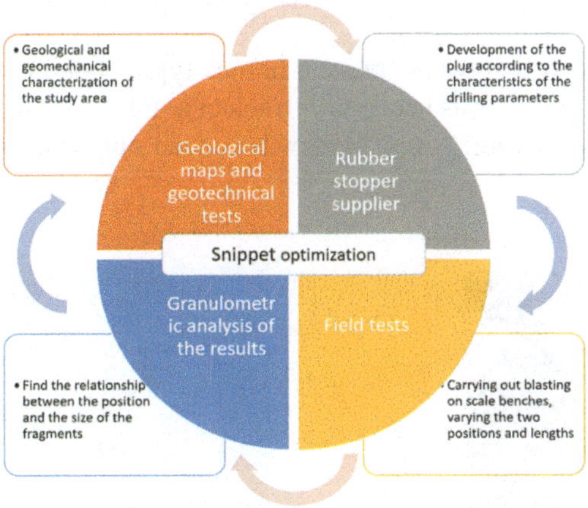

Fig. 1. Proposed model

Geological Maps. This resource allows us to previously know the scenario and all the studies that are had on it. The correct collection of bibliographic information and the validation in the field of the work previously carried out will allow the scenario to be accurately contextualized. Likewise, the geological formations and super units present in the study area influence the results of the research, it seeks to reduce the uncertainties that interfere with the veracity of the results.

Rubber Stopper. This variant in the material of the accessory that allows generating the air space inside the hole obeys similar behaviors to the already known conventional materials, with the benefit of its high elasticity and tensile strength, which qualifies it as a resistant material in the moment of blasting. Another of its characteristics is the impermeability and property of conforming to irregular spaces within the drill. This last property allows the air chamber to be further sealed.

Fragments. This research identified that there are several benefits that can be obtained by achieving the correct location of the bucket plug and its corresponding length in the bark, but it mainly focuses on reducing the size of the fragments in order to favor post-blasting processes. It can be inferred that the application of the air chambers technique reduces the size of the fragments compared to not applying it. However, factors such as the type of rock, the compressive strength must be taken into account.

4 Implementation

4.1 Baseline Scenario

The initial blasts were carried out at the Ccoriccocha mine in the mineralized zone, located in the Lurigancho district, Lima province and region; resulting in being located within the quadrants of Lurín-25j and Chosica-24j of the national geological chart of Peru, as can be seen, in effect 3 maps were used in order to observe the quadrants in greater detail.

These mineralized structures are present in rock types such as monzogranite, diorite and granodiorite. For this, it was necessary to select a reduced stage that would serve as a scale bench for the blasting. A test blast was carried out with the drilling of 8 holes (Figs. 2 and 3).

Fig. 2. Drilling the holes

Fig. 3. Drilled holes for test blasting

The parameters considered were applied in the same way as in the large open-pit drilling campaigns, in order to apply this same concept for all subsequent blasts and thus there are no variations in the blast variables.

The granulometry obtained served as a starting point for comparison with the granulometry of the fragments after the application of the rubber stopper. The results obtained after the image analysis are summarized in Fig. 4.

Finally, the P80 was calculated following an interpolation of the data obtained by the fragmentation. The concentration of the fragment size can be observed above 100 mm, with an established P80 of 7.30" (Fig. 5).

Implementation of the rubber stopper in the air chamber technique.

Size	Quantity	% Detained	% Detained Ac.	% Passant
214.68	7	4%	4%	96%
191.68	25	13%	17%	83%
178.3	14	7%	25%	75%
168.68	12	6%	31%	69%
145.58	35	19%	50%	50%
122.68	58	31%	81%	19%
99.68	19	10%	91%	9%
76.58	16	9%	99%	1%
72.42	1	1%	100%	0%

Fig. 4. Granulometric table of the base scenario

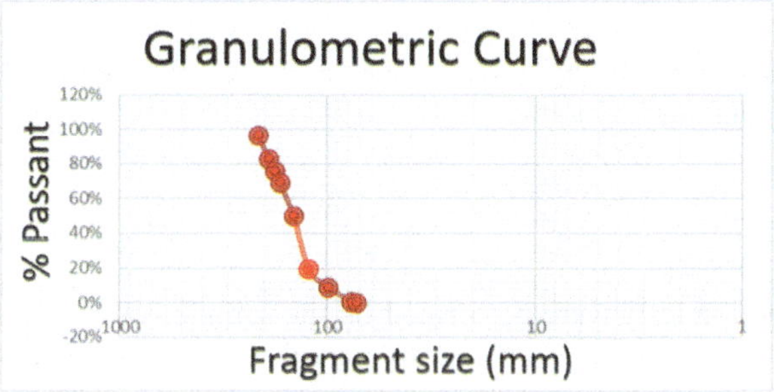

Fig. 5. Granulometric curve

The accessory or rubber plug was developed that will allow the creation of the air chamber or air deck in English, this has a concave circular design allowing the formation of the compression chamber in a blasting drill, its design allowed it to be easily installed with the help of an inert attacker (plastic or wood) without resorting to the need for additional efforts.

The plugs were requested according to the requirement with the appropriate dimensions of 1 ¼" in diameter and ¾" thick. The internal hollow structure that the circular cap has allowed the application with the etcher. Having carried out the necessary safety procedures to carry out the drilling and blasting process, the triangular mesh of 8 holes was marked and the drilling of these began.

The mesh of 8 holes was drilled for a calculated volume of 0.768 m3 considering a burden of 0.4 m., Spacing of 0.4 m. bench height 0.6 m. and 8 holes.

It was not necessary to place a mass of detritus on top of the plug. With the holes drilled as established, the rubber plug was placed considering a length of 0.1 m calculated from the bottom, said application was carried out with the support of an attacker to direct the plug and prevent it from pointing in another direction (Fig. 6).

The distance from the location of the plug is measured, this length corresponds to the length of the air chamber.

The plugs were placed in the considered production holes of the scale bench to perform the test. After having placed the plugs in the wells, the process continued with the loading of the explosive column; for this the UM. Ccoriccocha provided the

Fig. 6. Placing the plug in the holes

acquisition of explosives and accessories, which met the requirements indicated in the mining safety regulations.

5 Results

Uniaxial compression tests were performed on the samples taken from the field. From the laboratory tests carried out on the rocky massif to detail its characterization, it was obtained that the resistance to simple compression of the granodiorite present in the study area of the Ccoricocha mine is 114 MPa, a value very close to the minimum limit that it accepts the application of the technique of air chambers.

For the first blast with a length of 0.1 m in the lower position of the hole, the following results are obtained (Fig. 7).

Size (mm)	Quantity	% Detained	% Detained Ac.	% Passant
178.3	0	0%	0%	100%
165.1	1	1%	1%	99%
152.4	14	8%	9%	91%
139.7	14	8%	17%	83%
127	26	16%	33%	67%
114.3	45	27%	60%	40%
101.6	43	26%	86%	14%
88.9	16	10%	95%	5%
76.2	1	1%	96%	4%
63.5	5	3%	99%	1%
<63.5	2	1%	100%	

Fig. 7. Result table - Test 1

Likewise, the granulometric curve made graphs a curve with values close to 100 mm. of particle size (Fig. 8).

The P80 calculated for this blast delivered a value of 4.84 inches, evidencing a notable reduction in the size of the fragment compared to the base blast.

For the second blast, the position to the central part of the hole is varied. It is analyzed by digital images and the following results are obtained (Fig. 9).

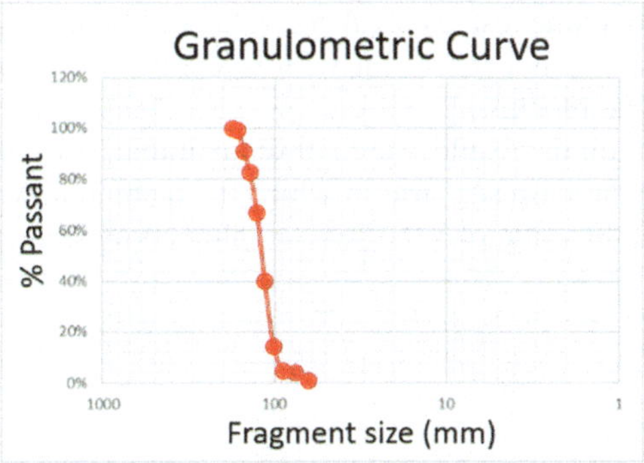

Fig. 8. Granulometric Curve 1

Tamaño (mm)	Cantidad	% Retenido	% Retenido Ac.	% Pasante
178.3	0	0%	0%	100%
165.1	1	1%	1%	99%
152.4	14	8%	9%	91%
139.7	14	8%	17%	83%
127	26	16%	33%	67%
114.3	45	27%	60%	40%
101.6	43	26%	86%	14%
88.9	16	10%	95%	5%
76.2	1	1%	96%	4%
63.5	5	3%	99%	1%
<63.5	2	1%	100%	

Fig. 9. Result table - Test 2

Likewise, a graph of the granulometric curve that behaves linear and with values close to 100 mm. (Fig. 10).

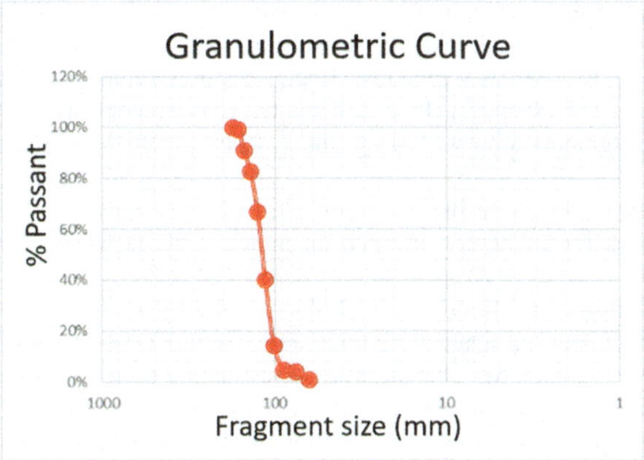

Fig. 10. Granulometric Curve 2

For this second technique test, a calculated value of P80 of 5.61 inches is obtained.

The third and last test is carried out with a length of 0.1 in the upper position, immediately close to the block.

For this last validation the results were not so favorable, since larger fragments were obtained compared to the base scenario in which the technique was not used. This test gives us a calculated P80 value of 7.91 fleas, a value above that calculated in the base scenario.

6 Conclusions

Compression tests validated the type of rock corresponding to the granodioritic intrusive. This calculated resistance of 114 MPa in a simple compression test and 110 MPa and 105 MPa in two validated sclerometric tests, allowed the application of the air chamber blasting technique in this scenario.

The results obtained in this research article responded to the reduction in the size of the fragments and irregular fragmentation, however, the reduction percentage for the lower positioning of 33.4% is significantly different when compared with trials and implementations of other authors that obtain a 20% reduction. The explanation could be given because the index of resistance to simple compression of the granodiorite present in the study area was close to the lower limit allowed for the application of the technique.

It is concluded that the recommended position of the air chamber according to the results obtained for this type of test scenario in granodioritic rock is in the lower part of the hole at a bottom length of 0.10 m.

References

1. EXSA, "MANUAL PRACTICO DE VOLADURA Edición especial," Lima, Perú (2019)
2. Faramarzi, F., Mansouri, H., Ebrahimi Farsangi, M.A.: A rock engineering systems based model to predict rock fragmentation by blasting. Int. J. Rock Mech. Min. Sci. **60**, 82–94 (2013)
3. Saharan, M.R., Sazid, M., Singh, T.N.: Explosive energy utilization enhancement with air-decking and stemming plug, 'SPARSH.' Proc. Eng. **191**, 1211–1217 (2017)
4. J. R. A. & Menon, R.: "Minas y canteras," Encicl. Segur. y salud en el Trab. **3**, 35–40 (2001)
5. Zhang, Z.X., Hou, D.F., Guo, Z., He, Z., Zhang, Q.: Experimental study of surface constraint effect on rock fragmentation by blasting. Int. J. Rock Mech. Min. Sci. **128**(May), 104278 (2020)
6. Babaeian, M., Ataei, M., Sereshki, F., Sotoudeh, F., Mohammadi, S.: A new framework for evaluation of rock fragmentation in open pit mines. J. Rock Mech. Geotech. Eng. **11**(2), 325–336 (2019)
7. Jhanwar, J.C., Jethwa, J.L., Reddy, A.H.: Influence of air-deck blasting on fragmentation in jointed rocks in an open-pit manganese mine. Eng. Geol. **57**, 13–29 (2000)
8. Balakrishnan, V., Pradhan, M., Dhekne, P.Y.: Investigating rock fragmentation in distributed spherical air-gap blasting technique. Powder Technol. **362**, 101–110 (2020)
9. Kabwe, E.: Improving collar zone fragmentation by top air-deck blasting technique. Geotech. Geol. Eng. **35**(1), 157–167 (2016). https://doi.org/10.1007/s10706-016-0094-7
10. Sazid, M., Singh, T.: Mechanism of air deck technique in rock blasting- a brief review. In: Fourth Indian Rock Conference, no. May, pp. 29–31 (2013)

11. Zhang, Z.-X.: Optimum fragmentation. Rock Fract. Blasting **2**, 411–423 (2016)
12. Saqib, S., Tariq, S., Ali, Z.: Improving rock fragmentation using airdeck blasting technique. Pakistan J. Eng. Appl. Sci. **17**, 46–52 (2015)
13. Ouchterlony, F., Sanchidrián, J.A.: A review of development of better prediction equations for blast fragmentation. J. Rock Mech. Geotech. Eng. **11**, 1094–1109 (2019)
14. Jang, H., Kawamura, Y., Shinji, U.: An empirical approach of overbreak resistance factor for tunnel blasting. Tunn. Undergr. Sp. Technol. **92**(July), 103060 (2019)

Shotcrete Elaboration Method Using Geopolymer Cement Product of the Alkaline Activation of Tailings for the Support of Underground Work in Polymetallic Mines of Peru

Jhoseph La-Fuente[1], Saac Huacho[1], Humberto Pehovaz-Alvarez[1], and Carlos Raymundo[2](✉)

[1] Ingenieria de Gestion Minera, Universidad Peruana de Ciencias Aplicadas, Prolongación Primavera 2390, Lima 15023, Peru
{u201614627,u201610038,pcgmhpeh}@upc.edu.pe
[2] Direccion de Investigacion, Universidad Peruana de Ciencias Aplicadas, Prolongación Primavera 2390, Lima 15023, Peru
carlos.raymundo@upc.edu.pe

Abstract. The objective of this article is to obtain geopolymer cement for the manufacture of shotcrete from the alkaline activation of mining tailings, which can improve the durability, resistance to compression, setting and fluidity of the shotcrete. For this, alkaline activators such as sodium hydroxide (NaOH) and sodium silicate (Na_2SiO_3) will be used. Likewise, the standardized test will be carried out to determine the compressive strength of concrete in cylindrical samples, in order to analyze the mechanical properties acquired by the shotcrete made with geopolymer cement of 12 and 15M NaOH concentration. The results highlight the possibility of obtaining an eco-efficient cement with compressive strengths of up to 18 MPa at 28 days, reductions of up to 73% in CO2 emissions from the production of Portland cement and a sustainable alternative for the use of tailings.

Keywords: Alkaline activation · Geopolymer cement · Tailings

1 Introduction

Currently, Peruvian mining is suffering from rockfall problems, many of which have caused accidents and the death of workers. According to OSINERGMIN [1], it was found that 20% of accidents with fatalities that occurred in 2019 were caused by geomechanical failures, due to an inefficient application of adequate support. In addition, the Ministry of Energy and Mines (MINEM) [2] indicated that 29% of accidents in the last 20 years were caused by rockslide, which refers to failures in support. That is why the mining sector has the need to design more efficient processes, methods and/ or techniques to reduce the problems caused by the instability of underground workings. Consequently, today a suitable and optimal type of support is being sought, since it is an essential

S. Trzcielinski et al. (Eds.): AHFE 2021, LNNS 274, pp. 254–262, 2021.
https://doi.org/10.1007/978-3-030-80462-6_32

process to protect personnel and equipment from accidents, it also encompasses value, because it is an additional high-cost work that reduces the speed of mining progress and/or production [3].

According to the Chilean Cement and Concrete Institute (ICH) [4], in underground mining, support has a great scope because, due to the nature of the work, all work carried out inside the mine is carried out in empty spaces, unstabilized product of the breakage of the rock or mineral extracted. To cover and minimize the problems of geomechanical instability, mining uses passive and active support, of these two types of support, shotcrete (passive) is the most used support in the Peruvian-mining industry.

2 State of the Art

2.1 Shotcrete Production Method

Since its successful application in the United States in 1907, shotcrete has quickly become popular in geotechnical engineering. During this period, the quality of the shotcrete, the working environment, the production efficiency and the mechanical properties have been significantly improved through the continuous reform of the technology and the improvement of the composition of its raw materials. However, these studies only focused on the composition, proportion and additives of shotcrete raw materials [5, 6].

2.2 Method of Making Shotcrete with Geopolymer Cement

According to C.S. Cundy, geopolymerization is any reaction that transforms a well-defined solid precursor, in the presence of an alkaline solution with or without soluble silicate, into aluminosilicate gels. The formation of geopolymers can be summarized in three steps: the first step consists of the dissolution of the Si and Al species in an alkaline medium; the second step is the condensation of the dissolved species in manometers and, finally, polycondensation occurs, which leads to the formation of an amorphous gel [7, 8].

2.3 Method of Making Shotcrete by Alkaline Activation of Mining Tailings

Currently, in alkaline activation in tailings containing lead, it has been shown that this compound is immobilized in the forms of lead in network (41.75%), lead in silicate of PbO-3SiO2 (3.89%) and PbO -7SiO2 (54.37%), respectively. On the other hand, by increasing the addition of Pb (NO3) 2 to 6%, more PbO-3SiO2 is formed, while the network lead fractions and PbO-7SiO2 decrease, due to the formation of a more geopolymer gel phase. Low and more lead oxides participate in the alkaline activation process. This was determined in an investigation on the implementation and improvement of the alkaline activation of mining tailings for the immobilization of lead in cementitious binders, which sought to reduce the impacts caused by mining tailings, this in terms of its ecological approach, since it it is known that tailings are composed of heavy metals that penetrate and damage soil and water resources [9–11].

3 Input

3.1 General Contribution

Figure 1 represents the flow chart of the process from a general point of view, divided into three fundamental phases. The process begins from the Analysis Phase, which consists of the collection of information and the taking of samples of tailings from the tailings deposit No. 2 of the Huancapetí mining unit of Compañía Minera Lincuna SA, in addition to the granulometric analysis and chemical analysis of said samples, which must meet certain conditions. In the same way, the Design Phase is presented, consisting of the design and preparation of the activator solutions at different molar concentrations of NaOH (12M and 15M), the dosage of the activator to carry out the alkaline activation of the polymetallic mining tailings and the design of the mixtures of shotcrete with geopolymer cement activated at different molar concentrations of NaOH. Finally, the evaluation and discussion phase is represented by the evaluation of the mechanical properties of the shotcrete with geopolymer cement by means of the simple compressive strength test of the shotcrete samples and the comparison of the different previously designed mixtures.

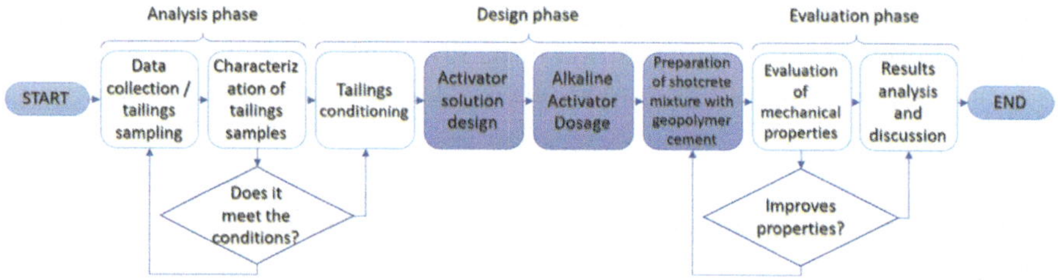

Fig. 1. General flow chart

3.2 Detail Contribution

Activator Solution Design. For the preparation of the activator solution, necessary for the alkaline activation of the tailings, Sodium Silicate and Sodium Hydroxide (NaOH) will be used, the first being a commercial alkaline activator (pH = 2) and of high efficiency, which will grant greater mechanical resistance to the geopolymer cement to be obtained, in addition to accelerating the activation process. On the other hand, NaOH is an activator that will be necessary due to its ability to activate a greater number of aluminosilicate species.

Alkaline Activation. For the alkaline activation of the mining tailings, the solid component (dry tailings) was progressively mixed with the previously prepared activating solution, both 12M and 15M NaOH in separate containers, for which two types of geopolymer cement were obtained. It is important to emphasize that the workability of the geopolymer mixture was determined by the amount of the activating solution that was used. Likewise, to obtain the ideal geopolymer cement, a solid/activating solution ratio of 1.5 was used, as a standard measure to obtain a pasty mixture, ideal for cementitious material.

Shotcrete Mixing Design. For the design of the shotcrete mixture, geopolymer cement was used as a replacement for Ordinary Portland cement together with the aggregate of settling type sand and steel fibers. For the pertinent calculations, the ACI 211 Standards were used, referred to the "Guide for the design of concrete mixtures", hand in hand with the criteria imposed by the ACI 506 R standards (Shotcrete Guide). The calculations were developed based on a design of 210 kg/cm2 and a volume of 1 m3, in addition to considering a particle size of the arid aggregate not greater than 3/8".

3.3 Indicators

Moisture Percentage
Calculate the water content present in the mine tailings sample. Calculation formula:

$$\% \, humidity = \frac{m_h - m_s}{m_s} \times 100 \tag{1}$$

Where:

mh: wet mass.
ms: dry sample mass.

Reference Level:

Good: Percentage of humidity less than 45%.
Medium: Percentage of humidity between 50%–75%.
Bad: Percentage of humidity greater than 75%.

Water/Cement Ratio
It allows to improve the workability or mechanical resistance of the designed concrete mix. Calculation: (Fig. 2).

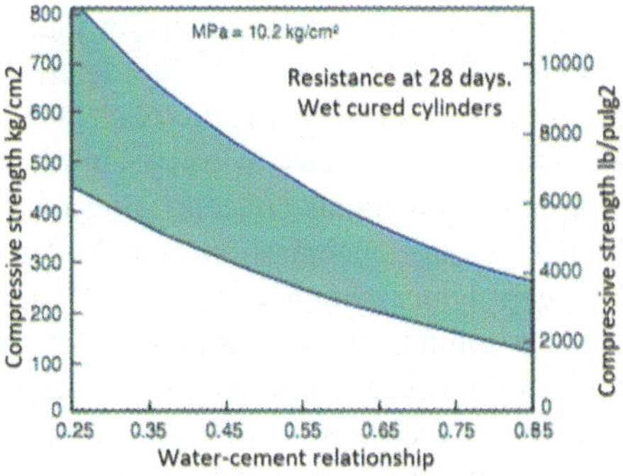

Fig. 2. Water/cement ratio

Reference Level:

Good: Complete hydration will be obtained with workability with a ratio greater than 60%.
Medium: Complete hydration will be obtained with greater mechanical resistance with a ratio between 40% and 60%.
Bad: Full hydration will not be obtained with a w/c ratio less than 40%.

Compression Stress. It allows to measure the resistance that a sample is capable of supporting.

Calculation formula:

$$\sigma_c = \frac{P}{A} \tag{2}$$

Where:

P: Applied load (N).
A: Cross-sectional area that supports the applied load (m2).

Reference Level:

Good: superior resistance (>5%) to samples with ordinary cement.
Medium: Equivalent strength (0%–5%) to samples with ordinary cement.
Bad: lower resistance to samples with ordinary cement.

4 Validation

4.1 Validation Scenario

The validation scenario took place in the laboratory of the LEM - ENGIL SRL facilities. To carry out the elaboration method, shotcreting with geopolymer cement product of the alkaline activation of mining tailings was used, polymetallic tailings from the tailings deposit No. 2 of the Huancapetí mining unit of Compañía Minera Lincuna SA were used, which were characterized physically and geochemically, in addition to being conditioned for this research through drying in a laboratory oven to determine its humidity. On the other hand, 2 activating solutions of sodium silicate (Na2SiO3) and sodium hydroxide (NaOH) were designed, which are necessary to carry out alkaline activation; Design 1 being an activator solution with a molar concentration of 12M NaOH and design 2 an activator solution with a molar concentration of 15M NaOH. Consequently, two mixtures of shotcrete with geopolymer cement called "12M" and "15M" were designed, corresponding to the activator solution used to obtain the geopolymer cement used in each of the mixtures. Then, 8 test pieces were prepared in cylindrical molds of 4 inches in diameter and 8 inches in height for each type of mixture, which were cured and subjected to the resistance test to simple compression on days 1, 7, 14 and 28 with respect to the cure time, with the use of a concrete dam.

4.2 Design of the Validation

Two shotcrete mix designs were established, based on the ACI 506 R (GUIDE TO SHOTCRETE) Standard. The first design used geopolymer cement obtained from tailings activation with an activator solution with a molar concentration of 12M NaOH, while the second design used geopolymer cement activated by an activator solution with a molar concentration of 15M NaOH. Table 1 indicate the amount (in Kg) of the components used for each of the shotcrete mix designs, differing mainly by the volume of geopolymer cement that was used in each design. The water/cement ratio used was 0.40 in order to give the shotcrete mixture greater mechanical resistance, but a reduction in workability.

Table 1. Shotcrete mix design with geopolymer cement.

Componente	Peso (Kg)	Densidad (Kg/m^3)	Volumen (m^3)
Agua	7.44	1000	0.007440
Cemento geopolímero (12M)	18.60	2860	0.006503
Cemento geopolímero (15M)	18.60	3000	0.006200
Agregados	47.41	2680	0.017690
Fibra de acero	0.95	2700	0.000352

A total of 16 cylindrical specimens of 4 inches in diameter and 8 inches in height, 8 shotcrete specimens with 12M geopolymer cement and another 8 shotcrete specimens with 15M geopolymer cement were prepared, in order to have two specimens of the same design for each day of failure to give reliability to the results obtained after the execution of the resistance test to simple compression. The test was applied on days 1, 7, 14 and 28 according to the curing time of each of the specimens to perform an analysis of the variation of the resistance of these shotcrete designs over time, in addition to analyzing the influence of activation alkaline to different molarities of NaOH to its mechanical properties.

Table 2 represents the results of the test of the resistance to simple compression of the shotcrete specimens with geopolymer cement at a molar concentration of 12M NaOH, called "P-12"; It is important to emphasize that only the results of those specimens that presented greater resistance than the 2 tested per day were represented. The load factor (F'c) was taken as = 20.594 MPa or 210kg/cm2, these results were obtained at days 1, 7, 14 and 28 of curing at room temperature, where a maximum of 88.3% resistance to compression with respect to the established F'c. Likewise, the samples presented a type 3 fracture, which means that vertical columnar cracks were evidenced in both bases and not well-formed cones.

Table 2 represents the results of the test of the resistance to simple compression of the shotcrete specimens with geopolymer cement at a molar concentration of 15M NaOH, called "P-15"; It is important to emphasize that only the results of those specimens that presented greater resistance than the 2 tested per day were represented. These results were obtained at days 1, 7, 14 and 28 of curing at room temperature, where a maximum of 92.4% compressive strength was reached with respect to the established F'c. Likewise, the samples of this design also presented a type 3 fracture.

Table 2. Results of shotcrete compression strength test with geopolymer cement.

Code	Break date	Age (days)	Specimen area (cm3)	Burden (kN)	Burden (Kg)	Strength (MPa)	Endurance	F'c (MPa)	Type of fracture
P-12-1	30/10/20	1	75.4	7.2	734.2	1.00	4.6	20.594	3
P-12-2	06/11/20	7	75.4	18.2	1856	2.40	11.7	20.594	3
P-12-3	13/10/20	14	75.4	64.8	6604	8.59	41.7	20.594	3
P-12-4	27/10/20	28	75.4	137.1	13977	18.18	88.3	20.594	3
P-15-1	30/10/20	1	75.4	7.6	775	1.00	4.9	20.594	3
P-15–2	06/1120	7	75.4	29.4	2998	3.90	18.9	20.594	3
P-15-3	13/10/20	14	75.4	69.4	7073	9.20	44.7	20.594	3
P-15–4	27/10/20	28	75.4	143.5	14632	19.03	92.4	20.594	3

As a final result, when making a comparison of the behavior of both designs with respect to time, it was obtained that the shotcrete design that used geopolymer cement obtained from the alkaline activation of the tailings with a solution of higher molar concentration of NaOH (15M) presented better mechanical properties on each of the test days.

5 Results

Two types of geopolymer shotcrete were obtained as a result of alkaline activation (12M and 15M NaOH), of which the 15M shotcrete has greater resistance with 18 MPa at 28 days of curing, for which it can be stated that it is found in the optimum range of shotcrete performance according to ACI regulations. In terms of percentage, the shotcrete reached 80% efficiency, being an average value among the various scenarios studied.

6 Conclusions

It is concluded that the formation of geopolymer concretes for the elaboration of shotcrete is possible using the alkaline activation of polymetallic mining tailings. The tailings from deposit No. 2 of the Huancapetí Mining Unit of Compañía Minera Lincuna S.A. It is a source of aluminosilicates that has a reactivity in highly alkaline media; in other words, this material facilitates the alkaline activation.

Likewise, the molar concentration of the sodium hydroxide solution directly influences the mechanical strength that a geopolymeric mixture will develop. As the concentration of the solution increases, the mechanical resistance that will develop increases, this increase would be given by the increase in amorphous material that can be dissolved and by a greater amount of positive ions that stabilize the geopolymer chains.

In addition, the use of sand type settling for the dosage of the mixture of shot-crete with geopolymer cement following the ACI 506 R (Specialized Guide for the preparation of shotcrete) regulation produced an increase in mechanical resistance as the amount of this arid in its composition. Likewise, the water/cement ratio has a double influence on the dosage of the geopolymeric mixture, on the one hand, it entails an increase in the workability of the mixture, but at the same time it produces a significant decrease in mechanical resistance as the amount of water increases.

Also, the curing process influences the development of mechanical strength of the specimens. According to the literature analyzed, a higher curing temperature increases the speed at which the geopolymerization chemical reactions develop, however, at room temperature it is possible to obtain good resistance in these days in 1, 7, 14, 28 days. materials.

Finally, the analysis of the mechanical properties of the shotcrete with geopolymer cement of 12M 15M indicates that the addition of an adequate amount of activating solution (NaOH + Na2SiO3), allows to achieve compressive strengths ranging from 3.15 MPa to 18.60 MPa approximately.

References

1. OSINERGMIN, "Boletín Estadístico de la Gerencia de Supervisi6n Minera," 30 April 2019. https://www.osinergmin.gob.pe/seccion/centro_documental/mineria/estadisticaseindicadores/accidentes-mortales/Boletin-GSM-Accidentes-Mortales-2019-04.pdf. Accessed 24 June 2020
2. Ministerio de Energía y Minas (MINEM), "Estadísticas de accidentes mortales en el sector minero," 2019. http://www.minem.gob.pe/_detalle.php?idSector=1&idTitular=170&idMenu=sub151&idCateg=170. Accessed 24 Jun 2020
3. Galan, I., Baldermann, A., Kusterle, W., Dietzel, M., Mittermayr, F.: Durability of shotcrete for underground support– review and update. Constr. Build. Mater. **202**,, 465–493 (2019). https://doi.org/10.1016/j.conbuildmat.2018.12.151
4. Instituto del Cemento y Hormig6n de Chile, "Guia Chilena de Hormig6n Proyectado," 2019. https://ich.cl/shotcrete/documentos-shotcrete/2da-edicion-guia-chilena-de-hormigon-proyectado/. Accessed 24 June 2020
5. Thomas, A.: Sprayed Concrete Lined Tunnels. CRC Press (2019)
6. Carmona, S., Molins, C.: Use of BCN test for controlling tension capacity of fiber reinforced shotcrete in mining works. Constr. Build. Mater. **198**, 399–410 (2019). https://doi.org/10.1016/j.conbuildmat.2018.11.229
7. Yan, D., Chen, S., Zeng, Q., Xu, S., Li, H.: Correlating the elastic properties of metakaolin-based geopolymer with its composition. Mater. Des. **95**, 306–318 (2016). https://doi.org/10.1016/j.matdes.2016.01.107
8. Cundy, C.S., Cox, P.A.: The hydrothermal synthesis of zeolites: history and development from the earliest days to the present time. Chem. Rev. **103**(3), 663–701 (2003). https://doi.org/10.1021/cr020060i
9. Cárdenas Ticlavilca, F.J.: "Propuesta de uso de relaves de mina polimetálica en la fabricación de unidades de albañilería - caso ex unidad minera Mercedes 3." Universidad Peruana de Ciencias Aplicadas (UPC) (2019). https://doi.org/10.19083/tesis/625225.

10. Perumal, P., Piekkari, K., Sreenivasan, H., Kinnunen, P., Illikainen, M.: One-part geopolymers from mining residues – effect of thermal treatment on three different tailings. Miner. Eng. **144**, 106026 (2019). https://doi.org/10.1016/j.mineng.2019.106026
11. Glid, M., Sobrados, I., Ben Rhaiem, H., Sanz, J., Amara, A.B.H.: Alkaline activation of metakaolinite-silica mixtures: role of dissolved silica concentration on the formation of geopolymers. Ceram. Int. **43**(15), 12641–12650 (2017). https://doi.org/10.1016/j.ceramint.2017.06.144

An Empirical Review and Implication of Globalization to the South African Automotive Industry

John M. Ikome[1(✉)], T. Laseinde[2], and M. G. Kanakana Katumba[2]

[1] Mechanical and Industrial Engineering (MIET) Department, University of Johannesburg, Doornfontein, Johannesburg, South Africa

[2] University of South Africa (UNISA), Pretoria, South Africa

Abstract. This paper examines the implication of globalization and the global competitiveness of South African Automotive Industry (SAAI). The South African automotive industry forms part of small and medium scale enterprise (SMEs). However this industry has been going through a number of constrains ranging from competitiveness and innovation problem, therefore getting an optimum solution to the global competitiveness of this industry is of great importance and the economic wellbeing.

Keywords: Competitiveness · Globalisation · Process control

1 Introduction

South African economic growth is influence by three main industries that are dominant which includes mining, agriculture and manufacturing. The manufacturing industry plays a vital role in the economic growth of the country, ranging from employment, domestic growth and exporting. This clearly shows that any improvement related to this industry is of great importance for the nation as a whole. However, despite the importance of this industry, growth in the automotive industry in particularly has been very slow in both domestic and international market. The automotive industry is a global industry and South African automotive industry must compete with other automotive industries globally in order to gain growth and competitive advantage. Internationalization is acknowledged and profoundly referring to the increasing linkage of national economies by way of international trade [1]. This perspective refers to the characteristic and trend of global production. Globalization and internationalization coexist, that they are interrelated terms pertaining to the geographical extension of economic activities across nations and state borders but however, includes transnational and functional integrations of dispersed activities [2]. This was further contended by a research scholar Gereffi G, were he argued that the search for new markets, raw materials and other resources including manufacturing exports for goods, started centuries ago and that internationalization is not a new thing [3]. Also according to another research work, multinational enterprise can be classified as an example of internationalization whereby

S. Trzcielinski et al. (Eds.): AHFE 2021, LNNS 274, pp. 263–270, 2021.
https://doi.org/10.1007/978-3-030-80462-6_33

they try to sell their products in a foreign country or aspires, through trade policies, and are obliged to establish production facilities offshore in order to gain access to new markets [4].

On the other hand, automakers were rationalizes as the first to globalize and whilst there was a significant difference between vehicle models produced in different countries, there was no reason for suppliers to follow [5]. The importance and origin of globalization was also analyzed by a number of scholars, were they stated that once similar vehicles were being manufactured by different automotive industries in different regions, it made economic sense for them to work with the same suppliers for standardization of parts purposed [5, 6].

This authors later added that " despite the opportunity and development that globalization in the automotive industry may create for developing countries, it may pose some limitations to local suppliers due to the preference of global suppliers of cheap automotive components [5, 6].

1.1 The Automotive Industry

Globalization is the term that is used to describe the world economic trends that is pulling previous economies of distinct nations together. Furthermore, many research noted the significance of globalization, by including the fragmentation of production to lower product units, dissatisfaction with the expensive systems and strategies of manufacturing cars for stock, and not to order, innovative modular construction of parts used to assemble cars by suppliers and possible switch to alternative energy power cars [7].

As a result of this, created room and opportunity for global economy participation by a number of suppliers in both the international and local market, According to Kaplinsky et al. [8], This also created a huge competition that has expanded significantly to the extent of surfacing of two major type of automotive industry suppliers and the competitors are no longer those merely down the street but now span across the world.

Direct investment across national borders has increased, which means that the economy now function on a global scale with severe competition, as a number of nations seek to attract new investment. The increased competition, promotes individualism as firms pursue increased profit and not really development, ultimately making the Multinational Corporation a threat to the local firms.

In recognizing the significance of South African automotive contributions to the national economy, they are factors affecting it competitiveness that have become a matter of prime concern [9]. According to Joshi et al. [10], this factors are often identified as Determinants of Competitiveness, and developing strategies that can improve this indicators, will result in sustainable competitive advantage. In a similar research work done by Gratama, et al. [11], states that globalization has created business opportunities for all nation, but however, the gap between developed and developing countries is alarming and making it difficult for developing countries to be independent genuinely.

In majority cases, the developing countries serve as hot countries to enterprises own by multinational cooperation. This leads to the host country benefiting, job creation, improvement of work force access to foreign market and capital. But in realistic, government and MNCs have different intentions. The government wants the benefits but requires strategies of minimizing the consequences of remaining merely a host with limited development. Figure 1 shows the Influence of key performance indicators on competitive priorities.

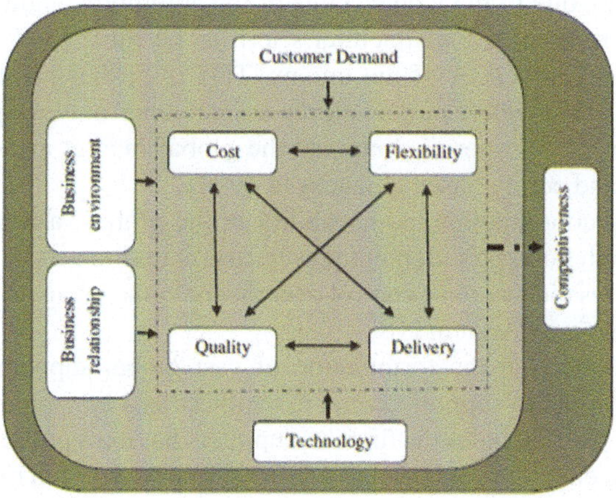

Fig. 1. Influence of key performance indicators on competitive priorities.

2 Original Equipment Manufacturers (OEM's)

In globalization, the identification of new markets and favorable production location is of key importance for an organization, and the OEMs plays a key role under this aspect by continuously searching for production locations and new favorable marketing locations that can minimize cost. As a result of this, there has been a global shift of production from developed to the less developed countries.

This shift has created a lot of opportunities, for developing countries ranging from employments, technology and skills transfer, economic prosperity and growth, increase trade relationship between industrialized and developing countries that can induce transfer of technology and innovation, leading to competition [12].

Globalization has brought great opportunities by breaking of economic, geographical and social boundaries and production growth, including finance and market shares, but however, it cannot be denied that not all developing countries are equally involved nor positioned to benefit from it. Empirical evidence of a research work done by Soubbotina [13] acknowledged East Asian economies' growth, being significantly boosted by globalization, but many other developing countries have been rather slow to integrate into the world economy particularly Africa.

2.1 The Global Automotive Sectors Drifts

According to Gastrow [14], the have been an increase of cross border trade and Foreign Direct Investment (FDI) since the 1980s facilitation and liberalization of trade and investment through the World Trade Organization (WTO) agreements. Brazil, China and

India offers surplus market with good economic growth due to low labor cost. Gastrow [15], further commented that, this countries suppler local market and export to emerging nation and due to this, there is a huge stimulation of Foreign Direct Investment (FDI).

Power [16], said, due to the iconic status of the automotive industry, usually result in political reaction as local manufacturer's tern to be threatened by lead firms and subsequent, political influences of the unions. This is further justified by a research work done by PWC [17], where he said eleven assemblers from European Union, Japan and the United States of America dominate the global market and manufacturing by enhancing unions and equity based alliances in 1990s.

The is a strong coordination and capability in the global value chain of the automotive industry, with huge procurement power, that is dominated by the top ten global automotive industries that exercise control over the manufacturing process including the supply chain management system [18].

This clearly shows that the formulation of vast global suppliers uses the global manufacturing networks to support several assemblers as a key trend in the automotive global value chain. This includes the first ties suppliers having a greater power within the supply chain management process as the assumed increased role in research, innovation and manufacturing: but despite this powers, the main control remains largely within the assembler's hands [19]. From a practical point of view, it is palpable that the suppliers that account for a manufacturing cost of 75% of a vehicle represent the assembler's greatest target for reduction of cost, which is a great impact of the supplier innovation strategies [20].

2.2 The New Automotive Supplier's Role

The automotive sector design has developed leverage in its design capacity in multiple numbers of its product that is sold in the global market. Sturgeon et al. [21], (2019) highlighted that the creation of two classes of automotive industry local and global suppliers, came as a result of globalization and that earlier, major automotive manufacturing industries either exported automotive components to off shore plants that assemble vehicles or depended on local suppliers in each manufacturing location.

However now our days, the global suppliers have been included in making the focused to change, compared to the past, and creating the trend to expand in regards to the customers and marketing segments of large supplier to the automotive industry. The European Union, Japan and United States of America major firms are the once benefiting from this supplier and in addition, the suppliers had to adapt to complex development approach of vehicle manufacturing in order to maintain the supply chain linkages. Table 1 illustrates the manufacturers of vehicles including their global manufacturing percentage.

The total number of out-put units is further illustrated in Fig. 2 and it shows that General motors, Toyota and Volkswagen group are topping the list of the total units produced.

According to the analysis of OICA [22], it shows that the global growth and manufacturing shares of emerging manufacturers has increased to 12% from the year 2005 to 2008. These details are analyzed in Table 2 and it shows the lead 11 OEM's from Germany, United States of America and Japan's production of over 2.5 million for each country, making a total contribution of 83% of vehicles manufactured globally. In 2011,

Table 1. Vehicle Production by company (2005–2017).

Firm	Out-put units 2005 (000 000)	Out-put nits 2008 (000 000)	Out-put nits 2011 (000 000)	Out-put nits 2014 (000 000)	Out-put nits 2017 (000 000)
General Motors	7.6	8.2	8.9	6.4	9.1
Toyota	6	6.2	8	7.2	8.1
Volkswagen Group	5.1	5	5.7	6.1	8.1
Honda	2.7	2.9	3.7	3	2.9
Ford	6.7	6.6	6.3	4.6	4.8
Hyundai	2.5	2.7	2.5	4.6	6.6
PSA group	3.1	3.3	3.4	3	3.5
Nissan	2.6	2.9	3.2	2.7	4.6
Chrysler		2.5	0.9		2
Renault	2.4	2.4	2.5	2.3	2.8
Fiat	2.4	2	2.3	2.4	2.4
Daimler AG	4.4	4.2	2	1.4	1.5
Total of above	45.5	46.4	51	44.6	56.4
Other manufacturer	10.1	13.1	17.3	14.9	22.4
Total	55.6	59.5	68.3	60.5	78.8

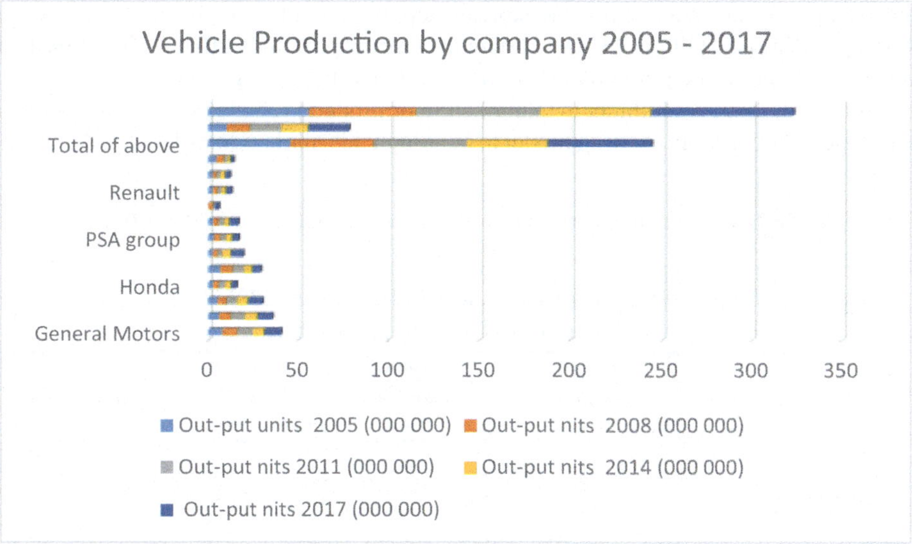

Fig. 2. Automotive production by companies 2005–2017.

the analysis shows that the situation was stable. However, in 2017 it shows a declined in global share of manufacturing percentage of 10.8% between 2011 to 2017 of the 12 leading OEM's.

3 The Manufacturing Trends of Automobile

Evaluating back from 1973, it shows that that total number of vehicles produced globally has tripled as from 33 to 78.9 million in 2007 (OICA) [23]. In a similar analysis, made by Sturgeon [24], he commented that the global automotive growth is as a result of new market segment in China and India. The establishment of new market in India and China has help to drive the growth space within the automotive industry, and while 7 countries accounted for 80% of global automotive production in 1975, the same number of shares were accounted for by 11 countries in 2005. The global Vehicle production and the geographical dispersion has grown from an average rate of 2% from 1975 to 1990, and a rise to 3% in 1990 to 2005, low motorization and huge rise of population resulted in a surge for new investments in India and China where market growth and manufacturing has increased rapidly. Gastrow et al. [25] further said from the 1980s, a huge number of industries have shifted from a series of isolated industries to a more assimilated industries. According to Power [18], global assimilation embeds firms in a larger region and manufacturing of global scale including innovation. Therefore, the automotive manufacturing industry is in the midst of a profound transition. The statistics on Table 3, illustrates the automotive industry's growth in emerging market segments, and as a result, this global growth is because of the following market: Latin America (Brazil and Mexico), Asia (India China and Korean Republic) and East of Europe (Mainly the Russian Federation) OICA [23]. According to Gastrow [26], India, Brazil and China that falls under large developing countries offers large market with good economic growth and as a result force assemblers to manufacture or produce vehicles specifically for this market segment or adapt existing model to be sold in this markets in order to make profit.

Table 2. Production of automotive vehicles in selected nations (2006–2011).

Motor vehicle production, selected countries, 2006-2011 in 000 units and in % for growth rate

	2006	% Growth Rate 2005 to 2006	2008	% Growth Rate 2007 to 2008	2010	% Growth Rate 2009 to 2010	2011	% Growth Rate 2010 to 2011
China	7 277	22	9 345	5,2	18 264	32,4	18 418	0,8
India	2 016	14,4	2 314	2,7	3 536	33,9	3 926	10,4
Republic of Korea	3 840	6,4	3 806	-6,8	4 271	21,6	4 657	9
France	3 169	-4,8	2 568	-14,8	2 229	8,9	2 242	0,6
Brazil	2 611	13,8	3 220	8,2	3 648	14,6	3 406	0,7
Mexico	2 045	2,4	2 191	4,6	2 345	50,2	2 680	14,4
Russia	1 503	10,4	1 790	7,8	1 403	93,5	1 988	41,7
Germany	5 819	6,8	6 040	-2,8	5 905	13,4	6 311	6,9
Spain	2 777	4	2 541	-12	2 387	10	2 353	-1,4
Canada	2 571	0,3	2 077	-19,4	2 071	39	2 134	3,2
Japan	11 484	1	11 563	-0,3	9 625	21,3	8 398	-12,8
United States	11 292	-4,5	8 705	-19,3	7 761	35,4	8 653	11,5

It was further outlined by Gastrow [26] that the shift of production and consumption from east to west is as a result of globalization and also that the shift will continue to reshape the automotive industry's activities globally to more advanced ways, but however, remains structured within the national and regional marketing segments that are nested under the global framework.

4 Conclusion

According to the empirical review, it is evident that the South African automotive industry is a very important industry to the economic growth of the nation, but of late, this industry has not been globally competitive and developing any strategy that will improve its competitiveness is of great importance.

References

1. Hartzenberg, T., Marudzikwa, S.: Transfer of Technology for Successful Integration into the Global Economy: A Case Study of the South African Automotive Industry Transfer of Technology. United Nations, New York (2002)
2. Choi, S.H., Jeon, B.N.: The impact of the macroeconomic environment on merger activity: evidence from US time-series data. Appl. Financ. Econ. **21**(4), 233–249 (2011)
3. Vahtra, P., Zashev, P.: Russian automo manufacturing sector-an industry snapshot for foreign component manufacturers. www.tse.fi/pei. Accessed 28 Apr 2013 (2008)
4. Kumaraswamy, A., Mudambi, R., Saranga, H., Tripathy, A.: Strategic Adaptation to deregulation in the Indian Auto Component Industry. Sloan Industry Studies: Working Papers, Alfred P. Sloan Foundation, New York (2008)
5. Department of Trade and Industry (DTI): South Africa's motor trade product. http://www.thedi.gov.za/econdb/raportt/rapprmocurent.html. Accessed 2 Feb 2013 (2005)
6. Gereffi, G.: International trade and industrial upgrading in the apparel commodity chain. J. Int. Econ. **48**(1), 37–70 (1999). https://doi.org/10.1016/S0022-1996(98)00075-0
7. Sturgeon, T.J., Van Biesebroeck, J.: Global value chains in the automotive industry: an enhanced role for developing countries? Int. J. Technol. Learn. Innov. Dev. **4**(1/2/3), 181–205 (2011)
8. Sutton, J.: The globalization process: auto-component supply chains in China and India. In: Annual World Bank Conference on Development Economics Europe 2005: Are We on Track to Achieve the Millennium Development Goals? World Bank and Oxford University Press, New York (2005)
9. Sturgeon, T.J., Memedovic, O., Biesebroeck, J.V., Gereffi, G.: Globalisation of the automotive industry: main features and trends. Int. J. Technol. Learn. Innov. Dev **2**(1/2), 7–24 (2009)
10. Castellacci, F.: Innovation and the competitiveness of industries: comparing the mainstream and the evolutionary approaches. Technol. Forecast. Soc. Change **75**, 984–1006 (2008)
11. Kao, C., Wu, W.-Y., Hsieh, W.-J., Wang, T.-Y., Lin, C., Chen, L.-H.: Measuring the national competitiveness of Southeast Asian countries. Eur. J. Oper. Res. **187**(2), 613–628 (2008)
12. Joshi, M., Dixit, S.: Enhancing competitiveness of the Indian Automobile Industry: A study using Porter's Diamond Model. Management and Change, vol. 15 (1&2). (2011). http://ssrn.com/abstract=1977444. Accessed 1 Apr 2013
13. Tripathi, G., Joshi, S.: Creating competitive advantage through sustainable supply chains: a theoretical framework for the assessment of practices, dynamic capabilities, and enterprise performance of manufacturing firms. Int. J. Recent Technol. Eng. **8**, 2277–3878 (2019). https://doi.org/10.35940/ijrte.D9501.118419

14. Gratama, J.-W., Kraan, J., Sutherland, D.R., Granger, V., Barnett, D.: Validation of the single-platform ISHAGE method for CD34(+) hematopoietic stem and progenitor cell enumeration in an international multicenter study. Cytotherapy **5**, 55–65 (2003). https://doi.org/10.1080/14653240310000083
15. Alfaro, A.L., Bizuneh, G., Moore, R., Ueno, S., Wang, R.: Microeconomics of Competitiveness. South Africa: Automotive Cluster. Harvard Business School, Kennedy School of Government (2012)
16. Subotica, T.P.: Beyond economic growth: an introduction to sustainable development (English). WBI learning resources series Washington, D.C.WorldBankGroup (2004). http://documents.worldbank.org/curated/en/454041468780615049/Beyond-economic-growth-an-introduction-to-sustainable-development
17. Gastrow, M.: Innovation, Skills Development and South African Labour Market Intelligence. Human Sciences Research Council, South Africa (2012)
18. J.D. Power and Associates: Global Automotive Outlook for 2011 Appears Positive as Mature Auto Markets Recover, Emerging Markets Continue to Expand (2011). http://content2.businesscenter.jdpower.com/JDPAContent/CorpComm/News/content/Releases/pdf/2011018-gao2.pdf. Accessed 20 Feb 2013
19. PricewaterhouseCoopers (PWC): Russian Automotive market-Is Russia the largest car market in Europe? (2008). www.pwc.com/bx/en/research-insights. Accessed 28 Apr 2020
20. Schwartz, M.: Trends in the Automobile Industry: Implications on Supply Chain Management, CISCO Internet Business Solutions Group. Scotia (2008)
21. Becker, H.: High Noon in the Automotive Industry. Helmut, Berlin (2006)
22. Damoense, M.Y., Agbola, F.W.: Economic analysis of the impact of policy reforms on the South African automobile industry. Int. J. Econ. Perspect. **3**(4), 285–296 (2009)
23. Sturgeon, T., Memedovic, O., Van Biesebroeck, J., Gereffi, G.: Globalisation of the automotive industry: main features and trends. Int. J. Technol. Learn. Innov. Dev. **2**(1), 7–24 (2009)
24. International Organization of Motor Vehicle Manufacturers OICA, (vide Organisation Internationale des Constructuers d'Automobiles) (OICA). (2013). Production and Economic 324 statistics 2012. http://oica.net/catergory/production-statistics/. Accessed 14 May 2013
25. International Organization of Motor Vehicle Manufacturers OICA, (vide Organisation Internationale des Constructuers d'Automobiles) (OICA). (2012) Production and Economic statistics 2011. http://oica.net/catergory/production-statistics/. Accessed: 14 Feb 2013
26. Gastrow, M.: A review of trends in the global automotive manufacturing industry and implications for developing countries. Afr. J. Bus. Manage. **6**(19), 5895–5905 (2012)

An Assessment of Logistics and Transport Subdivision as an Element of Globalization

Ayodeji Dennis Adeitan[1(✉)] and Clinton Aigbavboa[2]

[1] SARChl in Sustainable Construction Management and Leadership in the Built Environment, Postgraduate School of Operations Management, University of Johannesburg, Johannesburg 2028, South Africa
adeitandennis@gmail.com

[2] SARChl in Sustainable Construction Management and Leadership in the Built Environment, Faculty of Engineering and the Built Environment, University of Johannesburg, Johannesburg 2028, South Africa
caigbavboa@uj.ac.za

Abstract. The development of globalization has been recorded as one of the most meaningful basic system affecting logistics service providers, workers and investors. The procedure of transporting sets up the quality of being adequate of conveying items. The process of promoting in strategies and operational rules enhances the conveying freight, conveyance speed, process costs, quality of service, the facility's practice and saving energy. Transportation takes a crucial factor in the administration of logistics and supply chain in general. Assessing the ongoing circumstances, a solid scheme needs a reasonable fringe of logistics and an appropriate transport strategy to link the creating method. The target of this survey is to distinguish the basic factors impacting the transport segment and logistics through globalization procedure occurring in the global economy.

Keywords: Globalization · Economic development · Transportation · Logistics management

1 Introduction

The increasing global competition in businesses coupled with the globalization of the economy has made many logistics providers to explore new market outputs, to increase its trade turnover and flows of capital and services while using cheaper raw materials and labour which can be found in the country of the parent company or elsewhere. The movement of raw materials, labour, and services are characterized by covering distances which are only possible through the availability of effective and efficient transport section and provision of appropriate logistics functions. Over the years, implementation of modern technologies in transport system development such as haulage, transshipment, information flow in the supply chain, and implementation of innovative integrated supply chains solution has become a vital element in the globalization process of businesses. Therefore, key players in the logistics, transport, and forwarding businesses have to adapt to new developments required of them to improve in the reliability of services

S. Trzcielinski et al. (Eds.): AHFE 2021, LNNS 274, pp. 271–276, 2021.
https://doi.org/10.1007/978-3-030-80462-6_34

delivery in the globalized economy. According to [14], transport is an indispensable and important part of the process of economic competitiveness, that ensures important distribution for manufacturing, as well as effective and efficient transport which connects businesses to global markets. Hence, the technological progress in the design of transport networks has enabled it to transport a bulk load of raw materials over long distance while reducing transports costs. The globalization of the economy occurs in many areas such as outflow of product and services beyond borders, minimization of taxes and trade obstructions, technical, sociological, immigration, and technology flow [13]. The globalization of manufacturing and creation of complex chains of connections are the reason for the higher requirement for transportation and high-quality demand for the render services. It is accompanied by changes in transport systems which help in accelerating globalization processes across all transport modes. The creation of intermodal transport group are vital components of integrated logistics chains and means of understanding the essential aspect of logistics transport under globalization [7]. An effective, as well as efficient transportation network is the way at which big organizations can easily cope with the ever-changing demands of the global market and apply innovative methods of delivering bulky cargos within a limited time. The development of globalization has been recorded as one of the most significant basic system affecting logistics service providers and transportation is often referred to as one of the four key elements of globalization besides communications, international standardization, and trade liberalization [10]. Therefore, the aim of this study is to review the basic factors impacting the transport segment and logistics through globalization procedure occurring on the global economy.

2 Globalization and Transport

According to [4], globalization can be described as an ongoing process which can be noticed by the increasing number of cross border business, the capacity of worldwide financial flows, and increased labour flows. Presently, the global economy has changed greatly, it requires a new logistics direction, innovation in transport mode, and reduction in global trade costs. [10] mentioned that the four cornerstones of globalization are; worldwide standardization, liberalization of trade, communications, and transportation. Transport is referred to as the heart of globalization, it involves the capability to convey materials, equipment, and people all over the world. The opportunities for businesses to gain from business globalization are on the increase because of its operative impact on the transportation subdivision [9]. Business globalization impact on the transport subdivision is described in two ways and they are;

a) It leads to changes in transport needs, structure, and volume.
b) It leads to changes in procedures of entities arranging the transport services.

In developed nations, the transport needs are influenced by accelerated progress in the economy creating structure and the transport of processed products or freight which is the reflection of the economy that is visible in a volume of production, distribution structure, and material structure of the distribution. Manufacturing and trade globalization are some of the important characteristics of this present age, it comprises of economic scale activity, rapid change in technological advancement, lower manufacturing and trade costs, and higher work rate which have all contributed to the creation of resources today. According to [14], transport is a necessary feature of economic growth and competitiveness that support key circulation for manufacturing, vital movements of products, personal transportability, directly conjoining businesses and global markets. Therefore, it can be said that transport is the heart of globalization because the movement of goods and people around the world are made possible through transport. People and organizations will try to make a profit from globalization because of an increase in an efficient and effective transport network, as well as competitive, responsive and cost-effective transport sector that promotes trade. The creation of perfect conditions for efficient and cost-effective transport networks poses a great challenge to be addressed if the transportation sector is to fully contribute to globalization. An increase in globalization has contributed to a rise in global freight trade activity, firms and worldwide transport setup have been under rising pressures to support the increasing requirements of global market and the globalization associated with manufacturing and consumption [11]. Transporting bigger volume of freight and people quickly and more efficiently can be achieved through extensive technical development, since the efficiency of international distribution is improved by intermodal transportation with an expanding portion of it are cargo moving globally and are containerized. Also, functional growth and removal of bottlenecks in the transport division will considerably improve the efficiency of international transport networks and modes, reduce transit time, improve reliability and minimize the cost of moving goods. Hence, transportation can be referred to as an enabler of global trade, globalization could not have occurred without it.

3 Logistics and Transport Subdivision as an Element of Globalization

Logistics is referred to as a method of preparing, implementing, and regulating the efficient flow of goods, information, and revenues to adapt to the client's want while transport is an important part of logistics, moving commodities between distinct points in the supply chain. According to [12], globalization has been promoted by the expansion of current transport systems, from big freights to mini distribution trucks, the entire delivery system has become carefully merge connecting production activities with an international market. Therefore, it could be said that transport infrastructure and logistics industry play a substantial part in economic progression, integration of countries with the global economy, and serves as an element of globalization [5]. The minimization in lengthened transportation and communications costs has been a vital determinant of present globalization, it is supported by two transportation modes; maritime and air transportation [8]. Air transport has performed a vital part in supporting globalization and constantly improving to meet the necessity of the economic and social integration that bring about globalization. In order to grant the movement of commodities and people that aid productivity on a worldwide extent, air transport has already played a big part, and will continue this part in the future. Multimodal transport is an essential factor in the time and structural economic function of products and services which integrate the benefits of each mode to be efficient and effective. The current trend of transport and distribution logistics towards global expansion is to collaborate integrated logistics delivery, minimizing shipping costs, increasing flexibility and accuracy of freight. According [2], logistics importance's in the process of globalization includes; boosting modernization of economies, followed by the constant reduction of tariffs, and the opening of closed economies. The growth of the logistics industry promoted manufacturing, transportation, and worldwide competitive advantages to the firms and countries that invested in the field. The logistics industry and transport development has provided growth and aided globalization by broadening worldwide trade volume while forcing countries to increase their logistics volume.

4 Transport and Logistics as a Factor of the Global Economy

The activities of transport in global economy manage to be relevant because their involvement starts from the manufacturing process to distribution process to the end users and later turn into a vital element for attaining huge performance in international markets [1]. Transport and logistics have helped greatly in enhancing production and delivery processes, improve competitiveness and promote efficiency. However, transport occupies one-third of logistics costs and affects performance in international markets because it requires greater efforts to manage. Owing to increase in the magnitude of global production, transportation organizations readjust to modern requirements for shipping distribution. The distributions have to be well-timed and safety can be valuable as costs. According [5], the development of logistics and transportation infrastructure plays a key role in the worldwide economy by backing a series of product chains. It started with the advancement and efficiency in the transport system which helps to bolster the commodity chains, followed by minimizing the costs of telecommunication to establish control of the firm's operations, and lastly, to bring about developmental progression between commodity lines and different transport channels. Firms are finding ways of incorporating transport and logistics systems to improve their capabilities in the global competing market but the logistics systems and transport differs with variety of products and the geographical area of the market for the commodities, equipment and finished goods with the aim of procuring the proper product to the appropriate place at the proper time with reduced costs [2]. The crucial factors in global trade are time and capital cost of transport. The cost of global market for most goods is the resource cost of transport which varies with distance, handling during transit, size, and bulk density [3]. Time is another crucial element in the global business because it is needed to move goods from its origin to its final destination, also the loading and offloading it and to process the products and delivery trucks through customs. The transit time is of big concern to logistics companies than the direct transport costs because it brings about large stock holdings. [6] reported that one-day setback in freight can lead to an average cost of 0.8% in tariff and about 170% for industrialized countries. It is suggested that freighters must analyze the costs of several forms of transport to ascertain the amount of time that logistics organizations are willing to pay to minimize it. As a result of former developments, it is possible that the influence of transport and logistics processes enhance the global competitiveness of organizations in the microscale and domestic economies in the large scale, as well as the importance of transport infrastructure, and storage is understood. According to [9], the description of globalization processes studied form the global economic point of view is driven by the following features; a global network or system of transporting goods and services is formed, a global transport network development capable of increasing the speed and mass of transporting cargos, and the massive development of operations of a network of large banks in different nations. Economy progress requires logistics system to be advanced and integrated into one global network of products of commodity mass flows and it can only be achievable if international transport infrastructure creates and make use of the identical advanced technologies with identical transport system.

4.1 Conclusion

In this present era, the methods of a worldwide economy involve the unification of logistics services and transport subdivision which will act as an element of globalization. Transport is an indispensable and important part of the process of economic competitiveness, that ensures important distribution for manufacturing, as well as effective and efficient transport which connects businesses to global markets [14]. Therefore, logistics and transport have become a determinant that is helpful to globalization due to the utilization of modern technologies relating to haulage, transshipment and the flow of information between individual links in the supply chain. This has evidently provided acceleration of the transport process at every one of its stages, set up a network of links corresponding to the flows of goods, and an increase in the reliability of delivery. Lastly, the advancement in strategies and operational rules is likely to save energy, reduce transportation costs, increase competitiveness, and adjustments to global transport systems.

References

1. Barbosa, D.M.E.: Logistics and transport in Colombia: factors affecting the export performance. DOCUMENTOS DE TRABAJO FCEA (2015)
2. Chartterjee, L., Tsai, C.: Transportation logistics in global value and supply chains, working paper number: 2002G, Boston University (2002)
3. Deardoff, A.: The importance of the cost and time of transport for international trade. Paper in ECMT 2005, Time and Transport (2005)
4. Fischer, S.: Globalisation and its challenges. Am. Econ. Rev. **93**(2), 1–30 (2003)
5. Hayaloglu, P.: The impact of developments in the logistics sector on economic growth: the case of OECD countries. Int. J. Econ. Financ. Issues **5**(2), 523–530 (2015)
6. Hummels, D.: Time as a Trade Barrier, GTAP Working Paper No. 18, Global Trade Analysis Project, Purdue University (2001)
7. Kherbash, Q., Mocana, M.L.: A review of logistics and transport sector as a factor of globalization. In: 22nd International Economic Conference. Procedia Economics and Finance, vol. 27, pp. 42–47 (2015)
8. Kopp, A.: Summary of Discussions, in ECMT, Transport and International Trade, Round Table 130, ECMT (2006)
9. Kozlak, A.: Transport as a factor and subject of globalization. Medzinárodná Vedeckákonferencia Globalizácia a Jej SociálnoEkonomické Dôsledky 2008, pp. 273–278 (2008). ISBN 979-80-969745-1-0

10. Kumar, S., Hoffmann J.: Globalization: the maritime nexus. In: Handbook of Maritime Economics and Business, pp. 35–62. Informa, Lloyds List Press, London (2002)
11. Organisation for Economic Co-operation and Development: Globalisation, Transport and the Environment (2010). www.oecd.org/publishing/corrigenda
12. Rodrigue, J.-P., Comtois, C., Slack, B.: The Geography of Transport Systems. Routledge, London (2006)
13. Samimi, P., Jenatabadi, H.S.: Globalization and economic growth. empirical evidence on the role of complementarities. PLoS ONE **9**(4), e87824 (2014)
14. Yildirim, B.: A key element of economic growth. transport for a global economy. Challenges opportunities in the downturn. International Transport Forum, OECD, pp. 8–10 (2009)

Contextualizing Foreign Investments in the Nigerian Construction Industry

Mathew Ikuabe[1(✉)], Clinton Aigbavboa[1], Ayodeji Oke[1], Douglas Aghimien[1], and Wellington Thwala[1,2]

[1] cidb Centre of Excellence and Sustainable Human Settlement and Construction Research Centre, Faculty of Engineering and the Built Environment, University of Johannesburg, Johannesburg, South Africa

[2] SARChl in Sustainable Construction Management and Leadership in the Built Environment, Faculty of Engineering and the Built Environment, University of Johannesburg, Johannesburg, South Africa

Abstract. The contribution of the construction industry to the economy of any nation has been proven to be very important not just because it provides the infrastructural need which serves as an indicator of its development but also its immense contribution to the Gross Domestic Product (GDP) . This has placed the industry in a vantage position in calls for adequate financing. In developing countries such as Nigeria, the government is largely responsible for financing capital projects. Over time, it has been experienced that the low gross domestic savings of the nation has inhibited the adequate financing of capital projects, hence leading to a deficit in the provision of infrastructure. Foreign investment which could be either Foreign Direct Investment (FDI) or Foreign Portfolio Investment (FPI) serves as an alternative for capital investment financing and comes with its own benefits to the host nation. This study theoretically examines the infrastructural development in Nigeria and foreign investment in Nigeria with particular emphasis on the construction industry. Findings revealed that the inflow of foreign investment into the construction industry has been irregular and being inhibited by varying factors. The study concludes by emphasizing the need for collaborative efforts by government and other stakeholders to ensure the increase and stabilized inflow of foreign investment into the construction industry by creating a convenient investment environment and strengthening existing enabling investment laws.

Keywords: FDI · FPI · Construction industry · Infrastructure

1 Introduction

The construction industry is one of the major drivers in the economy of any nation because it is an important contributor to the process of development [1–3]. It produces the physical infrastructural features needed by all sectors of the economy. This includes schools, townships, offices, houses and other buildings; urban infrastructure (including water supply, sewerage, drainage); highways, roads, ports, railways, airports; power systems; irrigation and agriculture systems; telecommunications etc. [4] noted that the

S. Trzcielinski et al. (Eds.): AHFE 2021, LNNS 274, pp. 277–286, 2021.
https://doi.org/10.1007/978-3-030-80462-6_35

activities of the industry have a lot of significance to the achievement of national socio-economic development goals of providing infrastructure, sanctuary and employment. A sizeable significant proportion of the Gross Domestic Product (GDP) of both developed and under-developed countries is accounted for by the construction industry; hence its output is an integral and major output from the national perspective [5, 6].

The current state of infrastructure in Nigeria is poor to say the least. It becomes more worrisome because of the widening gulf between increase in its demand and limited resources to make this a reality. [7] asserted that in most developing countries it is the responsibility of government to provide infrastructure. The Nigerian government has over the years made efforts to ensure that the infrastructural needs of the country are met. The First and Second National Development Plans of 1962–1968 and 1970–1974 captured the construction projects that were of considerable benefits to the nation at large. In 2009, the Nigerian government with its vision 20:2020 initiated the idea with the aim of making the country attain the lofty heights of being among the twenty largest economies in the world by the year 2020. However, it is evident that policies initiated over the years have not attained the anticipated goals. One major challenge encountered by successive governments is the financing of capital projects. [8] asserted that the major source of capital formation in the construction sector that can spur growth and development in Nigeria is from the public sector, with the traditional approach in the major infrastructure procurement process of funding through annual capital budgetary provision. [9] noted that attempts have been made by past administration in trying to solve infrastructure deficit by seeking soft loans and grants from Multilateral Financial Institutions such as International Monetary Fund (IMF), World Bank and other lending nations and organizations. These loans and grants are normally characterized with conditionalities such as cut budgets in the social sectors; removed subsidies, exchange rate crisis, massive devaluation of local currency, terms of trade determination, foreign content and expatriate usage, unemployment and underemployment [10].

There is an urgent need for an alternative for capital investment financing as currently being done in the country. The Nigerian economy is one of such that domestic savings cannot adequately fend for the required infrastructural needs, hence the need for foreign investments in the form of Foreign Direct Investment (FDI) and Foreign Portfolio Investment (FPI). This is buttressed by [11], stating that Nigeria needs substantial amounts of foreign investment in the construction sector to speed up her economic growth most especially in the area of construction facility investment to promote development. With domestic savings not being adequate and the drawbacks from foreign loans, foreign investments present a long term avenue to meeting the infrastructural needs of the country. This comes with immense benefits such as technology transfer, human capital enhancement, trade investment and enterprise development [12, 13]. From the aforementioned constraints faced by the construction industry in Nigeria and the potential benefits it stands to get from foreign investments, this study is geared towards reviewing the current state of foreign investments in the construction industry in Nigeria.

2 Infrastructural Development in Nigeria

Since the attainment of independence, the Nigerian government has tried to consolidate on the infrastructural provision made by the British colonial masters. Pre-independence era experienced an outlay of infrastructural developments in various parts of the country albeit less coordinated. [14] noted that a ten-year development and welfare plan was introduced in 1946 initiated by the colonial government. This was an aftermath of the directive issued by Secretary of State for Colonies, thereby giving rise to the establishment of Central Development Board. However, much result was not attained in this plan as a result of the colonial government's needs rather than placing emphasis on the influence of the nation's economy. Irrespective of the drawbacks of the plan, it served as springboard to successive development plan for the country.

The first post-colonial development plan was initiated in 1962 with a view to steering up infrastructural development in the regions of the country. [15] noted that the objectives of the first national development plan were to expedite economic growth, enhance even distribution of national income, creation of savings for national development and improvement on the standard of living of the citizens. Some developmental strides were attained with this plan in place; however, it was short-lived due to the 30-month civil war that engulfed the country between 1967 and 1970. The second national development plan was birthed after the civil war and was termed one for national reconstruction and rehabilitation. Devastated areas were rebuilt and there was a rigorous drive for putting in place new infrastructures for the purpose of national development. The oil boom of the early 1970s helped spur on this developmental drive. [16] stated that the third national development plan was based on the injection of N30 billion into the national economy. The construction industry amongst other industries like agriculture, education and health was targeted to be a huge beneficiary of this plan. However, much was left to be desired in the implementation of this plan as the projected objectives were hardly met. Successive governments over the years initiated policies and programme to address infrastructural needs of the country; however, it is evident that much is still needed to be done to revamp the current state of infrastructure in the country.

Over the decades, the government of Nigeria has been the major financier of capital projects for national development. Budgetary allocations are made annually for forecast envisaged revenue earnings and also outlined capital expenditure under which the provisions of infrastructural projects are made. [17] reported that progress has been made over the years with respect to improving infrastructure in Nigeria when compared to a host of other Sub-Saharan nations. With quite a few advancements in road, power and information and communications technology (ICT) networks within a relatively large coverage in the country. However, even with these improvements in recent years, the huge infrastructural deficit is still quite evident. A report by [18] highlighted the current state of infrastructure in Nigeria by depicting the achievements and challenges on sectorial basis which includes air transport, ICT, ports, power, railway, roads, water resources, water and sanitation. In summary, the report listed among the achievements to include low-cost telecommunications coverage, market expansion of domestic airlines, geared efforts towards rural electrification, rail and road network expansion, ports concessions and progress on water resources institutional framework. The current challenges are tackling ICT market-efficiency gap, airport terminal concessioning, improvement of marine and land access to ports, improvement of customs delivery at the ports, improving rail service delivery so as to recapture traffic, adequate funding road maintenance, improving rural access roads, open defecation growth reversal for proper sanitation.

3 Foreign Investments in Nigeria

According to [19] and [20], foreign investment can be divided into foreign direct investment and foreign portfolio investment.

3.1 Foreign Portfolio Investment

[21] defined foreign portfolio investment (FPI) 'as an aspect of international capital flows comprising of transfer of financial assets: such as cash, stock or bonds across international borders in want of profit. It occurs when investors purchase non-controlling interests in foreign companies or buy foreign corporate or government bonds, short-term securities or notes. Accordingly, just as trade flows result from individuals and countries seeking to maximize their wellbeing by exploiting their own comparative advantage, so too, are capital flows as individuals and countries seeking to make themselves better off, moving accumulated assets to wherever they are likely to be most productive. FPI involves investment in equity shares, bonds, debentures of local companies or money market instruments of domestic economy by foreign entities. This form of investment is more often on a short term basis and very prone to rapidly exiting the country [22]. [23] stated that Investments in which the investor does not have an effective voice in the management of the enterprise are mainly portfolio.

FPI has become an increasingly important part of the world economy over the past three decades and many developed countries like China, United States of America, Japan etc. and developing countries like Nigeria are exploring it to develop their economies. The growing mismatch between capital requirement and domestic capital stock has consistently led to the call for foreign capital investment in developing economies. [24] noted that an important vehicle for augmenting the provision of funds for domestic investment is foreign capital inflow. Developing countries tend to provide incentives for prospective investors to lure them into investing in the capital market. [25] stated that foreign investors prefer developing countries to developed countries because of higher rate of return on investment. However, there are still glaring challenges posed to prospective foreign investors in these developing countries such as distorted investment incentives and irregular government policies.

In the face of the glaring need for foreign investment in the Nigerian capital market, it is evident that not much has been attained with respect to creating the enabling environment for foreign investors. The percentage of FPI in the Nigerian Bond Market is relatively small compared to domestic investments of Pension fund, Insurance companies, Merchant banks, Commercial banks and Discount houses even though there are opportunities to strengthen the market by attracting more foreign investors. Although steps have been taken over the years to encourage foreign capital inflow such as the abrogation of the Exchange Control Act 1962 in Nigeria has allowed foreigners to participate in the Nigerian capital market both as operators and investors. The internationalization of the Nigerian Stock Exchange which was part of the financial liberalization policy in Nigeria in the mid-2000 has also precipitated to a marginal increase in inflows of foreign portfolio investment into the Nigeria economy through the capital market [26]. The consolidation of such achievement by formulating policies by regulatory bodies is what is needed by capital market in Nigeria at the moment. However, [27] stated that expectations are high that there would be an increase in investments on Nigerian stocks with the rebasing of the nation's economy which projected its GDP at $510 billion.

3.2 Foreign Direct Investment

[28] defined Foreign Direct Investment (FDI) as a cross-border investment by a resident entity in one economy with the objective of obtaining a lasting interest in an enterprise resident in another economy. FDI are the net inflows of investment to acquire a lasting management interest (10% or more of voting stock) in an enterprise operating in an economy other than that of the investor (International Debt Statistics, n.d). Such investment may take the form of a green field investment or merger and acquisition (M&A), which entails the acquisition of existing interest rather than the new investment. [29] opined that the extension of enterprise involves flows of capital, technology, and entrepreneurial skills and, in more recent cases, management practices to the host economy, where they are combined with the local factors in the production of goods and services. As a result the inability of most developing countries national savings in financing their investments, the need for foreign investment in the form of FDI has been a viable tool for shoring up these inadequacies. In furtherance to this, the accelerated acceptance of FDI among economies of the world includes investment treaties on bilateral terms, enhanced technological progress and comparative advantage of global integrated production. This has led many countries taking active steps in encouraging FDI by proffering income tax holiday, guaranteed market preference, providing subsidies to foreign firms, exempting import duties and sometimes ensuring monopoly rights [30].

According to [31] Report, Nigeria experienced a flow of FDI worth USD 1.9 billion in 2018 indicating a drop from the previous year (USD 3.5 billion in 2017). This represented 25.1% of the nation's GDP while the countries making up the main investors include United States of America, China, United Kingdom, The Netherlands and France. Nigeria is third in the pack of nations in Africa with FDI flows closely following Egypt and Ethiopia, with investors' main interest in sectors like hydrocarbon, energy and construction. The vast FDI inflow in Nigeria has been greatly facilitated by the Nigerian Investment and Promotion Council Act of 1995. As expounded in the Act, foreign investors are allowed to maintain 100% ownership of their investment. And also, foreign investors in Nigeria are allowed to benefit a total profit and dividend repatriate without any hindrance [32].

4 Theoretical Views of Foreign Investment

The exogenous growth theory also known as the neo-classical growth model has long been an underpinning theorem with respect to foreign investment. [33] formulated the aggregate production function model. This is modeled against capital input (both domestic and foreign), labor input, and the rate of technological progress which changes over time. It is depicted through the framework that capital amassment has a direct contribution to economic growth in proportion to capital's share of the national output. This theory supports the view that there is an increase in capital stock as a result of foreign investment, which in turn has its impact on economic growth. [34] proposed the new endogenous growth models which identifies with long-run growth as a resultant of advancement in technology; it equally propagates a framework in which technology transfer, diffusion and spill-over effects are means in which foreign investment can perpetually increase rate of economic growth in the host nation. Foreign investments by multinational corporations bring about research and development (R&D), and also the human capital accumulation which yields growth spill-overs, and in turn impacts on the recipient country firms and economy [35].

[36] posited that the introduction of new technologies by foreign investments leads to increase in labour and capital productivity. This eventually brings about stabilized returns on investment and labour which has the potential to grow exogenously. This theory places emphasis on productivity as a result of foreign investment deployment of new technological input. Recipient nations tend to leverage on the importation of innovative technology which in turn aid in spurring new advancement in labour and capital productivity. [37] studied the determinants of foreign investment categorization. The study posited push or demand side and pull or supply side as the determinants of foreign investment flow to any nation. The pull factors as attributed by the study represent the micro and macro attributes influencing the flow of foreign investment in the host country. With particular reference to infrastructure, the size of the country's foreign reserve, borrowing cost, labour cost, inflation rate, Debt-GDP ratio, economic growth, industrial disputes and unemployment ratio. The push factors are the micro and macro features in the source countries that coerce investments out the recipient countries. These include drive for larger market, search for cheaper raw materials, trade conditions and tax policies of the source country.

[38] dwelt on inflows of foreign investment and their sources with emphasis on FDI. The study explained that the inflow of FDI can be influenced by output volatility in the source and host countries. And this having the potential to have negative or positive impacts on FDI flows. Due to non-synergy between the source and recipient countries in business cycle, positive effect may arise due to substitution while negative effect due to revenue impacts. The inflows of FDI to developing countries from developed countries have the propensity to move the reverse direction with respect to the business cycle of the source country [39]. Generally, these propositions and formulated frameworks of foreign investments can be classified into neo-classical (exogenous) growth models and the new endogenous growth models. The former assumes that technological upgrade is exogenous, while the latter posits that the stock of human capital and change in technology are the two main propelling factors for economic growth.

5 Foreign Investment in the Nigerian Construction Industry

The Nigerian construction industry has experienced its growth in leaps and bounds over the decades. With the nation experiencing unstable political regimes since independence, the construction industry like other industries in the economy of the nation has encountered irregular and unstable transitions in its contribution to the GDP of the country. [8] asserted that the contribution of the construction industry to the GDP of the country was between 3 and 6% from independence in 1960 to the 1980s, and then experiencing a slump to about 1% before the transition to civilian government in 1999. However, a significant improvement has been made since then with the industry averaging a contribution of 2.99%. Though the inherent challenges posed to the industry with respect to proper and adequate financing are still evident. This has led to the continuous call for foreign investment in the industry.

According to [40], since the switch to democracy in 1999 there has been significant increase in the flow of FDI into the construction industry in Nigeria. The implementation of enabling laws propelling the inflow of foreign investment can be attested to this amongst other factors. One of such is the Investment and Securities Act of 1999 which set up the Investment and Securities Tribunal as an appellate court for settlement of investment and securities issues [41]. This reposes the confidence in prospective investors by having in place the appropriate legal framework that gives guarantee and security of investments by setting up investment dispute mechanisms. Other enabling laws for foreign investment in Nigeria include the Nigerian Investment and Promotion Commission Act. No. 16 of 1995 which established the Nigerian Investment Promotion Council (NIPC) replacing the Industrial Development Coordination Committee (IDCC); the foreign (monitoring and miscellaneous provisions) Act No. 17 of 1995. These laws set up have over the years given the enabling environment for viable foreign investment by prospective foreign investors; however, the question yet to be answered is have much been achieved with these in place?

[42] revealed that there is a below average rating on the level of enhancement of the enabling laws for FDI in the Nigerian construction sector. This is as a result of the laws not conforming to international standards as some of them were enacted decades ago, therefore not appealing to current attractive terms for foreign investments. On the other hand, enforcement mechanisms of laws in Nigeria are quite weak. In Nigeria, efforts need to be geared towards effective implementation of policies and not just their formulation [43]. This perennial national setback must have hugely influenced the flow of foreign investment as no prospective investor would be confident to commit funds into a nation that is nonchalant when it comes to putting to work laid down policies and laws. One notable example is the National Construction Policy which the Nigerian government hasn't shown any sincerity of purpose towards its implementation [44].

Over the decades, the political instability experienced in Nigeria greatly affected foreign investments not just to the construction sector but other sectors of the economy. The lengthy rule of the military before the turn of the millennium greatly posed a serious challenge to convincing foreign investors to come do business in the country. However, there has been an improvement after the return to democratic rule in 1999 [11]. This is hugely influenced by the fact that a democratic set up has a more liberal inclination towards the tenets of the rule of law as compared to dictatorial one. [45] affirmed that nations with democratic governments are predictably likely to rake in as much as 70% higher in foreign investments against autocratic governments. With many years of military rule, making up for the lost ground of potential foreign investment in the construction industry in Nigeria might be difficult to attain as other countries are in stiff competition towards attracting foreign investors for economic development.

6 Conclusion

The study theoretically examined the position of foreign investments in the Nigerian construction industry. A review of the chronological events on infrastructural development from the colonial times to post-independence was outlined showcasing the state of infrastructure in the country which is still far below international standards. Theoretical propositions of foreign investment were analyzed by the study highlighting the benefits of the foreign investment of recipient nations. The inflow of FPI into Nigeria's stock market has proven to be low as evident in the data from the stock market; although, the country experienced a significant rise in the flow of FDI at the turn of the decade. However, a drop has been witnessed in recent years due to unfavorable and irregular government policies. Talking from the point view of the construction industry, outdated investment laws and weak law enforcement has inhibited the entrance of foreign investment. These factors lead to a lack of confidence on the part of prospective investors as there seem to be no guarantee of return on investments. As a result, the study recommends that a lot should be done on the part of the government in enacting laws that would propel and encourage investment from foreign investors as way of giving a guarantee of return of investment. Equally, law enforcement and legal frameworks should be strengthened as no investor would want to commit funds in a country where the rule of law is not upheld as the case is currently in the country.

References

1. Efem, N.: Federal Government plan to promote foreign direct investment in Nigeria, Thisday Newspaper, p. 4. (2010)
2. Oladirn, T., Ogunsemi, D., Aje, I.: Role of construction sector in economic growth: empirical evidence from Nigeria. FUTY J. Environ. **7**(1), 50–60 (2012)
3. Rangelova, F.: Fundamentals of Economics in Sustainable Construction. Bultest Standard Ltd., Bulgaria (2015)
4. Osei, V.: The construction industry and its linkages to the Ghanaian economy - polices to improve the sector's performance. Int. J. Dev. Econ. Sustain. **1**, 56–72 (2013)
5. Myšková, R.: Satisfaction of financial managers with accounting information system. In: Proceedings of the 3rd International Conference on Management, Marketing and Finances, WSEAS; Norwegian University of Science & Technology, pp. 76–81 (2009)
6. Šatanová, A., Krajüírová, L.: Application of accounting and reporting in a cost-oriented quality management in wood-processing companies. Drvna Industrija **63**(4), 283–289 (2012)
7. Adenikinju, A.: Analysis of the cost of infrastructure failures in a developing economy: electricity sector in Nigeria, AERC Research paper 148. African Economic Research Consortium, Nairobi (2005)
8. Isa, R.B., Jimoh, R.A., Achuenu, E.: An overview of the contribution of construction sector to sustainable development in Nigeria. Net J. Bus. Manage. **1**(1), 47–78 (2013)
9. Mogbo, C.T.: The quantity surveyor as a cost engineer. Paper Presented at the 21st Biennial Conference General Meeting of NIQS, November 25, held at Premier Hotel, Ibadan (2004)
10. Egolum, C.C.: Sustainable development in the midst of poverty in Nigeria: The options. Preface to the topic, delivered at the 5th Annual Lecture (2011) of Faculty of Environmental Sciences, 30th June 2011. Nnamdi Azikiwe University, Awka, Anambra State (2011)

11. Ebekozien, A., Ugochukwu, S., Okoye, P.: An analysis of the trends of foreign direct investment inflows in the Nigerian construction sector. Am. Int. J. Contemp. Res. **51**, 53–69 (2015)
12. Mustapha, B.: Nigeria ranks 19th in the world in attracting FDI. www.untad-export.by.htm. Accessed 5 Jan 2020 (2009)
13. Omagbeme, R.: Nigeria: Bayelsa and FDI. www.allafrica.com/storirs/201001140387.html. Accessed 16 Jan 2020 (2010)
14. Onah, F.O.: Managing Public Programmes and Projects, 2nd edn. Great AP Express Publishers, Nsukka (2010)
15. Iheanacho, E.N.: National development planning in Nigeria: an endless search for appropriate development strategy. Int. J. Econ. Dev. Res. Invest. **5**(2), 49–60 (2014)
16. Obi, E.A.: Development Administration. Bookpoint Ltd., Onitsha (2006)
17. African Development Bank.: An Infrastructure Action Plan for Nigeria: Closing the Infrastructure Gap and Accelerating Economic Transformation (2013)
18. World Bank.: Nigeria's Infrastructure: A Continental Perespective. http://documents.worldbank.org/curated/en/819441468077937967/Nigerias-infrastructure-a-continental-perspective. Accessed 23 Jan 2020 (2017)
19. Babalola, S.J., Dogon-Daji, S.D., Saka, J.O.: Exports, foreign direct investment and economic growth: an empirical application for Nigeria. Int. J. Econ. Financ. **4**(4), 95–105 (2012)
20. Orji, O.H.: Foreign direct and portfolio investment in Nigeria and selected African countries from 1980 to 2004. Paper Presented at the Senior Executive Course No. 26 of the National Institute for Policy and Strategic Studies, Kuru, Jos (2004)
21. Schneider, B.: Measuring Capital flight: Estimates and Interpretations, Working paper 194. Overseas Development Institute, London (2003)
22. Kolapo, Y.: Unveiling real economic pests. The Punch Newspaper, p. 12 (2010)
23. Fabayo, J.A.: Foreign direct investment and manufacturing industries in Nigeria: performance, prospects and problems. In: Nnanna, O.J., Okafor, C.M., Odoko, F.O. (eds.) Foreign Private Investment in Nigeria. Central Bank of Nigeria: Proceedings of the 12th Annual Conference of the Regional Research Units (2003)
24. Omisakin, O., Adeniyi, O., Omojolaibi, A.: Foreign direct investment, trade openness and growth in Nigeria. J. Econ. Theory **3**(2), 13–18 (2009)
25. Vita, G.D., Kyaw, K.S.: Determinants of FDI and portfolio flows to developing countries: a panel cointegration analysis. Eur. J. Econ. Financ. Administrative Sci. **13**, 161–168 (2008)
26. Ekeocha, P.C., Ekeocha, C.S., Malaolu, V.O., Moses, O.S.: Modelling the long run determinants of foreign portfolio investment in Nigeria. J. Econ. Sustain. Dev. **3**, 8–14 (2000)
27. Egwatu, P.: New GDP figures'll attract FDI, foreign portfolio investment, Vanguard Newspaper. http://www.vanguardngr.com/2014/04/new-gdp-iguresll-attract-fdi-foreign-portfolio-investment/ (2014)
28. OECD: Foreign Direct Investment for Development and Maximizing Benefits, Office for National Statistics, https://www.ons.gov.uk/statbulletin. Accessed 2nd Jan 2020 (2013)
29. Chenery, H., Stout, A.: Foreign assistance and economic development. Am. Econ. Rev. **55**, 679–733 (1966)
30. Bouoiyour, J.: Labour Productivity, Technological Gap and Spillovers: Evidence from Moroccan Manufacturing Industries. University of Pau (2003)
31. United Nations Conference on Trade and Development.: World Investment Report 2018. https://en.portal.santandertrade.com/establish-overseas/nigeria/investing. Accessed 23 Jan 2020 (2019)
32. Popoola, B.: Foreign direct investment in Nigeria. Am. J. Sci. Ind. Res. **2**(2), 45–53 (2013)
33. Cobb, C.W., Douglas P.H.: A theory of production. Am. Econ. Rev. **18**, 139–165 (1928)
34. Nair-Reichert, U., Weinhold, D.: Causality tests for cross-country panels: a new look at FDI and economic growth in developing countries. Oxford Bull. Econ. Stat. **6**(32), 153–171 (2001)

35. Barro, R., Sala-I-Martin, X.: Economic Growth. McGraw-Hill, Cambridge (1995)
36. De Jager, J.: Exogenous and Endogenous Growth. University of Pretoria ETD (2004)
37. Polat, B.: Determinants of FDI into central and eastern European countries: pull or push effects. Eurasian J. Econ. Financ. **3**(4), 39–47 (2015)
38. Nicet-Chenaf, D., Rougier, E.: Source and Host Country Volatility and FDI: AGravity Analysis of European Investment to Middle East and North Africa, LAREFI Working Paper, 2014-05 (2014)
39. Levy-Yeyati, E., Panizza, U., Stein, E.: The cyclical nature of north-south FDI flows. J. Int. Money Financ. **26**(1), 104–130 (2007)
40. Ebekozien, A., Abdul-Aziz, A., Jaafar, M.: A comparative study of foreign direct investment in Nigerian construction sector during military regime and democratic government from 1984–2017. Int. Trans. J. Eng. Manage. Appl. Sci. Technol. **9**(5), 425–437 (2018)
41. Wakil, G.H.: The role of foreign direct investment in sustainable economic growth: challenges for Nigeria. J. Invest. Facilitator **14**(3), 4–6 (2004)
42. Ugochukwu, S.C., Okoye, P.U., Ebekozien, A.: The challenges facing foreign direct investment in the Nigerian construction sector. Int. J. Eng. Bus. Enterprise Appl. **13**(1), 6–12 (2015)
43. Ugwuanyi, B.I., Chukwuemeka, E.O.: Obstacles to effective policy implementation by public bureaucracy in developing nations: the case of Nigeria. Singaporean J. Bus. Manage. Stud. **1**(8), 34–43 (2013)
44. Todero, R.: Infrastructure and economic development: the Nigeria experience. J. CBN Econ. Financ. Rev. **14**(3), 4–6 (2001)
45. Guerin, S.S., Manzocchi, S.: Political regime and FDI from advanced to emerging countries. Rev. World Econ. **145**(1), 75–91 (2009)

Business Management Engineers: Profile and Competencies of Generations X, Y and Z

Rodolfo Martinez-Gutierrez(✉), Carmen Esther Carey-Raygoza, Carlos Hurtado-Sanchez, Beatriz Chavez-Ceja, and Artemnio Lara-Chavez

Tecnológico Nacional de México/IT de Tijuana, Calzada Del Tecnológico S/N, Fraccionamiento Tomas Aquino, 22414 Tijuana, Baja California, México
{rodolfo.martinez,carmen.carey,carlos.sanchez,beatriz.chavez,alara}@tectijuana.edu.mx

Abstract. The 2017 report of the Organization for Economic Cooperation and Development (OECD) on competences in Mexico indicates the opportunity to rethink actions aimed at strengthening competitiveness, development and social progress, considering as a basis the impact from the capacities This is why this research is a series of methodological analyzes from the methodological perspective of the Fifth Systematic Helix (QHS) to determine the evolution level and the relevance needs of the educational programs in the area of economic-administrative sciences, in this research on Business Management Engineer of the Tecnologico Nacional de Mexico Campus Tijuana, considering the systematic approach of the QHS with representatives of the business sector, government, education, associations, consultants from the sectors subject of study and with this evaluate the prospects of generational competitiveness and the educational gaps in vocational training. In this research, the generations of graduates of generations X, Y and Z are taken as a sample of the universe. Considering representatives of the different sectors of employability such as the industrial sector, commerce, services, government or the entrepreneur. With the results of the research, the work situational analysis studies of each study program are proposed, to formulate initiatives of labor, professional and research competencies. Likewise, design and development of specialty strategies according to the sectorial feedback of generational competitiveness.

Keywords: Job competencies · Professional competencies · Research competencies

1 Introduction

At the Tecnologico Nacional de Mexico Campus Tijuana [1] the career of Engineering in Business Management is attached to the Department of Economic-Administrative Sciences (CEA) The comprehensive training of an Engineer in Business Management has the general objective of training professionals who contribute to the management of companies and process innovation; as well as the design, implementation and development of strategic business management, optimizing resources in a global environment, with

S. Trzcielinski et al. (Eds.): AHFE 2021, LNNS 274, pp. 287–291, 2021.
https://doi.org/10.1007/978-3-030-80462-6_36

ethics and social responsibility. The Business Management Engineer [2]. will acquire the knowledge, skills, attitudes and values that allow him to develop in the productive sector and in turn provide the development of human capital to achieve organizational objectives, within an ethical framework and a multicultural context; which leads him to promote the design, implementation and management of resources in an optimal and sustainable way. Scope to be covered by the graduate of the Engineering in Business Management career will have to develop in companies of the public and private sector of transformation and service, performing functions derived from the area of production, finance, human resources, marketing and auditing, as well as in the promotion and activation of professional services companies [3].

According to the research carried out by the OECD in turn, an analysis of their Graduate profiles of the best universities in the world, national ranking with the National Technology of Mexico campus Tijuana as a frame of reference for an institutional development project in the National Technology of Mexico Campus Tijuana for the formation of an Academic Body that has lines of research-oriented to the analysis and innovation of the needs of the study programs and their relevance for an effective process of professional integration and thus contribute to the sectoral development of their area of specialty.

2 Framework

The term generation [4] refers to an age group that shares throughout its history a set of formative experiences that distinguish them from its predecessors. Currently, in higher education institutions it is possible to see four generations coexisting: generation X (from 1965 to 1980), generation Y (from 1980 to 1994), and generation Z (from 1995 to 2010). Each generation responds, according to researchers on the subject, to different attitudes and expectations about work and their career.

Generation X individuals, according to this model, are nomadic archetypes. They have been independent since childhood. They have little reason to employ or adopt the visions of the postwar generation since they have rarely used their practical reality. Generation Y, offspring that make up the relief staff in the workplace, in today's generation, these are school children, the older ones are finishing postgraduate studies. They have grown up with planned lives. Generation Z is the first generation considered digital natives, that is, they were born immersed in digital culture. Its characteristics include the following: Experts in understanding technology; Multitasking; Socially open from technologies; Quickness and impatience; Interactive; and Resilient. Table 1 shows the main characteristics of each generation regarding technology.

The OECD highlighted that the main tool for conducting higher education was evaluation, accreditation of programs and institutions, information, relationships on formal and explicit agreements between the government and universities, as well as the distribution of financial resources. It also proposed the creation of a national accreditation system for university study programs that would not cross the official knowledge of a state body such as the Secretary of Public Education.

In this line of reconstruction, the OECD requested a group of researchers of different theoretical stature the Project Definition and selection of competencies: Theoretical and

Table 1. Main characteristics of each generation

Generation	Born	Technology	Digital competence
X	1965–1980	PC	Early digital adopters
Y	1981–1994	Smartphone	Digital natives
Z	1995–2010	AR/VR	Digital innates

Conceptual Foundations, whose main objective was to offer a resource for the process of defining, selecting and measuring the competencies that an individual needs to lead a successful and responsible life and for societies to face the challenges of the present and the future [5].

2.1 Model and Methodology

According to the educational profiles of the Tecnologico Nacional de México, the Business Management Engineer must cover the following competencies in his/her work, professional and research field, which are presented in Table 2.

Table 2. Competences and skills of the engineer in business management

Labor, professional and research skills	
•	Management skills
•	Analytical skills
•	Systemic approach
•	Technological knowledge
•	Research and critic skills

According to the regulations of the Tecnologico Nacional de Mexico, the Engineering in Business Management study program dates from 2009, when the following competencies were defined to be developed throughout the comprehensive training of the study plan. 1. Develop and apply managerial skills and engineering in the design, creation, anagement, development, strengthening and innovation of organizations, with a systemic and sustainable orientation for effective decision-making, 2. Design and innovate administrative structures and processes, based on the needs of organizations to compete efficiently in global markets, 3. Efficiently manage the organization's resources with a shared vision, in order to supply quality goods and services, 4. Apply quantitative and qualitative methods for the analysis and interpretation of data and systems modeling in organizational processes for continuous improvement, meeting world quality standards, 5. Design, evaluate and undertake new businesses and business projects that promote sustainable development and social responsibility in a competitive market, 6. Design and implement marketing strategies based on information collected from primary and secondary sources of the consumer or user of a product, according to market opportunities and threats, 7. Establish programs to strengthen safety and hygiene in organizations, 8. Manage comprehensive quality systems, exercising effective leadership and ethical commitment, applying the basic tools of engineering. 9. Interpret and apply legal norms that affect the creation and development of organizations, 10. Integrate, direct and develop work teams for continuous improvement and integral growth of organizations, 11. Analyze and interpret financial information to detect opportunities for improvement and investment in a global world, which affect the profitability of the business, 12. Use new information technologies in the organization to optimize communication processes and make decision-making more efficient, 13. Promote the development of human capital, for the achievement of organizational objectives, within an ethical framework and a multicultural context, 14. Apply research methods to develop and innovate systems, processes and products in the different dimensions of the organization, 15. Manage the supply chain of organizations with a process-oriented approach, 16. Analyze and interpret the global economy to facilitate decision making in the organization. For the purposes of this article, mention is made that it will be applied in the QHS and DCS research methodology as a sectoral linkage strategy and applied research [6–9].

3 Conclusions

When investigating the different universities that offer engineering degrees in business management, we observe that only the campuses belonging to the TecNM offer it, some universities have careers related to business or business and innovation engineering but, the subjects, objectives, and graduation profile are different. However, although this engineering is 11 years old, it is offered at 195 TecNM campuses, due to industry demands and the needs of engineers who can combine the knowledge of industrial engineering with the administration. Engineering in business management has allowed graduates to locate themselves in different areas, that's the importance of the analysis made, to glimpse the value of the profile and the necessary competencies in work centers. Within the aforementioned competencies the application of continuous improvement methodologies, cost reduction, and maximization of the capacity of industrial processes; give adequate follow-up to the projects that are presented in the different companies of the productive sector; understanding the use of diagramming tools and use them in process mapping or the creation of organizational diagrams within SMEs; In addition to applying the design of parameters, design of tolerances, continuous improvement programs, process control programs, preparation of processes and products for launch, they are considered essential for the Development of business management engineers.

References

1. Tecnologico Nacional de México Campus Tijuana (TecNM). https://www.tijuana.tecnm.mx/

2. Ingeniería en gestión empresarial: un híbrido de los institutos tecnológicos. https://anfei.mx/revista/index.php/revista/article/view/196/672
3. Ingenieria en Gestión Empresarial, Tecnologico Nacional de Mexico Campus Tijuana. https://www.tijuana.tecnm.mx/ing-en-gestion-empresarial/
4. Schroer, W.: Defining, Managing, and Marketing to Ge- nerations X, Y, and Z. The Portal, 10, 9 (2008). http://goo.gl/Fc40dB. Accessed 15 Feb 2015
5. Organización para la Cooperación y el Desarrollo Económicos (OCDE) (2017). Consultado el 15 de agosto, 2017. Disponible en. https://www.oecd.org/eco/surveys/mexico-2017-OECD-Estudios-economicos-de-la-ocde-vision-general.pdf
6. Martínez Gutiérrez, R.: Quinta Hélice Sistémica (QHS), un método para evaluar la competitividad internacional del sector electrónico en Baja California, México. Investigación Administrativa [en linea] 2012, (Julio-Diciembre): [Fecha de consulta: 20 de agosto de 2018] Disponible en: http://www.redalyc.org/articulo.oa?id=456045338003. ISSN 1870-6614
7. Martínez Gutiérrez, R.: Análisis de productividad y trazabilidad académica caso: Quinta Hélice Sistémica 2010–2020 [Reseña]. Administrar Lo Público, 8va. Compilación, (8)62–81, ISBN 978-9968-932-36-3 CICAP - Universidad de Costa Rica, San José (2020)
8. Martinez-Gutierrez, R.: Methodology of dictionaries of sector competences (DCS), to design standards of professional competences, research and labor. In: Nazir, S., Ahram, T., Karwowski, W. (eds.) AHFE 2020. AISC, vol. 1211, pp. 323–329. Springer, Cham (2020). https://doi.org/10.1007/978-3-030-50896-8_46
9. Martinez-Gutierrez, R.: Methodology of the fifth systemic helix for the development of public sector policies. J. Competitiveness Stud. **22**(3–4), 147–168 (2014)

Quality Evaluation of Measurement Service in Online Clothing Mass Customization Mode

Xiaoli Yu[1,2](✉) and Zhenzhen Huang[1]

[1] Faculty of Fashion and Art Engineering, Minjiang University, 200 Xiyuangong Road, Minhou, Fuzhou, Fujian, China
xiaoliyu@mju.edu.cn

[2] Fujian Provincial Research and Development Base for Garment Industry, Minjiang University, 200 Xiyuangong Road, Minho, Fuzhou, Fujian, China

Abstract. The measurement service of online clothing mass customization is one of the key factors to win customer recognition. In order to help apparel companies to objectively evaluate and optimize the quality of the measurement service, based on the SERVQUAL model and characteristics of measurement service, a service quality evaluation scale was designed from the perspective of user experience. Then, a set of questionnaire was designed according to the evaluation scale, and 373 consumers with online clothing customization experience were investigated. They were asked to evaluate the importance and satisfaction of each service quality evaluation index, and a straightforward IPA chart was drawn according to the evaluation results. Finally, according to the quadrant distribution of each quality factor, the factors with low customer satisfaction were found, the directions that apparel companies should focus on improving were clearly pointed out, and three targeted optimization suggestions were put forward.

Keywords: Mass customization · Measurement service · Service design · SERVQUAL · IPA

1 Introduction

Online clothing mass customization has developed rapidly in China in recent years. The measurement services provided by clothing enterprises not only include body measurement, but also fashion design and dress recommendations, such as fashion trends, style details, fabrics and accessories matching, etc. Dimension measurement is the technical support service with the help of various measuring instruments, and the fashion consultant is the emotional support service with the help of intelligent design software. Therefore, the measurement service in the online clothing mass customization mode is a comprehensive service.

A survey of relevant international academic literature reveals that the research results of body measurement in the garment industry are mainly about technology, which can be summarized into three aspects. The first is the study of anthropometric measurement techniques and data relationships. For example, some scholars have studied the mathematical relationship of various body size data in combination with different features of

S. Trzcielinski et al. (Eds.): AHFE 2021, LNNS 274, pp. 292–299, 2021.
https://doi.org/10.1007/978-3-030-80462-6_37

body types, so as to improve the quality of anthropometry, and also laid a foundation for the development of new standard size tables and size definitions [1]. The second is the research on human body data extraction equipment during measurement. For example, some scholars have proved through experiments that the data obtained by using 3D body scanning equipment is basically consistent with the manual measurement data, which can be used in the measurement service of mass clothing customization to improve the measurement efficiency [2]. The third is the research on how to manage the body data obtained by measurement. For example, some scholars have developed an APP software that can recognize language and voice, which can help service personnel store and manage human body data accurately and efficiently [3].

In summary, the research results on clothing measurement services from the perspective of user experience are relatively lacking. Classical studies on service quality are mostly carried out by European and American scholars based on the local marketing environment. Consumers in different regions have different shopping habits, consumer needs and aesthetic needs. In China, the quality of measurement services is one of the key factors for apparel companies to promote online apparel mass customization business. It not only determines whether customers will make up their minds to buy, but also affects customers' satisfaction with the final product. Therefore, it is necessary to study this important service link in depth.

2 Research Design

The online clothing mass customization model comes into being with the progress of technology and the upgrading of consumer demand. It is different from any previous business model. At present, there is no recognized perfect example. The measurement service of each clothing enterprise is constantly improved in the process of exploration. Therefore, this paper takes the classic SERVQUAL model as the basis [4], and then combines the characteristics of measurement service in the garment industry to design the service quality evaluation index. Then we carry out market research on customers with consumption experience to understand the importance and satisfaction of each service quality evaluation index in their mind, and then use the IPA chart to mark the customer's evaluation in the two-dimensional four-quadrant diagram [5, 6], so as to provide reference for garment enterprises to effectively improve customer service experience.

3 Evaluation Scale Design and Questionnaire Survey

3.1 Evaluation Scale Design and Pre-survey

In order to be more in line with the quality evaluation of door-to-door measurement services under the online clothing customization model, according to the SERVQUAL scale and the content of the measurement service, we redesigned an initial scale with 30 indicators, and then invited 4 experts in the clothing customization field to revise the expression of each indicator, merge indicators with similar meanings, and finally An evaluation scale with 5 dimensions and 19 indicators is formed, as shown in Table 1.

Table 1. Quality evaluation scale of measurement service

Evaluation dimension	Evaluation index	Serial number
Tangibility	Anthropometric equipment is advanced	No. 1
	Personalized design software is smart	No. 2
	Service staff are young and beautiful (handsome)	No. 3
	Service staff are dressed and made up appropriately	No. 4
Reliability	The service content is rich	No. 5
	The service process is normative	No. 6
	Service staff can record service data correctly	No.7
Responsiveness	Service staff arrive on time according to the appointment	No. 8
	Service content can be arranged according to customer needs	No. 9
	Customers' new needs can be dealt with in time	No. 10
	Service efficiency is high	No. 11
Assurance	The service process is highly informational	No. 12
	Customers can make appointments more freely	No. 13
	Service staff have high professional quality	No. 14
	Service staff are polite	No. 15
Empathy	Service staff can understand customer needs well	No. 16
	Service staff can consider the interests of customers	No. 17
	Service staff and customers have the same gender	No. 18
	The service process is comfortable	No. 19

According to the above service quality evaluation scale, we designed a set of questionnaire, which contains three parts: demographic characteristics, importance evaluation of service indicators and satisfaction evaluation of service indicators. The first part includes gender, age, monthly income, occupation, education level and the demands of choosing customized clothing. The second part requires the survey respondents to evaluate the importance of each quality index of the measurement service, using the Likert five-level scale, "1, 2, 3, 4, 5" respectively indicate "very unimportant, unimportant, general, Important, very important". The third part requires the survey respondents to evaluate the satisfaction of each quality index based on their own experience. The Likert five-level scale is also used. "1, 2, 3, 4, 5" respectively indicate "very dissatisfied, dissatisfied, general, satisfied, very satisfied".

The reliability and validity of the questionnaire must be tested before formal questionnaire survey is carried out. We first uploaded the questionnaire to "www.wjx.cn" and created the questionnaire link. Then, we contacted 40 customers by random sampling method, and they had at least one online custom clothing consumption within a year. After getting their consent, we issued the questionnaire link to them, and finally recovered 39 valid questionnaires. SPSS22.0 statistical analysis software was used to analyze the survey results. Cronbach's coefficient was 0.878, KMO value was 0.825 (>0.7), and Bartlett's Spherical Spherical Test Sig value was less than 0.05, which indicated that the questionnaire passed the reliability and validity tests and was reliable and meaningful.

3.2 Questionnaire Survey and Data Collection

During the formal survey, we contacted 400 customers who had experience in clothing customization and sent them the questionnaire link. 390 questionnaires were returned. After excluding the questionnaires with low authenticity, 373 were valid, with an effective rate of 93.2%. In order to ensure the reliability and stability of the survey data, we conducted reliability and validity tests on the recovered valid samples. The results show that Cronbach's coefficient is 0.864, KMO value is 0.803, and Sig value of Bartlett's spherical test is also less than 0.05. Therefore, it can be determined that the data of the questionnaire meet the reliability and validity requirements of the evaluation.

4 Result Analysis

4.1 Demographic Characteristics Analysis

Table 2 shows the demographic characteristics of the sample. There are significantly more male customers than women. It can be seen that men are a keen group of online custom clothing, which may be related to their willingness to try new things. Customers

Table 2. Demographic characteristics of the sample

Item	Attribute	Frequency	Ratio/%	Item	Attribute	Frequency	Ratio/%
Gender	Male	304	81.4	Occupation	Government and public institution personnel	113	30.4
	Female	69	18.6		Professional skill personnel	78	20.9
Age	<25	22	5.8		Trade, financial industry personnel	133	35.7
	25–35	65	17.3		other	49	13.1
	36–45	87	23.4	Education level	Senior high school and below	36	9.6
	46–55	95	25.6		Junior college	91	24.3
	56–65	73	19.7		Undergraduate	148	39.7
	>65	31	8.2		Postgraduate and above	98	26.4
Monthly income/Yuan	<5000	13	3.6	Demand for customized clothing	Unique style	73	19.5
	5000–10000	34	9.2		Good fit	89	23.8
	10001–15000	76	20.5		Special occasions	102	27.3
	15001–20000	130	34.8		High quality	68	18.2
	>20000	119	31.9		other	42	11.2

are mainly aged between 36 and 55, accounting for about 50% of the total number. Most of the customers are middle and high-income groups with a monthly income of more than 10,000 yuan, which is also in line with the main price positioning of major clothing customization companies. The customers' occupations are mainly government personnel, public institution personnel and business, trade and financial industry personnel. In terms of education level, undergraduate level personnel are obviously the most, and the proportion of high school and below is the least, which may be because the online customization process is relatively complex and requires a certain cultural foundation. There is no obvious difference in people's needs for choosing customized clothing. In general, the main consumer groups of online clothing mass customization model are middle-aged and young men with high income, decent jobs, good education, and willing to accept new things.

4.2 Analysis of Importance Perception and Satisfaction Perception

Using SPSS22.0 to analyze the survey results, the mean value of importance evaluation and satisfaction evaluation of each service factor can be obtained, as shown in Table 3. In the table, the mean values of importance evaluation are all greater than 3, ranging from 3.25 to 4.34, indicating that customers believe that each service factor is important. The five factors with the highest scores are: No. 14, No. 16, No. 8, No. 2 and No. 17. It can be seen that the professionalism of service staff and communication with customers are particularly important. The five factors with the lowest scores are: No. 18, No. 3, No. 7, No. 11 and No. 1. Accordingly, customers pay less attention to the gender and age of service staff, as well as service tools and service efficiency, and are not very sensitive to them.

The average value of satisfaction evaluation of 6 factors is greater than 4, indicating that customers are satisfied with the service quality in these six aspects, and enterprises should continue to maintain them. Arranged in descending order of scores, they are: No. 4, No. 18, No. 8, No. 15, No. 19 and No. 7. The scores of other factors are all lower than 4, indicating that customer satisfaction has not been achieved. In particular, special attention should be paid to the last five factors, arranged in order from low to high scores, which are: No. 17, No. 13, No. 14, No. 9 and No. 16. These five aspects are the weak links that need to be improved urgently, and need to be focused on by the business side.

Each service factor have a corresponding average value of importance evaluation (I) and a average value of satisfaction evaluation (P). If $I\text{-}P < 0$, it means that customers are satisfied with the experience of these service factors. On the contrary, if $I\text{-}P \geq 0$, it means that the customer's experience is not satisfactory.

I-P of only 7 factors is less than zero, while I-P of the other 12 factors is all greater than zero, indicating that the service experience of these 12 aspects fails to meet customer needs. Among them, the difference of 4 factors is quite prominent, which needs special attention. They are: No. 14, No. 17, No. 16, No. 13, indicating that these four aspects are in urgent need of optimization.

4.3 IPA Analysis

According to Table 3, we can calculate that the total average value of importance evaluation is 4.00, and the total average value of satisfaction evaluation is 3.87. Taking importance as the X axis, satisfaction as the Y axis, and (4.00, 3.87) as the intersection point, a two-dimensional four-quadrant chart can be drawn. Finally, we map the mean value of importance and the mean value of satisfaction of each factor to the chart to get the IPA chart, as shown in Fig. 1.

The first quadrant in Fig. 1 belongs to the dominant area, in which the five factors are: No. 4, No. 5, No. 8, No. 15 and No. 19. It shows that customers think these five aspects are very important, and the satisfaction of actual experience is also relatively high. Therefore, the enterprise should not only continue to focus on, but also consolidate and give full play to these advantages.

The second quadrant belongs to the maintenance area, in which the four factors are: No. 1, No. 7, No. 11 and No. 18. It shows that customers have no excessive demands in these four aspects, and their satisfaction is high. The service provided by the enterprise

Table 3. Importance, satisfaction and IPA Index of measurement service factors

Evaluation dimension	Serial number of evaluation index	Importance evaluation (I)		Importance evaluation (P)		I-P	Significance results of paired-samples T test
		Mean	Standard deviation	Mean	Standard deviation		
Tangibility	No. 1	3.83	0.978	3.94	0.899	–	Not significant
	No. 2	4.2	0.886	3.81	1.004	0.11	Significant
	No. 3	3.25	1.012	3.84	0.995	0.39	Significant
	No. 4	4.02	0.901	4.18	0.982	–	Not significant
						0.59	
						–	
						0.16	
Reliability	No. 5	4.11	0.994	3.98	1.027	0.13	Not significant
	No. 6	4.13	0.901	3.77	0.908	0.36	Significant
	No. 7	3.59	0.962	4.01	0.916	–	Significant
						0.42	
Responsiveness	No. 8	4.24	0.848	4.13	0.878	0.11	Not significant
	No. 9	3.98	0.997	3.68	0.965	0.30	Significant
	No. 10	4.06	0.902	3.78	0.949	0.34	Significant
	No. 11	3.77	0.957	3.88	0.964	–	Not significant
						0.11	
Assurance	No. 12	3.84	1.001	3.76	0.998	0.08	Not significant
	No. 13	4.05	0.893	3.47	1.008	0.58	Significant
	No. 14	4.41	0.879	3.63	1.017	0.78	Significant
	No. 15	4.03	0.916	4.08	0.879	–	Not significant
						0.03	
Empathy	No. 16	4.34	0.862	3.69	1.016	0.65	Significant
	No. 17	4.17	0.995	3.45	1.041	0.72	Significant
	No. 18	3.65	1.089	4.17	0.873	0.52	Significant
	No. 19	4.08	0.896	4.02	0.916	0.06	Not significant

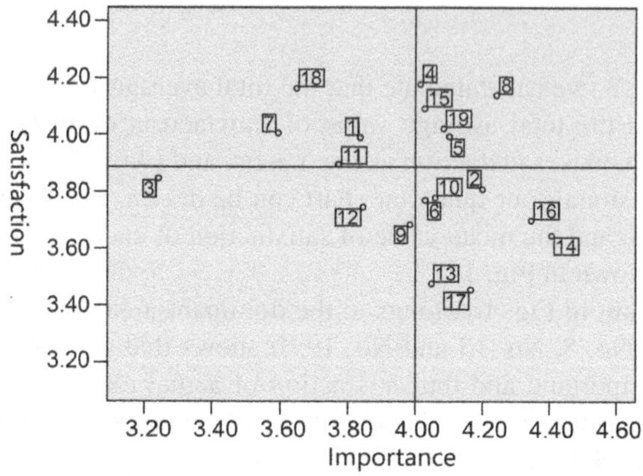

Fig. 1. IPA chart of Importance-satisfaction

has exceeded their expectations. Therefore, the enterprise does not need to invest too much energy in these aspects, as long as the original condition is maintained.

The third quadrant belongs to the secondary improvement area, in which the three factors are: No. 3, No. 9 and No. 12. According to the principle of IPA analysis, these three aspects have not attracted special attention from customers, but the experience satisfaction is relatively low. Therefore, enterprises can not ignore them, should find the reasons of low satisfaction, reduce the shortcomings.

During the survey, we found that many service staffs do not have a deep understanding of the fabric, version, craftsmanship and dress collocation of the clothing, and just take up their posts after a short intensive training. The service they can provide customers is only a uniform set of standardized content. But in fact, some customers are particularly concerned about fabrics or craftsmanship, and some customers are particularly concerned about colors or collocation. If enterprises can understand these different needs before service, and arrange more flexible and professional services in a targeted manner, it will greatly promote the improvement of service quality.

The fourth quadrant belongs to the key improvement area, in which the seven factors are: No. 2, No. 6, No. 10, No. 13, No. 14, No. 16 and No. 17. It shows that customers have high expectations and demands for these seven aspects, but the satisfaction after the actual experience is low. Therefore, enterprises should invest major resources and energy to improve the service quality in these aspects.

In the investigation, we found that there are two reasons why the personalized design software is not satisfactory. First, the resources are not rich enough, the style details, fabrics and accessories that customers can choose are very limited, and it is difficult to achieve the expected uniqueness. Second, virtual fitting cannot be realized. Although customers have seen the three-dimensional effect of their own designs, but still cannot experience trying on the clothes. In addition, some service personnel have poor professionalism and the service process is not standardized. In the process of communicating with customers, they cannot understand their real needs well, and even blindly pursue sales performance, and put forward some unreasonable design suggestions and dress suggestions. Customers generally want to make an appointment for door-to-door measurement service during non-working hours, but it is difficult for them to make an appointment during these time periods because the enterprise is short of manpower. Some customers responded that after the measurenment service, if they have new needs, it will be troublesome to communicate with service staff. If the enterprise can put the design software on the Internet and establish an online communication platform, the service efficiency will be higher.

5 Optimization Suggestions

Based on the above analysis results, in order to better help apparel companies develop online mass customization business, the following four suggestions are put forward. The first is to enrich the database of personalized design software to provide customers with more options. In addition, the virtual fitting function should be added, so that customers can see the fitting effect in advance, so as to judge the products more accurately. The second is to focus on building a professional service team, standardize the service process, modularize the service content, understand customers' needs in advance, and flexibly arrange personalized service content, also, rationally arrange the number of service personnel, improve the wage performance system, and balance the interests of the enterprise and customers. The third is to develop an online communication platform that can help companies improve service efficiency, improve follow-up services, and also maintain customer relationships and increase customer stickiness.

Acknowledgments. The work described in this paper was fully supported by the open fund of Apparel Industry Technology Development Base in Fujian Province (No. MJXY-KF-201901).

References

1. Balach, M., Lesiakowska-Jablonska, M., Frydrych, I.: Anthropometry and size groups in the clothing industry. AUTEX Res. J. **20**(1), 56–62 (2020)
2. De Raeve, A., Cools, J., Vasile, S.: 3D body scanning as a valuable tool in a mass customization business model for the clothing industry. J. Fashion Technol. Textile Eng. **S4**, 1–6 (2018)
3. Chunlan, W., Fengqi, R., Yunxia, C., etc.: Design and implementation of customized garment measurement platform based on speech recognition. In: Software Guide, vol. 18, no 9, pp. 163–166 (2019)
4. Yalan, W., Tainyi, L., Pin, L., Lu, H.P.: Contribution and trend to quality research-a literature review of SERVQUAL model from 1998 to 2013. Informatica Economica **19**(1), 34–45 (2015)
5. Martilla, J.A., James, J.C: Importance-performance analysis. J. Market. **41**(1), 77–79 (1977)
6. Haemoon, O.: Revisiting importance performance analysis. Tour. Manage. **22**(6), 617–627 (2001)

Production Management in Industry

Segmentation and Sequencing in CAx Instructional Videos

Lorena Niebuhr(✉), Nina Rußkamp, and Eva-Maria Jakobs

Department of Textlinguistics and Technical Communication, Human-Computer Interaction Center, RWTH Aachen University, Campus-Boulevard 57, 52074 Aachen, Germany
{l.niebuhr,n.russkamp,e.m.jakobs}@tk.rwth-aachen.de

Abstract. CAx systems (computer-aided technologies) contribute to making automated product manufacturing processes more efficient and flexible. Due to their high complexity, their usage requires both initial and recurring training. This paper addresses instructional videos as a potential means to train CAx users. In a literature study, design and quality issues concerning the segmentation and sequencing of instructions, in particular instructional videos, are investigated from a technical communication perspective. The qualitative and quantitative literature analysis reveals that CAx instructional videos and their design are rarely investigated topics. There is a need to learn more about how to design CAx video tutorials effectively in terms of segmentation and sequencing, depending on task complexity and worker type. Clear design principles would help small and medium-sized enterprises to produce effective videos for their employees, thereby reducing the duration and cost of vocational training.

Keywords: Instructional videos · CAx systems · Segmentation · Sequencing · Vocational training

1 Introduction

CAx systems (computer-aided technologies) are a key factor in making automated product manufacturing processes more efficient and flexible. *CAx systems* is the collective term for all computer-supported systems, whereas *x* is a variable for computer-aided processes such as *design* (CA*D*) or *manufacturing* (CA*M*) [1]. As CAx systems are highly complex, their usage requires both initial and recurring training [2]. CAx training is only a small part of the vocational education of engineers and shop floor workers [2]. Thus, CAx-related skills are typically acquired through on-the-job training or vocational training courses. The latter are time-consuming and costly for the companies. Instructional videos included in blended learning concepts are a promising alternative. They are seen as a communicative tool for solving problems – in this case, the lack of knowledge to operate CAx systems – and enable employees to acquire know-how and skills independently of time and place. The learning effect depends on the quality of the instructional video [3]. However, there has been little research on the quality and design aspects of CAx instructional videos, in particular [4].

S. Trzcielinski et al. (Eds.): AHFE 2021, LNNS 274, pp. 303–310, 2021.
https://doi.org/10.1007/978-3-030-80462-6_38

Guidelines such as [5] see segmenting of content and sequencing of segments as crucial factors for achieving good learning outcomes by instructional videos. This paper aims to give an overview of related literature, focusing on the segmentation and sequencing of content in instructional videos for CAx users.

The topic is investigated from a technical communication perspective within the scope of two research and development projects. In the project WerkerLab[1], a blended learning concept for teaching CAx related skills is developed in collaboration with partners from engineering and software developers. The concept is directed to small and medium-sized enterprises (SMEs). CAx instructional videos are a key component of this approach. In the project CAM2030[2], CAM (computer-aided manufacturing) systems are expanded by digital tools such as evolutionary algorithms, cloud computing, and artificial intelligence to support CAM users in processing highly complex tasks. In addition, the usability and user-oriented acceptance of CAM systems are investigated based on a model of action chains for CAM planning processes. The results are used to derive a training concept for CAM users.

2 Methodology

The literature search was conducted from September to November 2020 in the database of the RWTH Aachen University's library, in Scopus, Web of Knowledge, and Google Scholar. It used the keywords *segmentation, sequencing, instruction, instructional video, CAx systems,* as well as forward and backward snowballing. The search was limited to German and English publications. Irrelevant initial findings were cleaned out. The resulting findings (n = 120) were iteratively screened and reduced to papers discussing instructions in general, content segmentation, sequencing, and instructional videos (for CAx experts in particular). The analysis of the final corpus (n = 50) was driven by two questions: Q1: How are segmentation and sequencing discussed as a feature of instructions? Q2: Are there recommendations concerning instructional videos for CAx experts? The corpus was analyzed quantitatively and qualitatively. The quantitative analysis refers to the distribution of papers to disciplines, as well as the number of publications dealing with the sub-topics 'instructional videos (for CAx systems)', 'segmentation', and 'sequencing'. The qualitative analysis adapts [6]. The literature corpus was analyzed top-down – guided by the research questions, and bottom-up – guided by the material. The main categories of the category system are instructional videos, segmentation, and sequencing.

[1] This research and development project is funded by the European Regional Development Fund (ERDF) (funding number EFRE-0801580) and implemented by the LeitmarktAgentur NRW-Project Management Agency Jülich. The author is responsible for the content of the publication.

EUROPEAN UNION
Investing in our Future
European Regional
Development Fund

EFRE.NRW
Investitionen in Wachstum
und Beschäftigung

[2] This research and development project is funded by the German Federal Ministry of Education and Research (BMBF) within the "Innovations for Tomorrow's Production, Services, and Work" Program (funding number 02J19B081) and implemented by the Project Management Agency Karlsruhe (PTKA). The author is responsible for the content of this publication.

3 Results

In the following, selected results of the literature study are presented. Section 3.1 describes quantitative results, Sect. 3.2 qualitative results.

3.1 Quantitative Results

In the reviewed literature, the sub-topics segmentation (n = 32), sequencing (n = 25), and instructional videos (n = 38) are almost equally often discussed (partly separately, partly together), whereas few publications deal with complex software systems, i.e. CAx systems (n = 9). Publications on instructional videos frequently investigate the efficiency of knowledge transfer of paper-based versus video instructions. Regarding literature on segmentation and/or sequencing only, about a half considers instructional videos, the others focus on paper-based instructions or instructions in general. Authors' areas of expertise are diverse (sorted by descending number of publications: psychology, educational science, instructional technology, technical communication, computer science, and others). Recommendations on segmentation and sequencing in instructional videos are mainly discussed from the perspective of technical communication and instructional technology.

3.2 Qualitative Results

The first subsection describes instructing, the second relates to instructional videos. Subsections segmentation and sequencing deal with general findings as well as recommendations for instructional videos regarding segmentation and sequencing. The last subsection refers to instructional videos as a part of CAx training.

Instructions: *Instructing* is understood as a directive writing or speaking act [7]. The central goal of instructing is to impart procedural knowledge and enable the user to independently achieve a desired state.

[8] describes a theoretical framework for text-based software instructions that abstracts instructions to a sequence of states (desired, prerequisite, interim, unwanted) and actions (by a human or software system). The goal that is achieved by following a procedural instruction is defined as the desired state. The prerequisite state is the condition for moving towards the goal following the instruction. Interim states occur on the way to reach the goal and can be seen as subgoals. Finally, unwanted states such as errors or system malfunctions are considered, too. [8] distinguishes between human and system actions. Human actions are defined as actions users take to reach a goal. System actions are a response to those human actions. Additionally, external events outside the system have to be considered in an instruction. This framework can also be applied to instructional videos.

Instructional Videos: In an appreciable part of the literature on instructional videos, instructional videos for software training are compared to paper-based instructions [9–12], or classical classroom training [13]. Key findings show advantages of instructional videos, such as higher motivation and retention [5] as well as better accuracy in task performance [9]. Disadvantages of instructional videos mainly concern problems such as finding relevant information in the video [14] or the non-adjustable pace of the instruction/demonstration [9]. Some authors, however, stress that these disadvantages can be tackled by adjusting instructional videos' design [5, 14].

A small number of papers offer design guidelines and heuristics based on empirical findings and/or existing literature for instructional videos teaching (complex) software, e.g., [3, 5, 15]. They provide structured advice for improving the design of instructional videos as well as guidance on segmenting and sequencing content in instructional videos, which is detailed in the following subsections.

Segmentation: The term *segmentation* means breaking down complex units into their elements (segments). Segmentation is the sine qua non for sequencing.

In educational and psychological research, segmentation has mainly been investigated against the background of cognitive load theory. Related studies highlight that segmentation is a potential solution to cognitive overload, e.g., [16].

Both the length and the granularity of segments are seen as crucial factors for instructional design [5]. The literature dealing with paper-based instructions and classroom teaching highlights that segments should be short and appropriate. [17] say that a segment should comprise a maximum of five unknown chunks. [18] and [19] investigate how the granularity of a procedural instruction segment influences initial performance, learning, and transfer. A high granularity has a positive effect on initial performance but hampers learning and transfer. Thus, the authors recommend applying so-called fading to the instruction. Fading means continuously decreasing the granularity of single segments and combining small steps in one step.

Segmentation is often accompanied by inserting user-controls, which allow the user to time actions, e.g., start a segment, pause a segment, and go to the next segment. These actions enable users to determine the pace of learning. Studies on segmentation and user-paced learning show "that people learn more deeply when a multimedia message is presented in user-paced segments rather than a continuous unit" [20]. This effect is named the *segmenting principle* and is most likely to emerge if the learning content is complex, the presentation of the learning content is fast-paced, and the learner is inexperienced in the learning content [21]. In contrast, segmentation can have a negative effect on learning if the learner has a high level of prior knowledge. This effect is called the *expertise reversal effect* [22]. Thus, it depends on task complexity, presentation speed, and the learner's prior knowledge if segmenting the learning content eases learning.

Recommendations for Instructional Videos: Design guidelines and heuristics sensitize for the need of content segmentation in instructional videos [3, 5]. They give advice on the length of videos, video segments, and pauses between segments, and describe how to support user-paced learning. There is a broad consensus in the analyzed literature that videos should be as short as possible. Yet, the definition of 'short' differs between papers. Differences exist in the recommendation of the exact duration of a video or segment – ranging from 15–60 s [15] to six minutes [23]. If a defined procedure cannot be shown in the recommended duration, the procedure should be split into separate videos [5]. The procedure should be broken up into meaningful, "bite-size" [16] pieces with a clear beginning and end [3].

Segmenting the content within one video can support the retention process [24]. Pauses (two to five-second breaks) demarcate concept or event boundaries and support the cognitive organization of information [3]. A few studies stress the importance of enabling users to move through the video's content in a self-determined way to quickly find relevant information [5, 14]. Therefore, the content should be segmented, and the segments labeled. Other options to make segments directly selectable are an interactive timeline and/or an index linked to the segments [5].

Sequencing: The term *sequencing* refers to the determination of the segments' order [25]. Sequencing strategies differ in their application scope. Overall, they can relate to the organization of either several topics/ideas (macro-level strategies) or a single one (micro-level strategies) [25]. [26] offer an approach to categorize sequencing principles and rules. They distinguish between world-related, concept-related, inquiry-related, learning-related, and utilization-related sequences; they relate these types of sequences to principles and rules such as ordering segments chronologically, from observations to reasoning, from concrete to abstract. Sequencing principles are not mutually exclusive; sequencing approaches can employ more than one principle [27].

Recommendations for Instructional Videos: For the design of instructional videos, [3, 28], and [12] recommend the simple-to-complex pattern. The simple-to-complex-pattern refers to an increase of task complexity over time. Even on a low level of complexity, the instructions should refer to real-world problems instead of artificial scenarios [5]. Concerning task-based instructions, the order of sequences should correspond to the user's mental plan and the physical steps of task execution. The sequence prototypically comprises the goal or purpose as well as the actions and (intermediate) states that have to be passed through [5]. Different studies on instructional videos apply the simple-to-complex pattern as part of Demonstration-Based Training (DBT) approaches [3, 29] to investigate the role of supporting motivation, task performance, retention [30], and reviews [31]. According to [29], simple-to-complex sequencing facilitates understanding and contributes to retention and the learner's motivation.

Integration of Instructional Videos in (CAx) Software Training: The reviewed literature sparsely deals with the integration of instructional videos into software training. Only three studies [32, 33], and [34] discuss this topic.

[32] investigate the effect of practice in learning with instructional videos for software training: Students participated in one of three software training conditions (video only, video followed by practice, video preceded by practice); afterward, their learning outcomes were compared. The assumption that instructions should precede practice to support acquiring procedural knowledge (see [5]) could not be confirmed. Thus, it seems to be unclear if and when instructional videos should be followed by practice.

Sequencing can also refer to the integration of an instructional video into a higher-order training concept. Only [33] and [34] discuss the integration of instructional videos in CAx training. [33] investigate the use of instructional videos to teach undergraduate engineering students simulation methods. The course comprises four phases: theory, training, problem-solving, and presentation. In the training phase, the students learn to use the CAx system based on instructional videos. The videos guide them through exemplary and common processes step by step. The evaluation shows two main advantages of instructional video; the videos enable the learner to adjust the instruction to the personal learning pace and access the instruction independently from time and space. [33] recommend using videos in addition to classic media. They advise against completely replacing them.

[34] tailors training to the needs of CAD users and includes videos besides paper-based documentation and classic classroom training. The video has the purpose of giving an overview of the CAD system introducing different tools and functions. In addition, instructional segments of exemplary CAD processes are given in the video. Participants of the CAD course later evaluated the course, methods, and materials. They perceived the video and especially the instructional segments positively.

4 Discussion

Instructional videos are increasingly used in various areas of life, including industry, but there is little research that addresses their practical design. Often teaching-learning settings at school and university are considered, but not (further) vocational training. In particular, CAx instructional videos and their design are rarely investigated topics.

In the literature, recommendations for segmenting and sequencing instructional videos for complex software systems are given, yet no CAx-specific recommendations exist. It is unclear if it is sufficient to apply existing recommendations to the field of CAx systems without adjustments or if and how they have to be modified for CAx systems, the purposes of vocational training, and the contextual requirements of SMEs.

Additionally, the recommendations in the literature are often unspecific, ambiguous, and/or not detailed enough, e.g., statements concerning the granularity of segments as well as the duration of videos or video segments. A promising approach seems to be to insert breaks between subgoals. However, it is unclear according to which principle they should be placed depending on the segment type, the sequencing pattern, and/or the learner type. According to the literature, segments should be labeled to support users to find relevant information. Information is missing about how the labeling should be implemented, e.g., in terms of naming patterns.

Sequencing is mostly discussed concerning sequencing patterns. Coming from Demonstration-Based Training (DBT) approaches, simple-to-complex is the most commonly discussed sequencing pattern. Again, it is unclear how this principle should be used in practice, e.g., how many and which levels of increasing complexity are appropriate between the poles 'simple' and 'complex'.

5 Conclusion and Outlook

The industrial practice in Germany indicates an increasing interest in instructional videos for CAx systems that fit the requirements of software training and SMEs. In contrast to large companies, SMEs lack financial resources to invest in specialists for producing high-quality instructional videos. There is a need for further research to enable mechanical engineering experts as well as software developers in SMEs to produce their own instructional videos and incorporate them into CAx training courses. There is a lack of research on the interplay of factors i.e., the objective, subject matter, learner type, context of use, and design features of the instructional video such as segmentation and sequencing. User requirements for instructional videos need to be explored in detail. Studies in realistic environments with workers on the shop floor differing in their level of experience and knowledge are needed. Future research should learn from existing research and examine what can be transferred from the use of instructional videos in school and university classrooms into industrial settings. Finally, research is needed on integrating instructional videos in CAx training.

Limitations of this study can be the following: Firstly, relevant literature may have been overlooked. Secondly, the literature corpus is limited to German and English publications. Thirdly, publications on instructional videos for specific contexts other than software training have been excluded from this literature study.

References

1. Vajna, S., Weber, C., Zeman, K., Hehenberger, P., Gerhard, D., Wartzack, S.: CAx für Ingenieure: Eine praxisbezogene Einführung. Springer Vieweg, Wiesbaden (2018)
2. Jakobs, E.-M., Digmayer, C., Vogelsang, S., Servos, M.: Not ready for industry 4.0: usability of CAx systems. In: Ahram, T., Falcão, C. (eds.) Advances in Usability and User Experience. AHFE 2017 International Conference on Applied Human Factors and Ergonomics, pp. 51–62. Springer International Publishing, Cham (2017)
3. Brar, J., van der Meij, H.: Complex software training: harnessing and optimizing video instruction. Comput. Hum. Behav. **70**, 475–485 (2017)
4. Tandi, M., Niebuhr, L., Jakobs, E.-M.: How-to: instructional video. Recommendations for the design of software video trainings for production workers. In: 13th International Conference on Advances in Computer-Human Interactions, pp. 378–383 (2020)
5. van der Meij, H., van der Meij, J.: Eight guidelines for the design of instructional videos for software training. Tech. Commun. **60**, 205–228 (2013)
6. Mayring, P.: Qualitative Inhaltsanalyse. Grundlagen und Techniken. Beltz Verlag, Weinheim Basel (2010)
7. Ballstaedt, S.-P.: Sprachliche Kommunikation: Verstehen und Verständlichkeit. Narr Francke Attempto, Tübingen (2019)
8. Farkas, D.K.: The logical and rhetorical construction of procedural discourse. Tech. Commun. **46**, 42–54 (1999)
9. Alexander, K.P.: The usability of print and online video instructions. Tech. Commun. Q. **22**, 237–259 (2013)
10. van der Meij, J., van der Meij, H.: A test of the design of a video tutorial for software training. J. Comput. Assist. Learn. **31**, 116–132 (2015)
11. Donkor, F.: The comparative instructional effectiveness of print-based and video-based instructional materials for teaching practical skills at a distance. Int. Rev. Res. Open Dist. Learn. **11**, 96–116 (2010)
12. Kok, A.: Video-Tutorial versus textbasierte Anleitung: Eine vergleichende Untersuchung zur Interaktion und Lernerfolg am Beispiel von Einführungen in die Literaturverwaltungssoftware Citavi. Berliner Handreichungen zur Bibliotheks- und Informationswissenschaft 457 (2020)
13. Wyeld, T.: Using video tutorials to reduce cognitive load and frustration for early learners of 3D maya: a comparative case study discussion on knowledge visualisation modes. In: 20th International Conference Information Visualisation, pp. 209–215 (2016)

14. Ponzanelli, L.: Too long; didn't watch! extracting relevant fragments from software development video Tutorials. In: 38th International Conference on Software Engineering, pp. 261–272 (2016)
15. Plaisant, C., Shneiderman, B.: Show Me! guidelines for producing recorded demonstrations. In: IEEE Symposium on Visual Languages and Human-Centric Computing, pp. 171–178 (2005)
16. Mayer, R.E., Moreno, R.: Nine ways to reduce cognitive load in multimedia learning. Educ. Psychol. **38**, 43–52 (2003)
17. Niegemann, H.M., Domagk, S., Hessel, S., Hein, A., Hupfer, M., Zobel, A.: Kompendium multimediales Lernen. Springer, Berlin Heidelberg (2008)
18. Eiriksdottir, E., Catrambone, R.: Instruction use depends on the level of detail. In: Human Factors and Ergonomics Society Annual Meeting, pp. 2365–2369 (2014)
19. Eiriksdottir, E., Catrambone, R.: Procedural instructions, principles, and examples: how to structure instructions for procedural tasks to enhance performance. Learn. Transfer Hum. Factors **53**, 749–770 (2011)
20. Mayer, R.E., Pilegard, C.: Principles for managing essential processing in multimedia learning: segmenting, pretraining, and modality principles. In: Mayer, R.E. (ed.) The Cambridge Handbook of Multimedia Learning, pp. 316–344. Cambridge University Press, Cambridge (2014)
21. Mayer, R.E.: Segmenting principle. In: Mayer, R.E. (ed.) The Cambridge Handbook of Multimedia Learning, pp. 175–188. Cambridge University Press, Cambridge, New York (2009)
22. van Merriënboer, J.G.: An expertise reversal effect of segmentation in learning from animations. Comput. Hum. Behav. **27**, 46–52 (2011)
23. Guo, P.J., Kim, J. Rubin, R.: How video production affects student engagement: an empirical study of MOOC videos. In: 1st ACM Conference on Learning at Scale, pp. 41–50 (2016)
24. Lusk, D., Evans, A.D., Jeffrey, T.R., Palmer, K.R., Wikstrom, C.S., Doolittle, P.E.: Multimedia learning and individual differences: mediating the effects of working memory capacity with segmentation. Br. J. Edu. Technol. **40**, 636–651 (2009)
25. Reigeluth, C.M., Merrill, M.D., Wilson, B.G., Spiller, R.T.: The elaboration theory of instruction: a model for sequencing and synthesizing instruction. Instr. Sci. **9**, 195–219 (1980)
26. Posner, G.J., Strike, K.A.: A categorization scheme for principles of sequencing content. Rev. Educ. Res. **46**, 665–690 (1976)
27. van Patten, J., Chao, C., Reigeluth, C.M.: A review of strategies for sequencing and synthesizing instruction. Rev. Educ. Res. **56**, 437–471 (1986)
28. Steehouder, M., van der Meij, H.: Designing and evaluating procedural instructions with the four components model. In: IEEE International Professional Communication Conference, pp. 797–801 (2005)
29. van der Meij, H., van der Meij, J.: The effects of reviews in video tutorials. J. Comput. Assist. Learn. **32**, 332–344 (2016)
30. van der Meij, H., van der Meij, J., Voerman, T., Duipmans, E.: Supporting motivation, task performance and retention in video tutorials for software training. Educ. Tech. Res. Dev. **66**, 597–614 (2018)
31. van der Meij, H.: Reviews in instructional videos. Comput. Educ. **114**, 164–174 (2017)
32. van der Meij, H., Rensink, I., van der Meij, J.: Effects of practice with videos for software training. Comput. Hum. Behav. **89**, 439–445 (2018)
33. Brenner, G., Walter, H.W.: Individualized learning with instructional videos. In: Engineering Simulation Education SummerSim 2018, pp. 1–7 (2018)
34. Frome, F. S.: Course, video, and manual dexterity: tailoring training to CAD users. In: 22nd ACM/IEEE Design Automation Conference, pp. 226–231 (1985)

Data-Driven Determination and Plausibility Check of Requirement Profiles in Logistics

Markus Kohl[1](✉), Sandra Häring[2], Jens Lopitzsch[2], and Johannes Fottner[1]

[1] Chair of Materials Handling, Material Flow, Logistics, Technical University of Munich, Boltzmann Street 15, 85748 Garching, Germany
{Markus.Kohl,J.Fottner}@tum.de

[2] MAN Truck and Bus SE, Dachauer Street 667, 80995 Munich, Germany

Abstract. The introduction of new technologies driven by Industry 4.0 is transforming existing logistics processes. The changing tasks of the employees in this context require a systematic review of current requirement profiles (RPs) to appropriately bundle future employee tasks in homogeneous profiles. Thus, this paper develops a method for the determination of RPs taking Industry 4.0 in logistics into account. In this way, tasks are systematically transferred into RPs on the basis of similar characteristics. A data-driven method is chosen to reduce the subjectivity of the approach by using similarity factors. These reflect the central aspects of the socio-technical system and are derived from literature and practice. The method, which has been validated at a commercial vehicle manufacturer, helps in making decisions about RPs and can provide information about how eliminated, changed and new tasks affect the composition of employee RPs.

Keywords: Requirement profile · Logistics processes · Industry 4.0 · Competency · Clustering approach · Classification model · HRM

1 Introduction

The introduction of new technologies is changing logistics processes. Despite all these new technologies, employees and their skills are still of great importance for successful operational performance [1]. However, the changes in activities and requirements that will result from the fourth industrial revolution make sensible a systematic review of current requirement and task profiles for their suitability, as well as the early development of future profiles. Although there are already approaches in literature and practice to derive or determine requirement profiles (RPs)(s. chapter 2), a method to systematically check and determine employee RPs especially in logistics with a data-driven approach is missing. Consequently, the research goal is the development of a method for the determination of RPs by using similar characteristics to systematically transfer tasks carried out by employees in logistics into RPs. The profiles should be homogeneous within themselves and heterogeneous to each other. Furthermore, the approach should be able to validate the plausibility of already existing profiles. In order to achieve the objective, a brief overview of competencies and RPs as well as existing approaches

S. Trzcielinski et al. (Eds.): AHFE 2021, LNNS 274, pp. 311–319, 2021.
https://doi.org/10.1007/978-3-030-80462-6_39

to determine them is given. Based on this, the research gap is identified. The research approach for developing a method for the data-driven determination of RPs is presented in the third section, followed by the resulting method. For validation, the method is applied in a company of the commercial vehicle industry. The paper concludes with a summarizing discussion and an outlook on further steps.

2 Background

In order to cope with changing processes and tasks in the future, companies must develop qualification strategies and for these, competency models and profiles can be considered as essential. In practice, competency models combine the strategic and operational level of competency management [2]. Link and Rosenstiel [2] define a competency model as a detailed, behavior-specific description of skills and attributes that are seen as relevant to operate successfully in a workplace. Different types of competency models are distinguished based on their content and structure. As a competency model, job bundles can be a promising instrument for coping with increasing dynamization, complexity and uncertainty, since they contain fundamental tasks and requirements in a job bundle [3]. A job bundle is composed of a *job profile*, which is defined by the elements "organizational integration" and "objectives/core tasks" and the *requirement profile* including "professional requirements" and "personal behavioral requirements" of the same or similar jobs [3], such as those of a logistics planner or warehouseman.

Regarding the creation of RPs, various methods can be found in literature and practice. Basically, a requirements analysis is carried out to determine the target profile, which can be done either bottom-up or top-down [4]. In the bottom-up procedure, the requirements are derived from concrete tasks. In the top-down approach, the company's goals and strategies are used as a starting point. Based on the results, the objectives for the respective positions can be developed. In the next step, these objectives are transferred into concrete requirements for a position [5]. Building on the bottom-up and top-down approaches, Briscoe & Hall differentiate three methods to determine RPs [6]:

Survey-Based Approach: This method follows a bottom-up approach in which critical, success-promising behaviors are identified using, for example, the Critical Incident Technique (CIT) [6]. The CIT is used to collect success-critical work situations in which the behavior of high-performing and low-performing individuals can be clearly identified and distinguished. These identified behaviors are used as the basis for the requirements that are defined in the RP [7].

Strategy-Based Approach: In contrast to the survey-based approach, the strategy-based approach follows a top-down procedure. Goals and requirements are derived from the corporate strategy. Grote et al. assess the future orientation of this approach as an advantage, since competencies are derived from the future requirements [7].

Value-Based Approach: The value-based method also follows a top-down approach by using higher-level norms and values of the company as a basis for competence modelling [6]. Limitations consist of the less scientific and less systematic survey process. In addition, this approach may result in less task-oriented competencies.

Job-Bundle Approach: Another approach to determine RPs was developed by Becker [3] and follows a bottom-up approach by only including elementary and relatively permanent tasks and requirements in a job bundle. Based on the expected performance in a specific situation, relevant core tasks and basic requirements are determined and bundled. Core tasks, organizational integration, professional requirements, and personal behavioral requirements are taking into account in a job-bundle. As a competency model, job bundles can be a promising instrument for coping with increasing dynamization, complexity and uncertainty [3].

It becomes apparent that a wide variety of approaches for determining RPs is theoretically and practically possible. In the context of increasing dynamics in working environments with rapidly changing activities and tasks, however, a method should be applied that is both, future and task-oriented.

As each of the industrial revolutions, the fourth industrial revolution will also lead to changed task and qualification requirements [8]. Changing tasks and requirements demand instruments of competency management to identify new requirements, on the one hand, and to develop new RPs, on the other. The need for a re-conception of RPs results from the emergence of new occupational fields and task profiles [9]. With regard to the changing tasks of employees, a suitable method is needed to critically examine RPs and, if necessary, to create new profiles. In doing so, it is necessary to work out similarities and differences from the process and task descriptions and to take these into account when creating RPs. In addition, it is advisable to include new, changed and eliminated tasks in the creation of these profiles to build these RPs, detached from current structures and roles. Data-driven approaches are useful in the course of this providing traceable and transparent decisions when creating RPs.

3 Research Methodology

To develop a method to determine RPs, which incorporates a data-driven approach, a structured procedure is followed. Even though none of the existing approaches provides sufficient support, the established concept to determine RPs can provide as a base for adjustments. Therefore, first, the requirements on the method are derived from the objective and the findings of the state of the art. Then, the existing approaches are analyzed and their requirements fulfilment is assessed. As the selected method, job-bundles, is not able to fulfil all requirements sufficiently, an adaptation is necessary. The challenge is to satisfy the requirements on the method and to consider the data-based determination and the plausibility check of already existing RPs.

Requirements: The requirements on the method for the data-driven determination of RPs, which incorporates a determination and plausibility check, comprise two fields: The inclusion of task-specific aspects during the formation of the RPs and data-driven approach. The specific requirements are listed in Table 1.

Assessment: The evaluation of the approaches is based on a three-step scale. This reflects the degree of fulfillment of the requirements with the expressions "not fulfilled", "partially fulfilled", and "completely fulfilled". The classification is based on a literature review. The evaluation of the approaches indicates that none of the listed approaches for determining requirements profiles adequately meets all of the requirements imposed on the method. The best rated is the job-bundle approach by Becker as it is particularly suitable for determining RPs and is used as basic model for the new method. Nevertheless, this approach shows weaknesses with regard to the plausibility check of RPs and the derivation of the RPs is not reproducible on the basis of data.

Table 1. Defined requirements and assessment of the preselected approaches.

Requirement	Survey-based approach	Strategy-based approach	Value-based approach	Job bundle
Determining RPs	●	●	●	●
Plausibility check of RPs	○	○	○	○
Data-based determination	◐	◐	◐	◐
Task and process-oriented	●	○	○	●
Bundling of similar tasks	◐	◐	◐	●

○ not fulfilled ◐ partly fulfilled ● completely fulfilled

Adaptation: The analysis revealed that adjustments are necessary to fulfill all specified requirements and to eliminate the deficits. The core of the new method is a database consisting of a task-oriented process description and the determination of similarity factors (SFs) capturing the individual tasks in the processes. This database with the SFs is used to define the requirements that the respective tasks entail. In addition, it seems practical to consider not only future processes and tasks, but also current ones, in order to check the plausibility of existing RPs and structures. The assessment further indicates that previous approaches show deficits in the analysis and bundling of similar tasks. To remedy this with the method, the SFs are used to analyze common features and to bundle tasks with comparable characteristics into RPs.

4 Determination and Plausibility Check of Requirement Profiles

The method consists of three steps as displayed in Fig. 1. The data is generated by a process analysis taking the SFs into consideration. For future processes, overall core task classes are defined, which categorize tasks fundamentally. The collected information is then used to apply the classification methods, which weight the influence of the SFs and thus allow conclusions to be drawn as to which aspects of the process description are responsible for composing the RPs. These weighted factors are utilized in the clustering algorithm, determining the number and composition of the profiles. To obtain the final RPs, these are adapted by experts from technical and HR departments.

Step 1: The aim of the first part is to generate a database that will be used for analysis in the subsequent steps. Input data on future tasks is needed to determine future RPs. The

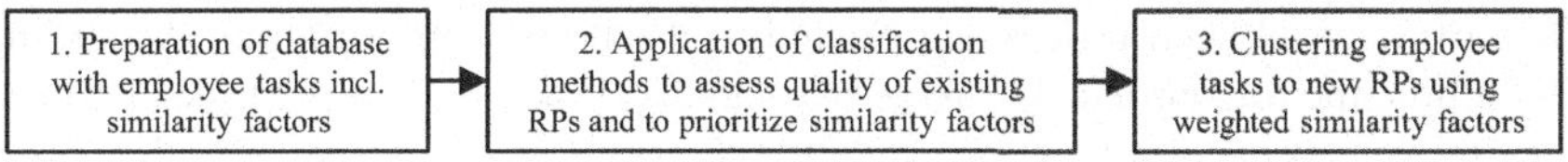

Fig. 1. Steps of the method for determining requirement profiles.

determination is carried out by means of workshops with experts. SFs are determined to compare the requirements of the tasks under consideration. Each SF represents a specific requirement, which is assigned to different levels depending on its characteristics. The result of the first step is therefore a database in which all tasks are recorded on the basis of the SFs. The data should be available in numerical form to enable calculations in the following steps.

Figure 2 lists all parameters that are used as SFs for the analysis. For clarity, these were divided into requirement groups based on considerations by Becker [3], including personal behavioral requirements, and Wilk [10], including degree of self-reliance, contact diversity, and complexity of tasks. In addition to the aforementioned groups, other subject-specific aspects were added. For logistics, an addition of knowledge material flow, knowledge information flow, process knowledge, IT complexity, and software and simulation appears to be appropriate. Thus, the flow-oriented view of logistics is covered in material and information flow with the interrelated processes. Furthermore, the groups IT complexity and software and simulation aim at new requirements resulting from digitalization and autonomization.

Degree of self-reliance	Decision on measures	Knowledge information flow (logistics software)	Number of systems
	Way of working		Frequency of use
	Dealing with malfunctions		Dealing with malfunctions
Contact diversity	Kind of interfaces		
	Number of interfaces		
	Range of communication	IT-complexity	Interpretation of data
Complexity of tasks	Problem cause		Number of data sources
	Alternative solutions		Condition of data
	Effects of decisions		IT-devices: frequency of use
	Dynamics of the working environment		IT-devices: dealing with malfunction
	Time pressure	Software & simulation	Programming
Knowledge material flow	Duration of training for assistive devices		Intelligent software & sim.: frequency of use
	Dangerous goods		
Process knowledge	Dealing with process disturbances		Intelligent software & sim.: dealing with malfunctions
	Connectivity of processes		

Fig. 2. Overview of the similarity factors.

Step 2: For the formation of new profiles, the tasks included in the data set are assigned to higher-level core task classes based on a qualitative consideration of content. This is done because the same superordinate core tasks and goals are considered an aspect of similarity according to the job-bundle approach [3] and thus represent a criterion for similar requirements. In this context, meaningful classes are first defined for the process under consideration. For the logistics area, these are based on the flow-oriented definition of logistics subdividing into tasks of planning and control as well as the processing of goods flows [11]. The processing can be divided into two types: processing related to the flow of materials and processing related to the flow of information. According to this definition, the four core task classes are planning tasks, controlling tasks, processing administrative tasks and processing operational tasks. For future scenario, the core task class of a key user, responsible for the IT systems, was also anticipated on the basis of the workshops. This need results from a future working environment that will be increasingly characterized by digitization and autonomization [12].

After assigning the activities to the core task classes, the classification procedure is carried out. Based on the defined higher-level core task classes, a classification algorithm is used to build a model specifically for the underlying data set. A supervised machine learning approach is used to predict the output value based on the different characteristics in the data set with the highest possible accuracy. The classification procedure can be used to analyze, on the one hand, whether the previously made assignment to core task classes can be reasonably reproduced and, on the other hand, which SFs are decisive for classifying tasks into different categories.

For this purpose, the data set is divided into a training set and a test set. As output, the classification model generates a confusion matrix, which provides information about the error rate, i.e., the quality of the model, for the test set. From the matrix, it can be seen how many activities from the data set can be correctly assigned to the previously specified higher-level task classes based on the rules formed by the algorithm. From this information, it is possible to determine which characteristics of the tasks in the data set are decisive for assigning activities to different classes.

Step 3: The identified SFs and their weighting are used to generate suggestions of clusters with similar tasks and requirements using a cluster algorithm. First, an optimal number of clusters or RPs is calculated, and then a clustering procedure is used to generate a proposal for the assignment of tasks to the profiles. In this work, it was decided to use a k-means clustering algorithm.

To determine the number of RPs, the Variance Ratio Criterion (VRC) and Adjusted Rand Index (ARI) are calculated using the weighted SFs. These two quantities provide information about the optimal number of clusters. The k-means procedure can then be performed with the determined number of clusters. The clustering manages to consider or group similar personal behavioral requirements and occupational requirements via the SFs. This produces as output an assignment of similar tasks to the same clusters.

The assignment of tasks to profiles proposed by the cluster algorithm is still subjected to a manual review by experts from the human resource and technical departments. This involves a critical assessment of whether the assignment made by the system is reasonable and feasible for practical application and the given organizational framework conditions. Furthermore, it is assessed whether clusters may need to be combined or subdivided again.

Plausibility Check: Within the scope of the method, it is also possible to check the plausibility of existing RPs. For this purpose, the data must also be available as described in step 1. The plausibility check is then carried out in step 2. The existing RPs serve as output values for the classification algorithm instead of the higher-level core task classes. Otherwise, the procedure is analogous to the method for determining new RPs. With regard to the output, the observation of the error rate is particularly interesting.

If this is below the specified value, the quality can be regarded as sufficient. Thus, the assignment of the already existing RPs seems to be plausible. An insufficient error rate can indicate a less than optimal assignment and recommends a new allocation of tasks to RPs. In addition, the classification procedure provides transparency as to which SFs are the most relevant for differentiation between the various profiles.

5 Case Study and Evaluation

The method was validated at a commercial vehicle manufacturer. Workshops were organized in all logistics areas to determine future processes, and the method was then used to determine RPs. A summary of the steps and the results is shown in Fig. 3.

In **step 1**, workshops were conducted for each working group with experts from different plants in the logistics network. The sub-processes and tasks in the respective areas were identified and described with regard to their future changes. In addition, the SFs were determined for each task. The experts classified which core tasks would be differentiated in the future and assigned the tasks to these.

This assignment was used in **step 2** to identify the appropriate classification method. The output of the Random Forest Model yielded the best result with an error rate of 3.2 percent. In addition, the Confusion Matrix indicates which individual tasks were incorrectly assigned by the model. For example, a task that was assigned to the processing administrative task class was shifted by the model to the processing operational task class. The classification procedure also provides information as to which the most important SFs for the classification are. In this case, the similarity factor "connectivity of processes" is the factor with the highest significance for differentiation.

The weighted SFs were used in **step 3** for the cluster analysis. The parameters VRC and ARI suggested a cluster number of 10 as reasonable, whereupon the k-means clustering procedure was performed for 10 clusters. The cluster result shows, for example, that the tasks of the goods receipt process have similarities to the tasks from the distribution process. It also suggests the split of the key user tasks for administrative and planning/controlling areas. The latter are managed separately due to the stronger focus on predictive analytics and simulation.

This is followed by a review of the results with the workshop participants and experts from the human resource department. The review and adaptation were carried out for all ten clusters of the proposal, so that ultimately eight future RPs for the logistics of the commercial vehicle manufacturer were derived (see Fig. 3).

1. Preparation of database with employee tasks incl. similarity factors	2. Application of classification methods to assess quality of existing RPs and to prioritize similarity factors	3. Clustering employee tasks to new RPs using weighted similarity factors	Resulting future RPs
• Conducting workshops in 7 areas • Identification of 70 tasks along the logistics process chain • Determination of the similarity factors for each task based on the assessment of the experts from different plants	• Identification of the core activity classes and assignment of the respective tasks • Selection of the classification method taking into account the core activity classes • Use of the Forest Model due to the lowest error rate • Obtainment of the similarity factors including their weighting in the classification procedure • Factors such as the number of data sources, the degree of interconnectedness of the process landscape, and the input state of the data are primarily responsible for the composition of the RPs	• Use of the weighted similarity factors to calculate the VRC and the ARI • Parameters yield 10 clusters as the optimal number of clusters • Performing the k-mean clustering algorithm for 10 clusters • Analysis of the results reveals cluster with homogeneous similarity factors • Expert interviews merge the clusters to 8 RPs	• Processing - administrative • Processing - operational • Coordinators (incl. dispatcher) • Logistics planner • Forecasting • Program planning • Key user planning/control • Key user administrative

Fig. 3. Steps and results of the data-driven determination method in the case study.

6 Conclusions

Existing logistics processes are undergoing a transformation in the course of Industry 4.0, which is accompanied by a change in the tasks of employees. In the course of this, it is necessary to review the current RPs and identify future profiles at an early stage. A corresponding method for this is presented in this work. The method for systematically checking plausibility and developing future RPs is based on similarity factors. These reflect the central aspects of the socio-technical system, consisting of people, organization and technology, and were derived from literature and practical application. Core tasks, technical requirements and personal behavioral demands from process descriptions are taken into account and validated by means of application in various logistics departments of a commercial vehicle manufacturer. The SFs, such as degree of independence, task complexity, process knowledge and IT complexity, are the basis for the data-based approach. The method consists of three steps: generating the data basis by a process analysis, processing a classification procedure to weight the SFs, and applying the clustering algorithm to determine the number and composition of the RPs. The validated method for the determination and checking of RPs is applicable to current and future logistics processes if sufficiently described. It helps with decisions on RPs and can provide information on how omitted, changing and new tasks affect the composition of RPs. A systematic integration of the method into the human resource management enables targeted personnel development and strategic planning towards Industry 4.0 in logistics. Further research is needed to include the application of the method in logistics areas of other companies outside the automotive industry. Furthermore, there is a need to adapt and test the method for other disciplines, such as production.

Acknowledgements. This research was funded in part by MAN Truck & Bus SE and in part by the Chair of Materials Handling, Material Flow, Logistics (fml) at the Technical University of Munich. All support is gratefully acknowledged. Any opinions, findings, conclusions, or recommendations expressed in this paper are those of the writers and do not necessarily reflect the views of MAN Truck & Bus SE or the fml.

References

1. Neumann, W.P., Winkelhaus, S., Grosse, E.H., Glock, C.H.: Industry 4.0 and the human factor – a systems framework and analysis methodology for successful development. Int. J. Prod. Econ. (2021)
2. Link, K., von Rosenstiel, L.: Strategisches HRM Kompetenzmanagement (2012)
3. Becker, M.: Personalentwicklung: bildung, förderung und organisationsentwicklung in theorie und praxis. Schäffer-Poeschel (2013)
4. Lorenz, M., Rohrschneider, U.: Erfolgreiche Personalauswahl. Gabler Verlag, Wiesbaden (2015). https://doi.org/10.1007/978-3-8349-4766-6
5. Lorenz, M., Ruessmann, M., Strack, R., Lueth, K.L., Bolle, M.: Man and machine in industry 4.0: how will technology transform the industrial workforce through 2025. BCG (2015)
6. Briscoe, J., Hall, D.: Grooming and picking leaders using competency frameworks: Do they work? An alternative approach and new guidelines for practice (1999)
7. Grote, S., Kauffeld, S., Denison, K., Billich-Knapp, M., Frieling, E.: Kompetenzen und deren management: ein Überblick. In: Grote, S., Kauffeld, S., Frieling, E. (eds.) Kompetenzmanagement. Grundlagen und Praxisbeispiele, Schäffer-Poeschel Verlag, Stuttgart (2012)
8. Kipper, L.M., et al.: Scientific mapping to identify competencies required by industry 4.0 (2021)
9. Majkovic, A.-L., et al.: IAP Studie 2017–Teil 2: der mensch in der arbeitswelt 4.0 (2017)
10. Wilk, G.: Stellenbeschreibungen und anforderungsprofile. Kompetente Unterstützung für erfolgreiche Personalarbeit, Haufe Group, Freiburg, München, Stuttgart (2018)
11. Pfohl, H.-C.: Logistikmanagement. Konzeption und Funktionen. Springer, Heidelberg (2016).https://doi.org/10.1007/978-3-662-48784-6
12. Bauer, W., Dworschak, B., Zaiser, H.: Weiterbildung und kompetenzentwicklung für die industrie 4.0. In: Handbuch Industrie 4.0. Springer, Heidelberg (2017). https://doi.org/10.1007/978-3-662-45279-0_36

Robotic Process Automation in Industrial Engineering: Challenges and Future Perspectives

Sebastian Schlund(✉) and Mathias Schmidt

Human-Machine Interaction, Institute of Management Science, Technische Universität Wien (TU Wien), Theresianumgasse 27, 1040 Wien, Austria
{sebastian.schlund,mathias.schmidt}@tuwien.ac.at

Abstract. Robotic Process Automation (RPA) is an emerging technology in the field of software automation, which is making rapid progress within efficiency-driven industries. Whilst it has found broad dispersion in the automation of office tasks, its potentials in manufacturing and industrial engineering are unexhausted. This paper explores the basic concepts and possible extensions of RPA, and identifies targets promising application areas in industrial engineering. For this purpose, a systematic literature review was carried out to deepen insights into the state-of-the-art, methods and research type facets. Over the past five years 21 scientific contributions have been identified and classified according to bibliometric data, their specific contribution (i.e., analysis, concepts, methods, metrics, tools) and the applied research type (i.e., evaluation, experience, solution, validation, vision). The results of literature survey reveals existing research gaps in particular emphasizing on strong demands of manufacturing and industrial engineering sectors to catch up on research work to leverage large potentials of RPA for future applications.

Keywords: Robotic Process Automation · Manufacturing · Industrial engineering · Literature review

1 Introduction

Robotic Process Automation (RPA) is an emerging technology in the field of software automation, which is making rapid progress within efficiency-driven industries. The term consists of three nouns that describe the automation of a process by a robot-like entity. Despite the notion of a robot as referring to a (physical) machine in the first place, the concept uses artificial/software robots (aka bots) to automate tasks, activities and entire processes. Main benefits are seen as the withdrawal of human intervention resulting in a decreased error rate and less operational efforts [1, 2]. In particular, RPA is a software-based concept to automate processes that underlie certain preconditions and rules. RPA is widely considered not to be a temporary trend, but a straightforward evolution of process automation technologies such as 'traditional' workflow automation or macros especially developing towards a competitive advantage in sectors like banking

S. Trzcielinski et al. (Eds.): AHFE 2021, LNNS 274, pp. 320–327, 2021.
https://doi.org/10.1007/978-3-030-80462-6_40

and insurance [3]. Essential precondition is the definition of appropriate processes that are characterized by a high frequency of occurrence, repetitive character, digital in- and outputs, the use of more than one software system and a well-defined, detailed process description [4]. As RPA implementations apply a wide range of machine learning tools, like optical character recognition, natural language processing and image analysis, it enhances the scope of traditional workflow automation well beyond strictly rule-based solutions. Therefore, RPA opens the perspective to automate non-routine tasks as long as they use the same set of platforms and features while not explicitly relying on human capabilities like manual dexterity or problem-solving skills. The immense interest and high expectations to RPA can be explained by benefits involved with process automation such as productivity gains, automatic documentation of processes and increased quality. Another benefit that usually appears after the first pilot implementations is the scalability due to transferability to comparable processes in other entities and in terms of the reuse of programming architectures and code. [5, 6].

Considering the above discussion, this paper explores the current dissipation of RPA implementations in industry. As most public attention and publications focus on back-office tasks in banking or insurance industry, the analysis specifically deals with applications in the manufacturing sector, especially regarding industry-specific processes such industrial engineering. The main research question therefore is:

What specific implementations and use cases of RPA exist within the application area of manufacturing and industrial engineering?

To appropriately address this question, the rest of the paper is structured as follows: Sect. 2 discusses the research methodology. Section 3 presents the results of the systematic literature review. Section 4 derives challenges and future perspectives in industrial engineering and is followed by a discussion (Sect. 5) and conclusion (Sect. 6).

2 Research Methodology

In order to reach that goal as stated in Sect. 1, a systematic literature review (SLR) referring to Okoli and Shabram [7] was conducted. The SLR was carried out in two phases, the research phase and the subsequent evaluation phase. In the first step, keywords were used, researching the databases Scopus, Springer Link, IEEE, Web of Science and ScienceDirect. Initially, the search covered the keywords ‘robotic’, ‘process’ and ‘automation’. As the results largely covered publications in the domain of ‘hardware’ robots, more specific keywords were used by an ‘AND’ linkage. Due to the novelty of the topic, no work older than five years was included. For an overview of the search strategy, the selected parameters and chosen criteria are listed in Table 1.

After identifying relevant works, a screening was carried out to evaluate the relevance of the contributions to a number scale from one to five, where five corresponds to the highest value. In order to increase objectivity in relation to the evaluation, criteria such as timeliness of publication, publication type, and objectivity of the literature were crucial. A backward search [8] was conducted via the highest rated papers, tracing the references to discover further relevant publications.

Table 1. Parameters of the SLR

Databases	Scopus, SpringerLink, IEEE, Web of Science, ScienceDirect
Search terms	Robotic Process Automation
Subset	Software-Implementation, Lifecycle, Digitalization, Data Mining, Framework, Classification, Mapping-study, Criteria, Business process, Manufacturing, Engineering
Inclusion criteria	Journal and conference papers published between 2016 and 2020
Exclusion criteria	Robotic automation referring to mechanical/physical robots

3 Results of the Literature Review

The geographical distribution of the studies did not show any specific results, since RPA is a topic of global interest. The total of 21 studies that met the final criteria spreads over 15 different countries, with a slight peak in Germany (4) and the USA (3). With respect to the contribution and research type facets as described in Wortmann et al. [9] and summarized in Tables 2 and 3 most papers treat methods (11), concepts (10) and metrics (10) from a solution (11), experience (11) or validation (10) perspective.

Table 2. Contribution (according to [9])

Analyses	Papers contributing investigations without constructive contributions	4/21
Concepts	Papers suggesting ways of thinking things, such as new metamodels or taxonomies	10/21
Methods	Papers suggesting ways of doing things	11/21
Metrics	Papers suggesting ways of measuring things	10/21
Tools	Papers presenting software tools	8/21

Table 3. Research type facets (according to [9])

Vision	Non-disruptive research agendas	5/21
Validation	Papers presenting novel techniques and experimenting with them	10/21
Solution	A novel solution is presented and argued for with case studies	11/21
Experience	Report of personal experiences, for instance on the application of a specific method	11/21
Evaluation	Papers evaluating existing techniques	6/21

An overall picture of the research field and its trends is conveyed by the keywords appearing in the papers. In a simple evaluation, the number of papers, out of 21 existing ones, containing terms listed in table 4 are presented. Although the term 'manufacturing' was mentioned nine times, in most cases it is used to describe difficult applications or missing studies.

Table 4. Distribution of keywords within the selected papers

Implementation	Software (18), Humans (18), Implementation (17), Training (15), Back Office (14), Routine (11), Lifecycle (9), Front Office (7), CoE (6), Jobs (6), KPI (5), Human Error (5), Knowledge Management (2)
Digital Model	Process Mining (13), Bot/Robot (12), Algorithm (12), AI (10), Self-learning (7), Logs (6), OCR (5), DTO (3)
Technology	Manufacturing (9), Pilot (7), IoT (6), 4.0 (3), Long Tail (2), Industrial engineering (1)
RPA objectives	Efficiency (16), Errors (14), Reduction (13), Problem (12), Limitation (10), Standards (9), FTE (5), Usability (4)

4 Challenges and Future Perspectives in Industrial Engineering

Anagnoste [10] provides an overview of the automation potential of various industries in relation to their respective processes. In the area of manufacturing, a high suitability is shown for finance-related areas, while industry-specific processes show the lowest overall potential. Most of the publications do not even include industrial engineering or manufacturing as promising application domains [11].

As manufacturing is considered to be a rather physical domain with a dominating share of manual work and a rather low IT integration, it does not straightforwardly fall into the precondition for efficient RPA implementation [4]. Mehta et al. [12] state further challenges as fragmentary model representations and simulation capabilities.

Despite the overall conclusion of the review that the use of RPA in domain-specific manufacturing tasks is not considered as application focus, industrial engineering contains several processes, which fulfil most of the criteria required for efficient implementation. Especially against the background of large digitization activities like the worldwide initiative 'Industry 4.0' [13], respectively 'Smart Factory' or 'CPPS' (Cyber Physical Production Systems) [14] the potential benefit of RPA is very likely to grow over the years to come. All of these initiatives are grounded on the scenario of self-organizing systems, digitization of products and processes (digital twins), over the entire life cycle. For highly automated industries such as semiconductor and chemical industry or even automobile industry, the respective processes already fit into the criteria pattern. Table 5 provides an overview over five common activities within industrial engineering that provide promising processes.

Process Planning such as the planning of assembly lines, new or adjusted workplaces is already carried out with the use of various IT planning, process simulation, ergonomics as well as project management tools. Due to the increasing trend towards changeable work

Table 5. Process attributes for automation for common industrial engineering (IE) processes

RPA Criteria \ IE Process	Frequent occurrence	Repetitive character	Digital I/Os	Multiple software systems	Defined process description
Process Planning	(X)	X	(X)	X	(X)
Planning and Scheduling	X	X	X	X	X
Authoring Tasks	(X)	X	X	X	(X)
Robot Programming	(X)	X	X	X	X
Incident Management	X	X	X	X	(X)

X: applicable (X): applicable to some extent–: not applicable.

systems and decreasing lot sizes the needed frequency of process (re)planning is very likely to increase. Nevertheless, it does not reach the frequency level of a 'traditional' back office process like sick leave or travel expense documentation.

Planning and Scheduling already consists of largely hybrid workflows as it usually combines ERP, MES as well as shop floor data and actions. As it is a frequent process it already uses automated IT systems and middleware interfaces. However, due to increasing changeability, the need for customizable and less invasive approaches of process automation opens up application potential for RPA.

Authoring Tasks are required by most assistance systems in manufacturing such as augmented reality applications or worker assistance systems for assembly [15, 16], maintenance or teaching tasks. By the increasing implementation of digital assistance systems [17] efforts to obtain, transform and move data are created to display the right information at the most useable representation, via the frontend of an assistance system (terminal, data glasses, other devices like projection systems). As these activities are currently emerging within industrial engineering, they are expected to grow massively in importance over the coming years.

Robot Programming is currently shifting from an activity carried out by robot experts in automation departments into a commodity activity within industrial engineering. With the rise of (fenceless) collaborative robots for direct human-machine interaction, multimodal teaching concepts and function-block based programming interfaces, robot programming already meets most of the RPA criteria [18]. First use cases show a large potential in productivity increase [19, 20].

Incident Management covers the workflows resulting by the occurrence of an unplanned event such as a failure of a machine, missing material, quality issues or unplanned maintenance tasks [21]. Besides the action of the problem solving these processes require documentation activities and information workflows that usually cover the criteria for efficient RPA application. Vast potential for optimization still exists in the recording or evaluation of existing logs. Clear inputs (data) result in an increase in efficiency through process mining. However, even in the case of unclean inputs, fuzzy logic can be used to remedy the situation [22].

5 Discussion

Over the past five years, RPA has been mainly used to automate back office tasks in service processes, mostly support activities without customer contact and with low cognitive requirements [1]. RPA however, can achieve much higher functionality by leveraging technologies such as and cognitive self-learning methods as described in [10, 23]. In line with growing digitization trends in manufacturing and, RPA provides large potential to have a significant impact on industrial engineering.

As mentioned earlier, 30–50% of attempted RPA implementations fail [11]. To discover the suitability of RPA to the dedicated processes, pilot tests at a small scale are suitable to deliver fast results and maintain motivation in early stages [3]. Some software vendors advertise the simplicity of their software, claiming little to no requirement of programming knowledge and therefore often suggesting a fast application of the respective RPA solutions. Nevertheless, the process should be analyzed very carefully, as it is a long-term commitment to change the system and carry the inherent economic risks involved [4].

Within the systematic literature research, almost no evidence on successful RPA applications in manufacturing and industrial engineering has been found. The most straightforward explanation is the difficult application of RPA to industry-specific processes and the novelty of the entire topic. As described in Ivančić et al. [6], it is often difficult to make meaningful statements since only a handful of case studies have been conducted and many aspects have not yet been elaborated. Ivančić et al.list case studies according to industry sectors. However, the evaluation revealed that two-thirds were conducted in the service or telecommunication sector, followed by finance and insurance. In sum, only 5.36% of the published papers from 2012 to 2020 actually applied implementations, while models or methods were proposed in 21.43% [3].

6 Conclusion

The global market of commercial RPA implementations already is considered to be about USD 3bn in 2021, generated primarily by the "Big Three" Automation Anywhere (USA), BluePrism (UK), and UiPath (Romania) [24]. As the break-even point for most companies is reached after just one year, a further increase in market share is likely [6]. A general worldwide trend can be observed to stray away from simple rule-based tasks to complex knowledge-based activities. This change is to be achieved by leveraging technologies such as AI, process-mining, cognitive computing and data analytics to further improve functionality of RPA.

Contradictory to the fact that manufacturing has always been very' in favor of (mechanical) automation, robotic process automation today is only used sporadically in industrial engineering. Considering recent developments in manufacturing and the impact on industrial engineering processes like discussed in Sect. 4, it is very likely that RPA implementations expect a significant rise over the years to come. However, the versatility of RPA can be seen by the application to everyday processes without high cognitive effort on the one hand and the search for the most intelligent system possible that can work with ambiguous information and processes on the other hand.

Acknowledgments. This research was supported by the Austrian Research Promotion Agency (FFG), the Austrian Ministry for Climate Action, Environment, Energy, Mobility, Innovation and Technology (bmk), the TU Wien through the endowed professorship "HCCPPAS" (FFG-852789).

References

1. Aguirre, S., Rodriguez, A.: Automation of a business process using Robotic Process Automation (RPA): a case study. In: Figueroa-García, J.C., López-Santana, E.R., Villa-Ramírez, J.L., Ferro-Escobar, R. (eds.) Applied Computer Sciences in Engineering, pp. 65–71. Springer International Publishing, Cham (2017). https://doi.org/10.1007/978-3-319-66963-2_7
2. Enríquez, J.G., Jiménez-Ramírez, A., Domínguez-Mayo, F.J., García-García, J.A.: Robotic process automation: a scientific and industrial systematic mapping study. IEEE Access **8**, 39113–39129 (2020)
3. Suri, V.K., Elia, M., van Hillegersberg, J.: Software bots - the next frontier for shared services and functional excellence. In: Oshri, I., Kotlarsky, J., Willcocks, L.P. (eds.) Global Sourcing 2017. LNBIP, vol. 306, pp. 81–94. Springer, Cham (2017). https://doi.org/10.1007/978-3-319-70305-3_5
4. Kirchmer, M., Franz, P.: Value-driven Robotic Process Automation (RPA). In: Shishkov, B. (ed.) BMSD 2019. LNBIP, vol. 356, pp. 31–46. Springer, Cham (2019). https://doi.org/10.1007/978-3-030-24854-3_3
5. Scheer, A.-W.: Unternehmung 4.0. Wiesbaden: Springer Fachmedien Wiesbaden (2020). https://doi.org/10.1007/978-3-658-27694-2
6. Ivančić, L., Vugec, D.S., Vukšić, V.B.: Robotic process automation: systematic literature review. In: International Conference on Business Process Management, pp. 280–295. Springer, Cham (2019).https://doi.org/10.1007/978-3-030-30429-4_19
7. Okoli, C., Schabram, K.: A guide to conducting a systematic literature review of infor-mation systems research, sprouts working papers on information systems. Amsterdam. **10**(26), 1–49 (2010)
8. Levy, Y., Ellis, T.J.: A systems approach to conduct an effective literature review in support of information systems research. Inf. Sci. J. **9**(2016), 181–212 (2006)
9. Wortmann, A., Combemale, B., Barais, O.: A systematic mapping study on modeling for industry 4.0. In: 2017 ACM/IEEE 20th International Conference on Model Driven Engineering Languages and Systems (MODELS), pp. 281–291. IEEE (2017)
10. Anagnoste, S.: Robotic automation process–the operating system for the digital enterprise. In: Proceedings of the International Conference on Business Excellence, vol. **12**(1), pp. 54–69. Sciendo (2018)
11. Taulli, T.: The Robotic Process Automation Handbook: A Guide to Implementing RPA Systems. Apress, Berkeley (2020). https://doi.org/10.1007/978-1-4842-5729-6
12. Mehta, P., Rao, P., Wu, Z.D., Jovanović, V., Wodo, O., Kuttolamadom, M.: Smart manufacturing: state-of-the-art review in context of conventional and modern manufacturing process modeling, monitoring and control. In: ASME 2018 13th International Manufacturing Science

and Engineering Conference. American Society of Mechanical Engineers Digital Collection (2018)
13. Bauer, W., Ganschar, O., Gerlach, S., Hämmerle, M., Krause, T., Schlund, S.: Industrie 4.0–flexiblere und reaktionsfähigere produktionsarbeit–ergebnisse der Industrie-4.0-Leitstudie des Fraunhofer IAO. Werkstatttechnik Online 104(3), 134–138 (2014)
14. Monostori, L., et al.: Cyber-physical systems in manufacturing. CIRP Ann. **65**(2), 621–641 (2016)
15. Rupprecht, P., Kueffner-McCauley, H., Schlund, S.: Information provision utilizing a dynamic projection system in industrial site assembly. Procedia CIRP **93**, 1182–1187 (2020)
16. Zigart, T., Schlund, S.: Evaluation of augmented reality technologies in manufacturing – a literature review. In: Nunes, I.L. (ed.) AHFE 2020. AISC, vol. 1207, pp. 75–82. Springer, Cham (2020). https://doi.org/10.1007/978-3-030-51369-6_11
17. Mayrhofer, W., Kames, D., Schlund, S.: Made in Austria 2019–survey results of the first industry panel on production work in Austria. Manage. Prod. Eng. Rev. (2020)
18. Ionescu, T.B., Schlund, S.: A participatory programming model for de-mocratizing cobot technology in public and industrial fablabs. Procedia CIRP **81**, 93–98 (2019)
19. Ionescu, T.B.: Leveraging graphical user interface automation for generic robot programming. Robotics **10**(1), 3 (2021)
20. Ionescu, T.B., Fröhlich, J., Lachenmayr, M.: Improving safeguards and functionality in industrial collaborative robot HMIs through GUI automation. In: 2020 25th IEEE International Conference on Emerging Technologies and Factory Automation (ETFA), vol. 1, pp. 557–564. IEEE (2020)
21. Bauer, W., Ganschar, O., Pokorni, B., Schlund, S.: Concept of a failures management assistance system for the reaction on unforeseeable events during the ramp-up. Procedia Cirp **25**, 420–425 (2014)
22. Agostinelli, S., Marrella, A., Mecella, M.: Research challenges for intelligent robotic process automation. In: Di Francescomarino, C., Dijkman, R., Zdun, U. (eds.) BPM 2019. LNBIP, vol. 362, pp. 12–18. Springer, Cham (2019). https://doi.org/10.1007/978-3-030-37453-2_2
23. Sönmez, Ö.E., Börekçi, D.Y.: A conceptual study on RPAs as of intelligent automation. In: Kahraman, C., Cebi, S., Cevik Onar, S., Oztaysi, B., Tolga, A.C., Sari, I.U. (eds.) INFUS 2019. AISC, vol. 1029, pp. 65–72. Springer, Cham (2020). https://doi.org/10.1007/978-3-030-23756-1_10
24. Anagnoste, S.: Setting up a robotic process automation center of excellence. Manage. Dyn. Knowl. Econ. **6**(2), 307–332 (2018)

Process Improvement for the Reduction of Rework Applying TPM and Kaizen in a Company in the Metalworking Sector

Angel Murga-Vasquez, Jennyfer Valenzuela-Garcia, and Percy Castro-Rangel(✉)

Ingenieria Industrial, Universidad Peruana de Ciencias Aplicadas, Prolongación Primavera 2390, Lima 15023, Peru
{u201616404,u201510319,pcahpcas}@upc.edu.pe

Abstract. This article shows the implementation of a Lean Manufacturing tools model based on Total Productive Maintenance TPM and Kaizen. The implementation of the TPM and Kaizen tools of the proposed model allowed the reduction of reprocessing rates in the CNC machining line in a company in the metalworking sector. First, an investigation based on success stories has carried out to corroborate that there is a significant relationship between the two tools. Finally, a model based on the TPM and Kaizen tools has proposed, consisting of four stages: system research, planning, implementation, and control. In the implementation stage, the Autonomous Maintenance pillar of the TPM and 5S is included. Indeed, successful results has obtained. Reprocesses decreased by 30% nd Overall Equipment Effectiveness (OEE) from 39.4% initial to 50.7%.

Keywords: Metalworking sector · Lean Manufacturing · 5S · TPM · Kaizen

1 Introduction

The company under study develops machining products and services for the mining and industrial sectors. The company manufactures spare parts made of bronze, cast iron, carbon steels, stainless steels, aluminum and non-metallic materials; manufactures the following types of spare parts: gear wheels in general, shape wheels, couplings, pins, idlers, bronze products, non-ferrous materials, CNC machine machining (Lathe and Center, metallic structures). The company had significant increases in production volumes in 2018 and 2019, showing efficiency problems in production processes with a significant increase in the number of reprocesses. This study proposes the application of Lean Manufacturing tools such as TPM and Kaizen in companies in the metalworking sector.

According to the literature reviewed in scientific articles and adaptation of successful models to the case study, the results are supported as improvements in maintenance indicators. The results obtained are a reduction in the number of Reprocesses in the CNC machining line from 13.5.0% to 9.5%, an improvement in the OEE of the CNC machining line system from 39.4% to 50.7%. The results obtained belong to a pilot area

S. Trzcielinski et al. (Eds.): AHFE 2021, LNNS 274, pp. 328–335, 2021.
https://doi.org/10.1007/978-3-030-80462-6_41

within the company, the same that will be implemented throughout the organization. Finally, we hope that the proposed model of this study will be replicated in companies in the sector.

2 Literature Review

Suarez et al. [1] define Kaizen as a comprehensive philosophy of personal, work, family and community development, which progressively seeks improvements and innovations that impact all the activities we carry out on a daily basis; that is, as a consequence of applying Kaizen, it should lead us to a spiral of improvement and innovation in our operational processes in which not a single day, not an hour, not a minute, not even a second should pass when not thinking about how to continue changing and growing.

Kongar et al. [2], indicates that Kaizen events generally consist of a small group of five to ten employees of all levels, representing various functions within the organization, all of whom are affected by the project in question, obtaining positive impacts on both business processes and human resources. However, Kaizen event evaluations rely heavily on quantitative analysis that aims to measure, analyze and evaluate the operational performance of the organization's activities.

2.1 Impact of TPM and Kaizen

Many studies in the literature show the success of the implementation of Lean Manufacturing tools. According to Sahoo [3], he found that the top five most widely implemented Lean practices are Cellular Manufacturing, Total Productive Maintenance, 5S, Work Standardization, and Quality Management Practices. Additionally, leadership and organizational culture were found to be crucial factors for the success of Lean Manufacturing.

Other authors propose the implementation of Kaizen with the aim of applying improvements in a complex environment such as medicine. After the development of the proposed Kaizen events, the average treatment time was reduced by 10% [4].

Cherrafi et al. [5], developed a study based on two success stories in the aerospace and automotive industries that through the application of Gemba and Kaizen, obtaining as a result in both cases significant cost savings of 5,000 USD in 90 days and 113,000 USD USD in one year respectively.

Hooi et al. [6] conducted a study of TPM implementation in the Malaysian manufacturing industries. The study identified that senior management commitment is critical in the initial stage of TPM implementation, as well as traditional maintenance, maintenance planning, correct execution (autonomous maintenance) and continuous improvement (Kaizen in teams), allow gradually the commitment in its entirety, obtaining better results in production indicators.

Bataineh et al. [7], propose a 13-step TPM model which covers basic aspects of an improvement system such as: planning, implementation, verification, corrective action and control. The results obtained were an increase in the OEE index of 62.6% in 9 months.

Singh et al. [8], propose a new concept of mobile maintenance based on TPM in the Indian industry. The results they obtained were a significant improvement in OEE and improvements in productivity.

3 Innovative Technical Proposal

According to the diagnosis of the problem in the case study (See Fig. 1), the pilot area for the application of the proposed model is the CNC machining line.

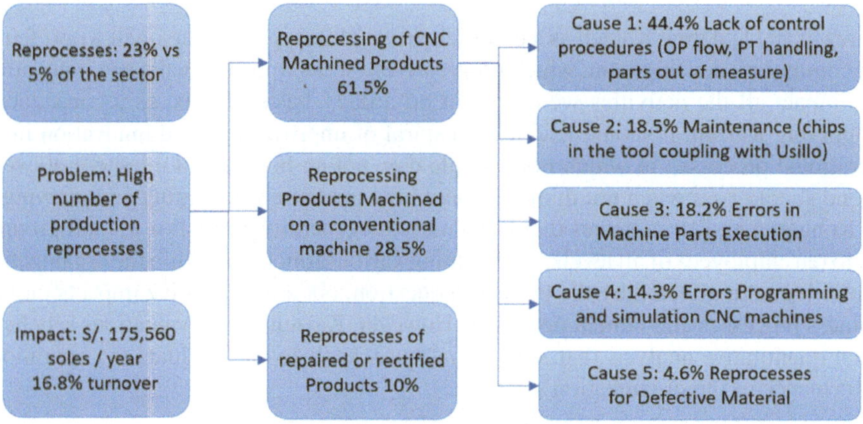

Fig. 1. Tree Diagram of the problem.

The innovative technical proposal consists of designing its own methodology for the implementation of TPM and Kaizen in the company under study. See Fig. 2.

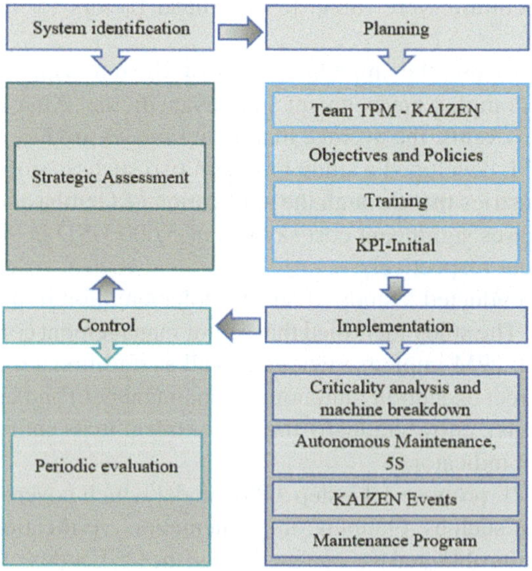

Fig. 2. Proposed model for the implementation of TPM - Kaizen.

Strategic Evaluation: the problem, the causes and the impact caused by the reprocessing are analyzed, also selecting the Pilot area.
Planning: the team that will lead the implementation is established, made up of company collaborators, objectives and policies are established. Finally, the training and the measurement of KPIs of the current situation.
Implementation: the criteria are established and the criticality analysis is developed. The breakdown of critical machines, autonomous maintenance, 5S, maintenance program will be carried out and Kaizen events will be carried out that include interviews with workers, identification of critical points with diagnostic tools and improvement action plans to obtain planned results.
Control: a periodic evaluation of the results of the KPIs, established in the proposed model, is carried out.

4 Results

4.1 Results Before the Intervention

According to Fig. 3, it was identified that reprocesses represent an average of 17.0% of parts produced in 2019, compared to the standard of 5.0% target. Of the total of reprocesses, the CNC machining line presents a rate of 61.5% of reprocesses of the total.

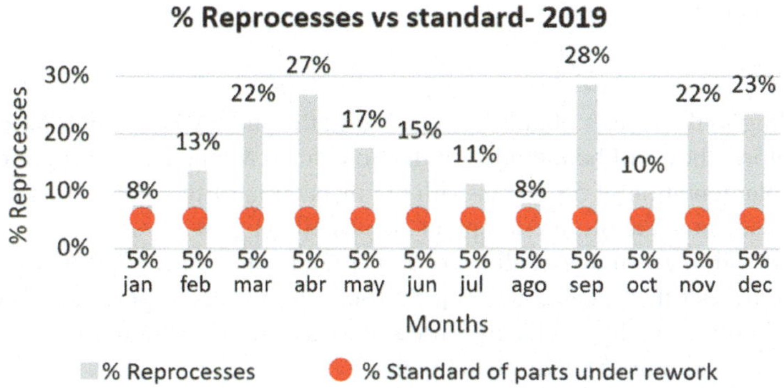

Fig. 3. Reprocesses (%) produced in 2019 vs the standard of 5%

In the planning stage, it was defined that the TPM-Kaizen team will be made up of three collaborators, the production manager, the plant supervisor and the quality supervisor. Policies and objectives were then established and training carried out:

Policies and objectives of the TPM / KAIZEN:

- Reduce the number of defects.
- Reduce customer complaints.
- Improve and optimize the capacity of the teams.

- Reduce maintenance cost.
- Have an optimal work environment.
- Developing employee competencies through training and education.

The Implementation phase begins with the criticality analysis and breakdown of each critical machine. The variables in the analysis are: 1. Effect on the Service provided; 2. Technical - Economic Value; 3. Failure affects; 4. Probability of Failure (Reliability); 5. Flexibility of the Equipment in the System; 6. Logistics Unit; 7. Dependence on Manpower; 8. Ease of Repair (Maintainability). See Table 1.

Table 1. Results of the criticity analysis

Item	Name of team	WEIGHT												Reference scale	Is it included in the PMP?
		Variable Type													
		1	2	3a	3b	3c	3d	4	5	6	7	8	Total		
1	Torno CNC DOOSAN Lynx 220	4	2	1	0	1	0	2	2	1	2	1	16	Critical	YES
2	Centro Smec LCV 650	4	2	1	0	1	0	2	2	1	2	1	16	Critical	YES

According to the results obtained, the CNC Lathe and SMEC Center have a critical scale, therefore, they will be included in the Scheduled Maintenance Program (PMP).

The implementation of 5S began with the inspection of the work areas of the CNC line, we can see how the production plant is currently (Fig. 4).

A classification was made of all the objects found out of place, for this we used three types of cards, and they placed the objects found in a specific place according to the type of card color. Finally, a cleaning program was established, and through an audit file, compliance with the 5S was guaranteed.

For Autonomous Maintenance a schedule of activities was established based on three points: cleaning, inspection and lubrication, which are developed by plant workers.

Among the proposed Kaizen events, is how to implement autonomous quality control (See Fig. 5). This format satisfies the largest number of OT work orders that are worked in the study area.

To carry out the transport of parts within the production area, supplies such as conditioned stretchers and carts for the parts were proposed.

The last step of this stage is the development of the maintenance program, for which at least three levels of critical machines were considered, the frequency of maintenance, execution time, specialty and the activities to be carried out.

Fig. 4. 5S implementation

Table 2. Autonomous maintenance activities

Type of activity	Activities	Frequency
Lubrication	Check the coolant level	Diary
Lubrication	Check the level of the track lubricant tank	Diary
Cleaning	Cleans chips on the track covers and inner bowl	Diary
Cleaning	Cleans chips from the cover turret, extension tube and swivel joint. Make sure the drive tube cap is installed, either on the rotary union or over the open part, of the hydraulic clamp plate	Diary
Inspection	Checks for proper drain operation at filter regulator	Weekly
Inspection	Check the air pressure gauge for 85 psi	Weekly
Cleaning	Clean exterior surfaces with a mild cleaner, but never use solvents	Weekly
Cleaning	Cleans chips out of the pan or reservoir in the coolant tank	Weekly

4.2 Results After the Intervention

In the Control stage, the OEE was measured to measure the impact generated by the implementation of the proposed model. The team will apply the improvement and will continue to have weekly meetings to reevaluate the maintenance plan and will continuously monitor the proposed KPIs. According to Table 3, an increase in OEE of 11.3% is observed.

	Quality Control Form	Annex 07090007				
		Rev. 2 Date 03/07/2015				
Description:		OP:				
Drawing / Sample:		Client:				
N° Pieces Produced:		N° of Champion pieces:				
DIMENSIONAL VALUES						
Measure	Nominal Value	Found value				
		Piece 1	Piece 2	Piece 3	Piece 4	Piece 5

Fig. 5. Quality control format

Table 3. OEE components calculated before and after TPM implementation

	BEFORE	AFTER
Availability	81%	89%
Efficiency	57%	62%
Quality	85%	92%
OEE	39.4%	50.7%

For the simulation of the implementation proposal, the Arena software will be used, in which we enter the company data obtained prior to the improvement project.

Simulation of Current Process. The simulation of the current process gives us the following results for a month of production: the parts produced are 185 units, the number of reprocesses is a total of 25 (13.5%). The simulation also provides the information that no waiting times are generated in quality control. On the other hand, the time per piece accepted is 23.43 min.

Simulation of Improved Process. The following is the simulation built in Arena after deployment. According to Fig. 12, the results obtained after the first month of implementation are shown, obtaining 231 pieces produced; Reprocesses are generated in a CNC Center machine (13), a CNC Lathe machine (4) and Quality Control (5), with a total of 22 reprocesses (9.5%). The time per piece accepted is 18.78 min.

Figure 12 Simulation of the improved process.

5 Conclusions

In the case study, the benefits of applying the proposed model based on the TPM and Kaizen tools have been detailed.

With the implementation of the proposed model, the impact produced by reprocessing in the company's CNC area was reduced by 30%. The results of the implementation were measured by calculating the OEE and the Arena software was used to simulate the results.

The results were an increase in the initial OEE from 39.4% to 50.7%. Availability increased by 8%, Performance by 5%, and Quality by 7%. Likewise, the decrease in production time was obtained by 19.8%.

References

1. Suárez-Barraza, M.F., Rodríguez-González, F.G., Stanley Hart, H.: Finding Kaizen core values in AACSB standards accreditation: a conceptual study. Total Qual. Manag. Bus. Excell. **30**(Sup1), S53–S73 (2019). https://doi.org/10.1080/14783363.2019.1665793
2. Vo, B., Kongar, E., Suárez-Barraza, M.F.: Kaizen event approach: a case study in the packaging industry. Int. J. Product. Perform. Manag. **68**(7), 1343–1372 (2019). https://doi.org/10.1108/IJPPM-07-2018-0282
3. Sahoo, S.: Assessing lean implementation and benefits within Indian automotive component manufacturing SMEs. Benchmarking **27**(3), 1042–1084 (2020). https://doi.org/10.1108/BIJ-07-2019-0299
4. Kapur, A., et al.: Improving efficiency and safety in external beam radiation therapy treatment delivery using a Kaizen approach. Pract. Radiat. Oncol. **7**(6), e499–e506 (2017). https://doi.org/10.1016/j.prro.2017.06.004
5. Cherrafi, A., et al.: Green and lean: a Gemba – Kaizen model for sustainability enhancement. Production Planning Control **30**(5–6), 385–399 (2019). https://doi.org/10.1080/09537287.2018.1501808
6. Hooi, L.W., Leong, T.Y.: Total productive maintenance and manufacturing performance improvement. J. Qual. Maint. Eng. **23**(1), 2–21 (2017). https://doi.org/10.1108/JQME-07-2015-0033
7. Bataineh, O., Al-Hawari, T., Alshraideh, H., Dalalah, D.: A sequential TPM-based scheme for improving production effectiveness presented with a case study. J. Qual. Maint. Eng. **25**(1), 144–161 (2019). https://doi.org/10.1108/JQME-07-2017-0045
8. Singh, J., Singh, H., Sharma, V.: Success of TPM concept in a manufacturing unit - a case study. Int. J. Product. Perform. Manag. **67**(3), 536–549 (2018). https://doi.org/10.1108/IJPPM-01-2017-0003

Improvement of the Global Efficiency of Mining Equipment Through Total Productive Maintenance - TPM

Orlando Lama, Josue Alayo, Ernesto Aparicio, and Cesar Nunura(✉)

Ingenieria Industrial, Universidad Peruana de Ciencias Aplicadas, Prolongación Primavera 2390, Lima 15023, Peru
{u201513240,u201518242,pcinvapa,pcincnun}@upc.edu.pe

Abstract. The model proposed in this article brings together the application bases of a Total Productive Maintenance (TPM) system as the beginning towards Lean transformation for a mining company. It takes as reference models and success stories that were previously implemented and is adapted to guarantee the success of its application. Likewise, it includes an optimal indicator to evaluate the efficiency of mining equipment, the Mining Production Index - MPI, which is an adaptation of the OEE to the mining context. This methodology can be applied to other mining environments since it is totally flexible in each of its steps. A case study is carried out to evaluate the potential application of the proposal, attacking each of the problems found in order to increase the efficiency of the mining equipment. Inside, the initial conditions of the mine were determined and how the proposed methodology could help improve it was evaluated.

Keywords: Efficiency improvement · Lean mining · Maintenance management · Mining equipment · TPM

1 Introduction

The mining industry constitutes a large part of the world economy, it is the main economic activity of most countries. A study carried out by a consultancy in 2018 [1] highlights that the global growth of Gold production will increase by 2.6% annually in the next 4 years due to a slight growth in the price of this metal, and the financial development of companies that will allow greater investment in the sector. One of the main engines of this global growth is mining production in Peru. Peru is one of the main mineral producing countries worldwide, it ranks second for the production of copper, silver, zinc, and sixth place in terms of gold production, being at the top in Latin America.

There is a constant search to improve the productivity and efficiency of mining operations, to maximize income, compete with rising prices, and stay out of the complex and heterogeneous business environment [2].

The maintenance area is the set of activities aimed at maintaining or restoring an asset to a state or conditions given for its safety and good use. The activities can be

S. Trzcielinski et al. (Eds.): AHFE 2021, LNNS 274, pp. 336–344, 2021.
https://doi.org/10.1007/978-3-030-80462-6_42

technical (performance in teams), administration (accounting, control) and management (decision-making) [3]. As equipment ages, operational failures are more frequent, leading to stoppages in operations. To correct this type of breakdown, the equipment is subjected to what is known as "corrective maintenance" or unscheduled. In this way, operating costs increase as the number of times this type of maintenance is performed on the machines increases.

The objective of this study is to improve the productivity and efficiency of mining equipment through the application of TPM principles, sequentially and measurably through techniques and tools that are efficient for this sector.

2 Literature Review

Given the importance of mining in the economy, this industry has paid attention to techniques such as Lean Manufacturing, which ensures the competitiveness of companies. The integration of Lean Manufacturing within the mining environment has been investigated for a little over 10 years [4–9]. Compared to the manufacturing industry, mining is much more exposed to natural and unstable conditions, so the implementation of Lean Manufacturing techniques does not always have the expected results.

Mining companies apply Lean Manufacturing in a certain way in their processes, either partially or by adapting these tools within the context. Of all the existing techniques in this philosophy, in mining, TPM, Kaizen continuous improvement, VSM, 5S, work standardization, or KPIS management stand out [6, 10]. The application of these already makes the company consider a change in its management and work philosophy. Godinho Filho et al. [11] studied the implementation and effect of lean manufacturing tools in small and medium-sized companies. His study highlighted that medium and small companies use lean techniques in a fragmented way, without having a holistic vision of LM, however, they managed to achieve improvements in their performance.

To make effective and efficient decisions, it is necessary that companies also establish metrics appropriate to their purposes and objectives [12]. In this way, the key metrics will be aligned with the objective of the companies, either by project or general. In the case of the TPM, one of the key indicators is the OEE or Global Equipment Efficiency. The OEE is considered one of the most important performance indicators, because it not only monitors productivity and production quality, but also because each of its elements lead to performance improvement solutions [9, 13].

3 Methodology

3.1 Model Sequence

This methodology consists of 11 steps which are described in Fig. 1.

This scheme follows a cycle of continuous improvement, the follow-up of which will guarantee to increase the efficiency of the teams and also the application of the six principles of the TPM in the organization.

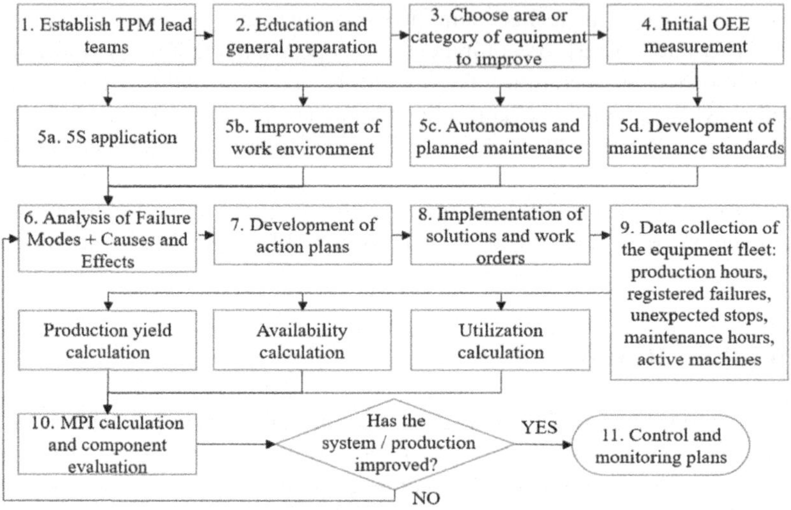

Fig. 1. Model for the implementation of TPM in mining

3.2 Key Indicator of Success

To find the most appropriate efficiency indicator, a comparison (Table 1) was made between three indicators that have been used according to the literature review, with the same purpose of measuring the efficiency of mining equipment.

Table 1. Comparison of oee variants

Variant	Characteristics	Key equipment
OEE_M	The indicator gives an individual view of the efficiency of each team. Affected by factors of quality of work and maintenance	Mining machinery and equipment
SEE	Applicable for excavators Measurement that considers excavator capacity, operator skill, and environmental conditions	Hydraulic and electric excavators
MPI	Indicates not only the efficiency of the equipment, but also its contribution to the availability, utilization, and performance of all equipment	All mining equipment and machinery

4 Validation in Case Study

4.1 Identification of Problems

The company under study has 11 assets, of which only 8 are operational. The 3 inactive ones are in the maintenance area and serve as a source of spare parts for the rest of the equipment in the event of a breakdown in them (Fig. 2). It was identified that there is no space in which these inoperative equipment can be parked, they are occupying a random space in the base area and without any registration (Fig. 3).

Fig. 2. Inoperative equipment used as a source of spare parts

Fig. 3. Current state of the repair shop.

Experts indicate that the ideal distribution of corrective-preventive maintenance is 20–80% respectively (Fig. 4). This can show that maintenance management for equipment is currently ineffective.

The availability values for each of these equipments are at very low levels, which impacts the OEE values. In this way, the efficiency of the equipment can be measured, which are effectively 56.7 for front loaders, 53.4 for excavators and 64.2 for dump trucks (Fig. 5).

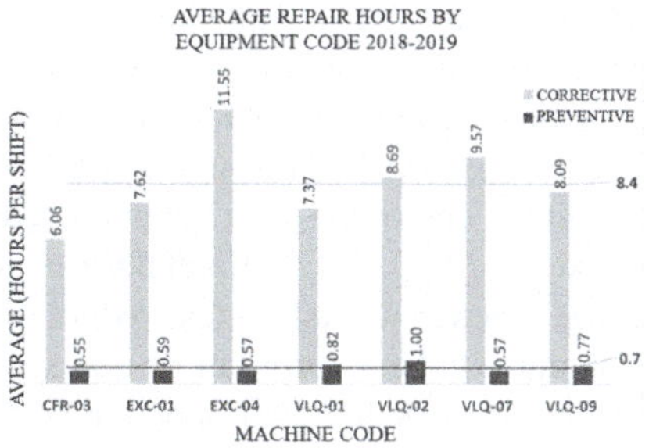

Fig. 4. Average repair hours per equipment

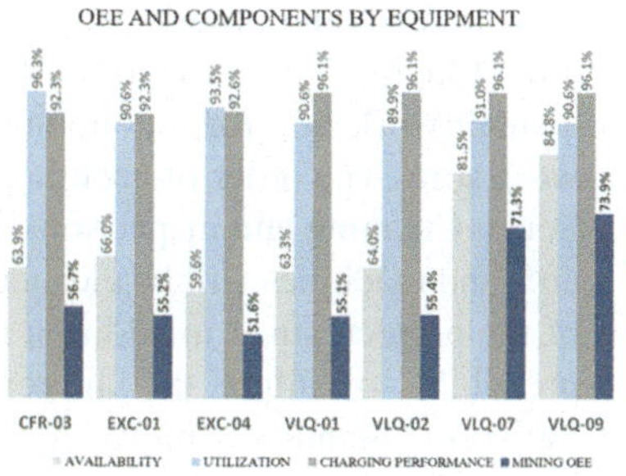

Fig. 5. OEE and components by equipment

4.2 Implementation Proposal

The application proposal in the case study will be sequential in accordance with the steps of the proposed scheme (Fig. 1), in order to successfully execute the new management plan. It is expected to have the support of the company and its workers to assume the new functions of the maintenance plan. Next, each step of the process and who will be part of it will be explained in detail.

1) **TPM Guiding Equipment.** The guide teams to be formed for the implementation of the TPM will be divided into 4 parts, the first being those in charge of the proposal, who will plan, implement, control and verify the TPM model; methodological advisor, who will provide his knowledge in the action plan; Finally, the chief of mechanics and chief of operations, these two positions are essential for the plan to be executed successfully, as they will be the leaders in their respective areas and will guide the entire team in the process together with those in charge of the proposal.
2) **Education and general preparation.** Training plans have been developed for the company's operators and mechanics twice a month for 5 consecutive hours, for a period of 6 months, these trainings will be interactive to make it more interesting for the workers.
3) **Equipment area to improve.** The equipment maintenance and operations area will be exclusively improved, with the aim of increasing productivity, availability, utilization, load performance and reliability of these.
4) **Initial OEE measurement.** The initial measurement of OEE and its equipment components was carried out thanks to the data provided by the company from the mechanics and operators reports for the year 2018 and 2019.
5) **Application of the proposal.** *a) 5'S Application:* The application of the proposal will begin with the Japanese methodology in the maintenance and production area, in order to achieve an efficient workplace, with order and standardization.
   Seiri (Organize and classify): The first step, all waste or objects that are not necessary in the mechanical workshop and in the production area will be eliminated, in other words, only what is necessary to fulfill the objective of the position will be kept.
   Seiton (Order efficiently): The tools and equipment of the workshop will be located in a strategic and orderly way, through shelves, work tables and shelves with the aim of facilitating the work of the mechanics, by the operators it will be provided a small workshop for autonomous maintenance to carry out before and after the work shift. The order of each implement will be established through labels.
   Seiso (Cleaning and inspection): Every day at the end of the shift the maintenance area and machines will be cleaned, this is a key factor for the optimal development of the project, since in addition to achieving a better workshop and reducing the probability of failure, the failure of the machine can be detected more quickly.
   Seiiketsu (Standardization): All processes in the workshop and operators will be standardized, from the beginning until the end of the day. When any autonomous repair or maintenance activity is carried out, the tools will be used and left in the indicated place.
   Shitsuke (Compliance or discipline): The heads of each area will verify in each shift that the processes have been carried out, in case they fulfill everything correctly, a bonus will be provided, to serve as an incentive.

b) *Improvement of the work environment:* New hand tools and workshop equipment will be purchased, a semi-open workshop with trenches will be built to facilitate the work of mechanics when repairing equipment; a washing area to keep the machinery clean; and a roof to work in the rainy season without problems. In addition, a small workshop for the operators, so that they carry out the autonomous maintenance of their equipment.
c) *Autonomous and planned maintenance:* Autonomous maintenance will be developed by the operator daily before and after the work shift, this will have a small one where they keep the basic equipment to be able to carry out this activity. The planned maintenance will be carried out by the maintenance manager and operations manager, they will analyze the behavior of the machines through the forms that the operators fill out each shift, and if necessary preventive maintenance will be scheduled, which will be developed outside the work shift of the machine, so as not to interrupt production.
d) *Development of maintenance standards:* This activity will be carried out by the maintenance manager, he will schedule preventive maintenance work, after a weekly evaluation of each equipment, he will request the necessary spare parts from the logistics area with an anticipation time of three days to the scheduled date of the repair, in the same way a safety stock of spare parts will be implemented that is usually changed frequently.

6) **Failure and Cause-Effects Analysis.** Through the new machinery and equipment control formats, information will be collected by the operators in order to predict the failure and thus be resolved before the failure occurs. This activity will be carried out by the maintenance and operations manager, who work as a team to analyze the causes of quantitative failures, through digital formats that we will provide and will be summarized in a Pareto chart. This way, they can solve the main cause of equipment failure.
7) **Development of action plans.** The information collected from the formats will be used to program and plan the change of the part and / or parts, this will be done outside the operation shift so as not to interrupt working hours.
8) **Implementation of solutions and work orders.** By an action plan developed by the chief of mechanics, the solution of the failure will be executed, this will be carried out through an analysis process of when would be the key moment of the change of the piece and / or pieces so as not to interrupt production, This will reduce the number of stoppages and improve the reliability of the equipment.
9) **Data collection.** The data collection will be through the new formats of the operators and mechanics, from this we will obtain the production hours, number of registered failures, unexpected stops, maintenance hours and loss of efficiency. These data will be stored in the company's system. Also, you will get the new indicators after implementation such as availability, utilization, load performance and reliability.
10) **MPI calculation.** The MPI will be calculated with the data obtained, if no improvements have been found then return to step 6 and develop the process again. This is based on the continuous improvement cycle.

11) **Control and monitoring plans.** The operations and maintenance manager will develop control plans for the equipment and its areas on a weekly basis. Likewise, an exhaustive monitoring of the implementation will be carried out through a maintenance audit.

4.3 Simulation of Results

The implementation of this model will have positive impacts on the company, on its work and management methods, and especially on the defined objectives. In order to validate that the proposal will really impact the main efficiency indicators, it will be evaluated through a simulation model, from which results will be obtained based on the operating time of the equipment. This was done in the Rockwell Arena Simulation 16 program.

a) Evaluation of results. 200 work shifts were replicated, for each of the dump trucks, in order to obtain optimal results that can be compared with the initial state of the dump trucks. The output of this procedure were the total operation times, corrective and preventive maintenance times and initial review. With these times, the calculation of the variables was carried out, for each of the equipment shown in Table 2.

Table 2. OEE and component evaluation

		EQUIPMENT			
Indicator	Status	*VLQ-01*	*VLQ-02*	*VLQ-07*	*VLQ-09*
Availability	Initial	63.3%	64.0%	81.5%	84.8%
	Obtained	**90.0%**	**90.0%**	**90.5%**	**90.3%**
Utilization	Initial	90.6%	89.9%	91.0%	90.6%
	Achieved	**86.8%**	**86.8%**	**87.7%**	**87.8%**
Performance	Permanent	96.1%	96.1%	96.1%	96.1%
OEE	Current	55.1%	55.4%	71.3%	73.9%
	Obtained	**75.1%**	**75.1%**	**76.3%**	**76.2%**
	Difference	**+20.0 pp**	**+19.7 pp**	**+5.0 pp**	**+2.3 pp**

The result of the simulation indicates that the increase in availability through the proposed methodology has a positive impact on the overall efficiency of the equipment. Operation times were increased, while corrective maintenance time was reduced. In the same way, the number of dumps increased and the number of failures fell. In this sense, the efficiency of the dump trucks increases by up to 20% points.

5 Conclusions

As mentioned in the literature review, Lean techniques were not originally intended to be implemented in other industries. However, many of these sought to adapt this philosophy because its implementation brings optimal benefits for companies. In the case of the mining industry, the registration and implementation studies of this technique are scarce due to the great differences between mining and other types of industries. Furthermore, the nature of the working conditions makes the implementation of these manufacturing techniques not entirely successful. The proposed scheme seeks to strengthen the efforts of other authors, who designed methodologies for Lean to be adapted in mining, starting with Total Productive Maintenance.

The results of the simulation, with the focus of increasing availability, show efficiency improvements by having an availability close to the objective and an increase in OEE. In addition to economic and operational improvements, the methodology also brings a very positive change in the workforce. This type of philosophy creates a safer environment, of autonomy and the morale of the workers increases. Thanks to this, the productivity of workers is directly increased.

The diagnosis of the case study showed that the company incurs very high costs related to inefficient maintenance activities that cause low efficiency in mining haulage equipment. The company performs below the international studies seen in the benchmark presented. The proposed model will allow the company to be led towards the objective of increasing its productivity and efficiency through the application of a Total Productive Maintenance model that will establish the bases for a transformation towards a culture of zero failures and waste. In the same way, an indicator adapted to the mining context is included, which will allow establishing a cycle of continuous improvement for the applied area.

References

1. INEI (2019). Informe técnico de producción nacional Diciembre 2018. Informe Técnico: Producción Nacional, 2, 65. https://www.inei.gob.pe/media/MenuRecursivo/boletines/informe-tecnico-de-produccion-nacional-febrero2019.pdf
2. Seifullina, A., Er, A., Garza-reyes, A., Nadeem, S.P., Kumar, V.: A lean implementation framework for the mining Industry. IFAC-PapersOnLine **51**(11), 1149–1154 (2018)
3. Fernández, C.S., Calvo, E.A.: Técnicas de mantenimiento en instalaciones mineras. Capítulo 1: Introducción al mantenimiento. ESCUELA POLITÉCNICA DE INGENIERÍA DE MINAS Y ENERGÍA, 1–21 (2018)
4. Oware, K.M., Samanhyia, S.: George Oppong Appiagyei Ampong (2015). Elimination of waste and inefficiencies through lean maintenance management in mining industries, Ghana. Int. J. Econ. Commerce Manag. United Kingdom **III**(10), 836–845 (2015). http://ijecm.co.uk/
5. Mostafa, S., Dumrak, J., Soltan, H.: Lean maintenance roadmap. Procedia Manuf. **2**(February), 434–444 (2015). https://doi.org/10.1016/j.promfg.2015.07.076
6. Lööw, J.: An investigation into lean production practice in mining. Int. J. Lean Six Sigma **10**, 123–142 (2019). https://doi.org/10.1108/IJLSS-07-2017-0085
7. Chlebus, E., Helman, J., Olejarczyk, M., Rosienkiewicz, M.: A new approach on implementing TPM in a mine - a case study. Arch. Civil Mech. Eng. **15**(4), 873–884 (2015). https://doi.org/10.1016/j.acme.2015.07.002
8. Mohammadi, M., Rai, P., Gupta, S.: Performance measurement of mining equipment. Int. J. Emerg. Technol. Adv. Eng. **5**(7), 240–248 (2015)
9. Korski, J., Tobór-Osadnik, K., Wyganowska, M.: Mining machines effectiveness and OEE Indicator. IOP Conf. Ser.: Mat. Sci. Eng. **268**(1). https://doi.org/10.1088/1757-899X/268/1/012010

10. Khaba, S., Bhar, C.: Lean awareness and potential for lean implementation in the Indian coal mining industry: an empirical study. Int. J. Qual. Reliab. Manag. **35**(6), 1215–1231 (2018). https://doi.org/10.1108/IJQRM-02-2017-0024Permanent
11. Godinho, M., Devós, G., Gunasekaran, A.: Lean manufacturing in Brazilian small and medium enterprises: implementation and effect on performance. Int. J. Prod. Res. **54**(24), 7523–7545 (2016). https://doi.org/10.1080/00207543.-2016.1201606
12. Nachiappan, R.M., Anantharam, N.: Evaluation of overall line effectiveness (OLE) in a continuous product line manufacturing system. J. Manuf. Technol. Manag. **17**(7), 987–1008 (2006)
13. Nallusamy, S., Majumdar, G.: Enhancement of overall equipment effectiveness using total productive maintenance in a manufacturing industry. Int. J. Performability Eng. **13**(2), 173–188 (2017)

Management System of Intelligent, Autonomous Environment (IAEMS). The Reference Model

Edmund Pawlowski(✉)

Faculty of Engineering Management, Poznan University of Technology, J. Rychlewskiego 2, 60-965 Poznan, Poland
Edmund.pawlowski@put.poznan.pl

Abstract. Intelligent Autonomous Environment (IAE) is a system of integrated industrial buildings, or a residential district, or a shopping district, etc. The full life cycle of such an environment may take several years or decades. Such an autonomous environment must have internal mechanisms to adapt to changing environmental conditions. Management of such an enterprise requires an integrated approach in terms of organizational structures, process management and support with modern management methods and techniques. The purpose of this paper is to present a reference model of such a management system.

Keywords: Management system · Intelligent autonomous environment · Reference model

1 Introduction

This paper is a part of a larger research project called "Innovative Management System of Designing, Building, Maintenance and Development of IAE". The project, undertaken at the construction enterprise WPIP Ltd. and Faculty of Engineering Management of Poznan University of Technology, started in 2018. The research framework of the project has been presented at the ISHE 2018 Conference [1, 2]. This paper is a summary of the research conducted between 2018 and 2021. The final effect of the research is a reference model of the IAE management system.

The paper consists of four parts:

1/ The Framework of the Research Project
2/ Description of the IAE management system reference model.
3/ Procedure for transforming the reference model into a management system design adapted to the specific conditions of the new IAE.
4/ Conclusions.

2 Framework of the Research Project

The research and development work is carried out in five stages.

S. Trzcielinski et al. (Eds.): AHFE 2021, LNNS 274, pp. 345–352, 2021.
https://doi.org/10.1007/978-3-030-80462-6_43

Stage 1. The result of the first stage was a methodology for designing an agile IAS management system including:

- development of the IAE management system agility concept,
- development of a methodology of designing an IAE management system.

The needs and expectations of the market of investors and users of smart facilities are changing and will continue to change with the changes occurring in the legal, economic, socio-demographic, technological and competitive segments of the business environment. Under such conditions, the IAE management system must be agile, that is, it must be able to take advantage of market opportunities, and thus able to meet the needs and expectations of investors building IAE and users of these facilities. The model of an agile enterprise is analysed in four dimensions [2, 3]: 1/ shrewdness of the enterprise, 2/ resource flexibility of the enterprise, 3/ enterprise's intelligence, and 4/ smartness of the enterprise. Agility should be investigated on four levels [3–6]: 1/ reacting for opportunities, 2/ creating opportunity, 3/ carrying out opportunities using virtual subsystems, 4/ carrying out opportunities by using virtual system – it means to become a broker of virtual network.

Management system design methodology includes organizational structure design and business process design. The methodological approach is based on the concept of multi-dimensional space of organizational structure design [7].

Stage 2. In the second stage, a model IAS management system was built consisting of five subsystems: a/ functional, b/ subject, c/ process, d/ methods and tools for fulfilling functions and processes, e/ information system.

The IAE management system will operate in external and internal conditions to which its functions, organizational units and processes must be adapted. For this purpose, a methodology was developed to study the macro and micro environment of IAE and to develop its forecast for five years. Then, models of management system were developed: functional (function tree), subjective (organizational chart), process (process maps) and a list of methods and tools supporting realization of management functions. The coherence of these components and their integration into the IAS management system was verified under laboratory conditions.

Stage 3. The final result of stage 3 is a prototype of the Intelligent Autonomous Environment Management System tested in near real conditions on a fragment of the existing WPIP enterprise organizational structure. The purpose of this phase was to check the feasibility of the structural model, process model and management methods. In particular, it checked:

1/ the completeness, integrity and scope of outsourcing of system functions.
2/ the separateness and inconsistency of the division of business processes and the division of competencies,
3/ workload of organizational units,
4/ completeness, integrity of the map of internal processes and relationships with suppliers and customers,

5/ internal logical correctness of processes,
6/ functionality of management methods (practical effectiveness of applications),
7/ labor consumption and cost of applying management methods.

Stage 4. In stage 4 particular subsystems of IAS management system were tested in operational conditions. The various subsystems of the model are made more detailed, including:

1/ Development of a detailed IAE management system function tree. The management system function model developed in Step 2 and verified under near real conditions in Step 3 was the starting point for developing the detailed IAE function tree.
2/ The subjective model of the management system developed in stage 2 was adjusted to the necessary level of system agility and takes into account the process features of the organizational structure.
3/ Mapping detailed management processes on the basis of maps of the main processes developed in Stage 2 and verified under laboratory conditions in Stage 3.
4/ Development of a demonstration version of an computer system of methods and tools to support IAE management.

Stage 5. The final result of stage 5 is a reference model of the full life cycle of the Intelligent Autonomous Environment Management System tested in real conditions.

3 Description of the Reference Model of IAE Management System

The reference model of the IAE management system consists of three groups of models (Fig. 1):

1/ Structural models of the management system.
2/ Process models management system.
3/ Models of method and tools supporting IAE management.

Reference Structural Model

The starting point for the structural model is the system function tree. The function tree answers the question of what the system should do, i.e., what business processes should be performed to achieve the intended effects of the created IAE system. The Reference Functional Model is presented in Table 1. It is a skeleton of the function tree, which is developed into a detailed tree for a specific investment case. On the main branch there are management functions concerning the enterprise as a whole: Strategic management, Developing a company management system, Human resources management, Economic control, and Operational management of IAE undertakings. The main functional subsystems correspond to the different phases of the IAE life cycle: Marketing and sales, IAE design, Construction of IAE, Exploitation of IAE, Development of IAE.

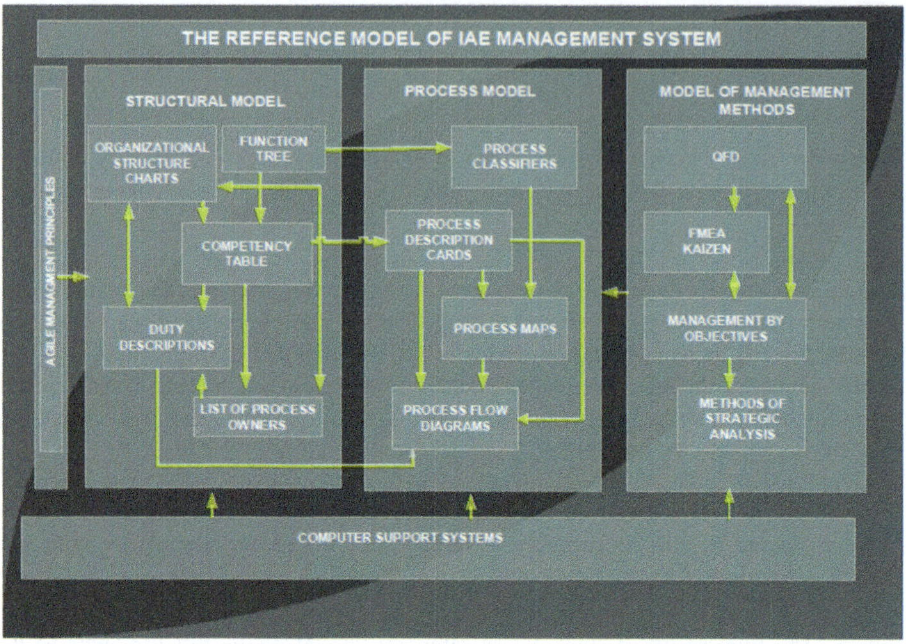

Fig. 1. The reference model of IAE management system

The Organizational Structure Chart answers the question what organizational units and how they are connected to each other are needed to realize the business processes defined in the function tree. The third element of the structural model are the Competency Tables that define who? (which organizational units) and to what extent they are responsible for realization of particular functions. On the basis of Competence Tables there are defined: Duty Descriptions and List of Process Owners.

Reference Process Model

Based on the Function Tree, a Process Classifier is developed, in which individual functions are described by indicating: input, process activation source, main processing operations, output, process owner, and determining the degree of process formalization. The processes most important for the whole system are described in Process Description Cards. Graphical presentation of processes is presented in the form of Process Maps and Process Flow Diagrams.

Reference Model of Management Methods

Methodical support for IAE managers are management methods. From among dozens of contemporary management methods and tools, 5 methods have been selected as the most useful for IAE management:

1/ Quality Function Deployment (QFD) – as a comprehensive analysis of the concept of developing the IAE system as a product and relating it to customers, suppliers, and competitors.

Table 1. The reference functional model of IAE Management System

No of function	Name of function
00	RATIONAL DEVELOPMENT AND ECONOMIC FULFILLMENT OF DEMAND FOR IAE
00.#.01	STRATEGIC MANAGEMENT OF IAE
00.#.01.01	Shaping the strategic management system of the Company
00.#.01.02	Strategic management of business portfolio (IAE)
00.#.01.03	Strategic management of other (apart from IAE)
00.#.01.04	Developing and implementing the Company Strategic Scorecard
00.#.02	DEVELOPING A COMPANY MANAGEMENT SYSTEM
00.#.02.01	Defining and implementing the concept of the management system for the Company
00.#.02.02	Developing management systems for IAE projects
00.#.02.03	Developing management systems for other (apart from the IAE) projects
00.#.03	HUMAN RESOURCES MANAGEMENT
00.#.04	ECONOMIC CONTROL
00.#.04.01	Economic control at the level of the Company
00.#.04.02	Economic control at the level of IAE projects
00.#.05	OPERATIONAL MANAGEMENT OF IAE UNDERTAKINGS
00.#.05.01	Operational management of the business portfolio of IAE projects
00.#.05.02	Operational lifecycle management of IAE
00.#.05.03	Operational management of IAE life cycle stakeholder relationships
01	MARKETING AND SALES
02	IAE DESIGN
02.01	Development studies
02.02	IAE concept design
02.03	Technical design of IAE investments
02.04	Legal and administrative preparation of IAE investments
02.05	Project documentation management of IAE
03	CONSTRUCTION of IAE
03.01	Sourcing subcontractors and entering into contracts
03.02	Procurement of technical and material resources
03.03	Current management of the construction process
03.04	Supervision of project execution
03.05	Final acceptance of the investment

(continued)

Table 1. (*continued*)

No of function	Name of function
03.06	Financial settlement of the investment
03.07	Construction documentation management
03.08	Supplier evaluation
04	EXPLOITATION of IAE
04.01	Shaping the IAE exploitation service system
04.02	Taking over documentation of objects from investments
04.03	Keeping records of IAE operation
04.04	Integrated support for use of buildings and premises
04.05	Maintenance of the technical condition of IAE facilities
04.06	Fire protection of the IAE
04.07	Maintenance of cleanliness and aesthetics
04.08	Facility security and property insurance
04.09	Lease administration
04.10	Proposing changes and improvements to the IAE
05	DEVELOPMENT of IAE
05.01	Adaptive shaping of the list of functions performed by IAE
05.02	Designing and implementing changes to IAE functionalities (technological, energy, intelligence, cubic capacity, etc.)
06	SUPPORT PROCESSES
06.01	Management of health and safety at work
06.02	Protection of intellectual property and information security
06.03	Management of information systems
06.04	Quality management
06.05	Securing legal services
06.06	Administrative support
06.07	Maintenance and protection of company property

2/ Methods of strategic analysis (SWOT / TOWS and Competitive Forces Model) - as methods preparing business strategy of investment in IAE.

3/ Management by Objective – as an effective management method in network business structures and turbulent environment.

4/ FMEA and KAIZEN – as verified methods for improving business processes.

4 Framework for Applying the Reference Model

The framework procedure for using each of the three types of reference models in developing an IAE project management system is generalized below. The procedure includes 8 steps:

1/ The analysis of internal and external conditions of the enterprise - using the methods of strategic SWOT analysis and micro environment analysis. The work of S. Trzcielinski [8] contains exemplary analyses of the macro-environment of IAE, whereas the scheme of the analysis of the competitive environment for the IAE sector is given by Kruszynski and Pawlowski [9].
2/ Develop a strategy for IAE contract implementation.
3/ Develop a function model for the contract - develop a function tree (for all phases of the cycle covered by the contract) based on the reference tree and predefined internal and external conditions of the IAE.
4/ Developing entity model for the contract - designing organizational units for IAE and integrating them into the organizational structure of the Company in particular phases of IAE cycle.
5/ Design of Competency Tables and definition of framework Job Duties for organizational units and management positions.
6/ Develop a Process Model for all phases of the IAE cycle.
7/ Selection and initiation of Management Methods for each phase of the IAE cycle.
8/ Complementary modeling of organizational structure and business processes when moving from one phase of the cycle to the next.

5 Summary and Conclusions

The research project called "Innovative Management System of Designing, Building, Maintenance and Development of Intelligent Autonomous Environment" began in 2018 and is currently in the final stages of completion. The goal of the final phase is to develop a Reference Management System Model that integrates all functional management subsystems and all phases of the IAE life cycle.

A reference model is a template, a generalized proposal for solving selected issues in a specific practical situation. Procedure for transforming the reference model into a management system design adapted to the specific conditions of the IAE depends on the specific business situation, including the type and size of the building, the phases of the object's life cycle covered by the contract, the specific requirements of the customer. The proposed management system modeling methodology is currently being verified for two concurrent IAE projects.

References

1. Pawlowski, E., Pawlowski, K., Trzcielinska, J., Trzcielinski, S.: Designing and management of intelligent, autonomous environment (IAE): the research framework. In: Ahram, T., Karwowski, W., Taiar, R. (eds.) IHSED 2018. AISC, vol. 876, pp. 381–386. Springer, Cham (2019). https://doi.org/10.1007/978-3-030-02053-8_58

2. Trzcielinski, S., Trzcielinska, J., Pawlowski, E., Pawlowski, K.: Methodology of Shaping the agility of the intelligent autonomous environment management system. In: Ahram, T., Karwowski, W., Taiar, R. (eds.) IHSED 2018. AISC, vol. 876, pp. 652–658. Springer, Cham (2019). https://doi.org/10.1007/978-3-030-02053-8_100
3. Trzcielinski, S.: Przedsiebiorstwo zwinne, Wydawnictwo Politechniki Poznanskiej, Poznan (2011)
4. Brander, S., Hinkelmann, K., et al.: Mining of agile business processes, artificial intelligence for business agility. Papers for AAAI Spring Symposium (2011)
5. Dove, R.: Engineering agile systems: creative-guidance. Frameworks for requirements and design. In: 4th Annual Conference on Systems Engineering Research (CSER), Los Angeles, CA, 7–8 April (2006)
6. Trzcieliński, S., Wojtkowski, W.: Toward the measure of organizational virtuality. Hum. Factors Ergon. Manuf. **17**, 575–586 (2007)
7. Pawlowski, E.: Designing the organizational structure of a company. A concept of multidimensional design space. In: Csatth, M., Trzcielinski, S. (eds.) Management Systems. Methods and Structures. Monograph. Publishing House of Poznan University of Technology, pp. 107–122 (2009)
8. Trzcielinski, S.: The influence of macroenvironment changes on agility of enterprise. In: Karwowski, W., Trzcielinski, S., Mrugalska, B. (eds.) AHFE 2019. AISC, vol. 971, pp. 111–121. Springer, Cham (2020). https://doi.org/10.1007/978-3-030-20494-5_11
9. Kruszynski, M., Pawlowski, E.: Management of intelligent autonomous environment in a context of competitive forces. The case of polish market. In: Mrugalska, B., Trzcielinski, S., Karwowski, W., Di Nicolantonio, M., Rossi, E. (eds.) AHFE 2020. AISC, vol. 1216, pp. 33–40. Springer, Cham (2020). https://doi.org/10.1007/978-3-030-51981-0_4

Application of Porter's Value Chain Model for Construing Potential Prospects and Lacunas in Industry 4.0 Adoption by 21st Century Manufacturers

Makinde Oluwafemi Ajayi[1(✉)] and Opeyeolu Timothy Laseinde[2]

[1] Faculty of Engineering and the Built Environment, University of Johannesburg, Johannesburg, Republic of South Africa
219090277@student.uj.ac.za

[2] Mechanical and Industrial Engineering Tech Department, University of Johannesburg, Johannesburg, Republic of South Africa
otlaseinde@uj.ac.za

Abstract. Manufacturing firms are lately experiencing a paradigm shift from mass production to customized production. As such, it becomes more pertinent for industrial entities to engage digital and intelligent systems of manufacturing. Industry 4.0, also known as the Fourth Industrial Revolution (4IR), is spurred by integrating digital technologies. This is embedded within an interoperable global value chain that guarantees sustainable operational efficiency and value optimization. This paper presents a review evaluating the potential opportunities and threats manufacturing firms experience with adopting the 4IR and the implication on value chains. An extensive literature review was conducted by analyzing accredited resources, using peer-reviewed articles, industry reports, and validated information obtainable in global repositories. The study generated a better understudying of Industry 4.0 technology adoption and their underlying impact (favorable and non-favorable) on industrial value chains. Significant findings reveal that the practice of industry 4.0 has a significant effect on organization cost, material resources, processes, products, systems, humans, and business models in manufacturing firms. The findings indicate that the impact of Industry 4.0 on manufacturing firms be studied holistically as against localized study of enterprises and country. As such, global views with inter-country economic peculiarities and realities must be considered.

Keyword: Industry 4.0· Opportunities· Threats· Manufacturing firms· Value chain

1 Introduction

In this 21st century, manufacturing industries have precipitously been exposed to the wind of change arising from global competition stirred by an unprecedented upsurge and innovation in technological inventions that have continuously made industries contend

S. Trzcielinski et al. (Eds.): AHFE 2021, LNNS 274, pp. 353–363, 2021.
https://doi.org/10.1007/978-3-030-80462-6_44

with the ever-changing demands of the market. Therefore, these criteria enable current manufacturing firms to be well acquainted with information technology opportunities and solutions to achieve their predetermined organizational goals and objectives.

The Industry 4.0 approach is very noteworthy, allowing manufacturing and retail activities to be streamlined while at the same time incorporating actors into the value chain of an organization in its entirety (producers and consumers). The use of Cyber-Physical System (CPS) generic concepts and the Industrial Internet of Things (IoT) have addressed some specialized necessities expected of an industrial manufacturing system in the 21st century.

The advent of the idea of Industry 4.0 is a universal term for the recent industrial paradigm that promotes a range of potential industrial developments. This development includes Cyber-Physical Systems (CPS), the Internet of Things (IoT), the Internet of Services (IoS), Big Data, Robots, Virtual Reality, and Cloud Manufacturing [1]. The implementation of these technologies is key to the advancement of other intelligent manufacturing procedures, such as machines, devices, production modules and products that autonomously exchange information, initiate actions, and control independently, thereby providing a smart manufacturing environment [2]. Therefore, the fourth industrial revolution gives rise to the interconnectivity of Cyber-Physical Systems assembling parts [3]. Cyber-physical system parts are interwoven arrangements that have been fragmented into parts. It ensures advanced connection, collection, and transmission of concurrent information that identifies, finds, trail, observes and improves the manufacturing course of action.

Industry 4.0 is largely driven by data. The application of industry 4.0 tools to make the best decisions has since become a competitive importance factor. Therefore, where competitive advantage originates from will not be limited to manufacturing in a synchronized manner or on an entirely new basis (such as additive manufacturing) and the integration of products with digital systems. For example, in case of a failure, the machine autonomously displays the replacement part required to be used, i.e., it allows companies to filter the relevant information from the data generated to make informed decisions [4, 5].

Evidently, in this 21st century, factory and production systems have been enhanced continuously by information technology support tools due to the need to control more complex technologies, the pressure of multiple-site manufacturing, and supporting logistic processes have become difficult tasks. The role of Information Technology at companies cannot be overstated. It has changed working conditions, level of efficiency, and its significance is unquestionable [6, 7].

In their work on Industry 4.0 readiness, Blanchet et al [8] noted that many developed countries in the world are still operating below the average in terms of manufacturing production versus GDP, as seen from indicators such as complexity of the manufacturing system, level of automation, workers readiness, and innovation strength. The emergence of Industry 4.0 supplies the relevant solution to the challenges associated with the fourth industrial revolution. The primary essence of Industry 4.0 is to guarantee improvements in arrears of automation and operational efficiency, as well as effectiveness [9].

Industry 4.0 today has amazingly become a contemporary issue amongst major players in the manufacturing sector. It is based on the unity of things, "smart" technologies with storing capability, and production facilities similar to the Cyber-Physical Manufacturing System (CPMS). This manufacturing order automatically allows the interchange of data, enable processes to start, and carry it out with independent control [10].

Industry 4.0 was founded on the notion of smart manufacturing. Smart manufacturing has introduced an entirely new method of manufacturing. Smart products are largely traceable to smart factories. The basic facts, prevailing conditions, and future program that are expedient to produce according to design are known at any timeA sound database system is of greatest importance for accessing prompt and correct information in order to filter through. The virtual systems have become so beneficial because it has provided the world opportunities to simulate different cases more so, the efficiency of processes, products, and the totality of supply chain is continuously optimized. Since the fourth industrial revolution, the cloud network has become inevitable for the corporation with an instrument called Cloud technology [10, 11].

Cloud technology makes ready smart data locations, products, services, and software to users (companies) to maintain lower costs and operations efficiency. This new method allows production based on customer's specifications. Many companies have now switched from bulky production to bulky customized products. The focus is to maintain a manufacturing system that can cope with the dynamism in business processes and practice. Such systems are flexible in nature, such that they can respond to interruptions from various angles. All these efforts are targeted towards operational efficiency and customer value creation, which underscores the value chain propounded by Micheal E Porter.

A value chain can be described as a predetermined order of activities being carried out by a firm operating in industry to present a unique product to its customers. Industry 4.0 provides an organization the avenue to manage the entire value chain of its product life cycle. Hence, the study aims to use Porter's value chain model as a tool to examine the underlining opportunities and threats the fourth industrial revolution has brought, thereby providing a template/model for application among manufacturing firms in developed, developing, and underdeveloped countries.

2 Review on Value Chain Approach: An Industry 4.0 Perspective

Nagy *et al.* [12] studied the effect of Industry 4.0 and the IoT, their role in developing Value Chain Business Strategy within a Hungarian economy. The research's main objective was to ascertain how corporate entities carrying out business activities in Hungary understand the Industry 4.0 phenomenon, the IoT tools used to enhance their operations, and those critical problems they faced during adjustment. The study adopted Porter's value chain model to illustrate the effect Industry 4.0 had on the case study firms, this model was considered valuable whenever special interest is given to corporate activities, which performs a vital role in promoting customer value creation. This study used two (2) methods in its investigation. Structured questionnaires were shared among manufacturing and logistics-based firms to find out the Internet of Things tools used and their challenges. Forty-Three (43) responses were retrieved which was further subjected to evaluation. Some interviews were further conducted with experts within the sector in order to gain deeper insights about their appliance, IoT tools development phases and other critical issues. Research findings showed that real-time data widely spread among organizations with possession of applicable analytical tools and procedures may have a significant effect on the company as a whole. Companies using CPS, CPPS, and Big Data Technologies evaluated were found to have higher level of logistic service, higher efficiency in processes with partners, enhanced cooperation amid specialized logistic functions, higher market and financial performance, and competitiveness. The study further submits that the application of better optimized production processes is key to attaining optimum manufacturing and economies of scale; it may also lead to high level economic sustainability.

Schonauer *et al* [13] also worked on the value chain approach under an Industry 4.0 perspective, focusing on conceptual development from the past industrial revolutions. The study aimed at illustrating the value chain concept holistically in developing business models and in all the four stages of the industrial revolution from the steam engine to electrical power and IT revolutions. Now, the Cyber-Physical System industrial change is being referred to as Industry 4.0. Furthermore, the study also focuses on analyzing the dynamics between rural and urban operations using social parameters from each stage of the industrial revolution with an emphasis on urbanization. Methodologically, the study used a third-level NUTS approach to consider three steps of statistics on the rural and urban studies, while the Eurostat website was used to visualize and analyze the collated data. The findings show that rural-urban value chains rely on the items generated, operations, services, distribution and trade, in the local, state and global space, connecting rural to global consumers from one stage to another,e.g., as smallholders, transporters, input suppliers or processors. Besides, it was revealed that developing countries have greater than 25% of their gross domestic product from agricultural value chains with more than 70% of its population dwelling in countryside arrears. The study finally submits that growth clusters of value-added activities generate substantial income for an economy.

Culot *et al* [14] used a Delphi-based scenario analysis to examine industry 4.0 and manufacturing company's prospects. The research adopted a value chain approach to evaluate the evolutionary trajectories peculiar to manufacturing firms. The study considered a rigorous selection of 76 experts with heterogeneous professional backgrounds. The result reveals that evolving development in manufacturing firms required the analysis of larger evolutionary dynamics extended beyond traditional industrial borders. The study concluded it was difficult to predict when the trend will stop, and the effect at the global economic level. Currently, selected responses must ensure that ecosystems and separate organizations overcome the shock on the short term. In the long run, new impetus will be injected into the economy aided by structural measures to curtail the gloomy possibilities of recession and employment. These structural measures include investment in inventions supported by public incentives to transform the spotlight on the 4IR trajectories.

Von Standen [15] worked on the impact of value chain management on Momentum's business performance. The study aimed to examine relevant literature and generate evidence to support whether value chain management has a significant impact on business performance. The research conducted an empirical study regarding value chain management models which was implemented by momentum for a specific group of customers.

With the aid of a well-structured questionnaire through a simple random sampling technique, information was elicited from a group of respondents who assumed different value chain responsibilities. Thereafter, data were retrieved and analyzed. The research findings from the empirical study show that Momentum's implementation of the value chain leads to an improvement in the individually measured component of business performance as well as an improvement in overall business performance. It was however, discovered that Momentum's implementation of value chain management did not improve business performance to the anticipated level. It was found that, on a decentralized basis, value chain management needs to be deployed in a separate business unit. The study, therefore, concluded that value chain management improves the business performance of Momentum Group Ltd. Conversely, the optimal implementation of value chain management requires a decentralized approach and creating a separate business unit.

Martínez-Olvera and Mora-Vargas [16] studied the framework for the analysis of 4IR Value Domains comprehensively. The study aimed to investigate the procedure of value creation in Industry 4.0. The study also had unique interest in the relationship existing between mass customization and corporate sustainability. In a bid to perform the research, related articles were reviewed and the research gap was identified. Therefore, the study proposed the Customer Product Process Resource (CPPR) 4.0 framework, considered a template to generate sustainability in business models that promote value creation with combined qualities of mass customization shift in an industry 4.0 environment and SMEs. The research result indicates that the effect the rate at which systems reconfigured has on the amount of produce far outweighs its impact on the level of customization. Hence, we can deduce that sustainability issues arising from developed business models with sustainable value propositions are resolved once high-level soft technology is realized.

Hopali and Vayvay [17] examined Industry 4.0 as the last Industrial Revolution and its opportunities for developing countries. The authors examined the responsibility of business partners and innovative technologies on the achievement of the 4IR. The enablers of smart factories have been extensively explored and how best they can be matched with supply chain partners in industry 4.0.

The main objective of the research was to make explicit procedures accessible to emerging economies that will generate sustainability and increase competitive advantage through the adoption of Industry 4.0, thereby enabling emerging economies to have a greater share of the global value chain of manufacturing with special significance to match two keywords for the first time (value chain partners and smart factory). The study findings show that industry 4.0 implementation has gone beyond smart factories but has started influencing people's daily living via value chain partners in different sectors. The study suggests there should be improvement in devices by satisfying and meeting the client's need in arrears of their utility anywhere and anytime. It further recommends that evolving countries should embrace industry 4.0 not only for efficient production but also for public wellbeing.

Sarc *et al.* [18] reviewed intelligent robotics and digitalization under the value chain perspective as it pertains to circular economy-oriented waste management. The research aimed to put in place systems that would be useful in treating waste machines or equipment in the future thereby making waste treatment process more efficient. Specifically, attention is focused on systems involved in sorting of (mixed) waste through robotics machineries. The study used a wide-ranging survey including international and national publications with an approximately 400 feedbacks. The online survey was performed via email using a questionnaire with nine questions and retrieved from companies in the DACH region of Austria, Switzerland and Germany. These companies in various sizes are practicing recycling technology and waste management. The research findings reveal that management of waste is inexorably evolving towards digital industrialization. Customer-adapted solutions should be coordinated with manufacturing firms to ensure effective implementation of the technologies. i.e. the recovery industry and the suppliers. Conclusively, it appears that smart technologies, especially the robotic-based sorting system in waste management has continued to concentrate more interest as a result of the one-sided substitution of human being or the introduction of robots. This eases humans of heavyweight physical sorting and saving them from non-optimal environmental pollution such as noise, air or water pollutants.

Strange and Zucchella [19] worked on Industry 4.0, global value chains, and international business. Their research aim was directed towards evaluating how the extensive implementation of latest digital technologies specifically; the IoTs, big data analytics, artificial Intelligence, and additive manufacturing also known as 3D printing, likely influence the situation and arrangement of events within the global value chains. To perform the study, various literature and sources were reviewed to understand the likely impact of the adoption of these emerging digital technologies otherwise known as fourth industrial revolution (4IR). The study further looked into differentiating these inventions from current ones and considering how these new technological inventions bring about new configurations amongst customers, suppliers, and the firm. Therefore, the findings showed that the emerging digital inventions can disrupt the method and the place where activities are situated and prearranged inside global value chains. The authors further submit that the 4IR is currently in its earliest period. Nonetheless, its consequences are evident in the arrears of competition and management strategy in many manufacturing firms. According to the authors, the study implication was to draw awareness to likely cyber-risks and possible consequences for an individual's confidentiality hence, the need for control.

Geissbauer *et al* [5] researched for PwC on Industry 4.0 and the building of digital enterprise. The research's primary aim was to evaluate the gains of digitized firms across their horizontal and vertical value chains while developing their digital product & service portfolio. To achieve this foretasted goal, the study conducted survey research covering more than 2,000 respondents from firms in 9 major industry sectors located in 26 countries. It went further to understanding corporate views and thoughts towards transformation into a digital business operation. Research findings from the survey depict that the 4IR concept has seized to be a 'impending trend' for lots of firms, it has now dominated their strategic and research plan. Corporations now link advanced connectivity with advanced automation, computer-powered processes, sensors and 3D printing, cloud computing, intelligent algorithms, connected capability, and IoTs services to change business processes and operations. The study, therefore, concludes that an enterprise that will change progressively with Industry 4.0 introduction, acquisition, and availability of digital abilities across these companies is quite germane. Hence, the process is time-consuming to grab or maintain a first-hand lead ahead of other competitors; it requires strategic management commitment and substantial investments that support its implementation.

Zangiacomi *et al* [20] examined the road to digitalization using more than one case study in the manufacturing industry. This work aims to present different case study analysis that shows an organizational perspective for the adoption and revolution path concerning Industry 4.0 in the production value chain. Methodologically, the study investigated many players' perception; a pre-selected group of companies from various firms in the production supply chain in northern Italy were considered. The final sample of 20 heterogeneous companies demonstrates its sufficiency. A survey research technique was used to capture the nature of the companies, the sector they belong, and their current stage in digitalization. Also, the method adopted by respective firms in adopting industry 4.0 technological know-how. The study results show these major challenges, common mistakes, and the most appropriate practices in the manufacturing industry regarding digital implementation.

It also offers a summary that acts as a guide that enables businesses to consider the most critical problems to be resolved whenever the introduction of new and emerging technology is addressed. Park and Heo [21] reviewed the change in electrical industrial value chain under the ICT scope. The study aimed at examining the various changes in value chain and further reviewed the process of transformation obtained in the electric power generating Industry resulting from the coming together of ICT in a systematic order. The study adopted a value chain transformation framework in the unity era, which is different from the current value chain element-centered method. Hence, the study reviewed relevant literature in an attempt to interpret again the changes that occur within the electric power generating industry in the area of counterflow, insertion, multiple dimension, and contraction or elimination of stages within the value chain. Data were retrieved from a scholastic database such as science direct, Google Scholar, etc. Therefore, the research findings reveal that the counter-flow value chain is equivalent to an increment in consumer influence. This implies the consumer role in the energy transition is increasing. The energy transition can be accelerated by the consumer's participation in expanding renewable energy combined with technologies that support eco-system. The study, therefore, concludes that heavy data and digitized platforms may enhance the usage of distributed resources resulting in a better and efficient operation of energy infrastructure.

3 Research Method

The study conducted desk research to examine the 4IR as regards the value chain approach. Different types of literature were collected from different outlets, such as: research papers, journals, documents, magazines, newspapers, business reports, government reports, consultants' reports, blogs, and online videos. All through the process of developing the study, the existing level of scientific published work was reviewed from time to time to know about any research being carried out in the related area. Initially, by defining the different ramifications of the implementation of the 4IR for the manufacturing sector in this 21st century, a measure was taken to consider the supply chain approach to 4IR.

A research on Industry 4.0 was subsequently carried out to clarify the possible risks and opportunities provided by the 4IR implementation to manufacturing companies using the supply chain approach. Our ultimate focus and basis of analysis was Porter's value chain model, which was considered able to capture pertinent areas that influence manufacturing companies operational efficiency. As soon as these areas are framed, inducing variables were used to analyze the consequences of various Industry 4.0 emerging technologies; the underlining threats and opportunities associated with each of these technologies were further discussed by connecting them with stages of the value chain under review. Conclusively, to support all the proposals made in the preceding stages, the study's research hypothesis was developed.

4 Review Analysis

This section presents the various approaches of the value chain towards Industry 4.0 and why Porter's value chain has gained more preference and relevance for manufacturing firms in the 21st century. The review presented methods used to analyze Industry 4.0, including Porter's value chain strategy, Delphi-based scenarios, Value Chain Management models, Consumer Product Process Resource 4.0 strategy, Value Chain Change System, Centric Approach to the Value Chain Aspect, and Global Value Chain Approach, as shown in Fig. 1.

From Fig. 1 above, we have seen the various approaches ranking at almost the same level except for CPPR 4.0 having a higher rate of usage which underscores the importance and need for concentration on value-added activities associated with the production process. The global value chain approach was ranked 12% which shows limited use of the approach towards the study of Industry 4.0. For example, in [19], the studies bone of contention was on how the prevalent implementation of latest digital and additive manufacturing (3D printing) likely affects the situation and arrangement of activities within the GVCs. In a similar vein, the value chain element-centric approach, value chain change framework, value chain management models, and Delphi-based scenario had received 12% rankings, suggesting that previous studies have not carried out much work using these approaches. Howbeit Porters value chain model which ranked more on a 15% point was considered beneficial while paying keen attention on corporate activities which often play an important role in client's value creation according to [12], and finally, the CPPR 4.0 approach was ranked 25% with the highest frequency. This method aims to explore the process of value creation under the fourth industrial revolution, with a keen interest in unraveling the relationship between mass customization and sustainability matters amongst manufacturing firms. Comparatively, it was observed that the areas of interest using porter's value chain model and Customer Product Process Resource 4.0 approach are similar. Both capture value creation activities in business organizations that underscore Porter's value chain model's preference to the study of Industry 4.0.

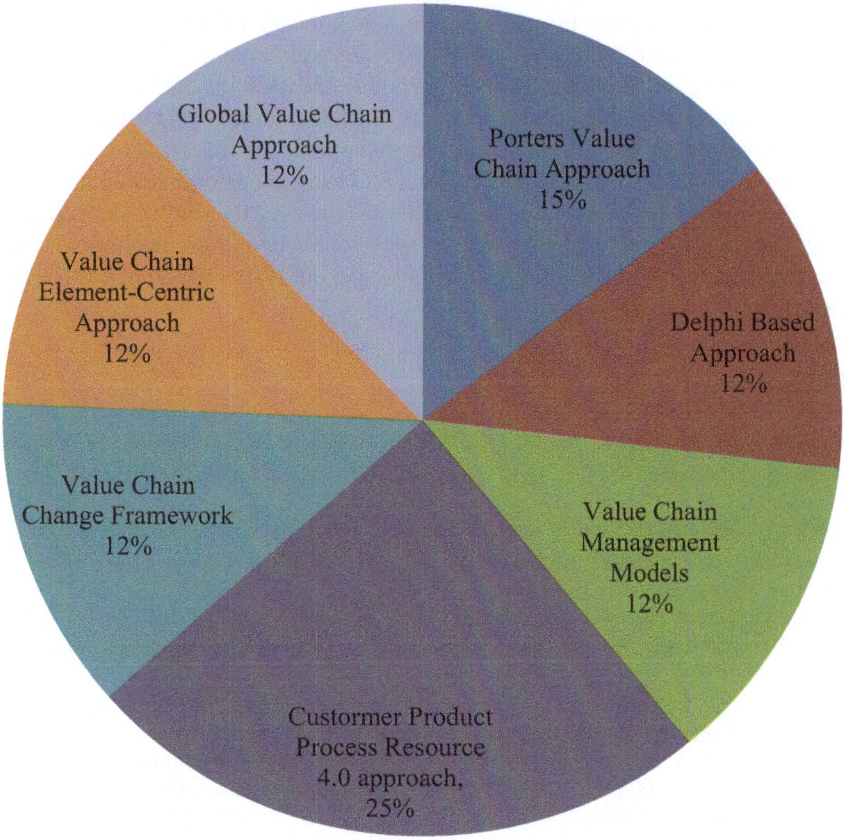

Fig. 1. Value chain approach analysis

5 Conclusion and Recommendation

The effect of industry 4.0 is overwhelming; it engulfs the entire business entity. Therefore it is necessary to understand how the component elements explore the various opportunities and threats that come with digitization. To fully grasp its details, there seems to be a necessity to develop a model that captures the fundamental process of a firm which is primarily creation of value for customer. The 4IR has its primary impact on the many components that determine value generation and it specifically cut across production. Those business activities that aid value generation and determine how useful they are with Industry 4.0 development must be recognized. Digitization has touched every aspect of the industry; it has increased efficiency in many areas, including logistics. Micheal Porter's value chain approach and Industry 4.0 perspective are quite worthy of note; it enables integration of business and production processes while integrating all players within the company's value chain, starting from the suppliers down to the customers. The appliance of the broad concepts of the CPSand the IoTs has addressed some procedural prerequisites expected of an industrial manufacturing system of the 21st century.

This review has therefore discussed findings on Industry 4.0 adoption in manufacturing firms using the value chain approach. This review addressed specific production units ranging from one case country to developing countries; the Industrial sector (SMEs and corporate organization), and circular economy down to international business operation. As clearly revealed, the value chain management approach in industrial intervention has brought about a unique and significant impact on business performance in the 21st century irrespective of their various lines of operation. However, the study lacks a crossbreed of views encompassing developed countries, developing countries, and underdeveloped countries. As such, the literature recognizes this gap, and it is recommended for further research.

References

1. Pereira, A.C., Romero, F.: A review of the meanings and the implications of the industry 4.0 concept. Procedia Manuf. **13**, 1206–1214 (2017). https://doi.org/10.1016/j.promfg.2017.09.032
2. Weyer, S., Schmitt, M., Ohmer, M., Gorecky, D.: Towards industry 4.0 - standardization as the crucial challenge for highly modular, multi-vendor production systems. IFAC-PapersOnLine **28**(3), 579–584 (2015). https://doi.org/10.1016/j.ifacol.2015.06.143
3. Chen, G., Wang, P., Feng, B., Li, Y., Liu, D.: The framework design of smart factory in discrete manufacturing industry based on cyber-physical system. Int. J. Comput. Integr. Manuf. **33**(1), 79–101 (2020)
4. Deloitte, A.G.: Industry 4.0 challenges and solutions for the digital transformation and use of exponential technologies. McKinsey Glob. Inst. **13**, 1–16 (2015)
5. Geissbauer, R., Vedso, J., Schrauf, S.: Industry 4.0: Building the Digital Enterprise. 2016 Global Industry 4.0 Survey. What We Mean by Industry 4.0/Survey Key Findings/Blueprint for Digital Success (2016)
6. Castelo-Branco, I., Cruz-Jesus, F., Oliveira, T.: Assessing industry 4.0 readiness in manufacturing: evidence for the European Union. Comput. Ind. **107**, 22–32 (2019)
7. Basl, J.: Companies on the way to industry 4.0 and their readiness. J. Syst. Integr. **9**(3), 3–6 (2018)
8. Blanchet, M., Rinn, T., Von Thaden, G., De Thieulloy, G.: Think Act - Industry 4.0, Rol. Berger Strateg. Consult. GMBH, no. March 2014
9. Ślusarczyk, B.: Industry 4.0 – are we ready? Polish J. Manag. Stud. **17**(1), 232–248 (2018). https://doi.org/10.17512/pjms.2018.17.1.19
10. Crnjac, M., Veža, I., Banduka, N.: From concept to the introduction of industry 4.0. Int. J. Ind. Eng. Manag. **8**(1), 21–30 (2017)
11. Sharp, M., Ak, R., Hedberg, T., Jr.: A survey of the advancing use and development of machine learning in smart manufacturing. J. Manuf. Syst. **48**, 170–179 (2018)
12. Nagy, J., Oláh, J., Erdei, E., Máté, D., Popp, J.: The role and impact of Industry 4.0 and the internet of things on the business strategy of the value chain—the case of Hungary. Sustainability **10**(10), 3491 (2018). https://doi.org/10.3390/su10103491
13. Schonauer, C., Iov, C.A., Petreanu, T.E.: Migration policies and EU enlargement as regional value Chains. Res. Sci. Today **18**, 116 (2019)
14. Culot, G., Orzes, G., Sartor, M., Nassimbeni, G.: The future of manufacturing: a Delphi-based scenario analysis on industry 4.0. Technol. Forecast. Soc. Change **157**, 120092 (2020). https://doi.org/10.1016/j.techfore.2020.120092
15. Von Staden, D.H.: The Impact of Value Chain Management on the Business Performance of Momentum. University of Johannesburg (2007)

16. Martínez-Olvera, C., Mora-Vargas, J.: A comprehensive framework for the analysis of industry 4.0 value domains. Sustainability **11**(10), 2960 (2019). https://doi.org/10.3390/su11102960
17. Hopali, E., Vayvay, Ö.: Industry 4.0 as the last industrial revolution and its opportunities for developing countries. In: Analyzing the Impacts of Industry 4.0 in Modern Business Environments, IGI Global, pp. 65–80 (2018)
18. Sarc, R., Curtis, A., Kandlbauer, L., Khodier, K., Lorber, K.E., Pomberger, R.: Digitalisation and intelligent robotics in value chain of circular economy oriented waste management – a review. Waste Manag. **95**, 476–492 (2019). https://doi.org/10.1016/j.wasman.2019.06.035
19. Strange, R., Zucchella, A.: Industry 4 . 0, Global Value Chains and International Business Article information, no. December 2017. https://doi.org/10.1108/MBR-05-2017-0028.
20. Zangiacomi, A., et al.: The management of operations moving towards digitalization : a multiple case study in manufacturing. Prod. Plan. Control **31**(2–3), 143–157 (2020). https://doi.org/10.1080/09537287.2019.1631468
21. Park, C., Heo, W.: Review of the changing electricity industry value chain in the ICT convergence era. J. Clean. Prod. **258**, 120743 (2020)

Human Capital Management in Industry

Analyzing the Dynamics of Work Accidents in Manufacturing to Understand "Reasonably Foreseeable Behaviors"

Lucia Botti[1(✉)], Riccardo Melloni[1,2], and Maniva Oliva[1]

[1] Interdepartment Research Center on Security and Safety (CRIS), University of Modena and Reggio Emilia, Via P. Vivarelli, 10, 41125 Modena, Italy
{lucia.botti,riccardo.melloni,maniva.oliva}@unimore.it
[2] Department of Engineering "Enzo Ferrari" (DIEF), University of Modena and Reggio Emilia, Via P. Vivarelli, 10, 41125 Modena, Italy

Abstract. This paper introduces a structured analysis of the leading causes of occupational accidents that occurred in the Italian manufacturing industry in the last decades. The aim of this research was to understand the human factors and the common causes of unsafe behaviors that may have determined the investigated accidents. The European Directive 2006/2/EC defines the essential health and safety requirements for the design of machinery for its intended purpose or for a purpose which can reasonably be foreseen. In this context, the present research supports the design of modern manufacturing systems, with the aim to investigate and understand workers' reasonably foreseeable behaviors and to provide directions for prevention strategies.

The results show the leading causes of such events as, for example, the adoption of an improper procedure or worker misplacement. The findings of this research will help practitioners and researchers understand the human behaviors and the limits of the machinery, including the intended use and reasonably foreseeable misuses thereof.

Keywords: Human factors · Reasonably foreseeable behaviors · Accidents in manufacturing · Occupational accidents

1 Introduction

The creation of healthy and safe working environments is a collective and social concern for the Institutions of industrialized countries and a fundamental factor for the improvement of people's quality of life. The benefits of improved occupational health and safety (OHS) are not limited to better health conditions and social benefits, but also include economic paybacks [1, 2]. The European Union legislation has been promoting the improvement of OHS in both private and public organizations since the publication of the Directive 89/391/EEC in 1989, on the introduction of measures to encourage improvements in the safety and health of workers [3]. This legislation determines the employers' responsibility to ensure healthy and safety workplaces, and the workers' duty

S. Trzcielinski et al. (Eds.): AHFE 2021, LNNS 274, pp. 367–375, 2021.
https://doi.org/10.1007/978-3-030-80462-6_45

to follow the employer's health and safety instructions, and report potentially dangerous situations. In 2014, the European Commission defined three major challenges on OHS for the policy agenda of the period 2014–2020: to improve implementation of existing health and safety rules; to improve the prevention of work-related diseases by tackling new and emerging risks without neglecting existing risks; to take account of the ageing of the EU's workforce [4]. The European Statistics on Accidents at Work (ESAW), i.e. the main data source for EU statistics relating to OHS, reports that 3.1 million serious accidents and 3,332 fatal accidents occurred in the EU-27 [5] in 2018, corresponding to approximately 940 serious accidents for every fatal accident [6]. Such data reveal a 0.3% increase of serious accidents and a 1.8% increase of fatal accidents from the previous year. Specifically, a fatal accident at work leads to the death of the worker involved, within one year from its occurrence. Serious accidents, also known as "non-fatal accidents", are those that result in serious injuries and a minimum number of four days of absence from work. The highest percentage of fatal accidents that occurred in 2018 in EU-27 was found in construction (21%), followed by transportation and storage (17%), manufacturing (15%) and agriculture, forestry and fishing (13%) [6]. These four sectors determined the 66% of all fatal accidents that occurred during work in EU-27, in 2018. Manufacturing registered the highest percentage of serious accidents at work in 2018 (19%), followed by wholesale and retail trade (12%), construction (12%) and human health and social work activities (11%). These four sectors determined more than half (54%) of all serious accidents at work in EU-27, in 2018 [6].

Manufacturers are required to design and construct machinery "so that it is fitted for its function, and can be operated, adjusted and maintained without putting persons at risk when these operations are carried out under the conditions foreseen, but also taking into account any reasonably foreseeable misuse thereof" [7]. Despite the strong efforts of manufacturers in producing safe equipment and complying with the requirements of safety regulations, statistics show that fatal and serious accidents still occur. More research is necessary to investigate people's behavior and to understand the role of human factors in the dynamics of accidents at work.

In 2020, an investigation was performed aiming to understand the apparent and root causes of serious occupational accidents occurred in Italy, in metal production industry [8]. Specifically, apparent causes are the determinants of the accident that can be derived from the event description in the accident reports. Root causes were identified following a hierarchical investigation process inspired by the approach proposed in Mosconi et al. [9] and based on the Five Whys technique [10]. The research revealed that the contact of the injured worker with fixed machinery for bending operations determined a significant portion of the accidents that occurred in metal production. The leading apparent causes were the voluntary adoption of an improper procedure, the failure or the improper use of equipment, and the lack of coordination. The authors experienced difficulties in understanding the root causes of occupational accidents due to incomplete descriptions and information about the dynamics of the events in the records. The authors also concluded that more efforts are required to investigate and to design proper prevention strategies that contrast the causes of occupational accidents [8].

This research aims to extend the investigation in [8], introducing the results of a structured analysis on the dynamics of more than 400 occupational accidents occurred between 2002 and 2016 in Italy, in metal production industry and in manufacturing of metal components, industrial plants and machinery. The ultimate aim was to recognize common patterns in accident dynamics and to identify workers' reasonably foreseeable behaviors on which prevention strategies should focus. The remainder of this paper is as follows: Sect. 2 shows the materials and the methodology adopted during this research; Sect. 3 introduces the results of the investigation; finally, Sect. 4 discusses the results and provides directions for future research.

2 Materials and Method

This research includes the analysis of occupational accidents that caused a serious injury to the workers involved in metal production and in manufacturing of metal components, industrial plants and machinery. The events occurred in Italy, between 2002 and 2016. Data and information on each event are from the webtool InforMO, i.e. the online database owned by the Italian National Institute for Insurance against Accidents at Work (INAIL) [11]. The INAIL database collects each accident report, which include a short description of the event, the accident year, the type and the location of injury, a limited number of personal information about the injured worker and about the company, and other information, e.g. the activity that the worker was performing when the accident occurred, the risk factors and the risk control measures available in the workplace. The registration of the accident in the database is usually demanded to the occupational physicians or other safety inspectors who intervened at the scene of the accident. A total of 452 workers was involved in the 441 occupational accidents included in this research. The occupational accidents occurred in metal production were 347, while 94 accidents occurred in manufacturing of metal components, industrial plants and machinery. The 3% of all the events included in this study caused the injury of more than one worker. Personal information was available for the 65% of the injured workers in the reference sample. Young and foreign workers appear to be the population categories majorly involved. The methodology adopted in this paper for the identification of the apparent causes of the occupational accidents is the hierarchical process proposed in [8]. Such methodology supports the structured analysis of the temporal succession of the events and their interactions in a formal logical hierarchy [8]. The investigator retraces the sequence of the events from the description and the data in the accident report. Then, following the hierarchical process, extends the temporal dimension to the analysis of worker's behaviors that lead to the injury. The result is the definition of the cause-effect relations and the identification of potential common patterns, which provide a quantitative and qualitative analysis of the accidents [12]. The investigation in this paper was limited to the identification of apparent causes of accidents, i.e. the determinants of the accident that can be derived from the description in the accident report.

3 Results

The 34% of the investigated accidents was due to the contact of the injured worker with the moving parts of machinery. Other frequent circumstances in which the accidents occurred involved the falling of objects from above (19%) and the sudden startup of machinery (11%). Table 1 shows the distribution of occupational accidents by type of accident and industry, i.e. metal production and manufacturing of metal components, industrial plants and machinery (also referred as manufacturing in this paper). The results in Table 1 by industry are similar to the results obtained for the entire sample. A further classification of the occupational accidents was possible based on the risk factor that determined the unfortunate event (Table 2). Table 2 shows that the most frequent risk factor for the investigated accidents was the use of fixed machinery for work, in both metal production (48%) and manufacturing (40%). The second most frequent risk factor was the use of fixed material transport equipment (e.g. conveyors, cranes, hoists, etc.). It was observed in the 14% of the occupational accidents occurred in metal production and in the 18% of the accidents occurred in manufacturing (see in Table 2).

Table 1. Distribution of occupational accidents in metal production and in manufacturing of metal components, industrial plants and machinery, by type of accident.

Type of accident	Metal production	Manufacturing of metal components, industrial plants and machinery
Contact of the injured worker with the moving parts of machinery	124 (36%)	28 (30%)
Falling of objects from above	66 (19%)	19 (20%)
Sudden startup of machinery	33 (10%)	17 (18%)
Contact of the injured worker with moving objects/vehicles	29 (8%)	6 (6%)
Solid projection	24 (7%)	4 (4%)
Falling from above	22 (6%)	8 (9%)
Uncoordinated movement of the injured worker (hitting an object)	13 (4%)	6 (6%)
Other types of accident	36 (10%)	6 (6%)

The findings of the investigation of apparent causes of the occupational accidents are in Fig. 1 and Fig. 2. The voluntary adoption of an improper procedure was the leading apparent cause of the investigated accidents. It caused more than half of the occupational accidents in both metal production and manufacturing. This is the case, for example, of a worker who got injured while making a maintenance intervention at an operating machinery, i.e. the company procedure required to switch off the machinery before starting any maintenance intervention.

The second leading apparent cause of accidents was the worker misplacement. Such apparent cause was found in the 29% of the occupational accidents occurred in metal production and in the 22% of the accidents occurred in manufacturing. An example is the case of a worker who hit an operating manual cross-cut saw and injured his hand while turning the tool off. The failure of the equipment or the use of improper equipment was the third leading apparent cause of occupational accidents in metal production. The case of a worker who used a regular forklift for lifting unstable barrels, causing the amputation

Table 2. Distribution of occupational accidents in metal production and in manufacturing of metal components, industrial plants and machinery, by risk factor.

Risk factor	Metal production	Manufacturing of metal components, industrial plants and machinery
Fixed machinery for metal, wood, rubber, plastic, paper, ceramics, etc	168 (48%)	38 (40%)
Fixed material transport equipment (e.g. conveyors, cranes, hoists, etc.)	47 (14%)	17 (18%)
Industrial trucks (e.g. forklifts)	32 (9%)	13 (14%)
Manual material lifting and carrying	31 (9%)	5 (5%)
Work at height (e.g. with lifting platforms, ladders, etc.)	17 (5%)	5 (5%)
Power tools (e.g. drills, screwdrivers, etc.)	14 (4%)	2 (2%)
Manual tools (e.g. hammers, amps, etc.)	10 (3%)	6 (6%)
Fixed heating systems and plants	8 (2%)	3 (3%)
Floors (slippery or damaged)	6 (2%)	
Lifting equipment for people or objects as elevators and hoists	4 (1%)	1 (1%)
Other risk factors	10 (3%)	4 (4%)

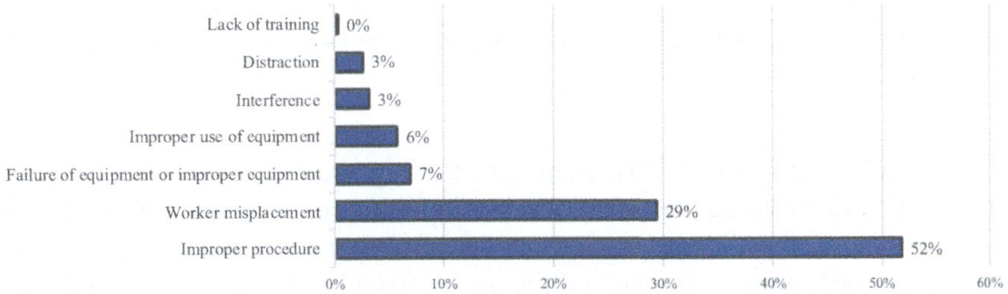

Fig. 1. Apparent causes of occupational accidents in metal production, occurred in Italy between 2002 and 2016.

of his finger, is an example. Finally, interference was the third leading apparent cause for occupational accidents occurred in manufacturing. In case of interference, the cause of the accidents is related to a further activity that was ongoing at the time of the event. An example of this situation is the case of a worker who was driving a lift truck, when he hit a second worker causing the colleague's leg fracture.

A deeper investigation was performed aiming to understand the leading apparent causes of occupational accidents, by risk factor.

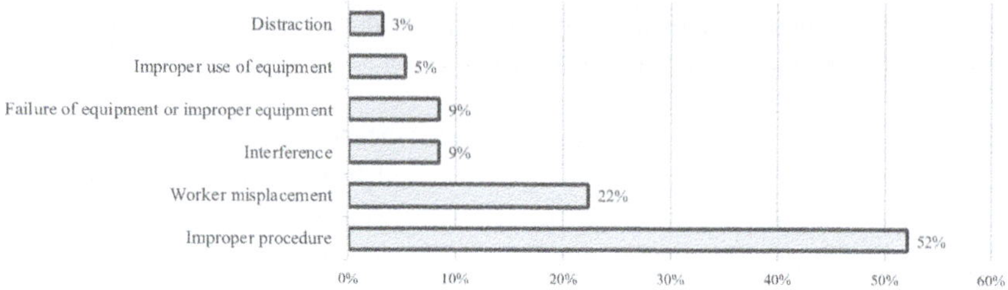

Fig. 2. Apparent causes of occupational accidents in manufacturing of metal components, industrial plants and machinery, occurred in Italy between 2002 and 2016.

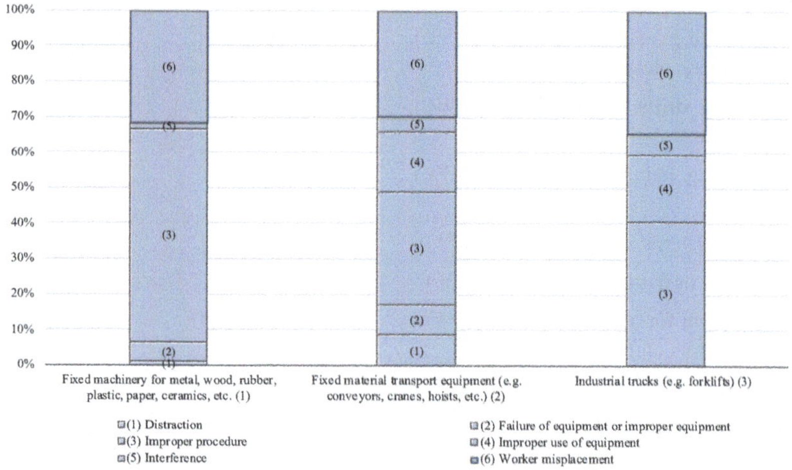

Fig. 3. Distribution of leading apparent causes of occupational accidents in metal production, occurred in Italy between 2002 and 2016.

These results show that the distribution of leading apparent causes varies with the risk factors that determined the event. The use of an improper procedure was the leading apparent cause of accidents in metal production, regardless the risk factor, and in manufacturing operations that involved the use of fixed machinery (Fig. 3 and Fig. 4). The leading apparent cause of occupational accidents occurred in manufacturing, during operations that involved the use of fixed material transport equipment, was the worker misplacement (Fig. 4). Finally, the improper use of equipment was the leading apparent cause of occupational accidents occurred in manufacturing, during operations that involved the use of industrial trucks.

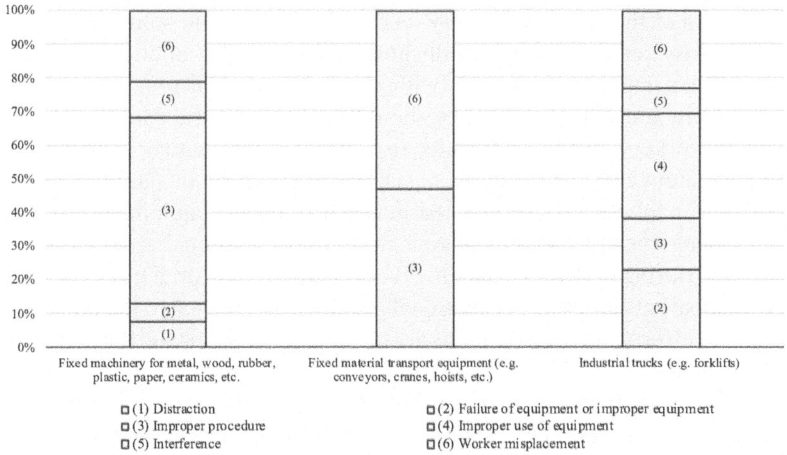

Fig. 4. Distribution of leading apparent causes of occupational accidents in manufacturing of metal components, industrial plants and machinery, occurred in Italy between 2002 and 2016.

4 Discussion and Conclusions

This study extended the results of the research in [8], including recent data on occupational accidents that caused serious injury, in the Italian metal production, and incorporating data on occupational accidents that occurred in the Italian manufacturing of metal components, industrial plants and machinery. The results revealed that the 34% of the investigated accidents was due to the contact of the injured worker with the moving parts of machinery. An example is the case of a worker who started a maintenance intervention on an operating machinery despite the company procedure required the machinery switch off before any intervention. The accident descriptions in the reports do not provide sufficient information to understand similar unsafe behaviors and other human factors. Hence, the investigation in this paper was limited to the identification of the apparent causes of occupational accidents. Other recurrent conditions in which the accidents occurred involved the falling of objects from above and the sudden startup of machinery. These findings are in line with the results in [8], which was limited to the investigation of accidents in metal production. The distribution of the occupational accidents by risk factors revealed strong analogies between metal production and manufacturing, i.e. the most frequent risk factors for occupational accidents was the use of fixed machinery for metal, wood, rubber, plastic, paper, ceramics, etc., followed by the use of fixed material transport equipment and the use of industrial trucks. The distributions of the apparent causes by risk factors revealed sensible variations in metal production and in manufacturing. The use of an improper procedure was the leading apparent cause of accidents in metal production, regardless the risk factor, and in manufacturing operations that involved the use of fixed machinery. An example is the case a worker who bypassed the two-hand control and activated the machinery with the pedal. The bypass of the machinery safety barrier, i.e. the two-hand control, costed the worker the arm amputation. The solution suggested on the accident report for such apparent cause, i.e. the improper procedure, is the application of an additional barrier between the risk factor, i.e. the fixed machinery, and the worker. This solution may result in higher complexity for the task accomplishment. Though, an additional barrier would encourage further improper behaviors, with no impact on their root causes. Hence, the focus of accident prevention strategies should be more on the causes rather than on the effects of workers' behavior. Results also revealed that worker misplacement was the leading apparent cause of occupational accidents occurred in manufacturing, during operations that involved the use of fixed material transport equipment. The improper use of equipment was the leading apparent cause of occupational accidents occurred in manufacturing, during operations that involved the use of industrial trucks. These results suggest that apparent causes of occupational accidents are related to the present risk factors. Still, more efforts and investigations are necessary to understand the apparent causes of occupational accidents.

Finally, these findings reveal that the leading apparent causes of the investigated accidents are related to workers' behavior, e.g. the worker misplacement or the voluntary adoption of an improper procedure. This may be due to the focus of the descriptions in the reports, which is mainly on the distribution of responsibilities for the accidents, rather than on the investigation of the root causes of workers' behaviors. Other frequent apparent causes are related to the failure of equipment or the use of improper equipment. Such causes may reveal lack of maintenance or the use of outdated and obsolete machinery. Future developments of this study will include the analysis of common causes of accidents that lead to the fatal injury of the workers involved.

References

1. Kim, N.K., Rahim, N.F.A., Iranmanesh, M., Foroughi, B.: The role of the safety climate in the successful implementation of safety management systems. Saf. Sci. (2019)
2. Botti, L., Mora, C., Regattieri, A.: Improving ergonomics in the meat industry: a case study of an Italian ham processing company. IFAC-PapersOnLine. **28**, 598–603 (2015)
3. Schütze, R.: Directive 89/391 on the introduction of measures to encourage improvements in the safety and health of workers at work. EU Treaties Legis, 612–621 (2018)
4. European Commission: EU Strategic Framework on Health and Safety at Work 2014–2020. EU – North Africa Semin. Strateg. better Occup. Saf. Heal. 28 (2015)
5. Eurostat: EU enlargements. https://ec.europa.eu/eurostat/statistics-explained/index.php?title=Glossary:EU_enlargements
6. Eurostat: Statistics explained: Accidents at work statistics (2020)
7. European Parliament, The Council of the European Union: Directive 2006/42/EC of the European Parliament and of the Council of 17 May 2006 on machinery, and amending Directive 95/16/EC (recast). Off. J. Eur. Union (2006)
8. Botti, L., Melloni, R., Mosconi, S., Oliva, M.: A Detailed investigation on apparent and root causes of accidents in manufacturing. In: Mrugalska, B., Trzcielinski, S., Karwowski, W., Di Nicolantonio, M., Rossi, E. (eds.) AHFE 2020. AISC, vol. 1216, pp. 18–25. Springer, Cham (2020). https://doi.org/10.1007/978-3-030-51981-0_2
9. Mosconi, S., Melloni, R., Oliva, M., Botti, L.: Participative ergonomics for the improvement of occupational health and safety in industry: a focus group-based approach. In: Perona, M. and Zanoni, S. (eds.) Proceedings of the XXIV Summer School "Francesco Turco" - Augmented Knowledge: A new era of Industrial Systems Engineering, pp. 437–443, Brescia, IT (2019)
10. Leino, A., Helfenstein, S.: Use of five whys in preventing. In: Proceedings of the 20th Annual Conference of the International Group for Lean Construction (2012)

11. INAIL: Infor.MO web. https://appsricercascientifica.inail.it/getinf/informo/home_informo.asp
12. Rogith, D., Iyengar, M.S., Singh, H.: Using fault trees to advance understanding of diagnostic errors. Jt. Comm. J. Qual. Patient Saf (2017)

Human Action Recognition on Exceptional Movement of Worker Operation

Chao-Lung Yang(✉), Shang-Che Hsu, Yu-Wei Hsu, and Yu-Chung Kang

Department of Industrial Management, National Taiwan University of Science and Technology, 43, Keelung Rd, Sec. 4, Taipei 106, Taiwan
{clyang,shang_hsu,m10701017,m10801012}@mail.ntust.edu.tw

Abstract. Time study is a technique to analyze human action based on ergonomic discipline and has been widely applied in various manufacturing sites. How to identify the standard operation by personnel and extract the operating time duration of each action is an emerging research problem in the human action recognition (HAR) research domain. In the real-world manufacturing, the operators actually conduct exceptional actions beyond the designed operation under standard operating procedures (SOP). These exceptional actions will lead to miss-recognition and disturb the measurement of action time. In this research, an exceptional action detection framework was proposed to detect exceptional actions that are not defined or should be avoided during HAR. First, the skeleton of human workers was computed, and Spatial-Temporal Graph Convolution Network model was implemented to provide the score of each action label. Then, the exceptional action can be classified by score correction with multiple support vector machine (SVM) classifiers. A dataset was created by simulating the actions which may occur during the drilling process. The preliminary experiment shows that the proposed framework is able to detect SOP and exceptional actions with up to 79.1% accuracy

Keywords: Time study · Human action recognition · Exceptional action · Manufacturing

1 Introduction

In a manufacturing site, time study is an essential technique to analyze human action based on ergonomic discipline. In traditional time study process, the method used to measure the operating time duration of each action is using timing device such as stopwatch [1]. However, using stopwatch to measure the working time with very short time is difficult. In order to avoid the human negligence error, it is essential to apply an automatic system. Golabchi et al. developed a comprehensive framework to utilize the simulation modeling, Predetermined Motion Time Systems (PMTS) and ergonomic and biomechanical assessment to demonstrate human action during working and ergonomic assessment [2]. However, the above-mentioned systems could only measure the time of the operation which should be assigned by human. If the system can identify the worker's action automatically, it will be more convenient for further analysis.

S. Trzcielinski et al. (Eds.): AHFE 2021, LNNS 274, pp. 376–383, 2021.
https://doi.org/10.1007/978-3-030-80462-6_46

Human action recognition (HAR) can be considered as a technology to obtain the information of human body by various types of sensing devices and then to understand the human behavior in computer vision. Applying HAR on a manufacturing site is an emerging research topic due to the dramatic progress of artificial intelligent technology. Based on the device to collect the information of body, HAR methods are classified into sensor-based HAR and vision-based HAR groups [3]. Sensor-based HAR utilizes the embedded sensors which are worn by subjects to capture human motion data. On the other hand, vision-based HAR uses video or image of subjects as the data for recognition. For instance, Shi et al. developed a two-stream adaptive graph convolutional network to recognize the action based on the skeleton information of subject from video [4].

Although the developed HAR model has achieved the promising performance, when applying it on the manufacturing site, one of major challenges: how to deal with the exceptional action which are not trained or learned in the model should be resolved. As known, most of HAR models are developed based on supervised learning methods. It means the model need to be trained by using tremendous historical action data with the labels which should be defined in advance. We can consider those action data contains actions that follow the production standard operating procedure (SOP). However, the exceptional actions which are not labeled or defined in SOP occurs frequently in the real-world production. For example, the worker might need to take a phone call, write up on the paper, drink water and so on which are not defined in SOP. When applying HAR to recognize human actions on production line, those "exceptional actions" will cause considerable recognition error. To distinguish exceptional actions from SOP, the exhaustive method is not reasonable because it is almost impossible to label or define the all possible exceptional actions in advance. Therefore, applying HAR model in the real-world operation sites will face the situation of distinguishing the exceptional actions, which might be mixed with SOP actions.

Recently, exceptional motion detection gradually becomes a popular research topic. More and more researchers attempted to solve the problem by anomaly detection. In 2017, Bouindour et al. proposed a two-stage approach by using pre-trained CNN for features extraction, and applying one-class SVM to learned the normal feature for detection and localization of spatial and temporal abnormal events in video [5]. In 2018, Amraee et al. presented a method by using two distinct one-class SVM for detecting anomalous events in crowded scenes [6]. In 2019, Ionescu et al. proposed a framework, Narrowed Motion Clusters, with a two-stage outlier elimination algorithm (k-mean clustering and one-class SVM) and CNN features, for abnormal event detection in video [7]. However, methods mentioned above are not used in human action recognition. Furthermore, these methods cannot recognize the unpredictable and untrained data.

In order to resolve exceptional action recognition problem, this research proposed a framework which combines HAR based model with the action-based one-class SVM model to recognize the SOP actions as well as the exceptional actions. The paper is structured as follows. In Sect. 2, the proposed framework is presented. Section 3, the experiment was performed to show a preliminary result. In Sect. 4, the conclusions were summarized.

2 Methodology

2.1 Model Framework

In this research, an image processing framework of classifying the exceptional actions from the real-world operation working video was proposed based on the results from GCN-based action recognition model. First, the video composed of operator's operation is considered as the input of the framework. Then, two deep learning models were utilized. One is human skeleton estimating tool which estimates the locations of the skeletons of the operator. The coordinates of 18 joints of human body can be retrieved. The second is the skeleton-based action recognition model which utilizes the coordinates of 18 joints of the operator's human body as the input. For each input frame containing 18 joint coordinates, a score will be generated by the action recognition model to represent the confidence of recognition for a particular action. It also means if there are k actions for recognition, k scores will be generated accordingly. Third, for each action, the confidence score of that action and the associated coordinates of upper body will be aggregated as a set of new features. In total, k feature sets will be generated to represent k action score. In the fourth step, for each action, a classification model was created to distinguish the exceptional action from normal actions. In the end, based on the results of k classification models, the decision boundary was determined to recognize the exceptional action.

2.2 Human Skeleton Coordinate Action Recognition

In order to recognize the human action, for each frame of a video, the skeleton information of human body is needed. In this research, the skeleton based human joints estimation was conducted by using open source OpenPose deep learning model. OpenPose was developed by CMU-Perceptual-Computing-Lab launched in 2017 [8]. Essentially, OpenPose equips with 2 CNN models; one is for human part detection with the confidence maps, and the other is for human part association with the part affinity fields. While the direction of connection of the adjacency detected points from the first model as the same as the detected part affinity field from second model, OpenPose assumes the skeleton connected with adjacency detected points come from same person. Then, the model goes through all detected points then provided the human pose estimation. The 2D pose estimation by OpenPose toolbox generates 18 human joints of a human in each frame. The pose estimation includes the coordinates x, y of each joint and the pose estimation confidence values.

After the coordinates of joints are generated, the human skeletons can be represented as a 2D coordinates of each human in each frame as a sequence of vectors. We can utilize the skeleton information by employing a spatial-temporal graph to construct all skeleton joints as a skeleton graph. In this research, spatial temporal graph convolution networks (STGCN) proposed in [9] was used to recognize the human action by using the skeleton graphs as input. Since STGCN can convolute the spatial and temporal information simultaneously, the model can fetch and determinate the video more comprehensively. The detail of the convolution of STGCN can be found in [9].

After STGCN computation, the score of each action to represent the confidence of the action can be computed. Based on OpenPose [8], there are 9 upper limb joints. In order to aggregate the skeleton information and score of each action, the 9 upper limb joints and the k score of k actions can be generated to represent a frame of the video. Assume the total frames of the video is T, then, a frame information matrix $\boldsymbol{X}$ can be generated as (1) where S_{ij} is the score of the ith frame represent to the jth action. J_i^x and J_i^y are the coordinates of x and y axis at the ith frame.

$$\boldsymbol{X} = \begin{bmatrix} S_{11} \cdots S_{1k} \; J_{10}^x \, J_{10}^y \cdots \quad \cdots J_{18}^x \, J_{18}^y \\ \vdots \\ S_{T1} \cdots S_{Tk} \; J_{T0}^x \, J_{T0}^y \cdots \quad \cdots J_{T8}^x \, J_{T8}^y \end{bmatrix} \tag{1}$$

Based on the result of STGCN, the maximum of the score would be selected to represent the action of the frame. A recognized action vector $\vec{A}$ can be generated as $\vec{A} = [a_1, a_2, \cdots, a_T]$ where a_i is the recognized action to represent the i frame.

2.3 Action Classification

As mentioned earlier, for each action j, a classification model was constructed to determine the decision boundary for distinguishing whether the frame is truly belonged to the action j.

Based on the recognized action vector $\vec{A}$, the one-hot encoding method was applied to expand $\vec{A}$ into a matrix AE as (2) where e_{ij} is 1 if the i frame is recognized as j action; otherwise is 0.

$$AE = \begin{bmatrix} e_{11} \; e_{12} & \cdots e_{1k} \\ e_{21} \; e_{22} & \cdots e_{2k} \\ \vdots & \vdots \\ e_{T1} \; e_{T2} & \cdots e_{Tk} \end{bmatrix} \tag{2}$$

According to the matrix X and the jth column of the matrix AE, Support Vector Machine (SVM) [10] was applied to construct a hyperplane or a set of hyperplanes to perform the classification on whether the frame is belong to the jth action. It also means for each action j, a SVM model will be constructed. If k actions could be recognized, k SVM models are needed. The detail of the constructing SVM can be found in [10].

Based on the results of all SVM models, if all of SVM models cannot distinguish a particular action in a frame, then the action in the frame will be distinguished as an exceptional action. If the frame belongs to more than one action, the highest confidence score from SVM classifier will be used to assign as the representative action.

3 Preliminary Experimental Results

3.1 Data

In order to evaluate the proposed framework, the video dataset recorded by simulating the human actions which commonly occur during drilling process was utilized. All videos were resized to a resolution of 848×880 with a frame rate of 60 frames per second (fps).

110 working-cycle videos were collected to simulate the standard of process (SOP) of drilling process which contains 4 actions: 1) loading 2) touching 3) drilling 4) unloading. To enlarge the diversity of the exceptional action, 4 exceptional actions such as using smartphone, drinking, head scratching, exiting were evaluated. The SOP actions and exceptional actions were illustrated in Fig. 1.

The total number of cycles in the training dataset and testing dataset were 100 and 10, respectively. Please note that only SOP actions are used in model training because the exceptional actions cannot be listed exhaustively. In the training dataset, the cycles number of training and validation was 80 and 20, respectively. For each working cycle in testing data, it was composed of SOP actions and exceptional actions.

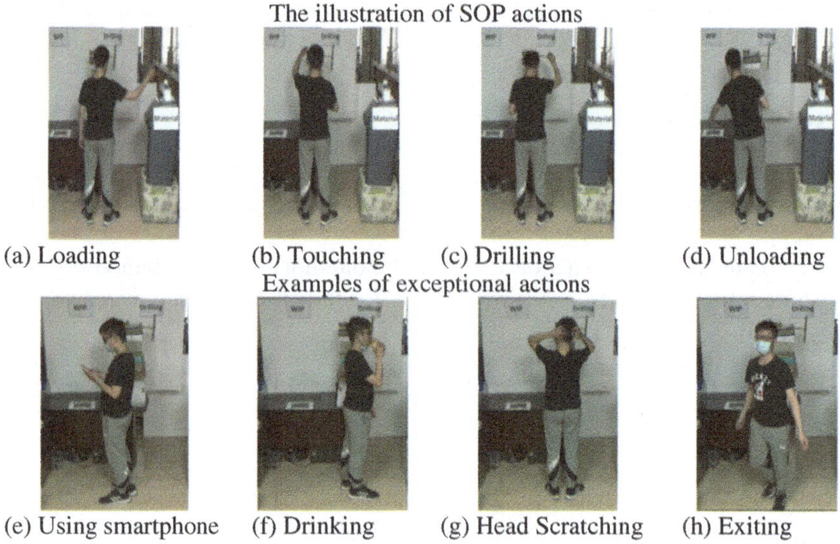

Fig. 1. The illustration of a SOP actions and examples of exceptional actions.

3.2 Implementation

In this research, the parameters of STGCN model was followed by the settings published in Yan et al.'s work in [10]. Essentially, the network model was composed of 9 layers of spatial temporal graph convolution operators. For the first three layers, 64 channels were used for output; for the following three layers, 128 channels were used for output, and for the rest of three layers, 256 channels were used for output. All of layers were computed with 9 temporal kernel size. The strides of the 4-th and the 7-th temporal convolution layers are set to 2 as pooling layer. A global pooling was implemented to generate the result tensor with 256 dimensional feature vector. At the end of the network, the SoftMax classifier was used to generate the feasible label for each human action. For the model training loss function, the cross entropy loss function was used to compute the training output. The GCN model with the minimum mean loss would be discovered at the sixtieth epoch when the parameter of GCN was set as following, the batch size to 64, window size to 80 and with a learning rate of 0.1, and the mean loss would be 0.22 afterward. For SVM kernel function, Gaussian radial basis function (RBF) was utilized to map the original data.

In order to evaluate the performance, the timing of the action label generated by the recognition model was compared with the timing of the label defined by the human expert. Basically, time discrepancy between the timestamp of HAR labels and the timestamp of manual labels can be calculated as following (3). Where $\boldsymbol{correct_i}$ is the discrepancy at the ith frame. The accuracy is the summary of the $\boldsymbol{correct_i}$. Where T is total frame of the video.

$$\boldsymbol{correct_i} = \begin{cases} 1, & \textit{The recognized action} = \textit{the actual action} \\ & 0, \textit{Otherwise} \end{cases} \tag{3}$$

$$\mathbf{accuracy} = \frac{\sum_{i=1}^{T} correct_i}{T} \tag{4}$$

3.3 Result

In the experiments, two results are presented: one is showing the recognition result of STGCN without utilizing action classification of SVM; another is the results of STGCN with SVM. Please note that exceptional actions were tested for both models to evaluate if the model can recognize them.

Table 1. The confusion matrix of STGCN model

Actual \ Predicted	Loading	Touching	Drilling	Unloading	Exceptional Actions
Loading	679	31	193	175	0
Touching	37	1211	16	83	0
Drilling	4	3	2015	288	0
Unloading	81	55	385	382	0
Exceptional Actions	851	300	663	298	0

Table 1 shows the confusion matrix of STGCN which was trained by only SOP actions. Obviously, it can only consider the exceptional actions as one of SOP actions. Therefore, the precision of Loading, Touching, Drilling and Unloading was 0.41, 0.76, 0.62 and 0.31, respectively which actually show poor action recognition. If no exceptional actions were used in the testing dataset, the accuracy of SOP actions ($\mathbf{accuracy}_{SOP}$) is 0.76 indicated in the grey box. However, after considering the exceptional actions, the accuracy was dropped to 0.54. It also means once the exceptional actions were presented in the working cycle, the accuracy of action recognition will be dramatically affected.

Table 2 shows the confusion matrix of STGCN-SVM. As can been seen, the STGCN-SVM is able to recognize the exceptional actions. The precision of Loading, Touching, Drilling and Unloading are 0.98, 0.71, 0.92 and 0.54, respectively. Compared with

STGCN only model, the prediction of Loading, Drilling and Unloading was improved. However, considerable frames of exceptional actions were recognized as "Touching" and "Unloading" because the hand skeleton sequences of "Using smartphone" and "Head Scratching" were quite similar to "Touching" and "Unloading". To evaluate the recognition of exceptional action of the proposed STGCN-SVM, the precision of exceptional actions 0.871 show promising result. It means that the framework could have good capability to recognize the exceptional action. Based on the Eq. (4), the accuracy of all of actions including exceptional actions was 0.79. Interestingly, if only considering SOP actions, the $\mathbf{accuracy}_{SOP}$ is 0.908 which is even better than the one (0.76) in STGCN only. It seems that the STGCN-SVM framework provides the capability to not only distinguish the exceptional actions but also improve SOP recognition.

Table 2. The confusion matrix of STGCN-SVM framework.

Actual \ Predicted	Loading	Touching	Drilling	Unloading	Exceptional Actions
Loading	749	112	38	154	25
Touching	0	1309	0	18	20
Drilling	3	0	2171	42	94
Unloading	0	0	0	893	10
Exceptional Actions	14	410	140	541	1007

The preliminary results show that the proposed STGCN-SVM framework is able to recognize the exceptional actions from working process video. Due to this capability, the proposed STGCN-SVM is invulnerable from the exceptional actions which usually cause recognition difficulty on HAR.

4 Conclusion and Future Work

Applying HAR model to recognize the human worker's operation has advantages of automatically identifying the human actions and measuring the operating time. However, the exceptional actions which are not defined in SOP cause the recognition error, especially when implementing in a manufacturing site. In this research, an image processing framework called STGCN-SVM for classifying either SOP or exceptional actions based on the real-world operation working videos was developed. Essentially, the STGCN model was trained by using SOP actions. Then, individual SVM classifier was further developed aiming to a particular action based on the confidence scores generated from STGCN. The decision boundary comparison process was performed to finally determine the action of an image frame.

In this work, the simulated videos of drilling environment were created to evaluate the proposed framework. Multiple exceptional actions were included in the testing dataset for evaluating the proposed framework. Based on the experimental result, the precisions of SOP actions are higher, since the framework is able to distinguish the exceptional actions from SOP. More importantly, with the precision of exceptional actions is 0.87, this framework can exclude the exceptional actions accurately, although some exceptional actions which are similar with SOP actions still causes troublesome on recognition. The work sheds a light on applying HAR system to the real-world operation for recognizing human operation without exhaustively label or define the exceptional action.

In the future work, improving the accuracy and precision of the HAR model is a continue work. Also, developing a filter which is able to identify the exceptional actions before going through the HAR model is another possible research direction.

Acknowledgment. We appreciate the financial support from the Ministry of Science and Technology of Taiwan, R.O.C. (Contract No. 109–2221-E-011 -101), and the Center for Cyber-Physical System Innovation" from The Featured Areas Research Center Program within the framework of the Higher Education Sprout Project by the Ministry of Education (MOE) in Taiwan. We also thank Wang Jhan-Yang Charitable Trust Fund (Contract No. WJY 2020-HR-01) for the generosity of financial support.

References

1. Andris, F., Benjamin, W.N.: Niebel's Methods, Standards, and Work Design, 13e. McGraw-Hill Education (2013)
2. Golabchi, A., Han, S., AbouRizk, S.: A simulation and visualization-based framework of labor efficiency and safety analysis for prevention through design and planning. Autom. Constr. **96**, 310–323 (2018)
3. Dang, L.M., et al.: Sensor-based and vision-based human activity recognition: A comprehensive survey. Pattern Recogn. **108**, 107561 (2020)
4. Shi, L., et al.: Two-stream adaptive graph convolutional networks for skeleton-based action recognition. In: Proceedings of the IEEE/CVF Conference on Computer Vision and Pattern Recognition (2019)
5. Bouindour, S., et al.: Abnormal event detection using convolutional neural networks and 1-class SVM classifier. In: 8th International Conference on Imaging for Crime Detection and Prevention (ICDP 2017) (2017)
6. Amraee, S., et al.: Abnormal event detection in crowded scenes using one-class SVM, **12**(6), 1115–1123 (2018)
7. Ionescu, R.T., et al.: Detecting Abnormal Events in Video Using Narrowed Normality Clusters. In: 2019 IEEE Winter Conference on Applications of Computer Vision (WACV) (2019)
8. Cao, Z., et al.: Realtime multi-person 2D pose estimation using part affinity fields. In: The IEEE Conference on Computer Vision and Pattern Recognition (CVPR). Honolulu, Hawaii, USA (2017)
9. Yan, S., Xiong, Y., Lin, D.: Spatial temporal graph convolutional networks for skeleton-based action recognition. In: Thirty-Second AAAI Conference on Artificial Intelligence. Hilton New Orleans Riverside, NewOrleans, Louisiana, USA (2018)
10. Wang, L.: Support vector machines: theory and applications, **177**. Springer Science & Business Media (2005)

Motion Analysis at Checking Phase in Automobile Repair Painting

Yuka Takai[1(✉)] and Shigeru Ikemoto[2]

[1] Osaka Sangyo University, 3-1-1 Nakagaito, Daito, Osaka, Japan
takai@ise.osaka-sandai.ac.jp
[2] Body Garage Ikemoto, 454-2 Shimofutamatacho, Higashiomi, Shiga, Japan

Abstract. Partial painting using metallic paint in automobile repair has three goals: (1) sufficiently coloring the damaged area, (2) preventing color unevenness because of the arrangement of aluminum flakes peculiar to metallic paint, and (3) obscuring the boundary between the color of the original body and the painting color. In (2) and (3), the degree of difficulty in painting and judging the finish is high. In this study, the checking phase for partial painting using metallic paint was targeted. In the checking phase, the finish of the painting is visually observed, and the shortage of painting and its quality are determined. It was clarified using three-dimensional motion analysis, from which an expert made a visual confirmation.

Keywords: Automobile repair · Painting · Metallic paint · Checking phase · Motion capture

1 Introduction

Painting work in automobile repair is carried out manually using a spray gun. Repair painting must be performed differently every time. This is because the paint must be blended and spray gun painted to match the color of the paint remaining on the body of the vehicle. In recent years, metallic and pearl coatings, which are highly modifiable, have become common for new car bodies, and the degree of difficulty in repair coating has increased. Additionally, because of the declining birthrate and aging population, the number of people entering the automobile repair industry is decreasing in Japan [1]. Consequently, it is necessary for the automobile repair industry to establish efficient and effective training methods for young engineers.

Therefore, we have been promoting the formalization of tacit knowledge by experts with the goal of creating digital teaching materials for automobile repair painting [2–4]; in this study, we focused on the checking phase. To visually check the color tone after painting and the texture of the painted surface, the experts place importance not only on the operation of the spray gun but also on the checking phase. It is important for young engineers to clarify the behavior of experts in the checking phase. The purpose of this study is to digitize the operation during the checking phase. The position where the expert checked the painted surface, and the distance between the painted object and painter were clarified using three-dimensional motion analysis.

S. Trzcielinski et al. (Eds.): AHFE 2021, LNNS 274, pp. 384–391, 2021.
https://doi.org/10.1007/978-3-030-80462-6_47

2 Methodology

2.1 Outline of Experiment

An auto repair painting working engineer repaired the damaged area at the front door of a passenger car. The painting repair work was measured using a 3D motion analysis system, and the checking phase was analyzed.

2.2 Subjects

Eleven painting repair technicians were included in this study. Subjects with more than 17 years of experience are referred to as experts, and those with less than four years of experience are non-experts. The average careers of the experts and non-experts were 31 years 7.8 months (±11 years 9 months) and 1 year 10.4 months (±1 year 5.4 months), respectively. The average height of the experts and non-experts was 1.68 m (±0.04 m) and 1.68 m (±0.04 m), respectively. This experiment was approved by an ethical review at Osaka Sangyo University. The experiment was fully explained to the subjects, and informed consent was obtained from all participants.

2.3 Materials and Tools

The object to be painted was the left front door of the Legacy made by SUBARU Co., Ltd. To make the situation more similar to the usual work in the actual car, the door was set up using an exclusive instrument in the painting booth. The lower left 100 mm square painted in gray, of the door painted in silver metallic (Color code 1F7) made by Toyota Motor Corp. was regarded as a damaged part. Figure 1 shows a photograph of the door with the damaged area depicted in red.

Three solvent-type paints made by Rock Paint Co., Ltd. were used for the repair painting. The first is a 1-liquid base coat (Product name: 077 Line ProTouch, 50% thinner dilution). The second is the top coat clear (product name: multi-top clear S series, 2: 1 type). The third is a gradation agent (product name: 077-P150 Nigori Clear P, 051-4F11 EcoMulti Blender, mixed at a ratio of 20:80). Each paint was mixed with a solvent according to the manufacturer's recommendations.

The spray gun used was a W-101 manufactured by ANEST IWATA Co., Ltd. Three spray guns were prepared for the three types of paint. The amount of paint discharged, pattern spread, and input pressure of the spray gun were unified to the same values for all the three spray guns. The subjects were instructed not to make spray gun adjustments.

2.4 Experimental Procedure

The subjects were instructed to paint the undamaged and damaged areas to minimize color differences and irregularities. After painting the damaged area with silver basecoat paint, the entire door was finished with clear paint.

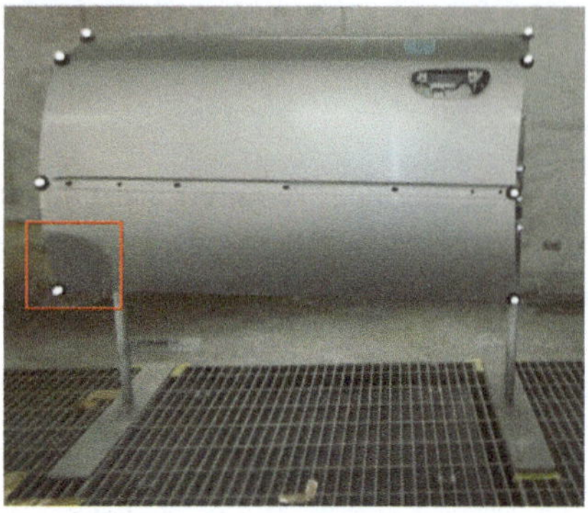

Fig. 1. Photograph of a door with a damaged part at the lower left 100 mm.

2.5 Record and Analysis

The subjects' movements were measured using an optical motion capture system MAC3D SYSTEM (Motion Analysis Co., Ltd). Infrared reflection markers were used as the reference points, and attached to the subjects at 21 locations and to the door at eight locations. The coordinate system was set as the x-axis for the left and right directions, the y-axis for the front and back directions, and the z-axis for the up and down directions for the examinee. The measurement conditions are shown in Fig. 2.

The analysis was performed while the subject performed the check. Because the checking phase was carried out several times by sandwiching another phase, the continuous checking phase was regarded as one checking phase. The trajectory of the infrared reflection markers on the top of the head and the distance between the top of the head and the door were calculated for each checking phase.

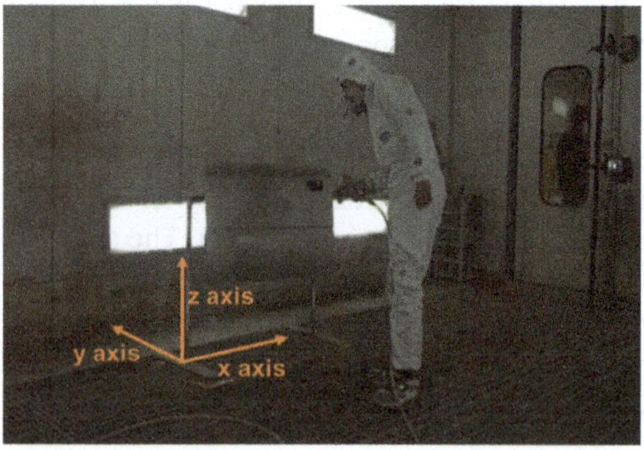

Fig. 2. Photograph of checking phase of expert.

3 Results

3.1 Times of Checking Phase

Table 1 lists the number of checking phases for each subject. Experts 3 and 5 had the highest number of checking phases (20 times). Expert 3 was the least common at seven times. The number of checking phases for the non-experts was 22 times higher for non-expert 6. Non-expert 3 was the least at seven times. The mean and standard deviation for each group showed no significant differences in the number of checking phases between the expert and non-expert groups.

Table 1. Checking phase times.

Subjects name	Times of checking (time)	Subjects name	Times of checking (time)
Expert 1	7	Non-expert 1	17
Expert 2	16	Non-expert 2	16
Expert 3	20	Non-expert 3	7
Expert 4	13	Non-expert 4	10
Expert 5	20	Non-expert 5	17
		Non-expert 6	22
Mean	15.2	Mean	14.8
S.D	5.4	S.D	5.4

3.2 Head Top Trajectory

Figures 3 and 4 show the trajectory of the top head of the expert and non-expert in the x-y plane. The x-y plane corresponds to the observation of the subject from the overhead. The red line connecting the red dots in the figure indicates the position of the door. The experts moved to the left and right relative to the door. Except for expert 1, the experts moved beyond the right end of the door. Additionally, experts 2 and 5 checked approximately 1500 mm to the left and approximately 2000 mm to the right from the damaged area. The non-experts had a smaller range of left and right movements to the door compared to the experts. The non-experts repeatedly moved to the left side of the door near the damaged area. All the non-experts moved within a range of approximately 500 mm to the left and approximately 1000 mm to the right from the damaged area. From these facts, it was deduced that the expert conducted the checking while moving widely from front to back and left to right around the damaged area.

Figures 5 and 6 show the trajectory of the head top of the expert and non-expert in the x-z plane. The x-z plane corresponds to observing the subject from behind. The red line connecting the red dots in the figure indicates the position of the door. The marker on the top head of the expert was often located about the z-axis at 1700 mm, and the damaged area was checked while standing. Particularly, the trajectory of the expert is located at the highest point on the z-axis when the trajectory significantly moves left and right on the x-axis; it can be inferred that the expert was standing when checking from a distance. The marker of the top of the head of the expert was seldom at the same position as the lower left of the door, which was a damaged area, and was located at a height of 1000 mm about the z-axis, which was at the top of the door, near the left end. The vertex marker of the non-expert was located about the z-axis at 1000 mm, which was near the upper end of the door. The vertex markers of the non-experts were closer to the lower left of the door as the damaged area, than those of the experts, and their faces were closer to the damaged area.

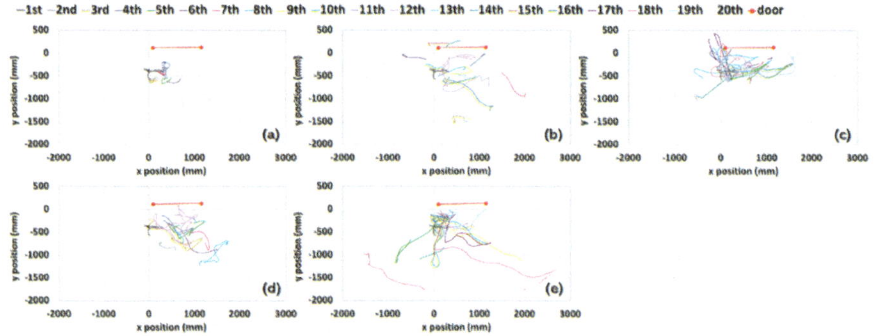

Fig. 3. Trajectory on x-y plane of (a) expert 1, (b) expert 2, (c) expert 3, (d) expert 4, and (e) expert 5.

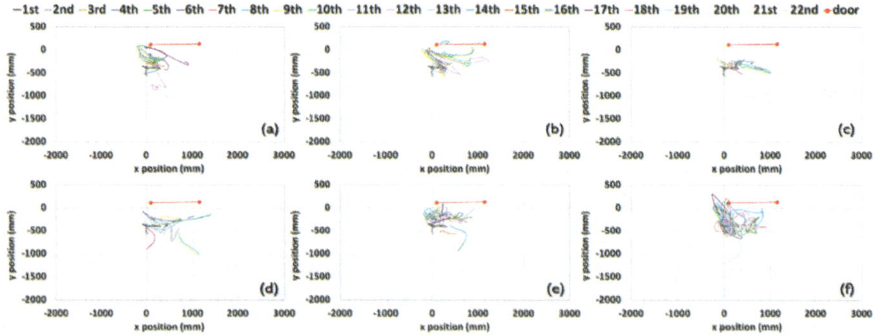

Fig. 4. Trajectory on x-y plane of (a) non-expert 1, (b) non-expert 2, (c) non-expert 3, (d) non-expert 4, (e) non-expert 5, and (f) non-expert 6.

3.3 Distance Between the Head Top Marker and Damaged Area

Figure 7 shows the linear distance between the vertex marker and damaged area for both the experts and non-experts. The average distance was 1200 mm and 779 mm for the

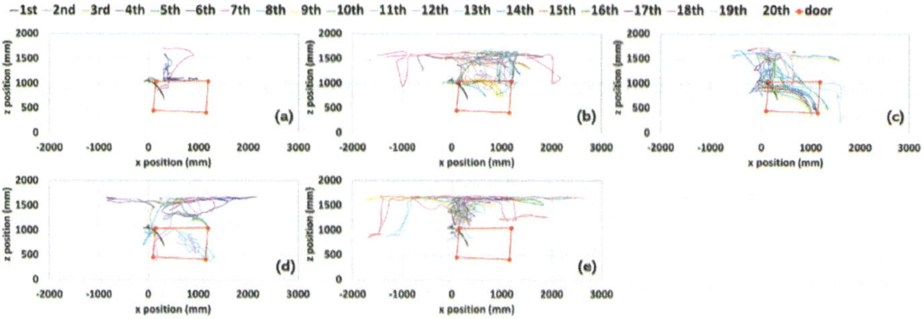

Fig. 5. Trajectory on x-z plane of (a) expert1, (b) expert 2, (c) expert 3, (d) expert 4, and (e) expert 5.

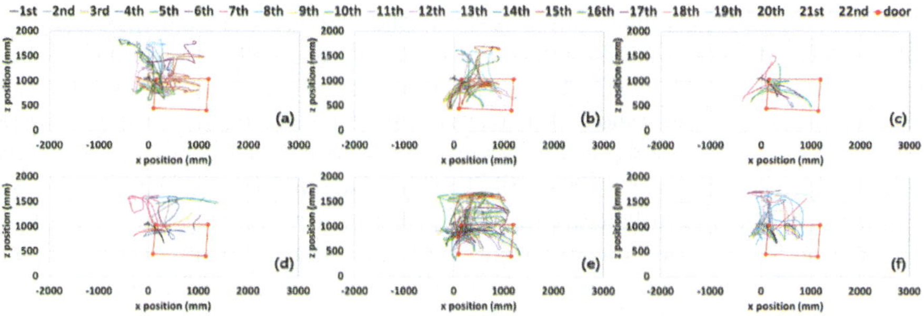

Fig. 6. Trajectory on x-z plane of (a) non-expert 1, (b) non-expert 2, (c) non-expert 3, (d) non-expert 4, (e) non-expert 5, and (f) non-expert 6.

experts and non-experts, respectively. The straight-line distance was long for experts 2 and 4, who made significant lateral movements along the x-axis. The straight-line distance was the longest for expert 4 among the non-experts. The value of the standard deviation tended to be larger for the experts than for the non-experts. In addition to the checking from a close location as in the case of the non-experts, the checking was also conducted from a distant location.

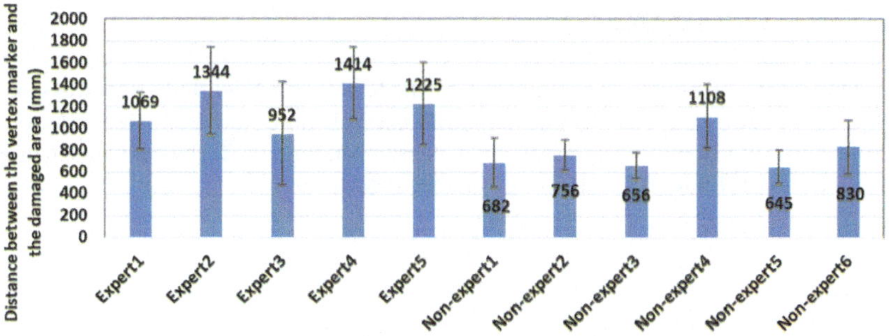

Fig. 7. Linear distance between vertex marker and damaged area.

4 Discussions

In the experiments conducted in this study, the subjects must achieve the following three main points.

(A) Adequate staining of the damaged area.
(B) Eliminate the occurrence of color unevenness caused by the arrangement of aluminum flakes peculiar to metallic paint.
(C) Obscure the boundary between the original body color and the painted color.

This will be a check to determine if these challenges are experienced. (A) can be solved by applying multiple coats. To achieve (B) and (C), it is necessary to have a technique for effectively operating a spray gun. From the results of the trajectory and linear distance, it became clear that the expert was away from the damaged area to the right and left, and stood and looked at the damaged area without lowering his head. Because the experts were aware of how the car user usually looked at the car body, the operation was probably to stand and check it. Car users often look at the car as they walk toward it, rather than squatting down. When this motion is reproduced, it is appropriate for the engineer to stand away from the damaged area and look at the damaged part to confirm whether it is sufficiently repaired. Additionally, it is necessary to check the overall balance of the color tone of the car body when coating metallic and pearl paints whose color appearance differs depending on the angle. When conscious of this, they started by looking at the door from a distance from a bird's eye view.

In contrast, the non-experts who have only seen the damaged area from the vicinity of the door cannot know what the damaged part looks like from a distance. Because of the behavior of the experts, we can instruct the non-experts on the necessity to check from a wide range from the viewpoint of the automobile user.

5 Conclusions

In the checking phase of automobile repair painting, the position from which the experts see the painted surface was clarified using three-dimensional motion analysis. The expert stood approximately 1200 mm away from the damaged area and looked at the damaged area without lowering the head.

Acknowledgments. This work was supported by the Kinki Aria Auto Body Repair Association, Kyoto Prefecture Auto Body Repair Association, Osaka Prefecture Auto Body Repair Association, Nara Prefecture Auto Body Repair Association, Nissan Automobile Technical College of Kyoto, Protorios Inc., and JSPS KAKENHI Grant Number 16K16329.

References

1. Ministry of Land, Infrastructure, Transport and Tourism: Report of Study group on securing and training automobile maintenance personnel (2016). https://www.mlit.go.jp/common/001127730.pdf

2. Ikemoto, S., Takai, Y.: Eye movement characteristics of engineers on a partial painting of metallic color. In: JSAE Annual Congress, No. 31-18, pp.1–4, Society of Automotive Engineers of Japan, Kanagawa (2018)
3. Ikemoto, S., Takai, Y.: Motion analysis of spray gun operation at base coat in automotive repair partial painting. In: JSAE Annual Congress, No. 47-19 pp. 1–4, Society of Automotive Engineers of Japan, Kanagawa (2019)
4. Takai, Y., Ikemoto, S.: Characteristics of expert in checking phase during partial painting in automobile repair, In: Mechanical Engineering Congress, Japan, No. S04109, p. 1, The Japan Society of Mechanical Engineers, Akita (2019)

Applications in Industry

Micro-influencers as an Effective Channel to Reach Customers Using Social Media in the Time of the COVID-19 Pandemic

Michal Trziszka(✉)

Faculty of Engineering and Management, Poznan University of Technology, Rychlewskiego 2, 60-965 Poznan, Poland
michal.trziszka@put.poznan.pl

Abstract. The coronavirus COVID-19 pandemic, which swept the world at the beginning of 2020, has radically changed the traditional marketing model. As a result of these changes, in the future, digital communication will be fundamental to gaining (and maintaining) consumer attention and confidence. More than ever, digital engagement will define consumer relationships and loyalty in a socially distanced reality. Recipients forced by lockdown to transfer their activity to the Internet - remote work, remote learning, shopping - necessarily focus more on the transmission from the Internet than before the crisis [1].

On the other hand, the sharp decline in organic reach on Facebook and ad blocker plugins make promoting content on social media more challenging. As the latest research shows, influencer marketing, which is currently one of the most popular forms of reaching potential customers, has become a means of effective promotion. Between 2013 and 2016, there was a 90-fold increase in the number of Google searches for "Marketing Influencer". In 2017 alone, this phrase saw a 325% increase in search results and is still growing [2]. Recent studies also show a new direction in influence marketing - an increase in the effectiveness of micro- and nano-scale influencers, who during a pandemic turn out to be much more effective in reaching recipients.

Keywords: Systems engineering · Cloud computing management · Influencers · Micro-influencers · Social media

1 Introduction

The subject of the study is to present the growing role of multidimensional communication models in internet marketing as an increasingly popular and appreciated by companies form of marketing influence in social media and the growing role of key entities in a given community influencing the consumer choices of buyers gathered around them, the so-called influencers.

Social media has revolutionized the way people communicate, where direct relationships have moved to the internet and at the same time have grown to a global scale. In the 1990s, when the first social networks were created, word of mouth marketing moved

S. Trzcielinski et al. (Eds.): AHFE 2021, LNNS 274, pp. 395–400, 2021.
https://doi.org/10.1007/978-3-030-80462-6_48

to the Internet, mainly to forums. The roots of influencer marketing can also be found in e-PR tactics. On Facebook, Instagram and Tweeter, two-way communication has turned into mass communication. Brands took advantage of this phenomenon, seeing it as a marketing potential. Marketing specialists have found that the use of influencer popularity translates into an increase in brand popularity and is much more effective than the so-called word of mouth marketing [3]. Influencer marketing is based on the authority mechanism. We are happy to buy those products recommended by friends who are authority figures for us, we identify with them or want to look like them, have what they do, look like them, so companies are looking for influencers who will be intermediaries between the client and the brand [4].

During the COVID-19 pandemic, when social activity shifted mainly to the Internet [5], digital engagement has become the main way of interacting with the customer, and staying close to customers and their individual needs are paramount. This required a radical overhaul of your marketing strategies, putting the latest customer data at the center of every interaction.

In a customer-driven marketing approach, brands create programs that authentically communicate with people using an committed customer service approach. To do this right, marketers need to understand consumer goals and use campaign elements such as message, data and channel selection that are most relevant and contextual at the time of engagement. In this pandemic crisis, consumers are turning more than ever to brands they can trust in times of uncertainty. The priority is to build a relationship over the transaction value. Paradoxically, the pandemic restrictions on advertising budgets of companies have become an opportunity for micro- and nanoifluensers [6], who have much lower rates than celebrities, but at the same time their message is more authentic, so inevitably convincing.

2 Influencers and Influencer Marketing

One of the reasons why influencer marketing is considered an effective brand promotion tool is the fact that it is based on the authenticity of the influencer. The second thing is trusting an influencer - if a company says it's great, it doesn't sound convincing enough. It's definitely easier for us to believe an influencer. Another aspect is measuring the results and changing the perception of the advertising message by recipients who have become resistant to traditional advertising. Authenticity, personalization of communication, and the ability to measure the effects of a campaign are the main advantages of influencer marketing over traditional advertising.

An accurate definition of influencers is provided by K. Sammis [7], who describes them as experts in a specific field who share their knowledge about products. According to this definition, they are therefore not so much brand ambassadors, but opinion leaders, due to the amount of information they have in a given industry. Audiences trust them because they treat them as equals. Often, influencers are not world-famous celebrities, but "ordinary" people who have a certain specialist knowledge and share it on their blogs. Among the 20 most popular Polish influencers, there are no actors, models, popular musicians, singers or television people.

Internet users approach the blogger as a friend, thanks to which they trust him and the content he publishes. An influencer is therefore considered an objective person who does not seek to maximize the company's profit, but to disseminate information that he has about given products or product categories.

For marketing specialists, it is extremely important to choose an influencer that will guarantee him the best match of the character of his profile to the promoted product and brand, and thus greater reach and engagement of recipients (followers). "Authentic, unique and credible content that arouses consumer curiosity is the greatest value that cooperation with an influencer can bring. Consistent implementation of such a strategy - e.g. in the form of cyclical inspired tasks planned every year - can permanently contribute to changing the perception and increasing brand awareness." [8] For this reason, it's worth giving the influencer a free hand to build a story that shows the product's features. He has a very good sense of his audience and knows what and in what form the message will arouse the greatest interest.

J. Baer on the portal Sprawny Marketing [9] cites an example of a study conducted by TopInfluence, which followed an influencer marketing program for Silk Almond almond milk and other products of this brand. The company found 258 influencers related to the dietary market and each of them was asked to create content about Silk milk as part of the "Meatless Mondays" initiative. No paid advertising was used, the posts of these people appeared on social media as organic content. A special application was used to track performance. It found that households that were exposed to influencer marketing purchased 10% more Silk products than the control group, and each of the 1,000 people who experienced influencer marketing purchased Silk products worth $ 285. higher than the control group. The click-through rate of blog posts alone is 11 times higher than the ROI of banner ads after 12 months [9].

Influencers, however, are not a homogeneous group. Depending on the number of followers (followers) of a given person's profile in a social network, we distinguish micro-influencers (from 1 thousand to 100 thousand followers), mid-tier influencers (from 100 thousand to 1 million followers) [9] and top-tier influencers (over 1 million followers). The effectiveness of micro-influencers is supported by the fact that many consumers are more likely to trust his opinion than celebrities, doubting the authenticity of the latter's recommendations.

Campaigns with influencers prove that they constitute an important marketing value for brands. Well coupled with the strategy of presence in social media and content marketing, they can result in great results, including sales. Influencer marketing shows that it is worth focusing on authenticity and more personal communication in reaching customers [10].

3 Increase in the Value of the Message of Micro-influencers During a Pandemic

According to research conducted in April 2020, the number of brands cooperating with influencers fell by 37%, which was directly related to companies' budget cuts due to the crisis caused by the Covid-19 pandemic [11]. At the same time, there was a sharp 41% drop in Influencer Marketing Efficiency - the ratio of average interactions in an influencer post mentioning a brand compared to a post published by the brand itself - compared to April 2019 [12].

Reduced marketing budgets and a large drop in the effectiveness of influencer communication forced marketers to completely change their strategy in reaching customers. Experts realized that popular influencer marketing doesn't necessarily mean influencer. Thus, they focused on more precisely reaching smaller, but more attractive for the brand groups of customers focused on lifestyle and ideas. Therefore, the most important value of influencer marketing is the level of fan engagement - it decides what effects the company will achieve with the money invested. "The results of the research conducted by Dentsu Aegis Network Polska and ARC Rynek i Opinia have shown, for example, that in the category of women's cosmetics, fashion and lifestyle bloggers have a much higher impact than cinema and television stars who mainly advertise such products." [13].

As it turns out, cooperation with micro-influencers also gives a better engagement rate [14] - an indicator that more and more marketers consider a more adequate measure of the campaign's success than its range. The average level of engagement generated by micro-influencers on posts is even 5 times higher than in the case of influencers with large reach. By engaging several micro-influencers, you can get the same reach as one top-tier influencer with a higher Instagram engagement rate [15].

Research by Indahash shows that a statistical micro-influencer is a woman who shares her posts one to three times a day [16]. A large part of the micro influencers surveyed is open to barter cooperation with enterprises, and often also to cooperation without remuneration. Micro influencers most often work in the field of fashion, sports, mobile applications and the beauty industry. By using micro-influencers in their advertising campaigns, brands can count on greater audience engagement. Statistics show that micro-influencers generate 60% more interest than average campaigns in social media. Campaigns based on this cooperation model are also characterized by a higher effectiveness rate and generate 2.2 times more conversions [16].

In April 2020, a month after the announcement of the global SARS-CoV-2 pandemic, 33.3% of all Instagram influencers working with brands were former nano-influencers. 31.9% of all #ad posts also came from them. This is the highest percentage since June 2019. Of all the brand collaborations that Socialbakers analyzed [21], nearly 94% are micro-influencers with less than 100,000 followers. This demonstrates brands' belief that positive campaign results can be achieved by having smaller, more authentic influencers who are closer to their audience [17].

4 Conclusion

Traditional marketing messages have a smaller and smaller impact on recipients. The revolution brought about by social media significantly changed the nature of relationships between people, and thus the form of communication. Internet users also use a more direct language, and messages addressed to them should be individualized, and not to an undefined mass target group [18]. A person works best in a simple, natural, authentic language, that is, using stories, metaphors and stories. In such a situation, the influencer becomes an intermediary, a medium through which the brand can establish relationships with consumers. Consumers, in turn, cease to be passive recipients of advertising messages. They participate in their creation, are responsible for distribution, interact, engage others, and really influence the decisions of people around them.

During the COVID-19 pandemic, the time spent by consumers on the Internet has increased worldwide. This situation has created opportunities for brands to expand their reach or engagement. At the same time, due to the crisis, most companies cut their advertising budgets, which made them look for less expensive channels to reach customers. This situation not only strengthened, but also accelerated the trend that has been going on for some time, related to the perception of micro-influencers as high-value resources, bringing good results without huge expenses. Following this path, it will be more profitable to engage a few microfinluencers to cooperate with them, who enter into dialogue with fans more often, shorten the mythical "distance" with users, and their channels constitute a much more strongly targeted group. As the presented research results show, the distribution of content on their channels will have a stronger impact on recipients who have greater trust in micro-influencers. Smaller influencers are also more likely to engage in non-standard actions that require more involvement on the part of the influencer.

Despite the economic crisis caused by the Covid-19 pandemic, as outlined in the article, brands continue to invest in influencer marketing to reach their target audience, but with a clearly different approach. The changes that have taken place in 2020 show that it is time for the digital marketing industry to evaluate the best approach and commit to a whole new set of customer needs and expectations. The challenge for marketers now is to understand the fundamental changes that are needed to adapt to new consumer priorities. What this new normality will look like after the pandemic is unknown yet, but focusing on customer-centric marketing is a way to overcome the uncertain and complex reality of Covid-19. Rather than starting with a product or communication channel, this strategy focuses primarily on individual consumers. With customer-oriented marketing, marketers implement commitment strategies based on trust and empathy, and their good carriers are the micro-influencers with whom the recipients identify the most. The signs are that brands are likely to continue to expand their collaboration with these less influential creators as part of a smarter social media strategy, also after the pandemic is over [19].

References

1. Internet source: https://forsal.pl/praca/aktualnosci/artykuly/7996051,jedna-trzecia-polakow-bedzie-pracowala-z-domu-podczas-drugiej-fali-pandemii.html. 3 November 2020
2. Kaczmarczyk, I.: Nowe trendy w influencer marketingu. In: https://whiteanchor.pl/nowe-trendy-w-influencer-marketingu. 28 October 2020
3. Enberg, J.: Global Influencer Marketing. In: https://www.emarketer.com/content/global-influencer-marketing. 29 October 2020
4. Quote. Infuencer marketing, Encyklopedia zarządzania. https://mfiles.pl/pl/index.php/Influencer_marketing. 28 October 2020
5. Internet source: https://www.socialbakers.com/blog/influencer-marketing-report. 1 November 2020
6. Internet source: https://www.travelmindset.com/why-influencer-marketing-now/. 2 November 2020
7. Sammis, K., Lincoln, C., Pomponi, S.: Influencer Marketing for Dummies, Wiley, New Jersey (2015)

8. Influencer marketing – pobudzanie "brand desire": https://friendsandbrands.pl/publikacje/influencer-marketing-pobudzanie-brand-desire/. 12 October 2020
9. Baer J., Influencer Marketing – szokujący wskaźnik ROI. https://sprawnymarketing.pl/influencer-marketing-roi/. 12 October 2020
10. Makuch, K.: Mikroinfluencerzy w kampaniach reklamowych w 2018 r., Social Media Manager nr4 (2017)
11. Cupriak, A., Fabijańczyk, J.: Influencer marketing praktycznie, e-book (2016). whitepress.pl
12. Full report: https://www.socialbakers.com/blog/influencer-marketing-report. 1 November 2020
13. Internet source: Co to jest engagement rate na Instagramie i jak go podnieść w 2019 roku? https://www.wildrocks.pl. 13 November 2020
14. Full report: https://semahead.pl/blog/influencer-marketing-warto-inwestowac-mikroinfluencerow.html. 14 October 2020
15. Women are the New Media przygotowanego przez firmę w 2017 r. Wynika z niego, że 68% influencerów w social mediach to kobiety: indahash.com. 14 October 2020
16. Internet report: https://www.socialbakers.com/blog/influencer-marketing-report. 1 November 2020
17. Internet source: https://morebananas.pl/blog/marketing-w-dobie-pandemii-trendy-2020. 2 November 2020
18. Internet source: https://marketingtechnews.net/news/2020/jun/29/the-state-of-influencer-marketing-amid-covid-19-concern-but-micro-influencers-show-path-forward/. 1 November 2020
19. Internet source: https://www.press.pl/tresc/61582,w-wyniku-epidemii-zyskaja-mikroinfluencerzy-kosztem-gwiazd. 3 November 2020

Correlations Linking Macroergonomic Premises with Key Features of Intelligent Buildings

Leszek Pacholski(✉)

Faculty of Engineering Management, Poznan University of Technology, J. Rychlewskiego 2 Street, Poznan, Poland
leszek.pacholski@put.poznan.pl

Abstract. The subject of this study is the processes of macroergonomic improvements of intelligent buildings. The macroergonomic premises involve both the technological-economic and political-legal component as well as subjectively treated the human factors – users of these buildings. The main research stream of this study regards the correlations linking the four types of the above-mentioned macroergonomic premises with the seven key features of the intelligent buildings. The presented expert opinions concerning the analyzed correlations are case studies and are based on inventive approaches. The Google Forms software was used to collect these opinions. Clarifying the joint opinions of the team of experts is based on the Delphi method. For the purposes of verifying the inference, the fuzzy formula of inference is used. The results of the research, are shown in the form of a table using the method of morphological analysis.

Keywords: Macroergonomics · Intelligent buildings · Fuzzy infernence

1 Introduction

Enterprises realizing the operation management of the intelligent buildings are forced to add to their range of action issues of its simultaneous use and improvements. Such situation requires searching for long-term efficiency of management in the company, based on balance between exploitation and exploration. It is possible to try to achieve this balance by realizing enumerated actions in different organizational units of the enterprise that puts into practice the new variant of operation management improvements. Such actions must be integrated and connected on a strategic level of company's management. This requires eclectic organizational structure, with co-existing units having various professional competencies, different organizational culture and separate motivation systems and leadership [1]. The macroergonomic premises for the process improvements of the both exploitation and exploration of the seven key characteristics of the intelligent buildings focuses on the particular role of their "essence", which is a composition of the subjectively treated human communities with the intelligent buildings technological-economic system innovative solutions [2, 3].

The main research stream of this study regards the correlations linking the four types of the macroergonomic premises with the seven key features of the intelligent

S. Trzcielinski et al. (Eds.): AHFE 2021, LNNS 274, pp. 401–408, 2021.
https://doi.org/10.1007/978-3-030-80462-6_49

buildings. Reports from research results are just correlations. They are not cause-effect relationships. The presented expert opinions concerning the analyzed correlations are case studies and are based on inventive approaches. The Google Forms software will be used to collect these expert opinions. Clarifying the joint opinions of the team of experts will be based on the Delphi method. For the purposes of verifying the inference, the fuzzy formula of inference will be used. The results of the research, are shown in the form of a table using the method of morphological analysis.

2 The Macroergonomic Premises

In seventies of the last century, the author of this article made an attempt of universal defining the subject of the research and ergonomic applications [4]. He has determined this subject as a so-called macroergonomic [5–8] interface (system, network, hybrid) combining human societies and the technological-economic, cultural-social and political-legal components. If to treat historically the problem of the subject of research and ergonomic applications, then, according to the above-mentioned universal proposal, one could say about four stages of the evolution of this interface. The first stage would refer to pre-industrial societies and the technological-economic and political-legal component in form of methods and tools for hunting or simple agricultural techniques and tools. The second stage encloses industrial societies, which have machines, energy and electronics on their disposure. The modern, third stage, information societies use computer networks and new media. The further, fourth stage might refer to creative and empathic societies and the technological-economic and political-legal component in form of nanotechnology applications and of artificial intelligence [9]. Current technological-economic, cultural-social and political-legal "partners" of human communities can be also identified more precisely on the basis of the concept of the macroeconomic cycles. These "partners" usually have the basic form of following so-called "technological fundamental innovations" (the steam machine and railways, the electricity and the internal-combustion engine, aviation, energetics and electronics, computer networks, pro-ecologic solutions, artificial intelligence and nanotechnology). These innovations, constituting the technological, economic, cultural-social and political-legal reaction to the human-centered evolution of "cultural-social imperatives and needs" (like: improving the efficiency of work and trade, the availability and mobility of assets, increase if the standard of life, energy networks, travelling, ecology, knowledge networks, psychological and social fitness, human health and quality of life).

Democracy and market freedom have caused that human societies where fascinated by the concept of the "association" network, but next they started protesting the functional, "tribal" order. Today, democracy and market are ruled by the research of the "buyer". Human societies leave traditional hierarchies and authorities in favor of a so-called „guru", who does not have to convince us, he only needs to be popular. There is chaos appearing within the human factors (the contemporary, "pro-society" order of the organization is being replaced by a formation called a "swarm"). However, this state of affairs does not change the fact that today we can still talk about the four types of the macroergonomic premises: technological and organizational, economic, socio-cultural and political-legal. The complexity of macroergonomic decision-making situations has forced to change the current research and application methodology based on two-valued logic. Fuzzy inference methods [10] seem to be more effective when necessary syncretic problem solving of the technological and organizational, economic as well as socio-cultural and political-legal improvements.

3 The Key Features of Intelligent Buildings

The intelligent buildings means a full system functionality of a sustainable creation of a construction technology (building) that is equipped with self-steering subsystems that also have abilities to adjust to the environment. Such adaptation concerns both an exploitation's reaction to changes in the external and internal environment and the ability for improvement along the process of the technological and social progress. "… It is a relatively autonomic environment – it has all the necessary functionalities built in. Its principal characteristics are hybrid connections, like: Smart Economy, Smart People, Smart Governance, Smart Mobility, Smart Environment and Smart Living. Such combinations are creating a completely new perspective on building objects as a whole" [2]. Intelligent buildings can have various designations. They can be residences, hotels, offices, banks, hospitals, schools, etc. Intelligent neighborhoods or intelligent cities must fulfill similar requirements.

Their design, production (construction) and exploitation (use and service) and exploration (improvements) requires expanding the classic engineering from the range of construction, technology and organization.

Energy efficiency and thermal isolations are also key features of intelligent buildings, just like sustainable use of natural resources, safety of the user, protection from environmental parameters' levels that can be harmful for human health (noise, air pollution, microclimate, and vibrations), accessibility, low costs of use and service and pro-environmental performance during the entire life cycle of the construction.

The intelligent building's should meet six following criteria exchanges [2]:

- provide full protection and steering of the building's assets,
- ensure the required technical and communication infrastructure of the building,
- provide the control of the functioning of the entire,
- enable an efficient connection with the exterior,
- maintain stable parameters of the material internal environment independently from changes outside and facilitate the adjustments of these parameters to changing needs of the user,
- meet demands of the ecological rationality of resource management (electric energy, heat energy, water energy) and the criterion of the human factor (health, safety, identity, visual comfort, esthetic design, intelligent management of usage costs).

In practice, the realization and improvement of enumerated criteria in an intelligent building is steered by three principal groups of technological systems [2]: security system, system of monitoring and steering technical installations, computing system. The first system usually encloses: the system of fire security, the transmission of fire alarms system and defects signals, audible warning system, smoke and heat spread control, fire suppression and extinguishing system, a system integrating fire protection devices, evacuation system, air quality monitoring and ventilation for underground garages, systems for removing smoke from evacuation paths, intrusion detection system, access control system, closed circuit television system. It is possible to divide a system of monitoring and steering technical installations into two subsystems: power supply and electricity control, and comfort steering. The electric subsystem includes: the division of energy, measuring and energy accounting, the monitoring of technical installations, the electric installation cabling, lighting control inside and outside of the building, emergency power supply, photovoltaic subsystem. The comfort subsystem includes: control of sources of the heat and coolness (including renewable energy sources), air conditioning and ventilation steering, heat steering, lighting and shading steering (shutters), horizontal and vertical transport subsystem, audio-video subsystems, sound system steering, subsystem of parking facilities [2].

Intelligent building have a computing system the encloses: a subsystem of structural cabling and of active devices, computing networks, data transmission devices, security subsystem, links with external networks, vision and sound transmission subsystem, telephone exchanges. The four sub-systems of operation management system ensure an efficient functioning of technical installations in intelligent buildings are following: - a system for power supply management and for steering electricity, comfort, information and communication systems, - security system for the building, - global system of management and steering in the building (its goal is gathering and analyzing information from the entire object and exchanging data with all cooperating systems and subsystems), - system of control of the energy supply in the building. All mentioned systems and automation and management processes in the building require determined norms and standards. Standardization topics enclosing issues in the design, realization, exploitation and improvement of intelligent building's is largely diversified. It refers to areas like: construction, ergonomics and health protection, fire security, telecommunication, informatics, energy, automation, data transmission in control systems, electronics, property protection [2].

4 Research Methods and Results

The main research stream of this study regards the correlations linking the four types of the macroergonomic premises with the seven key features of intelligent buildings. Conclusions from research results are just correlations. They are not cause-effect relationships. The research was conducted in the second half of 2020. The presented expert opinions (the team of seven experts- practitioners familiar with the concept of intelligent building) concerning the analyzed correlations are case studies and are based on inventive approaches. The Google Forms software will be used to collect these expert opinions. Clarifying the joint opinions of the team of experts will be based on the Delphi method. For the purposes of the research on the correlations linking the four types of the macroergonomic premises with the seven key features of intelligent buildings, four degrees of Delphi expert assessment of the strength of relations (five-level numerical scale from 1 to 5) were adopted for twenty eight of their potential variants. Taking into account each of the seven expert's opinion, it was assumed that in the next three rounds the experts narrowed the number strength of the relation. If the interquartile range and the new expert opinion was outside this new relation, its justification was expected. In this way, it was forced to move experts undecided and without adequate arguments to the majority group, i.e. to the group whose opinions are in the interquartile ranges.

For the purposes of verifying the inference, the fuzzy formula of inference will be used. As part of the fuzzy formula of inference ([10, 11]) concerned the research on the macroergonomic premises and the key features of intelligent buildings, it was assumed that the probability of high, medium and low strength of the correlations will be tagged [12]: p^h_{ai}, p^m_{ai} i p^l_{ai}.

The determination of the above probabilities was entrusted to the afore mentioned seven experts. The probability boundary conditions were defined as follows:

$pa^h_i + pa^m_i + pa^l_i = 1$ and $pa^h_i \geq 0$; $pa^m_i \geq 0$; $pa^l_i \geq 0$, where: i ∈ 1. The linguistic variable z (five-level scale of expert assessments) was introduced to describe the strength of the analyzed correlations and the fuzzy scale. For the level 1 (expert rating: no correlation) the linguistic scale was $0 < z < 0{,}333$. For the level one (expert rating: correlation strength slight) the linguistic scale was $0{,}167 \leq z < 0{,}5$. For the level three (expert rating: indirect relation) the linguistic scale was $0{,}333 \leq z < 0{,}667$. For the level four (expert rating: strength of the correlation significant) the linguistic scale was $0{,}5 \leq z < 0{,}833$. For the level five (expert rating: dominant correlation) the linguistic scale was $0{,}667 \leq z \leq 1$. The designation dj was further introduced as the result of the j-th expert's assessment, where: (dj ∈ 1, 2, 3, 4, 5). According to the fuzzy set theory, the membership of the set N was expressed by the membership function for the interval [0,1] of the form: μN(x). Each result of the dj expert assessment was assigned to the following five fuzzy sets: $N = \{(\beta 1, \mu N(\beta 1)), (\beta 2, \mu N(\beta 2)), (\beta 3, \mu N(\beta 3))\}$; for $j = 1, 2, 3,.., E$; where: $E = 7$ is the number of experts assessing the relationship. The normalized fuzzy set distribution (source: [11, 12]) of the scale of correlations linking four macroergonomic premises with the seven key features of intelligent buildings, was taken into account: (0, 0.333) - insignificant correlation, (0.333, 0.667) - relatively significant correlation and (0.667, 1.0) - a very important correlation. Then, fsf was introduced as an element of the fuzzy set s and msf as a membership function for the corresponding element of fuzzy sets and the probabilities of the generalized evaluation of seven experts were calculated (in two steps) as follows:

- Calculation of the average Pi probability score (where: i = 1, 2, 3) reflecting the frequency of selection made by the experts:

$$P_1 = \sum_{f=1}^{3} f_{1f} m_{1f} \; P_2 = \sum_{f=3}^{5} f_{2f} m_{2f} \; P_3 = \sum_{f=5}^{7} f_{3f} m_{3f}$$

- Calculation of pi (where: i = 1, 2, 3) corresponds to the probability of significance of the relation:

$$p_i = P_i / \sum_{i=1}^{3} P_i$$

The results of the research, based on the fuzzy formula of inference and concerning the correlations linking the four macroergonomic premises with the seven key features of the intelligent buildings are shown in Table 1.

This table uses the method of morphological analysis. A characteristic feature of this approach is the combination of intuitive and analytical elements, and the basic criterion for consideration is the principle of targeted searches according to specific procedures. Such an approach may offer a greater degree of technical character of action, because the analytical effect is obtained at the expense of limiting spontaneous creativity, which is the driving force of inventive methods (based on free associations). The use of the method of morphological analysis was aimed at creating such a picture of the problem situation in which all of the structural correlations, linking the four types of the macroergonomic premises with the seven key features of intelligent buildings, would be visually included.

Table 1. The morphological analysis of correlations linking the macroergonomic premises with the key characteristics of the intelligent buildings

	Technological	Economic	Socio-cultural	Political-legal
Energy efficiency	0,714	0,833	0,714	0,500
Sustainable use	0,500	0,833	0,714	0,500
Safety of the user	0,583	0,500	0,833	0,500
Protection	0,583	0,583	0,714	0,500
Accessibility	0,583	0,714	0,714	0,714
Low costs	0,714	0,833	0,714	0,500
Pro-environmental	0,583	0,833	0,833	0,500

(Source: own study)

Legend:
Energy efficiency - Energy efficiency and thermal isolations
Sustainable use - Sustainable use of natural resources
Protection - Protection from environmental parameters' levels (noise, air pollution)
Low costs - Low costs of use and service
Pro-environmental - Pro-environmental performance during the entire life cycle of the construction.

5 Conclusions and Summary

The numerical probabilities of the significance of correlation in Table 1 form the conclusion matrix consisting of four columns and seven rows. An additional fifth column and an additional eighth row show the calculated arithmetic averages (Fig. 1).

The arithmetic averages of these columns and rows can form the basis for summarizing the results of the research presented in this article. The values of these arithmetic

0,714	0,833	0,714	0,500	**0,690**
0,500	0,833	0,714	0,500	**0,637**
0,583	0,500	0,833	0,500	**0,604**
0,583	0,583	0,714	0,500	**0,584**
0,583	0,714	0,714	0,714	**0,681**
0,714	0,833	0,714	0,500	**0,690**
0,583	0,833	0,833	0,500	**0,687**
0,609	**0,733**	**0,748**	**0,531**	

Fig. 1. The conclusion matrix of numerical probabilities of the significance of correlation linking the macroergonomic premises with the key characteristics of the intelligent building. An additional columns and rows showing the calculated arithmetic averages (source: own study)

averages are as follows: - for the first column 0,609; - for the second 0,733; - for the third 0,748 and for the fourth column 0,531. For the following rows of the matrix, the averages are respectively: 0,690; 0,637; 0,604; 0,584; 0,681; 0,690 and 0,687. According to the criterion of the arithmetic averages and preferring fifteen most important correlations (out of a potential twenty eight). With two highest numerical probabilities of the significance of correlations (0.833 and 0.714 - see Table 1), the specific ranking in terms of the four macroergonomic premises for the key features of intelligent buildings can be distinguished.

The ranking of macroergonomic arguments in favor of the implementation of intelligent building opens socio-cultural premises with a set of the seven highest correlations (concerning the key features) and the arithmetic averages indicator 0.748. Such a situation, meaning for the implementation of intelligent buildings, the priority of macroergonomic socio-cultural premises over economic, technological and organizational premises, seems to be the most interesting conclusion from the presented research. Experts participating in the research were assigned the five highest correlations (concerning the key features) and the arithmetic averages indicator 0,733 to economic premises. For the technological and organizational premises of intelligent building, the experts surveyed awarded the arithmetic averages indicator 0,609 and two highest correlations concerning the key features of intelligent buildings. The political and legal premises obtained in this ranking the arithmetic averages indicator 0,531 and only one highest correlation concerning the key features of intelligent buildings. Within the seven key features of intelligent buildings, the following four attributes of smart construction have the highest correlations with macroergonomic premises: energy efficiency and thermal isolations, low costs of use and service, pro-environmental performance during the entire life cycle of the construction and accessibility. Just behind them is the argument about the sustainable use of natural resources. In summary, it should be to repeat, with particular emphasis, that the presented expert opinions concerning the analyzed correlations are case studies and are based on inventive approaches. Conclusions from research results are just correlations. They are not cause-effect relationships.

Acknowledgment. The research was carried out as part of the research project of the Polish National Research and Development Center (POIR.01.01.01–00-1229 /17) for the research and teaching needs of the Faculty of Engineering Management at the Poznan University of Technology.

References

1. Csaszar, F.A.: An efficient frontier in organization design: organizational structure as a determinant of exploration and exploitation. Organ. Sci. **24**(4), 1083–1101 (2013)
2. Kalkowska, J., Pacholski, L.: The conditioning of macroergonomic design and improvement of the intelligent building's use and service. In: Goossens, R., Murata, A. (eds.) Advances in Social and Occupational Ergonomics. AHFE 2019. Advances in Intelligent Systems and Computing, vol. 970, pp. 165–172. Springer, Cham (2020). https://doi.org/10.1007/978-3-030-20145-6_16
3. Kałkowska, J., Siemieniak, M., Pacholski, L.: The future of macroergonomic manufacturing systems. In: Goossens, R. (eds.) Advances in Social and Occupational Ergonomics. AHFE 2018. Advances in Intelligent Systems and Computing, vol. 792, pp. 365–374. Springer, Cham (2019). https://doi.org/10.1007/978-3-319-94000-7_37
4. Pacholski, L.: Methodology of ergonomic diagnosing in the enterprise, Rozprawy, vol. 81, pp. 1–148. Publishing House of Poznan University of Technology (1977). (In Polish)
5. Hendrick, H.W., Kleiner, B.M.: Macroergonomics: Theory, Methods and Applications, Lawrence Erlbaum Associates Publishers, Mahwah NJ (2002)
6. Pacholski, L., Jasiak, A.E.: Macroergonomic effects of rapid economic renewal. In: Scott, P.A., Bridger, R.S., Charteris, J. (eds.) Global Ergonomics, pp. 855–858. Elsevier, Amsterdam - Oxford – New York (1998)
7. Pacholski, L., Pawlewski, P.: The usage of simulation technology for macroergonomic industrial systems improvement. In: Goossens, R. (ed.) Advances in Social & Occupational Ergonomics. Advances in Intelligent Systems and Computing, vol. 487, pp. 3–14. Springer, Cham (2017). https://doi.org/10.1007/978-3-319-41688-5_1
8. Pacholski, L., Piotrowski, K.: Political ergonomics, macroergonomic battles. Hum. Factors Ergon. Manuf. **18**(5), 515–524 (2008)
9. Kollmann, T., Kuckertz, A., Stockmann, C.: Continuous innovation in entrepreneurial growth companies: exploring the ambidextrous strategy. J. Enterp. C. **17**(3), 297–322 (2009)
10. Pacholski, L.: Fuzzy logic application in ergonomic renewal of multiagent manufacturing systems. Cybern. Syst. Int. J. **29**, 715–728 (1998)
11. Kałkowska, J., Kozlov, A.V.: Decision making process for the knowledge-based enterprise: fuzzy sets theory application to strategic management. In: Świątek, J., Borzemski, L., Grzech, A., Wilimowska, Z. (eds.) Information Systems Architecture and Technology: Proceedings of 36th International Conference on Information Systems Architecture and Technology – ISAT 2015 – Part III. AISC, vol. 431, pp. 163–174. Springer, Cham (2016). https://doi.org/10.1007/978-3-319-28564-1_14
12. Kalkowska, J.: Proexploitation Approach for Public Transport Vehicles Developing Process, Rozprawy, pp. 1–164, Publishing House of Poznan University of Technology (2018). (In Polish)

Management Model Based on the Application of the DEA (Data Envelopment Analysis) and 5s Method to Improve the Efficiency of Production Processes in the Coffee Agronomic Sector

Fernando Bellido-Arcas[1], Cesar Ramirez-Valdivia[1], and Carlos Raymundo[2](✉)

[1] Ingenieria de Gestion Empresarial, Prolongacion Primavera 2390, Lima 15023, Peru
{u201518142,pcgecram}@upc.edu.pe
[2] Direccion de Investigacion, Prolongacion Primavera 2390, Lima 15023, Peru
carlos.raymundo@upc.edu.pe

Abstract. The purpose of the following article is to describe how by using efficiency control tools it is possible to obtain an improvement in the coffee production process. To complement and ensure a good result of the management model, the guidelines proposed in the lean methodology will be used, applying efficiency improvement tools such as the 5s method and continuous improvement. For the purposes of this study, the use of resources that generate a reduction in productivity and that are treated within inefficient production processes, such as the use of water, fertilizers and machinery, was used as information. Once the indicators to be taken into consideration for decision making were identified, the pertinent improvements were made, achieving an increase in the Soil Production Index from 65% to 75%, generating a greater product for the coffee growers, as well as the Production Index of work went from 68% to 75%.

Keywords: Benchmarking · Data envelopment analysis · Efficiency · Management model · 5s

1 Introduction

Peru is currently the ninth coffee producing country in the world, and the seventh exporter [1]. This activity also generates an economic support for more than 223 thousand producer families and occupies an area of more than 420 thousand hectares in the country, supporting the economic support of families living in very poor areas. However, for the same reason that this activity takes place in areas of economic scarcity [2], work practices do not take into account the efficient use of resources or the impact that these can have on the environment. In addition, among the main direct causes why the Peruvian coffee market has low competitiveness and social/environmental sustainability are: the low level of productivity and technical management of the crop without considering efficiency criteria, the great inconsistency of the coffee quality, the limited access to financial services that coffee-growing families have, the weak image and positioning of

S. Trzcielinski et al. (Eds.): AHFE 2021, LNNS 274, pp. 409–416, 2021.
https://doi.org/10.1007/978-3-030-80462-6_50

Peruvian coffee in national and international markets, the precarious social and economic conditions of the coffee-growing areas and the weak governance for the management of coffee production [3]. Consequently, there are not too many documented experiences of the use of resources or how they are treated within the production processes, and the issue of efficiency is still far behind. However, there are some important advances focused on implementing an efficient production policy that allows advances in the matter. It is believed that this is the way to achieve that coffee is consolidated as a market and becomes part of the group of products by which Peru is identified.

2 State of the Art

2.1 Lean Manufacturing

As Lean Manufacturing is one of the most widely implemented methodologies in the world when it comes to making improvements in the efficiency of production processes, there are many ways in which it can be applied depending on the sector in which it will be applied. The most used practices have been identified through a comparative matrix, designing an implementation methodology for small companies and for people with little experience [4, 5]. It is also understood that great experience is not needed to apply improvements within a process, as there are simple and agile alternatives to achieve a successful implementation of lean manufacturing. However, it is also necessary to analyze its application over time and how companies that have already implemented lean manufacturing tools move once the improvement has been made. With all that has been learned by companies and applied case studies, it is possible to develop a methodology with adaptation facilities to be used by industries [4, 6, 7].

2.2 Data Envelopment Analysis

For the development of the efficiency model, it is necessary to identify a terminal that allows identifying the necessary inputs to perform a comparison, grouping as much information as possible to compare [8, 9]. The more information it is possible to obtain from the sector, the more accurately the distance to achieve the ideal efficiency can be determined [8, 10]. In this way, each market needs to carry out an analysis to determine what will be the input and output elements to use for the organizations. In the case of cooperative companies, it has been determined that the variables to be considered will be land, work, fertilizers and machinery [10]. In addition, it is concluded that the land factor is the most important input when small farmers develop their activity, followed by the intensity of the fertilizer and the work of the machinery. Also, the labor factor, despite being relevant, maintains an inverse relationship with the producer, due to the small number of hectares they own and that increasing workforce will not increase their coffee production [11, 12].

3 Contribution

3.1 Proposed Model

In Fig. 1, the contribution made in the research is appreciated, where the DEA tool is used to evaluate the technical efficiencies that coffee cooperatives have and from said analysis of technical efficiencies, understanding the use of the microeconomic framework and Methodology on the stochastic frontier to analyze the production per productive unit per hectare of the Peruvian coffee sector. Thus, the relationship between the quantity of inputs and the production of coffee under conventional and flexible functional forms, the 5S tool is applied, having a quality management adjusted to the reduction of inputs and achieving efficient indicators in production and marketing of coffee.

Regarding the use of the lean methodology, they point out that the use of Lean tools in the manufacturing sector, such as the 5S tool, helps to order, clean and classify products or supplies in the warehouse or workplace, establishing a standard process and an action procedure for the activities to be developed and placing elements of information, signage, among others, in order to increase the satisfaction of internal or external customers, reducing or eliminating the waste they generate to optimize the management of the Cooperative.

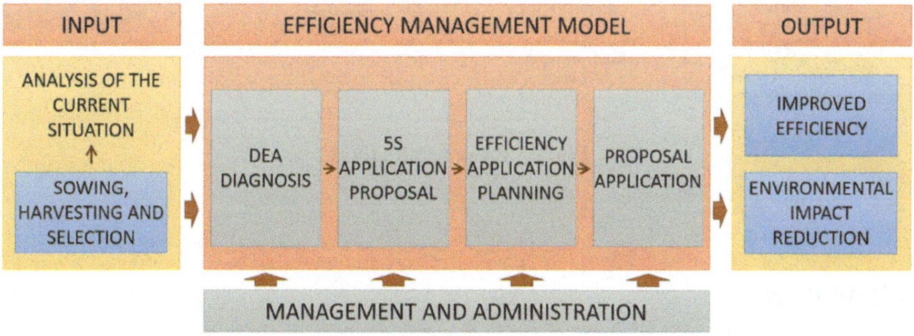

Fig. 1. Proposed model

The proposed model is composed of 3 components: DEA diagnosis, improvement proposal based on efficiency and continuous improvement planning.

Diagnosis of DEA, this component originates from the current analysis of the cooperative that has inefficient production processes. The purpose is to identify which variable (Land, labor, fertilizers, machinery and economy of scale) should be prioritized to optimize the results in production and costs.

Proposal for Improvement Based on Efficiency, in this component, an improvement plan will be carried out supported by the 5s methodology according to the variable previously identified.

Planning of Continuous Improvement, the objective is to plan the appropriate processes to optimize efficiency in the planting, harvesting and selection of coffee.

3.2 Proposed Method

In Fig. 2 the steps for the implementation of the efficiency management tool during coffee processing are detailed. In it, the phases are detailed, which are divided by the application of the DEA, diagnosis of the 5S, planning of the proposal, execution of the proposal and finally, the planning of the improvement.

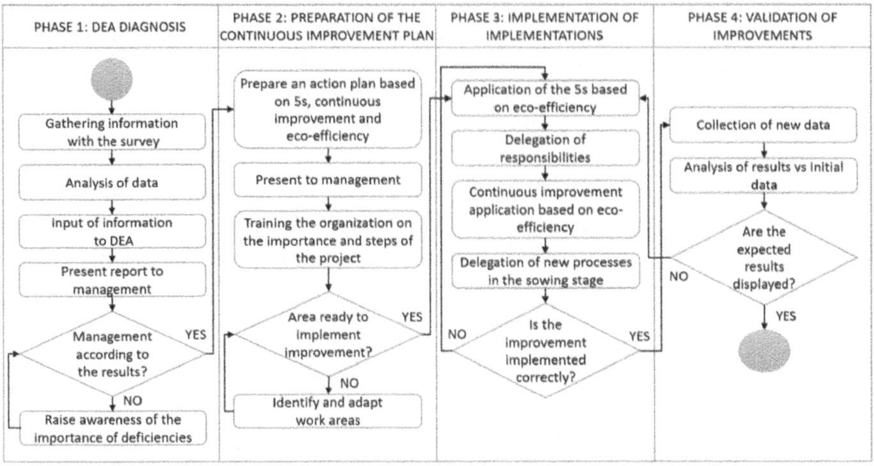

Fig. 2. Proposed method

3.3 Indicators

The indicators used before and after the implementation of the methodology are mentioned and explained below.

Soil Production Index (PTS), It establishes a value of the production for the surface with aptitude for irrigation, as an indicator of the performance achieved with the level of technology applied in the agricultural year.

$$PTS = \frac{total\ production\ (kg)}{irrigable\ area\ (ha)} \tag{1}$$

Labor Production Index (IPTR), It is presented as an indicator of the production in weight of the annual per capita harvest of the rural population.

$$IPTR = \frac{total\ production\ (kg)}{rural\ population\ (inhabitant)} \tag{2}$$

Unused Production Index (IPNU), It is proposed as an indicator that determines the amount of waste generated during the process.

$$IPNU = \frac{waste\ (kg)}{total\ production\ (kg)} \tag{3}$$

4 Validation

4.1 Description of the Case Study

The implementation of the efficiency management model was carried out in the Selva Alta coffee plantation in the agricultural coffee sector.

4.2 Development of the Contribution

Table 1 shows the evaluation of the effect that each variable has on the productivity of the coffee crop.

Table 1. Efficiency of the cooperative

Independent variables	Constant elasticity of substitution	Cobb-Douglas	Translogarithmic (Minflex Laurent Tranlog)	Leontief generalized (Diewert)	Quadratic
Land	0.02	0.89	0.81	2.73	24.88
Job	0.01	0.47	−0.3	−0.04	0.17
Fertilizers	0.0024	0.03	0.69	5.61	24.21
Machinery	0.0014	0.03	0.13	0.03	0.003
Economies of scale	0.0338	1.42	1.33	8.33	49.263

It can be seen that labor or labor has an indicator of −0.3, which indicates that labor does not impact the efficiency of small coffee farmers, this because as they are small coffee growers their land extensions are small, making use only of the family's own staff. Likewise, land has an impact on cultivation efficiency with 0.81, but in our efficiency analysis it is removed because small coffee growers do not have the option to buy farmland, therefore, it is removed from efficiency to seek in the cooperative.

Regarding fertilizers, it has an indicator of 0.69 which indicates that it is the most important factor after the cultivated land, in it the fertilizer plays an important role, since it represents the highest costs in the production of coffee, as well as vital input to increase productivity and technical efficiency, therefore, it will be used to optimize the efficiency of coffee growers, especially in the use of organic fertilizers, since it is the main input to produce organic coffee, thus being a sustainable and efficient, since it reduces electrical energy and water consumption.

To implement the 5S tools, the following artifacts were used according to stage 2 of the model.

Cleaning check-list, it is appreciated that the cooperative carries out before and after, where signage is added, that is, a name to each element for easy identification, due to its visual character, as well as the places or types of products. In the case of order, the signage served to keep the supplies in order and stored as appropriate. Which led to a training, for a greater knowledge and understanding of the processes, steps or procedures to apply in each case.

In Table 2 the daily jobs, framed within the 5S, in the concept of Cleaning.

Table 2. Daily work plan in the rooms

Activity	Schedule		Resources	Person in charge
	10–11 h	12–15 h		
Cleaning in cultivation (Check out)		×	Cleaning supplies	Collaborator 1
Cleaning when the plant grows and is pruned	×		Cleaning supplies	Collaborator 2
Harvest cleaning	×		Cleaning supplies	Collaborator 3

In it, it can be seen that there are three moments to be able to clean and that the plant grows well, for this reason a person is established for each activity. Indicating the times that these tasks must be carried out, to then inspect and know that the plants are growing healthy and that they do not have pests and that they have enough light or shade in their growth stage, maturity and fruit harvest.

Given this, the training plan was established, in it we see that the cooperative needs a training plan to train its staff in techniques to help them improve their productivity. For this reason, a person responsible for each technique is established, with cleaning, plant care and storage being the main techniques, since these directly influence the technical efficiencies of coffee growers.

After the training, the tasks foreseen in the 5S were carried out, photos were taken to demonstrate the status of the cultivation, warehouse and dispatch areas, establishing an execution plan of those responsible for areas, for which established responsible for each area. They are responsible for ensuring that their respective areas are clean and orderly, carrying out inspections and supervision of tasks from Monday to Sunday, this indicates that there must always be a person responsible for cleaning and maintaining order, including Sundays and holidays.

Then, with the control checklist, it is possible to take note of the observations within the daily activities carried out by the operators. Thus, corrective measures will be carried out and those responsible will be identified to know with whom to carry out the intervention at the appropriate time.

5 Results

The result obtained at the time of applying the improvements with respect to the indicators is established in Table 3.

According to the statistical analysis (Table 4), it can be seen that regarding the media production index it presents an optimal result after the improvements implemented in the proposal.

Table 3. Indicators after implementation

Indicator	Before	After
Soil production index	65%	75%
Labor Production Index	68%	75%
Production index not used	15%	10%

Table 4. Statistical analysis by scenario and indicator

	Soil production index	Work production Index	Production Index not used
High jungle	75.00%	75.00%	10.00%
Partner 1	81.00%	78.00%	9.00%
Partner 2	76.00%	79.00%	11.00%
Partner 3	77.00%	80.00%	8.00%
Average	77.25%	78.00%	9.50%
Median	76.50%	78.50%	9.50%
Dev. Est	0.016330	0.010759	0.026458
Max	81.00%	79.00%	11.00%
n	4	4	4

On the standard deviation of the production index, it turns out that the variability has a greater dispersion compared to the mean.

In the case of the unused production index, the average value is less than 9.5%, which reflects a great result.

The average in the coffee production index shows a successful result of 75%. On the other hand, the standard deviation in this indicator is higher compared to the other indicators, which implies a greater dispersion present.

When analyzing the various scenarios bearing in mind different sizes of land, it can be concluded that the proposal could be applicable and successful in any cooperative, regardless of the size of hectares it has.

6 Conclusions

The use of the DEA allows to know the technical efficiencies from the variables of the production of coffee cultivation, therefore from the comparative analysis of efficiencies and the production model, finding out which is the greatest input of incidence on the technical efficiency and increase the productivity of the coffee crop of the coffee cooperative. Which the Translogarithmic index of production gave that 0.69 of the fertilizers, indicates that it is the main input after the land 0.81, to improve the technical efficiency of coffee cultivation.

The use of Lean Manufacturing 5S tools and continuous improvement, allowed to improve the working conditions of the operators by cleaning, order and classification, of the inputs in the management of the warehouse and in the cleaning of the crop in all its phases.

The soil production index went from 65% to 75%, generating more product for the coffee growers.

Likewise, the Labor Production Index went from 68% to 75%.

The unused Production Index was reduced, from 15% to 10%, which means that the coffee bean is being processed more efficiently.

References

1. HELVETAS. Estudio de Mercado del Café Peruano (Primera Edición) (2017). https://camcafeperu.com.pe/admin/recursos/publicaciones/Estudio-de-mercado-del-cafe-peruano.pdf. [Último acceso: 25 April 2019]
2. Agraria.pe, Agraria.pe (2018). https://agraria.pe/noticias/actualmente-casi-el-35-de-los-productores-de-cafe-en-nuestro-17624. [Último acceso: 15 May 2019]
3. Ministerio de Agricultura y Riesgo (2018). Plan Nacional de Acción del Café 2018–2030. https://www.undp.org/content/dam/peru/docs/Publicaciones%20medio%20ambiente/PNA-Cafe%20(pliegos)%2018Oct2018%20(1).pdf. [Último acceso: 30 April 2019]
4. Sarria, M.P., Fonseca, G.A., Bocanegra, C.C.: Modelo metodológico de implementación de lean manufacturing. EAN **83**, 51–71 (2017)
5. Gobinath, S., Elangovan, D., Dharmalingam, S.: Lean manufacturing issues and challenges in manufacturing process– a review. Int. J. ChemTech. Res. **8**(1), 44–51 (2015)
6. Dauch, K.A., da Silva, J.E.A.R., de Souza Jabbour, A.B.L.: Evaluation of the implementation of the 5S methodology in a manufacturing company: analysis of steps, benefits and barriers. Exacta. **14**(2), 285–302 (2016)
7. Vargas-Hernández, J.G., Castillo, M.T.J., Muratalla-Bautista, G..: Lean manufacturing. a system improvement tool of production. Ingenieria Ind. Actual. y Nuevas Tend. **17**(5), 153–174 (2017)
8. Arieu, A.: Eficiencia técnica comparada en elevadores de granos de Argentina, bajo una aplicación de análisis de envolvente de datos. La situación del puerto de Bahía Blanca. Consorcio de gestión del puerto de Bahía Blanca. Universidad Tecnológica Nacional. Argentina (2002)
9. Abdulai, S.: Assessing the technical efficiency of maize production in northern Ghana: the data envelopment analysis approach. Cogent Food & Agriculture, Turquia (2018)
10. Jorge Andres, P.C., Darrell Lee, H.: Funciones de producción, análisis de economías a escala y eficiencia técnica en el eje cafetero colombiano: una aproximación con frontera estocástica, MPRA Paper 37179, University Library of Munich, Germany (2011)
11. Tatari, O., Egilmez, G., Kurmapu, D.: Socio-efficiency analysis of highways: a data envelopment analysis. J. Civ. Eng. Manag. **22**(6), 747–757 (2016)
12. Jiang, H., et al.: Sustainability efficiency assessment of wastewater treatment plants in China: a data envelopment analysis based on cluster benchmarking. J. Clean. Prod. **244**, 118729 (2019)

Lean and Agility and Enterprise

The Impact of Changes Perceived in the Environment on the Agility of the Enterprise

Stefan Trzcielinski(✉)

Faculty of Engineering Management, Poznan University of Technology, Rychlewskiego 2, 60-965 Poznan, Poland
stefan.trzcielinski@put.poznan.pl

Abstract. The concept of agile manufacturing and an agile enterprise was created in order to effectively compete in the conditions of increasing volatility in all segments of the macro-environment of firms. The turbulent environment generates not only threats, but also opportunities, the perception and use of which favours of gaining a competitive advantage on the one hand, and on the other hand is a manifestation of agility. The aim of this article is to investigate whether and how managers assess changes perceived in macroenvironment and how the changes affect the implementation of solutions increasing the agility of the enterprise. An agility is analysed in cross section of its four dimensions i.e. brightness, flexibility, intelligence and shrewdness of enterprise. For this dimensions indicators/symptoms of agility were elaborated. The perceived changes are treated as independent (explanatory) variables when the indicators represent the dependent (explained) variables. The empirical research was conducted in January 2020 in 100 medium and large enterprises operating in Poland. The results show that the assessment of macroenvironment's changes has rather weak influence on implementation the concept of agile enterprise.

Keywords: Business environment · Agile enterprise · Opportunity recognition · Opportunity creation · Brightness · Flexibility · Intelligence · Shrewdness

1 Introduction

The concept of agile manufacturing and an agile enterprise was popularized in literature in 1991. It is aimed at effective management in conditions of turbulent changes taking place in the environment of enterprises. Such environment generates threats but also generates opportunities [1]. The literature on the subject emphasizes, above all, that the agility of a company consists in its ability to quick response to emerging opportunities [2, 3]. It also presents a dispute as to whether, as Kirzner claims [4], opportunities exist in the environment and to use them, they must first be discovered, or, as Schumpeter stated much earlier [5], opportunities are created because of innovative activities. We believe that this dispute is apparent and that opportunities are both created and discovered. In both of these cases, we are talking about opportunity recognition. Taking this point of

S. Trzcielinski et al. (Eds.): AHFE 2021, LNNS 274, pp. 419–427, 2021.
https://doi.org/10.1007/978-3-030-80462-6_51

view, we believe that a company's agility is not only about quick response. In order to recognize opportunities, an agile company must be [6]:

- bright - to perceive the situation in business environment which are favour to achieving its goals,
- flexible - so that its resources can be used in various favourable situations,
- intelligent - to develop its knowledge and adapt its resources and organization to changes that may occur in the future,
- shrewd - to achieve its goals in a practical way in a given situation with the use of available resources.

Observation, analysis and evaluation of changes taking place in the environment of enterprises is therefore necessary for both creating opportunities focused on these changes (Schumpeterian model) and for identification of the opportunities that are created by events taking place in the environment (Kirznerian model). On this basis, a hypothesis can be made that there is a relationship between how changes in the business environment are perceived and assessed and whether enterprises implement the strategy of agility. Checking this hypothesis is the aim of this paper, however the scope of considerations is limited only to changes in the macro-environment of enterprises.

2 The Business Environment and the Agility of the Enterprise

The business environment is usually divided into the macroenvironment and the industrial environment. The macroenvironment covers the following segments: political and legal, economic, socio-demographic and technological, whereas the industry environment consists of customers, suppliers and competitors. Enterprises have a much greater impact on their industrial environment than on the macroenvironment. Therefore, creating opportunities (proactive model) concerns primarily the industrial environment, while discovering opportunities (reactive model) concerns both the industrial environment and the macro environment.

Agility is understood in different ways. It is related to the organization's ability to survive [7], managing and applying knowledge effectively [8] or the ability of organizations to quick response to unexpected changes in business environment [9]. These abilities are usually embedded in the context of counteracting threats and especially of taking advantage of opportunities that arise precisely because the environment is changing. We treat the ability to take advantage of opportunities as a defining feature of an agile enterprise. There can be two cases here. The first consists in reacting to opportunities that arise in the environment because of a series of events occurring in its segments. Their use is therefore reactive. The second case is the creation of opportunities, for example through innovative activities that create a demand in the market. Taking advantage of these opportunities follows the pro-active model. Creating opportunities does not take place in isolation from changes and their trends in the business environment. This means that, both in the reactive and proactive model, it is necessary to observe the environment in order to see a situation conducive to achieving the goals of the organization using the resources available to it. This ability to notice changes that may result in opportunities recognition is called brightness and we treat it as the first dimension of the company's agility [6, 10]. The second dimension is flexibility. This is a feature of the resources available to the enterprise that enables it to take advantage of a wider repertoire of opportunities. By available resources, we mean both the company's own resources and its network resources. The third dimension of enterprise agility is intelligence. It is the ability to develop knowledge (among others by learning) and prepare the organization for changes that take place in the environment, so that it can take advantage of future opportunities. The fourth dimension is shrewdness. Its essence is the ability of the organization to use opportunities in a practical and profitable way. These four dimensions (Fig. 1), after operationalization, are adopted to describe the agility of the enterprise and treated in the research model as dependent (explained) variables.

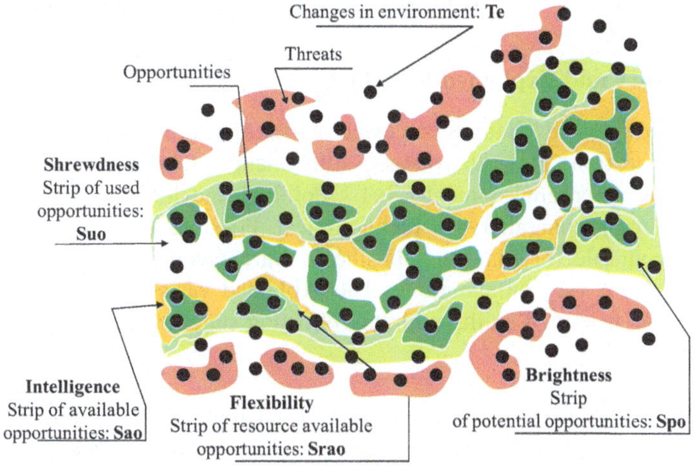

Fig. 1. Dimensions of enterprise's agility Source: [10]

3 The Research

3.1 Methodology of Research

In order to check whether the changes in the environment perceived by managers affect the application of the agility strategy, empirical studies were carried out in 65 medium and 35 large sized enterprises in Poland, belonging to 19 major production sectors, for example food, pharmaceuticals, paper, furniture and motor vehicles. The research was conducted in January 2020.

The data was collected using a closed-ended questionnaire using the CAPI (computer-assisted telephone interviewing) method. The respondents were presidents or owners of companies, managing directors, technical directors, production directors / chiefs, production engineers, including 13 women and 87 men.

The survey questions operationalize the perception of changes by managers (independent variable) and the agility of the enterprise in terms of its four dimensions, i.e.

brightness, flexibility, intelligence and shrewdness (dependent variable). The reliability of the measuring scales was checked using the Cronbach's alpha test. To obtain acceptable reliability, some questions were omitted in further analyzes. For the independent variable, the test value is 0.54, so it is poorly acceptable, while for agility (dependent variable) it is 0.79 and it is good, with the first three dimensions mentioned above respectively 0.75 (good), 0.53 (poorly acceptable), and 0.72 (good). The fourth dimension is described with just one question.

Since the responses to survey questions were expressed on the ordinal scale, the relationship between explanatory (independent) and explained (dependent) variables was examined using Spearman's rank correlation test. For statistically significantly correlated independent variables, it was checked whether and what their levels have a statistically significant effect on the dependent variables using the Kruskal-Wallis test and the post hoc test. These analyzes were performed with Statistica v. 13.3 software.

3.2 Explanatory and Explained Variables

Explanatory variables i.e. changes perceived in the environment have been operationalized with the question "How do you assess the changes taking place in the following segments of the macroenvironment": Iv1 - Political and legal; Iv2 - Economic; Iv3 - Social and demographical; Iv4 - Technological. Possible answers were as follows: 5-are very favourable for running a business; 4-are favourable; 3-are neutral; 2-are unfavourable; 1-are very unfavourable. The explained variables have been described in particular dimensions of agility with the following statements:

1. Brightness
   In the company, we recognize and analyze changes in environmental segments: DB1 - Political and legal; DB2 - Economic; DB3 - Social and demographical; DB4 - Technological; DB5 - Customers; DB6 - Suppliers; DB7 - Competitors. They can take the values: (5) very systematically; (4) systematically; (3) occasionally; (2) sporadically; (1) never.
2. Flexibility
   DF1 - In our firm, the operations are: (5) fully automated; (4) human work occurs, but automation / robotization dominates; (3) automation / robotization is in balance with human work; (2) there is automation / robotization but human work dominates; (1) there is no automation / robotization.
   DF2 - Our firm's policy towards machines that communicate themselves depends on: (5) we have such machines and we will increase their number; (4) it has been decided to introduce them within two years; (3) it has been decided to introduce them within five years; (2) Maybe we will introduce in the undefined future; (1) we will not introduce them.
   From the view point of the enterprise development the following competencies:
   DF3 - the employees' openness to changing the way and place of performing the job;
   DF4 –Multi-job of employees,
   are (5) very important; (4) important; (3) neutral; (2) not important; (1) definitively unimportant.

From the view point of use of employees' abilities the following forms of work organization:
DF5 - Job rotation;
DF6 - Flexible forms of employment,
are (5) very important; (4) important; (3) neutral; (2) not important; (1) definitively unimportant.

3. Intelligence
From the view point of the enterprise development the following competencies:
DI1 - innovativeness and willingness to propose new ideas;
DI2 - abilities to analyze data;
DI3 - readiness to learn,
are (5) very important; (4) important; (3) neutral; (2) not important; (1) definitively unimportant.
4. Shrewdness
DS1 - From the view point of the enterprise development the adaptation of employees to collaborate with robots is (5) very important; (4) important; (3) neutral; (2) not important; (1) definitively unimportant.

3.3 The Results

The correlation between the perception and assessment of changes in the environment and agility is presented in Table 1. The distinguished correlation coefficients are significant with $p < 0.05000$. The perception of changes in the political and legal segment of macroenvironment (Iv1) is positively correlated with flexibility (DF1) and shrewdness (DS1). In the economic segment (Iv2) it is positively correlated with flexibility (DF1) and negatively with flexibility (DF3) and Intelligence (DI2) and (DI3). Also in social and demographical segment (Iv3) it is negatively correlated with brightness (DB2).

For statistically significant correlations, we checked whether and what levels of explanatory variables on the Likert scale have a different impact on the agility of the enterprise. The result of these analyzes is presented in Table 2.

The value levels of the variable (Iv2) - "perception of changes in the economic segment" have different relevance for flexibility (DF1) and the variable (Iv3) - "perception of changes in social and demographic segment" has different significance for brightness (DB2).

4 Discussion and Conclusions

Explanatory (independent) variables - the perception of changes in the macroenvironment segments: political and legal (Iv1), economic (Iv2) and social and demographical (Iv3) are statistically significantly but relatively weakly correlated only with the six explained (dependent) variables. The technology segment variable (Iv4) is not correlated with any dependent variable.

The existing relationships between the variables are not obvious; therefore, we explain them through a possible, but certainly not the only possible interpretation.

A positive assessment of changes in the political and legal segment (Iv1) as well as in the economic segment (Iv2) is conducive to making decisions by company management boards to invest in technological development, which increases the level of automation and robotization of operations (DF1). As a result, the flexibility of enterprises is improved. In addition, a positive assessment of the trends in changes in the political and legal segment (Iv1) gives a sense of the stability of legislative regulations, which is conducive to shaping the investment development strategies of the company, an element of which may be an improvement of employees' adaptability to collaborate with robots (DS1).

Table 1. Correlation.

Variable		Independent variables			
		Iv1	Iv2	Iv3	Iv4
Brightness	DB1	-0,084	-0,021	-0,052	-0,182
	DB2	-0,087	-0,172	-0,290	-0,139
	DB3	0,140	-0,011	-0,055	-0,022
	DB4	0,156	-0,069	-0,083	0,067
	DB5	-0,124	-0,111	-0,089	-0,027
	DB6	-0,071	-0,190	-0,058	0,007
	DB7	0,053	-0,107	-0,030	-0,015
Flexibility	DF1	0,262	0,220	0,071	0,132
	DF2	0,174	0,049	-0,056	0,061
	DF3	0,128	-0,209	-0,076	0,067
	DF4	0,130	-0,027	-0,008	0,040
	DF5	0,174	-0,089	-0,121	0,106
	DF6	-0,067	-0,145	-0,012	-0,119
Intelligence	DI1	0,155	-0,192	-0,047	0,053
	DI2	0,166	-0,270	-0,067	-0,057
	DI3	0,033	-0,232	0,022	-0,062
Shrewdness	DS1	0,251	-0,015	0,033	0,065

The variable (Iv2) is also negatively correlated with the flexibility variable (DF3) and with the enterprise intelligence variables (DI2) and (DI3). Improving the assessment of changes in the economic segment of environment strengthens the feeling of managers about stabilization of the economic situation, which may lead to a reduction an interest in including in the development strategy of such employee competences as openness on change of the way and place of the job (DF3), ability of data analysis (DI2) and willingness to learn (DI3).

The perception of changes in the social and demographical segment of macroenvironment (Iv3) is also negatively correlated with brightness of enterprise defined as recognition and analysis of changes in economic segment of the environment (DB2). This may be due to the reduced interest of the boards of companies of conducting analysis of economic segment (DB2) along with improvement of the situation on the labour market and better assessment of the changes in the market (Iv3). Such attitude may exist when the situation on the labour market favours achieving the economic objectives of the firm.

Table 2 shows that the value levels (Iv2) have different effects on the variable (DF1). In particular, the better the situation in the economic segment of the macroenvironment is assessed, the greater its impact on the agility of the enterprise in its flexibility dimension

Table 2. Kruskal-Wallis and post hoc test.

Dependent: DF1	Value p for multiple comparisons; DF1 Independent variable (grouping): Iv2 Kruskal-Wallis test: H (4, N= 95) =15,40534 p =,0039				
	1	2	3	4	5
1		0,202614	0,016514	0,503593	0,048829
2	0,202614		1,000000	1,000000	0,740330
3	0,016514	1,000000		1,000000	1,000000
4	0,503593	1,000000	1,000000		0,778458
5	0,048829	0,740330	1,000000	0,778458	
Dependent: DB2	Value p for multiple comparisons; DB2 Independent variable (grouping): Iv3 Kruskal-Wallis test: H (4, N= 95) =15,63668 p =,0035				
	1	2	3	4	5
1		0,077393	0,004390	0,767584	0,618088
2	0,077393		0,824796	1,000000	1,000000
3	0,004390	0,824796		1,000000	1,000000
4	0,767584	1,000000	1,000000		1,000000
5	0,618088	1,000000	1,000000	1,000000	

by increasing the automation of operations (DF1). This means that there is a qualitative difference in the influence of levels 1, 3, and 5 of the variable (Iv2) on the variable (DF1). This relationship seems obvious, because the better the economic situation, the more it affects the company's investment decisions.

In the case of variables (Iv3) and (DB2), if changes in the social and demographical segment are rated no worse than neutral, they affect the agility of the enterprise in its brightness dimension by reducing the interest in looking for opportunities by recognizing and analyzing changes taking place in economic segment (DB2). The favourable situation on the labour market (value 3 or more of Iv3) may more radically replace the need to identify macroeconomic factors influencing the performance of the enterprise than value 1 or 2.

The obtained results allow formulating several conclusions.

Since the correlation between the independent and dependent variables is weak and not frequent, it follows that the perception and assessment of changes in the macroenvironment has little impact on the implementation of agility strategy in enterprises. The reason for this may be a relatively poor knowledge of the concept of agile enterprises, which is implemented mainly in the ICT sector in the field of agile software manufacturing.

To a marginal degree, independent variables (Iv) affect brightness (DB) and shrewdness (DS). The weak correlation between (Iv) and (DB) raises the suspicion that the analyzes of the macroenvironment segments and the assessment of the changes taking place in them are sketchy - carried out intuitively and not methodically. Previous studies [11–13], and [14] confirm this. This may also indicate the occurrence of the "dormant" effect, consisting in the fact that if the macroeconomic situation is good, and it was good in the period before the research, less attention is paid to opportunity recognition using both the reactive and proactive model. The greatest influence (Iv) posses on the agility in the flexibility dimension, the effect of which mainly relates to the level of automation of the operations.

The results in this article have some limitations. The research was conducted in January 2020, i.e. before the effects of the Covid 19 pandemic were revealed. It can be expected that the pandemic results in more in-depth analyze of the macroenvironment segments. For this reason, the representativeness of our results is of a historical nature. It is reasonable to suspect that the Covid 19 pandemic will change, relatively permanently, the attitude of managers to methodical research of the business environment. This suspicion indicates the need for future comparative studies.

The weak and sparse correlation between (Iv) and the dimensions of the agility of the enterprise may also indicate poor reliability of the measurement scales of these dimensions. As it results from the Cronbach's alpha coefficient, the reliability of the scale for flexibility adopted by us is poor. This indicates the need for study on symptoms of enterprise's agility and its particular dimensions.

References

1. Tilman, L.M., Jacoby, C.: Agility: How to Navigate the Unknown and Seize Opportunity in a World of Disruption. Missionday (2019)
2. Dove, R.K.: Response Ability: The Language, Structure, and Culture of the Agile Organization. Wiley, Taos (2001)
3. Alberts, D.S.: The Agility. A Survival Guide for Complex Enterprises and Endeavors. CCRP (2011)
4. Kirzner, I.M.: The Alert and Creative Entrepreneur: A Clarification. IFN Working Paper No. 760. Research Institute of Industrial Economics, Stockholm (2008)
5. Schumpeter, J.: The Theory of Economic Development: An inquiry into profits, capital, credit, interest and the business cycle. Harvard University Press, Cambridge (1934)
6. Trzcieliński, S., Trzcielińska, J.: Some elements of theory of opportunities. Hum. Factors Ergon. Manuf. Serv. Ind. **21**(2), 124–131 (2011)
7. Alberts, D.S.: The Agility. A Survival Gude for Complex Enterprises and Endeavors. CCRP (2011)
8. Dove, R.K.: Design Principles for Highly Adaptable Business Systems, With Tangible Manufacturing Examples. In: Maynard's Industrial Handbook, McGraw Hill (1999)
9. Nejatian, M., Zarei, M.H.: Moving towards organizational agility: are we improving in the right direction? Glob. J. Flex. Syst. Manag. **14**(4), 241–253 (2013). https://doi.org/10.1007/s40171-013-0048-3
10. Trzcieliński, S.: Agile Enterprise: Concepts and Some Results of Research. IEA, Madison (2007)
11. Trzcielinski, S.: The influence of knowledge based economy on agility of enterprises. In: Proceedings of the 6th International Conference on Applied Human Factors and Ergonomics (AHFE 2015), pp. 4780–4788. Elsevier (2015)
12. Trzcielinski, S.: Agile enterprise - research on flexibility. In: Proceedings of the 10th International Workshop on Robot Motion and Control, RoMoCo 2015, pp. 213–216. IEEE, Poznań (2015)

13. Trzcielinski, S., Trzcielińska J.: How enterprises identify market opportunities: research results and finding. In: Trzcieliński, S. (ed.) Advances in ergonomics of manufacturing: managing the enterprise of the future. Proceedings of the AHFE 2017 International Conference, pp. 116–128. Springer, Cham (2018)
14. Trzcielinski, S.: Prior knowledge and opportunity recognition. In: Trzcieliński, S., Mrugalska, B., Di Nicolantonio, M., Rossi, E. (eds.) Advances in Manufacturing, Production Management and Process Control. Proceedings of the AHFE 2018 International Conference, pp. 191–200. Springer, Cham (2019)

The Concept of Management by Objectives and Kaizen Implementation on the Basis on Production Intelligent Facility

Joanna Kałkowska(✉) and Hanna Włodarkiewicz-Klimek

Faculty of Engineering Management, Poznan University of Technology, Poznan, Poland
{joanna.kalkowska,hanna.wloadarkiewicz-klimek}@put.poznan.pl

Abstract. At present, enterprises are functioning in dynamically changing environment which from the one hand it forcing into introducing numerous of innovations and on the other hand inducing to achieve the competitive advantage among others through the implementation of contemporary strategies and management methods. It constitutes condition sine qua non not only for adapting the enterprise into appearing changes as well as the quick respond for the market requirements but also to increase enterprises' efficiency. The goal of this paper is to present the conception of implementing Management by Objectives (MBO) and Kaizen methods on the basis on production intelligent facility. According to this, the phases and stages as well as whole methodology of the implementation process of these both methods will be presented. The proposed methodology seems to be universal and can be applicable and modified in a different type of manufacturing enterprise.

Keywords: Management by objectives · Kaizen · Intelligent facility

1 Introduction

Nowadays, the enterprises are searching for competitive advantage concerning quality of products, application of modern information and telecommunication technologies (ICT), implementation of modern concepts and tools of management as well as proper support of decision making process and knowledge management. To achieve this, enterprises need to possess the ability of implementation of dedicated management techniques for continuous improvement to increase efficiency of carried out processes. Regardless of what management system the enterprise introduces, appropriate methods, tools and techniques are needed to introduce and, most importantly, maintain a management system [1]. It is also very important to be aware that we live in a world of dynamic changes and we need to adapt to them. Constant commitment in development, willingness to constantly improve both the quality of the product and the company enforces a strategic approach using, for example, management by objectives and kaizen. The purpose of this paper is to present the conception of implementing Management by Objectives and Kaizen methods on the basis on production intelligent facility.

S. Trzcielinski et al. (Eds.): AHFE 2021, LNNS 274, pp. 428–436, 2021.
https://doi.org/10.1007/978-3-030-80462-6_52

2 Intelligent Facility

Intelligent facility is a relatively autonomic environment with all the necessary functionalities built in. Its principal characteristics are hybrid connections, like: Smart Economy, Smart People, Smart Governance, Smart Mobility, Smart Environment and Smart Living. Such combinations are creating a completely new perspective on building objects as a whole. Intelligent buildings can have various designations. They can be residences, hotels, offices, banks, hospitals, schools, etc. their design, production (construction) and exploitation (use & service) and exploration (improvement) requires expanding the classic engineering from the range of construction, technology and organization. Energy efficiency and thermal isolations are also key characteristics of intelligent buildings, just like sustainable use of natural resources, safety of the user, protection from environmental parameters' levels that can be harmful for human health (noise, air pollution, microclimate, and vibrations), accessibility, low costs of use and service and pro-environmental performance during the entire life cycle of the construction [2]. At present, the rapid changes of technology together with the business globalization is considered as a basic factor of an enterprise development and improvement [3] and it requires the constant implementation of modern management techniques for continuous improvement of such intelligent facility.

3 Management by Objectives (MBO) and Kaizen

The main idea of Management by Objectives (MBO) is to concentrate the management and subordinates on jointly defining and negotiating goals, jointly defining measures of the desired final results and joint periodic reviews and evaluation of the results obtained. It is assumed that the motivation, initiative and activity of employees, especially of the management staff, are among the most valuable and least used resources of an enterprise, which determine its results [4, 5]. Employees who set their own goals or participate in setting them obtain better results than those who are imposed on these goals. Participation in setting goals, as well as consciously jointly defining the measures of their achievement, corresponding to the motivation remuneration system, increases the probability of their effective implementation. In the MBO concept, the main and central importance of the goal, which we can define as a valuable and desired state of affairs which determines the direction and structure of future action. The objectives should meet the following requirements and conditions: (1) The goals of each cell and employee are derived from the goals of the company and create a consistent system within the organization. (2) Objectives are related to strategic areas of activity and key results. (3) The strategic goals of enterprise are disaggregated by division into sub-specific goals of a lower order, until the formulation of goals for each employee (goal tree). (4) Targets must be ranked in order of importance (necessity to assign weights). (5) Goals are flexible (redefining and adapting to environmental conditions). (6) A goal is assigned to a person (team) and includes his individual contribution to achieving the company's goals. (7) Goal setting is consultative. (8) Each goal defined and assigned to the implementer must be formalized (assigned to the responsibilities of the implementer). (9) Each goal has a specific measure of effectiveness (measures of effectiveness may be: financial, qualitative, quantitative, mixed). (10) The goals assigned to the employee must be weighted (the sum of the weights of all goals is 100%). (11) The achievement of goals must be systematically checked. The control process focuses first on self-control, and then on periodic reviews and the results obtained (objective effects of work).

Setting and cascading goals is carried out according to the Balance Score Card technique in the following four perspectives: finance, brand, competences, systems (work procedures, management systems, IT, etc.). The use of the BSC concept in formulating, implementing and measuring goals in an organization enables a logical and multidimensional presentation of the strategy in the form of a set of measurable goals necessary to implement the company's mission [6, 7]. From this perspective, there is a tool that enables the implementation of strategic assumptions. It allows you to translate strategic goals into specific activities, and then cascade the goals into lower organizational units of the company, and then into key employees. The measures for assessing the achievement of goals, determined at each level, along with specific activities allowing for their fulfillment, at the same time complete the entire concept of achieving the strategic goals of the company.

Kaizen is understood as a philosophy which is based on continuous improvement. The assumptions of this concept of work were first presented by Masaki Imai in the book Kaizen - The Kay to Japanese Competitive Success, published in 1986 [8]. In Japanese, Kaizen means continuous improvement, improvement, refinement focused on the involvement of all employees implementing improvement processes in the enterprise (all employees should be a source of ideas for introducing improvements aimed at positive changes) [9]. The Kaizen philosophy assumes that a person's life path (personal and professional) should be focused on continuous improvement. Kaizen is also a tool enabling individual employees to organize and manage their own work. It leads to standardization, which introduces repeatability and order in the work performed. As a result, employees can easily identify problems and pinpoint correct solutions. Kaizen assumes introducing small, gradual improvements, but continuously and usually with high frequency, which in the long run leads to significant and noticeable results [10]. Kaizen focuses on eliminating losses in all processes and is equally about improving devices and improving work procedures. The starting point for the Kaizen philosophy is the appropriate definition of the concept of waste in the enterprise and its limitation in the implementation of processes in the enterprise. Traditionally, waste is understood as a negative financial result of the enterprise, and so, for example, in the case of a production process, waste occurs when the operation of the process does not bring profit. In relation to Kaizen, waste is therefore any activity that, from the customer's point of view, does not add value to the product or service created by the company. Eliminating or limiting all waste and its causes also improves the use of the company's potential. However, it will not be possible without identifying and understanding the processes implemented in the enterprise, maintaining constant awareness and searching for ways to improve work performance [11]. According to Hamel, Kaizen philosophy can be synthetically presented using the eleven principles: (1) Implement the thinking in line with the PDCA and SDCA cycles. (2) One should go to the place where the activities of a given process are performed in order to observe and document reality. (3) 5xwhy should be used to identify the root causes. (4) Never be content with the status quo. (5) What is essential must be improved. (6) Be action-oriented. (7) You should improve with the method of small, frequent steps which results in large and lasting changes for the better. (8) Use creativity, not money. (9) Kaizen is the task of all members of the organization. (10) There is no transformation without leadership transformation. (11) Whatever you do, do it with humility and respect for the individual [11].

4 Management by Objectives – Implementation Concept

Both concepts Management by Objectives and Kaizen can be fully implemented in the intelligent facility of construction branch. Concerning Management by Objectives the following organizational functions will be supported by the MBO method in the selected enterprise: (1) Shaping and implementing the Strategic Scorecard, (2) Shaping the tasks organizational units in the field of maintaining databases on business activity measures, (3) Shaping the system of business activity measures, (4) Training employees in the field of strategic goals and their operationalization through the business measurement system, (5) Determining the values of measures in the adopted planning perspectives, (6) Ongoing monitoring and recording of achieved values of measures, (7) Ongoing preparation of analyzes business activities in numerical and graphic terms for the Management Board and Service Managers, (8) Motivating employees. The expected results that the enterprise will achieve with the use of MBO method can be distinguished as following: (1) Transparent view of the activity of the enterprise, its units in the totality of goals and methods of achieving them. (2) Logical division of goals in the organization, which connects them with measures and activities defined in their implementation. (3) It is a useful tool in controlling the achievement of strategic goals of the company. (4) It can be used throughout the company through organizational unit achievement cards as well as achievement cards for individual employees, which allows synchronization of activities of all cells and individual employees in order to achieve strategic goals of the company. (5) It has a positive effect on employees, motivating them to act effectively.

The essence and logic of proceeding in the MBO method in connection with supporting the organizational functions can be implemented in four following stages:

(1) The members of the Management Board defines a maximum of 10 strategic goals, indicating at the same time: the main executor of the goal - a person responsible for the coordination and effects of achieving goals, organizational unit of the main goal executor, organizational units jointly responsible for achieving the goal, expected date of goal completion, the expected level of goal achievement expressed in the form of a measure (the measure should be expressed numerically or descriptively, the measure is scalable: expected level, satisfactory level, unacceptable level).
(2) The main executor of the strategic goal decomposes it into goals in four implementation perspectives, and at the same time indicates the cause-effect relationships between individual goals. Goals F1, F2 - goals in financial areas, K1, K2 in customer areas, etc. The concept of decomposition of strategic goals in perspectives is presented in Fig. 1.

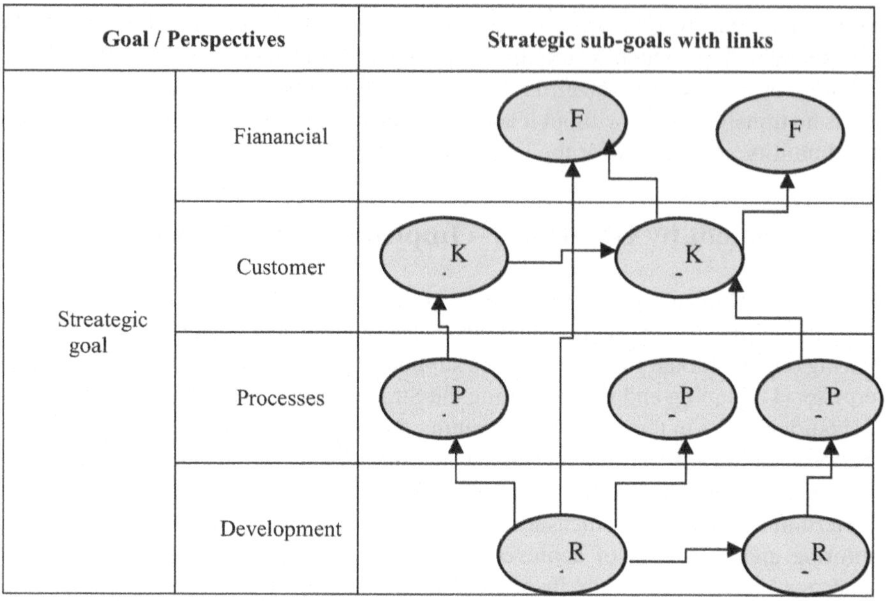

Fig. 1. Decomposition of strategic goals, prospects for implementation (financial, customer, processes and development), source: own study

(3) For partial strategic goals in four perspectives, the person responsible for the main goal determines strategic actions/skills (STEP 1), which will allow for full implementation of the set goal. Then, operational goals are determined for organizational units and then for key employees from a given unit (STEP 2), which allow the achievement of the main goal and activities describing the achievement of the main goal. Then, the manager of the organizational unit and the person assigned to achieve the sub-goal determines the deadline for achieving the goal (STEP 3) and measure (STEP 4). The designated deadline and measure must be approved by the person responsible for the main goal, which includes the sub-goal (STEP 5). The concept of the implementation of individual stages in the MBO concept using the strategic scorecard is presented in Table 1.

(4) The next stage of MBO is the combination of setting and measuring strategic and operational goals with the individual evaluation of the employee and determining the financial reward (financial bonus, other bonuses).

Table 1. Stages of implementation of activities in the MBO concept using the strategic scorecard

The way of interpreting the goal described in point (1)	Goal decomposition point (2)	STEP 1	STEP 2	STEP 3	STEP 4	STEP 5
Main strategic goals	Strategic goals in particular perspectives	Strategic activities	Operational goals	Deadline	Measure	Acceptance of the person responsible for the main goal
1… (we repeat the presented layout for each main strategic goal)	*financial* • …	• … • …	• … • …			
	customer • …	• … • …	• … • …			
	processess • …	• … • …	• … • …			
	development • …	• … • …	• … • …			

Source: own study

5 Kaizen – Implementation Concept

As well as in the case of Management by Objectives method, Kaizen will support the following organizational functions in the same enterprise: (1) Human Resources management, (2) Employee evaluation, (3) Motivating employees, (4) Employee development (including training, career paths). The use of the Kaizen approach will enable the stimulation of innovation, proper protection of organizational resources, cost reduction, quality improvement, increase in work efficiency, employee development, improvement of occupational health and safety as well as improvement of work processes in the enterprise.

The essence and logic of proceeding in the Kaizen method in connection with supporting the organizational functions includes mainly the individual idea improvement generation as well as assessment of the improvement generated by employee. Improvement ideas should be generated in dedicated IT system by filling application form of improvement. For the effective implementation of the Kaizen program, a Kaizen team and a Kaizen Leader should be established in the company. The team proposes the amount of cash prizes for the implemented improvement. The team should define the frequency of reporting improvements for individual organizational units/workstations (e.g. 1 improvement/month). For reported and positively verified but not implemented improvements, can be received points that add up in a given settlement period, which results in a cash prize.

The concept of an individual Kaizen supported by an IT system should: (1) Automatically remind/force the introduction of a new improvement in a specific period of time,

(2) Allow viewing of all reported improvements in the company, (3) Provide the user with monitoring of reported improvements, showing the status of the improvements, the decision to accept or reject the idea, (4) Force control of the implemented improvements by indicating/verifying the benefits of the improvement in a specific period of time.

The concept of the application form and the implementation of an individual Kaizen is presented in Table 2.

Table 2. The concept of Kaizen application form

Application form improvement				
3Name/title of idea				
Organizational unit submitting the idea				
Area of improvement/Branch				
Idea description				
Description of the proposed solution to the problem and the expected results				
Verification of the results achieved after implementation				
Submitted idea assessment form - evaluates the team appointed by the Kaizen Leader				
Evaluation of an idea due to its importance to the company (scale from 1 to 5)				
1	2	3	4	5
Assessment of the possibility of implementing the idea (scale from 1 to 5)				
1	2	3	4	5
A financial possibility to implement the idea (scale from 1 to 5)				
1	2	3	4	5
A technology possibility to implement the idea (scale from 1 to 5)				
1	2	3	4	5
An organizational possibility to implement the idea (scale from 1 to 5)				
1	2	3	4	5
Expected benefits				
Financial benefits (scale from 1 to 5)				
1	2	3	4	5
Technology benefits (scale from 1 to 5)				
1	2	3	4	5
Organizational benefits (scale from 1 to 5)				
1	2	3	4	5
Decision concerning idea implementation				
YES		NO		

Source: own study

6 Closing Remarks

At present, the rapid changes of technology together with the business globalization is considered as a basic factor of an enterprise development and improvement and it requires the constant implementation of modern management techniques. They are aimed at increasing the efficiency and effectiveness of processes occurring in a given enterprise. There are many different methods of organization and management that allow to effectively run a business. Certainly, every decision to introduce changes has a number of consequences. For example, when introducing Kaizen to the company, it should be remembered that it is not a simple method that can be introduced routinely. It is a concept that integrates many aspects of the company's operation. Its proper implementation requires the involvement of top management and all employees. Employees have to look for new solutions and improvements every day, keep their workplace in order and avoid waste. Many Japanese companies have achieved global success using Kaizen management, so European companies should follow their example. Kaizen allows enterprises to meet the requirements of producing products and services of the highest quality, low cost and exactly on time. The whole process should take place with the support of the highest levels of management, whose task is to create an appropriate work culture conducive to continuous development. Such conditions are conducive to the introduction of Kaizen in the described enterprise. Management by Objectives (MBO) and the Balanced Scorecard are management systems that align objectives with an organisation's vision. The main idea of introducing MBO in intelligent facility is to ensure that the employees get a clear understanding of their roles and responsibilities for better understanding the relation of their activities to the overall success of the enterprise. The great advantage of using these methods together is the direction of the enterprise's development activities, cost reduction and increasing of work efficiency. As the company belongs to construction branch, the use of these methods seems to be justified.

Acknowledgment. The research was carried out as part of the research project of the Polish National Research and Development Center (POIR.01.01.01–00-1229/17) for the research and teaching needs of the Faculty of Engineering Management at the Poznan University of Technology.

References

1. Wlodarkiewicz-Klimek, H.: Agility of knowledge-based organizations. In: Proceedings of the AHFE 2016 International Conference on The Human Aspects of Advanced Manufacturing (2016)
2. Kałkowska, J., Pacholski, L.: The conditioning of macroergonomic design and improvement of the intelligent building's use and service. In: Advances in Social and Occupational Ergonomics:/red. Richards H.M. Goossens, Atsuo Murata: Springer, pp. 165–172 (2020)
3. Trzcieliński, S.: Changeability of environment – enemy or ally? Management sci-ence in transition period in South Africa and Poland. Cracow University of Econ-omy, Cracow-Stellenbosch (2013)
4. Drucker, P.: Management by objectives: SMART (1996). http:// E:\Management by Objectives. (Drucker)_ SMART.htm

5. Rodgers, R., Hunter, J.E.: Impact of management by objectives on organizational productivity. J. Appl. Psychol. **76**(2), 322–336 (1991)
6. Kaplan, R.S., Norton, D.P.: The balanced scorecard—measures that drive performance. Harvard Business Review, pp. 71–79 (1992)
7. Kaplan, R.S., Norton, D.P.: The Balanced Scorecard. Harvard Business School Press, Boston (1996)
8. Imai, M.: KAIZEN, The Key to Japan's Competitive Success. McGraw-Hill, Inc., United State of America (1986)
9. Imran A., K.: KAIZEN: The Japanese strategy for continuous improvement. In: VSRS International Journal of Business & Management Research, vol. 1 (3), 2011, pp. 177–184 (2011)
10. Recht, R., Wilderom, C.: Kaizen and culture: on the transferability of Japanese suggestion systems. Int. Bus. Rev. **7**(1), 7–22 (1998)
11. Hamel, M.R.: Warsztaty Kaizen: praktyczny poradnik, jak prowadzić skuteczne warsztaty doskonalenia procesów. Wydawnictwo Lean Enterprise Institute Polska, Wrocław (2014)

Productivity Improvement of a Footwear Manufacturing Company Through Lean Tools

Jose Castillo-Castañeda[1], Brian Tarazona-Lopez[1], Claudia Leon-Chavarri[1], and Luis Cardenas[2](✉)

[1] Ingenieria Industrial, Universidad Peruana de Ciencias Aplicadas, Prolongación Primavera 2390, Lima 15023, Peru
{u201316389,u201313030,pcincleo}@upc.edu.pe
[2] Direccion de Investigacion, Universidad Peruana de Ciencias Aplicadas, Prolongación Primavera 2390, Lima 15023, Peru
luis.cardenas@upc.edu.pe

Abstract. This research focuses on the application of the primary tools in industrial engineering with regards to companies and sectors within a country. This analysis focuses on small and medium enterprises in the footwear sector. Small companies currently form a nucleus in the world economy, and thus, research sources include companies in the footwear sector in Lima, Perú. With the aim of continuous improvement, the objective of this research is to implement and control the processes used by the sector through Lean methodology with the following indicators: optimization of the effective time of operators by 13%, production increment by 20%, and finally reduction in reprocessing by 63%. The impact on work is based on changes in machines, stations, functions, tools, and workshop designs to increase productivity in the footwear industry by 23.62%. Therefore, applying operations as tools is essential to effectively analyze and control production.

Keywords: Footwear · Productivity improvement · SMEs · Kaizen · 5s

1 Introduction

The production of small- and medium-sized enterprises (SMEs) the footwear sector in Perú has increase by 5.7% in recent times (2017). In April 2018, the production in the manufacturing sector increased by 20.33%, which was a positive result, and thus, Perú now ranks fourth in footwear production in South America.

According to the National Society of Industries (SNI), by the end of 2018, 3,669 companies were engaged in footwear manufacturing in Perú. These companies comprise the following:

- Micro companies having a production capacity of approximately 40 pairs per day and representing 88% of all companies in this sector. They constitute 24% of production.
- Small companies having a production capacity of approximately 250 pairs per day and manufacturing 36% of the total production.

S. Trzcielinski et al. (Eds.): AHFE 2021, LNNS 274, pp. 437–445, 2021.
https://doi.org/10.1007/978-3-030-80462-6_53

- Medium-sized companies having a production capacity of approximately 700 pairs per day and constituting 40% of national production.

The SMEs form consortiums to improved performance in productivity to ensure supply for the corresponding demand when required and subsequently fulfill client requirements [1–3].

Therefore, to implement any type of lean tool, identifying the event under study and the qualitative and quantitative analyses of the root cause of the problem should be a priority because the most significant characteristic of the lean philosophy is its application in companies of varying sizes (i.e., small, medium, and large) in addition to the area and/or sector in which it is focused and developed. [4, 5].

Innovative approaches to be studied, applied, and implemented must be effective and easy to apply to ensure its standardization and use in the study environment (footwear companies here), and a new culture of applicability can be created. Consequently, the improvement proposal to be implemented is a combination of different engineering tools (i.e., 5S, VSM, and Kaizen) because as observed in other countries, the use of the tools in question, together with the primary causes identified, is feasible. The use of these tools has yielded positive results that are reflected by their economic impact on the company, in addition to increased productivity, which has an initial technical gap of 0.417 (Par/HH). The application of the tools yielded an increase of 21% with a value of 0.505 (Par/HH). Therefore, lean thinking and lean manufacturing aim to eliminate waste, which is enhanced through the optimization of resources. The primary motivation of this research is to capture the results obtained over a course of the time required to conduct the project in consideration with the primary causes of the problem that SMEs in the footwear sector encounter.

2 State of the Art

Lean philosophy aims to optimize resources with the available tools and to ensure the maximum use of resources for obtaining long-term results. Therefore, a sample analysis jointly conducted by various researchers recommends the following 17 practices for the use of lean tools in SMEs [6–8].

Therefore, when lean tools are used in production processes, it must be one of the philosophies always used by engineers when proposing continuous improvement. Such improvements are registered by the process and operators in such a manner that the operators can immediately make decisions and subsequently save time in the process [9–11].

When applying this philosophy to the process, workstations contributing the greatest delay and impacting the productivity of the company should be identified; the identified station is equipped and stopped with a total of 3 H-H [12].

The 5S tool in the lean manufacturing philosophy is responsible for reorganizing the process and optimizing the used resources. Thus, it has five phases, with the organization and cleaning stage being the most influential phase because being organized and having set points for each workstation are crucial. With the application of the 5S tool at workstations, a reduction gap of 61.97% was yielded, thereby validating the high impact of the implementation [13, 14].

The Kaizen tool aims to constantly improve production by eliminating waste from processes. The Kaizen method is a management system oriented to the continuous improvement of processes to subsequently eradicate all those inefficiencies of a production system. The application of this Japanese technique under high rates of defective products has helped reduce defective products by 33.3% [15, 16].

The order of the use of lean tools allows the process to be divided by sections to ensure that the development in the plant and the performance of each operator are evaluated.

3 Contribution

3.1 Case Study

The starting point is to develop a perspective of the economic sector directly influencing the country's gross domestic product, through which the opportunity for improvement as well as the causes with a direct influence and repercussions on the country's economy can be identified. The footwear sector, which represents 30.8% of the Peruvian economy, was considered for this case study. An analysis revealed that in Perú, SMEs represent 95.4% of the footwear manufacturing industry, and thus, an efficient solution was sought.

For impactful results, an established process must be proposed because SMEs typically cannot provide the necessary resources, such as budget, workforce, time, and the knowledge of experts, when implementing and adhering to lean practices.

SMEs represent more than 90% of the production industries in countries, such as Perú, Brazil, and Mexico (Fig. 1).

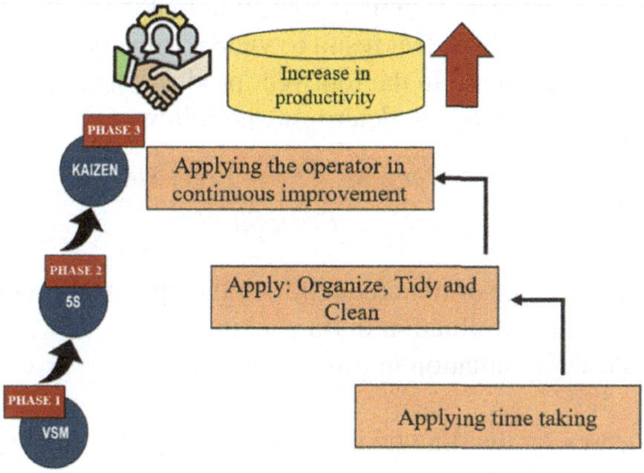

Fig. 1. Proposal

After certain studies of similar casuistry in other countries, of which the center and/or nucleus are SMEs in the footwear manufacturing industry, the lean tools that are to be used and show effectiveness in medium and large companies. The novel approach of this research is to apply them to small companies. The primary reason why SMEs in the footwear manufacturing sector are always unstable is their low productivity because of various factors. However, an exhaustive analysis revealed the following three primary causes with the greatest impact: Deficiency in work methodology, lack of quality control in manufacturing shoes, and delay in production because of a lack of product standardization, which constituted for a total of 38.64%, 31.82%, and 29.55%, respectively. The study and analysis further assisted in identifying the technical gap representing low productivity, which represents 0.417 pair/HH in the SMEs in the footwear manufacturing sector. The approach used in this research aims to increase it by more than 21% because the industrial conditions allow it.

The technical analysis gap represents a loss of S /. 47 016 soles per year, which can be increased to more than 80% using the correct tools.

The purpose of implementation is to increase the primary indicator, which is reflected in the economic situation of MSEs of the footwear manufacturing sector.

The engineering indicators that evaluated the primary causes in the first instance as follows:

- Operators optimization indicator:

$$\%\,Red.\,Ope. = \frac{Sub.\,Sit. - Prev.\,Sit}{Prev.\,Sit.} \times 100 \tag{1}$$

  - % Red. Ope: Percentage reduction of operators in the process
  - Sub. Sit.: Subsequent situation in the number of operators
  - Prev- Sit.: Previous situation in the number of operators

  This first indicator essentially indicates by how much the number of operators can be reduced through time-taking and simultaneous time-making comparisons and then by analyzing a previous and subsequent result to yield a value in terms of the percentage reduction of personnel required in the footwear production process.
- The second indicator is intended to determine the following:

$$\%\,Red.\,Time. = \frac{Sub.\,Sit.\,(T) - Prev.\,Sit\,(T)}{Prev.\,Sit\,(T)} \times 100 \tag{2}$$

  - % Red. Time: Percentage of time reduction in the operators production process.
  - Sub. Sit. (T): Subsequent situation in time taking.
  - Prev. Sit. (T): Previous situation in time taking.

  This indicator focuses on the time taking before and after applying the lean tools in such a manner that the changes being made can be measured.
- Reprocess percentage indicator

$$\%\,R = \frac{R.\,Pairs\,(Month\;\;number)}{P.\,Pairs\,(Month\;\;number)} \tag{3}$$

  - %R: Reprocess percentage
  - R. Pairs: Reprocess pairs
  - P. Pairs.: Pairs produced

- N: Number of the month under evaluation.

This indicator aims to control the amount of reprocessing that occurs in the operation to monitor the amounts of reprocessing being conducted in the company.

These indicators are based on the measurement of the primary causes, and thus, determining that to obtain these indicators, the use of the aforementioned lean tools is necessary with consistency and guidelines on the order and way of measurement are followed to ensure that the results obtained in the urban footwear production process can be observed and analyzed [17–19].

4 Validation

4.1 Diagnostic Value Stream Mapping

To implement the tools, a previous diagnosis is first required to evaluate the production performance and establish the cycle time rhythms; a work plan can be subsequently developed.

With the analysis on low productivity, resulting in the technical gap of 0.417 torque/HH, determining that the stations requiring the most time for task completion are as follows: the stacking station and finishing station requiring 3 h each.

Analysis of Time: During time analysis, the Westinghouse table was used, whose function is to apply highly repetitive operations during the shoemaking process.

The following factors were considered for the analysis:

The company manufactures more than 1,000 pairs of footwear annually. The manufacturing of a pair of shoes involves five processes (i.e., cut, fit, dress, finish, and assemble). Manufacturing time is 67.05 min. The operation is then categorized based on the five processes, and a value of 13.5 min is obtained. It is then converted into hours, that is, 0.225 h. The Westinghouse table is reviewed and five pilot observations are presented based on the results.

After obtaining the results of the pilots, the cycle analysis was conducted to determine the percentage difference. The times will be divided based on the results; that is, red indicates that the expected results are not obtained, whereas blue indicates that the work is optimal.

However, before implementing this approach into practice, the two missing tools (i.e., 5S and Kaizen) must be used because the combination allows the pilot plan to be performed optimally. The use of these tools will further broaden the panorama of analysis in low productivity in the MSEs in the footwear manufacturing sector.

4.2 5S Implementation

The 5S tool was implemented pre and post audit with the following evaluation criteria (Table 1):

The data in the tables reveal that the trends before and after the implementation can be observed, and the trends of each situation and/or criterion can be analyzed using the radar graph.

Table 1. Comparison of situations of the medium and small footwear sector

Wasteful audits	Initial situation	Subsequent situation	%Improvement
Result	33	88	63%

Waste Audit: In addition to detailing the required times, a waste audit was performed for further reinforcement because productivity lies in the amount of waste obtained during the process.

During the implementation, the waste control system in the production workshop was reduced by 58.93%. This checklist must be used for supervision by each operator on a monthly basis (Table 2).

Table 2. Comparison of the wasteful situations of the MSE footwear sector

Wasteful audits	Initial situation	Subsequent situation	%Improvement
Result	59	28	53%

Positive results were obtained, since the improvement made was of 53%, decreasing by more than 50% the waiting time for the item.

4.3 Kaizen Implementation

This phase aims to achieve a continuous improvement of the process to standardize the operational levels of quality in the products and the production of defective products.

To complete the implementation with the Kaizen tool, the use and awareness of continuous improvement by the operators were applied, and thus, a record implementation was performed, as shown in Fig. 2.

Continuous Improvement Format		Code:	XX_MC_2020
		Date:	-- / -- / --
Surnames and Names:			
Work Station:			
1. What observation was found at the work station?			
A. Plan (Establish ideas)		B. Do (Put to test the idea)	
C. Check (Identify ideas)		D. Act (Substitute idea)	
2. Suggestions:			

Fig. 2. Continuous improvement format

With the implementation format, workstations were able to reduce their lag times:

- Reduction of the time at the cutting station by 17.91%.
- Reduction of the time at the fitting station by 35%.
- Reduction of the time at the dressing station by 10.41%.
- Reduction of the time at the assembling station by 14.5%.
- Reduction of the time at the finishing station by 18.78%.

What Kaizen seeks is for operators to have the ability to observe opportunities for improvement and thus make decisions to ensure that the process and production can be continued without any bottlenecks, thus achieving process standardization [19].

4.4 Pilot Implementation

Fourth Improvement: In the fourth improvement, the results in the cutting and pairing workstations exhibited a reduction of 0.33% and 1.53%, respectively, that is, the expected results are achieved. With respect to the other workstations, a decrement can be noted; therefore, in the following improvement, the complete expected results may be obtained (Table 3).

Table 3. Pilot improvement 4

Process	Time (minutes)		Percentage difference
	Real	Proposed	
Cut	9	9.03	−0.33%
Fit	5.8	5.2	10.34%
Dress	15	15.23	−1.53%
Assemble	20	18.79	6.05%
Finish	8	7.31	8.63%
Total	57.8	55.56	3.88%

Fifth Improvement: In the fifth improvement, it can be seen that workstations reveal significant decreases, and thus, the percentage difference of each workstation has a negative result. Finally, a comparison using the table of expected results reveals that the anticipated results were achieved. Thus, the implementation enabled the combination of tools to reduce the times proposed, which subsequently enhanced production as is reflected in productivity (Table 4).

Table 4. Production time

Process	Time(minutes)		Percentage difference
	Real	Proposed	
Cut	8.5	9.03	−6.24
Fit	4.78	5.2	−8.79%
Dress	15.1	15.23	−0.86%
Assemble	18.63	18.79	−0.86%
Finish	6.5	7.31	−12.46%
Total	53.51	55.56	−3.83%

5 Conclusions

According to percentage indicator reduction of operators, a reduction of 13% was obtained; that is, the necessary number of operators for the production of footwear is reduced by one operator. Thus, the result was positive because the cost of the process is optimized.

Regarding the second percentage indicator, reduction in production time, an optimization of 17% was achieved, with a significant positive impact on the process because optimizing the process with few resources is possible.

Finally, regarding the third percentage indicator, reprocessing percentage, an improvement of more than 48% was achieved, an thus, an improvement was noted in delivery compliance compared with the amount estimated at the beginning in the urban footwear production process.

References

1. DIARIO PERU (2018). [En linea]. https://peru.info/espe/comercio-exterior/noticias/7/31/-como-esta-el-sector-manufacturaen-el-peru
2. Miñáz, W.: "Producción de calzado cae a tasas de dos dígitos por noveno mes", Diario Gestión, p. 18, 19 March 2019
3. "Perú produce más de 50 millones de pares de calzado de cuero al año y eso atrae a Brasil". Diario Gestión, 8 June 2017. Recuperado de https://gestion.pe/economia/peru-produce-50-millones-pares-calzado-cuero-ano-atrae-brasil-136833
4. Majava, J., Ojanperä, T.: Lean production development in SMEs: a case study. Manag. Prod. Eng. Rev. **8**(2), 41–48 (2017). https://doi.org/10.1515/mper-2017-0016
5. Teli, S.N.: Productivity Improvement by Value Stream Mapping, pp. 2349–5162 (2015)
6. Heese, H.S.: Inventory record inaccuracy, double marginalization, and RFID adoption. Prod. Oper. Manag. **16**(5), 542–553 (2009)
7. Venkataraman, K., Vijaya Ramnath, B., Muthu Kumar, V., Elanchezhian, C.: Application of value stream mapping for reduction of cycle time in a machining process. Procedia Mater. Sci. **6**, 1187–1196 (2014). https://doi.org/10.1016/j.mspro.2014.07.192
8. Rose, A.M.N.: Lean manufacturing best practices in SMEs, Kuala Lumpur, Malaysia, pp. 22–24, January 2011

9. Melton, T.: The benefits of lean manufacturing: whatlean thinking has to offer the process industries. Chem. Eng. Res. Design **83**(6), 662–673 (2005)
10. Suhardi, B.: Implementation of standardization work to improve productivity in indonesian furniture industry (2016)
11. Aulet, W.: A tale of two entre-preneurs: understanding differences in thetypes of entrepreneurship in the economy (2013)
12. Chaman, G., Gahagan, S.M.: Adding value to value stream mapping: a simulation model template for VSM, Institute of Industrial Engineers (2012)
13. Tinoco, O.: Application of the 5S to improve the perception of quality culture in textile manufacturing microenterprises in the Northern Cone of Lima (2016)
14. Gupta, C.: Can application of the 5S concept to organize the workplace in a scientific instrument manufacturing company (2015)
15. Abu, S.: Court shoe production line: improvement of process cycle efficiency by using lean tools (2017)
16. Gordian, B.: KAIZEN as a Strategy for Improving SSMEs performance: assessing its acceptability and feasibility in Tanzania (2014)
17. Ahmad, M.: Kanban in software development: a systematic literature review. In: Proceedings of the 2013 39th Euromicro Conference on Software Engineering and Advanced Applications, pp. 9–16 (2013)
18. Chaple, A.: Status of implementation of Lean manufacturing principles in the context of Indian industry (2014)
19. Nazri, M.: Facility layout for SME food industry via value stream mapping and simulation. Procedia Econ. Financ. **31**, 797–802 (2015)
20. Sivananda Devi, K., Arunachalam, V.P., Gunasekaran, N.: Lean manufacturing concepts in wet grinder assembly line through value state mapping. Int. J. Serv. Oper. Manag. **30**(3), 357 (2018). https://doi.org/10.1504/IJSOM.2018.092608

Warehouse Management Model Under the Lean Warehousing Philosophy to Reduce Product Returns in the Marketer

María Isidro[1], Angel Susaya[1], Gino Viacava[1], and Wilder Namay[2(✉)]

[1] Ingenieria Industrial, Universidad Peruana de Ciencias Aplicadas, Prolongación Primavera 2390, Lima 15023, Peru
{u201520154,u201520159,pcingvia}@upc.edu.pe

[2] Direccion de Investigacion, Universidad Peruana de Ciencias Aplicadas, Prolongación Primavera 2390, Lima 15023, Peru
wilder.namay@upc.edu.pe

Abstract. In this work, the high rate of return of orders is studied in mypes and medium import companies in Peru. For this, two papers were taken into account for the design of our warehouse management model under the lean warehousing philosophy with three stages called "creating stability", "creating flow" and "making flow". These are intended to ensure quality of the processes, increase the productivity of the warehouse and the operators and improve the reliability of the inventories respectively. As a result of the implementation, a reduction of the picking time from 3.5 to 1.5 h was obtained, an increase in the reliability of inventories of a 75 to 95% (ERI), an increase in operator performance from 50 to 85% and a reduction of expired products from 19 to 2%. In this way, the rejection level was reduced by one 14.1 to 7% in the company under study.

Keywords: Returns · Lean warehousing · Management of warehouses

1 Introduction

One of the most recurrent problems in trading companies worldwide is the product return rate [1]. In recent years, growth in returns ranges from $ 520.2 to $ 642.6 billion worldwide. In some countries, such as Peru, this problem in medium-sized companies has an average rejection percentage of 12.45% and a minimum of 5% in leading companies in the sector [2]. Currently, trading companies face significant operational and logistical challenges due to the increase in returns, since extra resources such as personnel and warehouse space have to be allocated to receive the merchandise and identify whether

S. Trzcielinski et al. (Eds.): AHFE 2021, LNNS 274, pp. 446–454, 2021.
https://doi.org/10.1007/978-3-030-80462-6_54

the items should be resold or discarded, thus incurring financial losses [3]. This problem is generally due to the absence of inventory control, inadequate distribution of spaces in the warehouse, poor order preparation process, delivery of products ready to expire and inefficient planning of the supply system [2].

As there is a high rate of return of products in medium-sized marketers, the present project aims to optimize the preparation time of the orders, since this process depends on the products being delivered to customers in a satisfactory manner. In various studies and proposals focused on solving the problem, different combinations of Lean Warehousing tools are developed. Some of the most relevant studies mention the following. For [4] the use of ABC multicriteria analysis helps to determine the best combination of products and thus establish an adequate distribution and a correct inventory count: In [5] it is verified that the method is highly precise, since 95.4% of articles were classified according to the proposed level of criticality. On the other hand, in [6] the inventory precision index was improved, as it increased from 67% to 91% and showed that using cyclical inventory can control the stock, in addition, inventory costs had a significant reduction, of annual losses, from 11% to 3%. Another study details that the application of the SLP managed to reduce the cycle time by 40.8% and the average travel distance by 5.5% [7].

2 State of the Art

2.1 First Expired First Out (FEFO)

FEFO is a storage policy which establishes that products are shipped in order to their expiration date regardless of the date of entry or acquisition. By this method, logistics controls are more efficient with respect to the useful life of the items. This results in a decrease in product loss and out-of-stock levels, while increasing product quality confidence and customer satisfaction [8].

2.2 Process Standardization

Standardization of Procedures (SOP) is an ideal tool to increase operator performance, improve operational efficiency, and prevent human error. It also contributes to improving the productivity of operations.

This tool details the work processes to be carried out in an orderly manner and in a sequence of activities. Previously, it must have been identified that processes do not add value in order to eliminate or mitigate them [9].

2.3 Value Stream Mapping (VSM)

The value stream map is a visual diagnostic tool, which is used to identify those activities without added value [10].

This tool is divided into a series of systematic steps, these are: Identify the product family, draw the current state of the process, identifying the inventories between operations, material flow and information; at this stage the initial VSM is drawn. The next stage is to analyze the vision of how the future state should be, at this stage the takt time and lead time must be calculated. In the penultimate stage, the future VSM is drawn. Then, to finish, an action plan must be drawn up and implemented. It is important to mention that the VSM must identify bottlenecks, identify where products are wasted and where resources are lost [11].

2.4 Systematic Layout Planning SLP

Systematic design planning is a lean tool used for warehouse restructuring. The SLP will have a direct and indirect effect on the efficiency of the circulation flow of the products, it will affect the times of reception and preparation of the merchandise, as well as influence the efficiency of the operators. Consequently, this will have a significant influence on the level of service of the company.

According to [12] this technique is used to determine the specific locations of the reception and shipping areas, storage areas, office, and equipment areas in the warehouse based on the prioritization of the proximity relationship.

2.5 ABC Multicriteria Classification

The ABC multicriteria classification is a tool which aims to classify the merchandise according to various criteria such as rotation level, lead time, cubic space, expiration, among others. This varies according to the characteristics of the product and the policies that the business manages [13].

This tool allows better inventory control, having as main objectives: Improving the allocation of resources, knowing the real value of your assets, identifying the right time to restock the merchandise.

3 Input

3.1 Proposed Model

The model is divided into three stages. The first is called "creating stability", which aims to improve and ensure the quality of the processes. For this, lean tools such as FEFO visual management and process standardization are used. The second stage is "create flow", which seeks to increase the productivity of the operators. For this, systematic design planning or PFS and ABC multicriteria analysis are used. Finally, the third stage is called "making the flow", which focuses on improving the reliability of inventories. For this, the cyclical inventory is used as a tool. It is worth mentioning that there is a previous step that is to carry out a VSM to identify possible problems in the warehouse. The proposed model is observed in detail in Fig. 1.

3.2 Detail

Next, the detail of each tool to be used by each stage is presented.

Stage 1: Create Stability

The basis of this stage is to improve the quality and operation of the process of products and workers. The following lean tools to apply are:

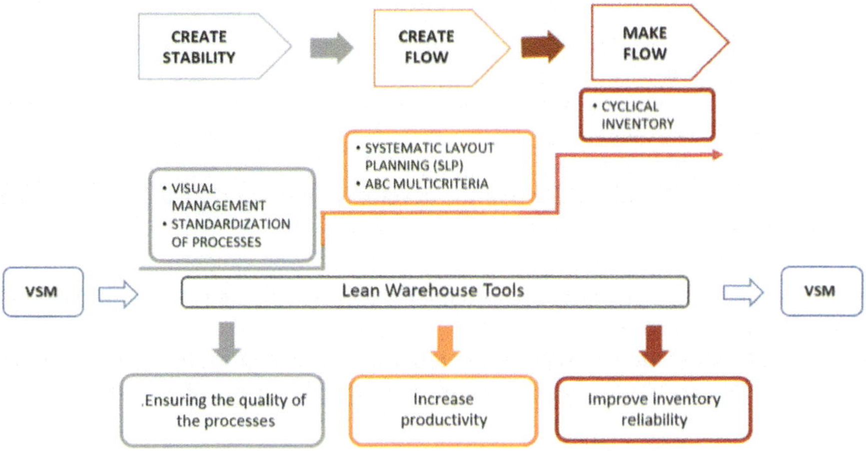

Fig. 1. Model proposal

Standardization of Processes: In this tool it will be divided into 5 steps. First, a study of critical processes will be carried out using a value-added analysis matrix. In it, activities that do not add value to the product will be identified in order to subsequently eliminate them. Then, we proceed to develop a new process that seeks to reduce times and increase efficiency.

Subsequently, the procedures manuals will be made where the activities for each process, the objectives of each of them and who will be responsible for their compliance will be established. Finally, a bimonthly audit of the process will be carried out using productivity and compliance indicators.

Visual Management FEFO: For a correct implementation of this tool. First, the expiration intervals of the products are identified. Then an identification card must be designed (Table 2) using the traffic light method, taking into account the red color for products that are about to expire and prioritize their dispatch, the amber products must have a constant review and avoid expiring and for green products the decision is that they should only be dispatched if there are no red and yellow items. Subsequently, work groups must be formed and trained. Finally, a schedule will be drawn up where you will identify those responsible for reviewing the merchandise and performing the new product categorization. This review will be carried out every two months at the beginning of each month (Fig. 2).

Produced by:			
PRODUCT SPECIFICATIONS	MARK THE COLOR		
Subfamily:	4 to 6 months	7 to 12 months	12 months or more
Product:			
Lot:			
Quantity:			
Date:			
Comments or observations:			

Fig. 2. Identification card.

Stage 2: Create Flow

The first step in the "create flow" stage is to ensure that the products are distributed in an efficient warehouse layout.

Systematic Layout Planning SLP: For the development of this tool the following steps were carried out.

- Analysis of the flow of materials: Evaluation of the path of materials within the process (use of the effort matrix).
- Analysis of the relationships between activities: This step evaluates the communication and flow between areas, defining who should be together and who should not.
- Designate spaces: Identify and calculate the total area of each department.
- Alternative design: The redistribution of the products is carried out based on the results of the ABC multicriteria analysis.

Stage 3: Make it Flow

This stage seeks to make the system flow, where not only 3Ms are minimized or eliminated (muda, mura, muri), but also a process flow that adapts to the operators and owners. The tool to use at this stage will be:

Cyclical Inventory: To carry out this tool, first, the level of inventory accuracy must be determined. This value is 72%. That is, there are 28% differences between physical and virtual stock. Then the merchandise is categorized using the ABC classification. Subsequently, criteria and counting frequencies must be defined. For this, it was established that products "A", "B" and "C" will have a monthly, bimonthly and quarterly counting frequency respectively. Likewise, 30.51 and 74 units will be taken as samples for the count in the order mentioned. Finally, the working groups will be held and the respective counts will be made.

4 Validation

4.1 Process Standardization

For the validation of this tool, cases of studies similar to the context of the present investigation were identified (Table 1).

Table 1. Increase in operator performance.

Indicator (5)	First Operator	Second Operator
Operator performance (5)	30%	17%

In the first case study, an improvement in the performance of the first operator of 30% and of the second by 17% was identified.

Table 2. Increase in operator performance.

Indicator (6)	Operator
Operator performance	14.7%

For the 2nd case study, an increase in operator performance was identified by 14.7%, focused on order preparation time.

4.2 Visual Management FEFO

Based on success stories, it can be validated that the implementation of FEFO visual management through the traffic light method contributes to reducing delivery delays from 38% to 10%. Likewise, in a 2nd case, a reduction of expired products from 20 to 0% was obtained. Therefore, it was estimated that for the development of the company under study, a reduction of 19 to 2% of expired products will be obtained. As shown in the following table (Table 3).

Table 3. % of expired products.

Expired products	Case	Company under study
Before	20%	19%
After	0%	2%

4.3 Systematic Layout Planning SLP

To validate this tool, the multicriteria ABC data was used, then the merchandise was ordered and stored by area classified in A, B, C. This was done in order to reduce distances and optimize times in the picking process. These results can be seen in Fig. 3 and Table 4.

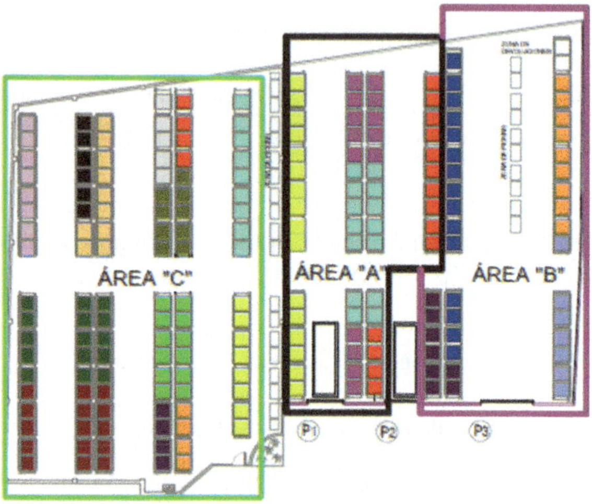

Fig. 3. Organized warehouse

Table 4. Product flow results in the new distribution

Product flow	Time (min)	Distance (m)
Before	44	130
After	25	105

In this image you can see that the merchandise was ordered by product family and by color in 3 respective areas.

In Table 5 it can be seen that the time (min) of the picking process in the warehouse was reduced by 43%. That is, from 44 to 25 min and the distance (m) was reduced from 130 to 105. That is, there was a reduction of 19%.

4.4 Cyclical Inventory

For the validation of this tool, cases of 2 success stories were analyzed. In the first, an improvement in inventory accuracy% was obtained from 67 to 91%. That is, an improvement of 24%. In the 2nd success story there was an increase in inventory reliability from 86% to 98% for the first customer and from 85 to 96% for the second customer.

Table 5. Analysis of results

Stages		Create Stability		Make Flow	Create Flow
Indicator		Operator performance	Expired products	ERI	Picking time
Unit		%	%	%	Hours
Status	Low	≤40	≥20	≤60	≥3.5
	Regular	40–69	20–11	60–84	3.5–1.6
	Acceptable	≥70	≤10	≥85	≤1.5
Before		50	19	75	3.5
After		85	10	95	1.5

4.5 Analysis of Results

5 Conclusions

It is concluded that with the warehouse management proposal under the lean warehousing philosophy, which used tools such as visual management, process standardization, systematic design planning, ABC multicriteria and cyclical inventory contribute to the reduction of returns in a medium-sized company in the sector. commercial.

With the proposal proposed for the case study, it is possible to reduce picking times by 57%, it is possible to optimize the performance of the operators from 50 to 85%, the% of expired products is reduced from 19% to a 2% and inventory reliability is improved from 75 to 95%. All of this impacts on the reduction of returns, reducing the ratio from 14.1% to 7%. In this way, annual savings in cost overruns are achieved with a total of S /. 189,727.41.

The designed model was based on 2 investigations. The first in a mass consumer multinational in India, the second based on an import company in Peru. This model consists of three stages, which are "create stability", "create flow" and "make flow." The present article aims to complete and implement a third stage following the guidelines of the aforementioned investigations. Based on the proposed model, it is necessary to carry out other studies considering different warehouse configurations and routing methods, consider a WMS to have greater control over inventory quantities and expiration dates and consider the impacts that the sales area has with respect to the number of returns.

References

1. Schrotenboer, A.H., Wruck, S., Roodbergen, K.J., Veenstra, M., Dijkstra, A.S.: Order router with product returns and interaction delays. Int. J. Prod. Res. **55**(21), 6394–6406 (2016). https://doi.org/10.1080/00207543.2016.1206982
2. Bonilla-Ramirez, K.A., Marcos-Palacios, P., Quiroz-Flores, J.C., Ramos-Palomino, E.D., Alvarez-Merino, J.C: Implementation of lean warehousing to reduce the level of returns in a distribution company. In: 2019 IEEE International Conference on Industrial Engineering

and Engineering Management (IEEM) (2019). https://doi.org/10.1109/ieem44572.2019.8978755
3. Cui, H., Rajagopalan, S., Ward, A.R.: Predicting product return volume using machine learning methods. Eur. J. Oper. Res. (2019). https://doi.org/10.1016/j.ejor.2019.05
4. Zhou, P., Fan, L.: A note on multi-criteria ABC inventory classification using weighted linear optimization. Eur. J. Oper. Res. **182**, 1488–1491 (2007). https://doi.org/10.1016/j.ejor.2006.08.052
5. Molenaers, A., Baets, H., Pintelon, L., Waeenbergh, G.: Criticality classification of spare parts: a case study. Int. J. Prod. Econ. **140**(2), 570–578 (2012). https://doi.org/10.1016/j.ijpe.2011.08.013
6. Rodriguez, P., Pacheco, D.: Proposal of an approach for the implantation of a cyclical inventory in dispatch sectors, vol. 35, pp. 4–8 (2015)
7. Tippayawong, K.Y., Sopadang, A.P., Patitad, P.: Improving Warehouse Layout Design of a Chicken Slaughterhouse Using Combined ABC Class Based and Optimized Allocation Techniques, vol. 1 (2013)
8. Sazvar, Z., Mirzapour Al-e-hashem, S.M.J., Govindan, K., Bahli, B.: A novel mathematical model for a multi-period, multi-product optimal ordering problem considering expiry dates in a FEFO system. Transp. Res. Part E Logist. Transp. Rev. **93**, 232–261 (2016). https://doi.org/10.1016/j.tre.2016.04.011
9. Basumerda, C., Rahmi, U., Sulistio, J.: Warehouse server productivity analysis with objective matrix (OMAX) method in passenger boarding bridge enterprise. IOP Conf. Ser. Mater. Sci. Eng. **673**, 012106 (2019). https://doi.org/10.1088/1757-899x/673/1/012106
10. Dotoli, M., Epicoco, N., Falagario, M., Costantino, N., Turchiano, B.: An integrated approach for warehouse analysis and optimization: a case study. Comput. Ind. **70**, 56–69 (2016). https://doi.org/10.1016/j.compind.2014.12.004
11. Carmignani, G., Zammori, F.: Lean thinking in the luxury-fashion market. Int. J. Retail Distrib. Manag. **43**(10/11), 988–1012 (2015). https://doi.org/10.1108/ijrdm-07-2014-0093
12. Sutari, O., Satish, R.U.: Development of Plant Layout Using Systematic Planning Layout (SLP) to Maximize Production. In: Proceedings of 7th IRF International Conference, vol.1, pp. 124–127 (2014). https://doi.org/10.9790/3021-021030107
13. Chen, Y., Li, K.W., Liu, S.F.: A comparative study on multicriteria abc analysis in inventory management. In: IEEE International Conference on Systems, Man and Cybernetics, no. 4811802, pp. 3280–3285 (2008). https://doi.org/10.1109/ICSMC.2008.4811802

Process Improvement Model Based on Lean Manufacturing and Plant Distribution to Reduce Production Times

Josue Arias-Castañeda[1], Ruth Condori-Gonza[1], Victor Aparicio-Lora[1], Alejandro Barbachan-Callirgos[1], and Wilder Namay[2(✉)]

[1] Ingenieria Industrial, Universidad Peruana de Ciencias Aplicadas, Prolongación Primavera 2390, Lima 15023, Peru
{u201520283,u201520113,pcinvapa,pcinabac}@upc.edu.pe
[2] Direccion de Investigacion, Universidad Peruana de Ciencias Aplicadas, Prolongación Primavera 2390, Lima 15023, Peru
wilder.namay@upc.edu.pe

Abstract. This work focuses on solving the problem of high production times in textile SMEs, with the aim of analyzing this problem and reducing the impact generated, by designing a model for optimizing production times, which consists of in 2 phases focused on the production process and personnel, using Lean tools and plant distribution. The model will be applied in a Peruvian SME textile company in its quilt and sheet production process. As a result of the implementation, a reduction in production time of 23.13% of the current time was obtained in the case of quilts. Likewise, in sheets the standard time was reduced by 52.48%. In the same way, the productivity of the operator was improved, increasing by 54.31% in the case of quilts and 53.37% in sheets.

Keywords: Lean manufacturing · Textile industry · Time

1 Introduction

Today, companies face constant competition and under pressure to reduce costs in the market and increase their productivity [1 , 2]. According to [3] and [4] they argue that companies have to focus on improving quality, increasing efficiency and increasing their operational performance. At the local level, the manufacturing sector is considered one of the most important sectors in the country's economy [5]. Likewise, the textile industry comprises a significant number of this, since it represents around 11% of manufacturing companies and 98.6% of them are Small and Medium Enterprises (SMEs), which are an excellent means to promote economic development [6]. Textile SMEs face the challenge of competing inside and outside the country with products of foreign origin, particularly those of Chinese origin [7], so improving competitiveness is a challenge for this industry, since doing so would allow them to enter and position themselves in markets and encourage production.

S. Trzcielinski et al. (Eds.): AHFE 2021, LNNS 274, pp. 455–463, 2021.
https://doi.org/10.1007/978-3-030-80462-6_55

Research carried out in the textile industry shows relevant results in the application of LM, such is the case of [8], which proposed a time optimization model based on 5s. This made it possible to reduce clutter and have more clarity on the state of the product in the process, obtaining an increase in productivity of 33.33%, with a decrease in production time of up to 20%.

In this article, this problem is analyzed and focuses on reducing the impact generated by designing a model for optimizing production times. This model consists of 2 phases: the first focused on operational processes and the second on personnel, which will use tools such as Value Stream Mapping (VSM), 5s, Sistematic Layout Planning (SLP), Standardized Work and Workshop Kaizen. These seek to solve the root causes such as clutter, lack of cleanliness, non-existent training and non-established work methods present in this company. The model is validated in a Peruvian textile SME company in its quilt and sheet production process.

2 State of the Art

In this second section, a review of the literature of the Lean tools used in said research is carried out, through a search and analysis of articles that allow defining the factors to take into account.

2.1 Lean Manufacturing (LM)

According to [1], LM consists of the following principles: define and identify the value chain, eliminate all unnecessary steps in the entire value chain, create a value stream so that the whole process flows smoothly and efficiently, from the raw material to the consumer. In the case study carried out by [9], a stage design based on LM was proposed to analyze the efficiency of the production process. The results showed an increase in productivity of more than 50%, and the cycle of a batch was reduced from 240 to 112 min, equivalent to 53.3% less.

2.2 Value Stream Map (VSM)

The value flow map is an LM technique, which consists of showing the operations of the production process in a graphical way, to identify waste, eliminate waste and manage the material flow with the information of the cycle time, the downtime, inventories [2].

2.3 5's

The 5S technique consists of five different elements, which begin with 'S' in the Japanese language: Seiri (classification), Seiton (order), Seiso (cleanliness), Seiketsu (standardization) and Shitsuke (discipline). According to [4], its main purpose is the organization and liberation of space in the workplace, as well as the construction of a quality environment with desired standards and continuous improvement in organizations. The research of [10] reveals that the application of 5S increases the efficiency and productivity of the company, obtaining a clean and orderly place, with a good organizational climate, industrial safety and quality. Its performance improved from 39% to 76% and waste was reduced by 71.42%.

2.4 Sistematic Layout Planning (SLP)

The Sistematic Layout Planning (SLP) technique consists of the physical order of the industrial and commercial elements, this order includes both the spaces required for the movement of materials, storage, indirect workers or operational services in order to have an optimal design of plant distribution in the workplace according to research by [11]. The research carried out by [12] in a textile company when applying plant distribution methods, a cost reduction of 36.82% was achieved.

2.5 Workshop Kaizen

This LM technique focuses on continuous improvement, through the participation of people and the elimination of waste. According to [13], the word Kaizen comes from two words: "Kai" which means change and "Zen" which is for the better, that is, always change for the better. This strengthens the ability of team members to work together, solve problems, and builds trust among team members [14]. The research carried out by [13] applied this technique in a textile industry, where there was a reduction in production time of 20%, which represents a decrease of $ 2.55 per piece, the quality of the product increased by 70% completed and the participation of employees in the company improved by 80%.

2.6 Standardized Work

This technique focuses on performing a task in an efficient, easy and safe way through a manual of established processes [15]. The standardized work aims to increase productivity and reduce times, through clear activities, optimization of movement, reaching balance in production. In the study by [15], this technique was applied in a Brazilian assembly company, where it defines the time required for each unit and the sequence of activities. Among the results obtained, there is evidence of a 15% reduction in cycle time, and a 34.5% decrease in operator movement.

3 Input

3.1 Basis

The proposed model uses the LM methodology to eliminate waste, improve the visual management of the work area, and improve the knowledge, skills and abilities of the company's staff. Likewise, the SLP tool is added, since it complements the objective, since, according to [16] the plant distribution proposal improved the efficiency of 90.43% in production (Table 1).

Table 1. Comparation

Factor	Time reduction in the production process	Less time for implementation	Greater adaptation of staff	Greater feasibility in SMEs	Total
Lean six sigma	3	5	3	3	14
VSM	5	5	3	5	18
5S	5	3	5	5	18
Standardized work	5	3	5	5	18
TPM	1	5	1	3	10
Poka yoke	1	5	3	3	12
Kanban	1	3	1	3	8
Kaizen	3	3	5	5	18
Jidoka	1	3	1	3	8
SLP	5	3	5	5	18

Where:

Low	1	The methodology is not very viable with the factor
Medium	3	The methodology is moderately feasible with the factor
High	5	The methodology is quite viable with the factor

3.2 Proposed Model

The proposed model focuses on two phases: The first is the approach to operational processes and the second is focused on personnel (Fig. 1).

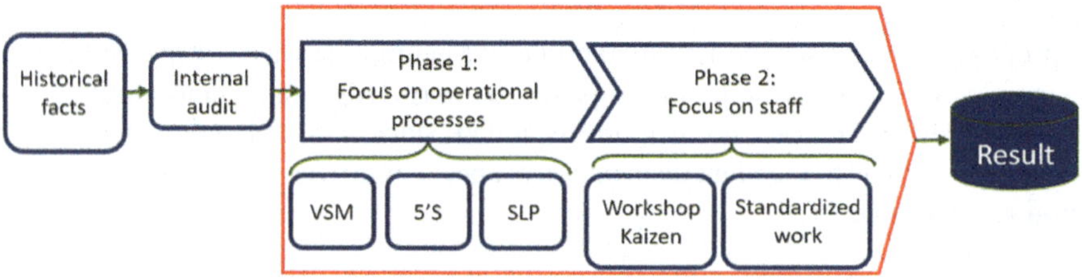

Fig. 1. Proposal model

3.3 Model Components

PHASE 1: Focus on Operational Processes
This phase is responsible for improving the operational processes of the production area, using tools that allow reducing the activities that do not generate value to the product, as well as the information within the process such as the Value Flow Map (VSM). In the same way, we want to improve the visual management and comfort of the staff in the work area through the implementation of 5'S. Finally, the third tool to use in this phase is Systematic Layout Planning (SLP), which seeks to maximize the work flow with an adequate plant distribution, minimize the movement of personnel in production.

PHASE 2: Focus on Staff
This second phase focuses on the operators, who are the main resources in an organization. The aim is to improve the knowledge, skills and abilities of the staff, implementing the Kaizen Workshop tool, which is the key to change for all types of organizations, as it allows the work team to identify and implement a significant improvement in a process, through workshops. Likewise, the second tool to use is work standardization, which focuses on selecting the best work practices, establishing methods and procedures to be carried out in each activity.

3.4 Indicators

Indicators are presented to control the model, which are used in each tool used. The main ones are:

1) *Metrics for VSM:* **Lead Time:** The objective is to calculate the time from when the production process begins until the product is delivered to the customer.

$$LT = TA + TCC \tag{1}$$

LT: Lead time (days).
TA: Provisioning time (days).
TTC: Total cycle time (min).

2) *Metrics for 5'S:* **Reduced space:** It allows to know how much the available space has increased after order and cleaning.

$$A = L.\, a \tag{2}$$

A: Production area.
L: Length of gap.
a: Width of space.

3) *Metrics for SLP:* **Distance Traveled:** It is possible to determine how many meters the material travels before and with this tool how much it has reduced.

$$D = D2 - D1 \quad (3)$$

D: Total distance traveled.
D1: Initial distance.
D2: Final distance.

4 Validation

4.1 Describe the Initial Validation Scenario

The first scenario represents the current situation of the company, where the data that was collected was for the entire year 2018. The second scenario represents the simulation in ARENA after the implementation of the proposal. It should be noted that for this simulation 1000 data were collected for each process, with the aim that the result is as close as possible to reality.

4.2 Initial Diagnosis or Previous Studies

According to previous studies carried out in the textile sector in 2013 by researchers [17], the average production times of sheets and quilts were determined, which are 9.0 min and 36.3 min respectively.

4.3 Design of Validation

- **VALUE STREAM MAPPING (VSM)**
  The following Table shows the expected results after the implementation of the VSM tool (Table 2).

Table 2. Expected results of VSM implementation.

Indicators	Product family	Current situation	Future situation	Variation
Lead time (days)	Quilts	1.103	0.963	- 12.69%
	Sheets	1,038	0.905	- 12.81%

- **5'S**
  The following Table shows the expected results after the implementation of tool 5's (Table 3).

Table 3. Expected results of the implementation 5's.

Indicator	Current situation	Future situation	Variation
Reduced space in production areas (m^2)	150	133	- 11.3%

- **SISTEMATIC LAYOUT PLANNING (SLP)**
  The third tool used in the company was SLP, from which three plant distribution proposals were proposed, which are Multiproduct Diagram, Transfer Diagram and Matrix Tables. The expected results of proposal 1 are shown below (Table 4).

Table 4. Expected results of SLP implementation.

Indicator	Current situation	Future situation	Variation
Unnecessary transportation (m)	71.3	55.5	- 22.2%

- **EVALUATE THE SUCCESS OR FAILURE OF THE PROJECT**

From the Table 5 it can be stated that for every S/. 1 invested there will be a profit of S/. 4.78.

Table 5. Expected results of the project evaluation.

Indicator	Formula
Cost benefit ratio	$RBC = \frac{137,814}{28,316} = 4.87$

5 Conclusions

The proposed model of this research work is based on two pillars, which are focus on operational processes and focus on personnel. In the first approach, the tools VSM, 5's and Plant Distribution (SLP) were used. In the second approach, the Workshop Kaizen and Standardized Work tools were used. The expected results of the present model is a decrease of 23.13% from the current time in the case of quilts. Likewise, in sheets the standard time was reduced by 52.48%.

The results obtained after the implementation of the improvement allowed the reduction of 12.75% of the average lead time in the company. Also, the average value-added time was reduced by 26.65%.

The research model being unique, had difficulties to perform the validation. Therefore, it was carried out in 3 production lines of the company, additional to the production line of quilts and sheets, which were the production line of towels, production line of pillow protectors and production line of mattress protectors. Through the statistical analysis that on average the model would reduce indicator 1 by 9.09%, indicator 2 by 18.70%, indicator 3 by 10.2%, indicator 4 by 16.36% and indicator 5 would increase on average 38.2%

References

1. Chauhan, G., Chauhan, V.: A phase-wise approach to implement lean manufacturing. Int. J. Lean Six Sigma **10**(1), 106–122 (2018). https://doi.org/10.1108/IJLSS-09-2017-0110
2. Carvalho, C.P., Carvalho, D.S., Silva, M.B.: Value stream mapping as a lean manufacturing tool: a new account approach for cost saving in a textile company. Int. J. Oper. Prod. Manage. **7**(1), 1 (2018). https://doi.org/10.4995/ijpme.2019.8607
3. Panwar, A., Jain, R., Rathore, A.P.S., Nepal, B., Lyons, A.C.: The impact of lean practices on operational performance – an empirical investigation of Indian process industries. Prod. Plann. Control **29**(2), 158–169 (2017). https://doi.org/10.1080/09537287.2017.1397788
4. Randhawa, J., Ahuja, I.: Evaluating impact of 5s implementation on business performance. Int. J. Prod. Perform. Manage. (2017). https://doi.org/10.1108/IJPPM-08-2016-0154
5. PROMPERÚ: Peruvian textile industry surprises the world. FOREIGN TRADE (2018). https://peru.info/es-pe/comercio-exterior/noticias/7/32/industria-textil-peruana-sorprende-al-mundo
6. From Althaus Peru's labor legislation threatens the prosperity of SMEs. LAMPADIA (2018). https://www.lampadia.com/analisis/gobernanza/no-hay-medianas-empresas-en-el-peru/
7. Godinho, M., Ganga, G.M.D., Gunasekaran, A.: Lean manufacturing in Brazilian small and medium enterprises: implementation and effect on performance. Int. J. Prod. Res. **54**(24), 7523–7545 (2016). https://doi.org/10.1080/00207543.2016.1201606
8. Ochoa, B.: Proposal for the Implementation of the 5s Tool in the Textile Company Zoga (Undergraduate thesis). University of Azuay, Ecuador (2013)
9. Trojanowska, J., Kolinski, A., Galusik, D., Varela, M.L.R., Machado, J.: A methodology of improvement of manufacturing productivity through increasing operational efficiency of the production process. In: Hamrol, A., Ciszak, O., Legutko, S., Jurczyk, M. (eds.) Advances in Manufacturing. LNME, pp. 23–32. Springer, Cham (2018). https://doi.org/10.1007/978-3-319-68619-6_3
10. Lamprea, E.J.H., Carreno, Z.M.C., Sánchez, P.M.T.M.: Impact of 5s on productivity, quality, organizational climate and industrial safety in caucho metal ltda. Chil. Eng. J. **23**(1), 107–117 (2015). https://doi.org/10.4067/S0718-33052015000100013
11. Ojaghi, Y., et al.: Production layout optimization for small and medium scale food industry. Proc. CIRP **26**(1), 247–251 (2015). https://doi.org/10.1016/j.procir.2014.07.050
12. Montero, Y., Calderón, O., Lorente, L., Herrera, I., Machado, C., Saraguro, R.: Improvement plant layout of production line in textile company: a case study. Proc. Int. Conf. Ind. Eng. Oper. Manage. **1**(3), 741–750 (2019)
13. Fontes, E., Loos, M.: Application of the kaizen methodology: a case study in a Brazil's central-western textile industry. Spaces. **38**(1), 6–18 (2017)

14. Lantz, A.: Participative work design in lean production. J. Work. Learn. **27**(1), 19–33 (2015). https://doi.org/10.1108/JWL-03-2014-0026
15. Fin, J., Vidor, G., Cecconello, I., De Campos, V.: Improvement based on standardized work: an implementation case study. Braz. J. Oper. Prod. Manage. **14**(3), 388–395 (2017). https://doi.org/10.14488/BJOPM.2017.v14.n3.a12
16. Ojaghi, Y., Khadmi, A., Mohd, N., Ghorbani, R., Syed, A.: Production Layout Optimization for Small and Medium Scale Food Industry, 26(12), 247–251 (2015). https://doi.org/10.1016/j.procir.2014.07.050
17. Inca, U., Muyulema, E.: Proposal for the start-up of a company for the manufacture of sheets and quilts for the association of women from the citadel camilo ponce in the city of Riobamba (2013)

A Lean Manufacturing and RCM-Based Production Process Improvement Model for Increasing the Production Capacities of Carbonated Beverage Bottling Companies

Esthefani Capcha-Huamali[1], William Vila-Huaman[1], Gino Viacava-Campos[1], Juan Quiroz-Flores[1], and Luis Cardenas[2(✉)]

[1] Ingenieria Industrial, Universidad Peruana de Ciencias Aplicadas, Prolongación Primavera 2390, Lima 15023, Peru
{u201322377,u201420521,pcingvia,pciijqui}@upc.edu.pe
[2] Direccion de Investigacion, Universidad Peruana de Ciencias Aplicadas, Prolongación Primavera 2390, Lima 15023, Peru
luis.cardenas@upc.edu.pe

Abstract. The beverage manufacturing industry is an extremely competitive sector. Small- and medium-sized businesses usually encounter order fulfillment issues and have high operating costs, negatively affecting their profitability. In this study, we consider a small soft drink production company to design an improvement model. The proposed design is based on the combination of lean manufacturing tools and other tools that may be required depending on the situation. In this sense, the proposed model is based on the 5S methodology, workstation redesigning, process standardization, and reliability-based maintenance. The implementation of the proposed design will increase the efficiency of the production process by 8.8%.

Keywords: Lean manufacturing · SMEs · Efficiency · Standardization · Beverage production

1 Introduction

Currently, the beverage industry is a sector reporting the highest sales volume in the world. Between 2009 and 2018, this industry sold 696,100–994,100 million liters of beverages [1]. In Peru, the beverage sector is constantly growing. Further, it is considered to be an economic source for the country because it represents approximately 2% of the national gross domestic product (GDP) and generates more than 180,000 jobs [2]. According to the Ministry of Production, soft drinks are the dominant beverage within the beverage industry, with a 39.6% market share. In addition, their sales have experienced an increase of 5.1% when compared with historic sales [3]. Approximately 68% of the beverage market is dominated by three main brands, with Coca Cola leading sales at 27% but closely followed by Inca Kola at 24% and Kola Real at 10%. Similarly, the

S. Trzcielinski et al. (Eds.): AHFE 2021, LNNS 274, pp. 464–472, 2021.
https://doi.org/10.1007/978-3-030-80462-6_56

remaining market has also witnessed considerable competition, especially in case of SMEs trying to withhold their market position against such large chains [4].

In addition, most medium-sized businesses are often afflicted by inefficient production processes, which not only hinder their growth but also negatively impact the service levels offered to their customers, further affecting their profitability. The problem identified in the beverage sector arises from the deficient management of the production process [5] or deficient management and production controls [6]. Majority of the case studies reported in the literature focus on managing the availability of machines because they are considered bottlenecks of the production chain; a machine stoppage translates into lost time, resources, and revenue.

2 State of the Art

2.1 Lean Manufacturing

In the case studies found, lean manufacturing is considered to be a work philosophy, which defines the way in which the production system may be optimized, focusing on eliminating all types of waste and increasing the productivity of companies. The methodology has been applicable in SMEs, where the most widely implemented lean manufacturing tools were VSM, TPM, 5S, work standardization, and quality management practices. Thus, SMEs can adopt management initiatives to reap benefits [7, 8]. Finally, the research articles indicate that lean manufacturing improves performance and quality, minimizes waste, and improves processes and reduces company expenses [9].

2.2 5S Methodology

To remain competitive, companies must reduce the additional time invested in machine preparation operations, increase the frequency of useful changes, and be more flexible in accordance with product diversification [10, 11]. Therefore, appropriate item classification, organization of all the necessary elements according to the frequency of requirement, and maintaining the workplace clean and neat directly contribute to productivity and the engagement of operators in the implementation process.

2.3 Process Standardization

The formats used to standardize the activities of the company must at least include information regarding who prepared it, who approved it, the version number, and the effective date of the document [12, 13]. The process variations in the beverage case study can be attributed to the many identified failures. Therefore, the current process times and failures are assessed to propose solutions. The resulting operator times measured were 1 min with faults and 0.22 min without faults for the bottle filling station and 1.15 min with faults and 0.45 min without faults for the unloading process. Based on the results, the authors proposed standardizing processes based on optimal times, eliminating idle time, conducting toolbox meetings about the machines, and completing processes in a timely manner. Thus, the process times were reduced [14].

2.4 Reliability-Centered Maintenance

The reliability-centered maintenance (RCM) tool can be used to measure the reliability of the components of each machine. RCM aims to schedule and plan system preventive maintenances based on a reliable design and the production quality. Preliminary analysis is a critical requirement associated with the implementation of RCM. This involves detailed analysis of the data and deep exploration into each fault. The need for a failure mode and effects analysis (FMEA) also extends the implementation time [15, 16].

3 Contribution

3.1 Proposed Model

The model is divided into four interrelated stages, with the objective of improving the production process management efficiency. As shown in the following image, the 5S methodology will be considered as the foundation for adapting other tools. This will be especially helpful when reorganizing the workstation layouts because the job performance increases when the workstation adapts to the workers. Additionally, a standardization of the working method will be applied in critical processes such that an adequate flow is established to eliminate all types of waste, including time, unnecessary movements, material waste, and even defective products. Finally, the RCM tool is proposed to address machinery issues because machines are essential production components (Fig. 1).

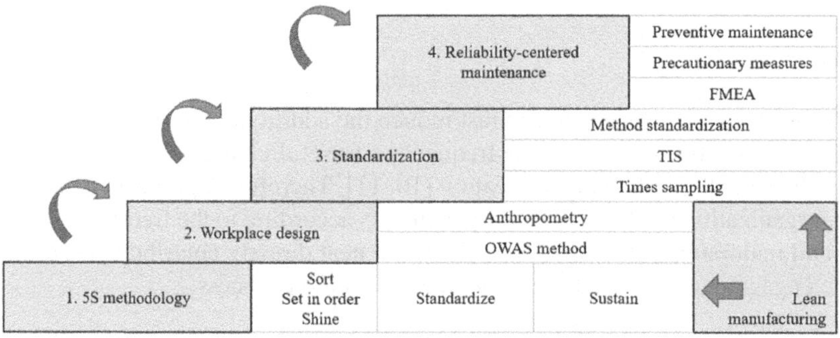

Fig. 1. Proposal model

Component 1: 5S Methodology. The 5S methodology is related to work conditions because it aims to create an efficient and productive work environment. This will allow the tasks to be executed in an organized, orderly, and clean manner, creating an efficient and production work environment [17]. For this component, we conducted a 5S audit. The company scored 2.5 out of 5, indicating that the company is in poor condition.

The development of this phase begins with selecting and eliminating the elements that do not add value to the company. Subsequently, the necessary elements are ordered in their proper place; further, cleaning is integrated as part of the daily work. Standards, such as schedules and policies, are developed.

Component 2: Workplace Design. It is a tool that aims to create a better workplace design for the operators to improve their efficiency and productivity in the company. The workplace must be designed to meet both the needs of the company and the requirements of the operator performing each task. According to a research paper, a healthy environment helps workers to perform their functions in the most comfortable way possible. Therefore, a properly organized workplace adapted to the specific needs of each operator would simplify tasks and prevent annoyances that make work difficult [18].

Component 3: Process Standardization. The proposed model uses the process standardization tool to establish best practices so that the operators can standardize the activities that they perform at their workstations. The objective of this tool is to reduce the cycle time of these activities and remove the different types of waste generated [19]. In this stage, each critical company process is assessed, and the movements of the operators when performing their activities are recorded. In addition, time is taken to find and eliminate the subactivities that do not add value to the final product.

Component 4: Reliability-Centered Maintenance. The RCM tool is used to develop a maintenance plan for critical production process machines and provide greater focus on the equipment components [16]. Therefore, the number of stops reported for each machine is initially used as the basis to select the most critical machine. Subsequently, a list of all the machine functions and failures is prepared. An FMEA is conducted, wherein the root cause, sequence, and risk of each fault are analyzed. Therefore, a list of spare parts is prepared to reduce the repair times, and a maintenance plan is also developed. The indicators used to assess the machine performance and availability for this component are the waste levels and the OEE, MTTR, and MTBF values.

3.2 Indicators

After implementing the components, indicators will be assigned to monitor the project improvement.

Efficiency: It is the relation between the result achieved and the resources used. Resources are optimized to achieve the defined objectives.

$$Efficiency = \frac{Useful\ time}{Total\ time} \tag{1}$$

Defective Products: This indicator will be used to determine how many noncompliant products will be discarded.

$$PD = \frac{Bottles\ with\ dents}{Month} = \frac{Bottles}{Month} \tag{2}$$

Lost Time: This is the idle operator time within the process because it increases the cycle times (%).

$$\frac{\%\ Downtime\ in\ spraying\ processes\ due\ to\ aches\ and\ pain}{\%\ Drying\ time} \tag{3}$$

Waste: The waste of a number of inputs or resources during the production process directly affects the stock of raw materials, generating shortages.

% of labels used in the process

$$\frac{Discarded\ labels}{Month} = \frac{Label}{Month} \tag{4}$$

% of content liquid lost during filling process

$$\frac{Burnt\ bags}{Month} = \frac{Bags}{Month} \tag{5}$$

Overall Equipment Effectiveness (OEE): This is a vital indicator that represents the real defect-free production capabilities, the actual process performance rates, and the equipment availability rates (%).

$$OEE = Availability \times Quality \times Performance \tag{6}$$

4 Validation

The actual company data used for the development of this project were obtained in 2019. Based on these data, we were able to identify management issues associated with the production process. This affects 18% of the revenue generated from sales. The root causes determined for this process are inadequate workplace design, inadequate work method, and poor maintenance management. These problems are evident in two specific processes: the spraying area and the packaging area.

4.1 Component 1: 5S Methodology

The assessment was scheduled on a weekly basis, aiming at seven days per week, to obtain a more accurate representation of the activities performed at the company. First, the sorting stage selected the items in good condition. The remaining items were either discarded or recycled. As shown in the implementation, 70% of the items deemed as unnecessary were discarded. After this step, signs were posted to instruct the operators to return all the items that they use at their corresponding place. However, because of the lack of space, the concept of setting in order was not achieved as expected, which can be attributed for its score of 2.8. Next, the operators strived to keep their workplace clean. In addition, the provision of a temporary container in each work area was satisfactory because the operators no longer had to approach the general bin located near the main door.

4.2 Component 2: Workplace Design

To realize the pilot of the workplace design, tables were exchanged with tables in better condition (tables from the dining room and warehouse), especially in the spraying area. The number of bottles dropped decreased when these tables were used. In addition, the operators began to use a better working method. During the implementation weeks, the number of products discarded in the spraying area decreased. Since the change was implemented, the number of dented bottles was reduced to an average of 37%.

The simulation result indicates that the risk level was reduced from 3 to 1, which means that the new workplace does not generate any harmful effect on the musculoskeletal system.

4.3 Component 3: Process Standardization

This tool was applied in four processes, i.e., lotting, labeling, bottling, and spraying.

Lotting Process: Operators spend a long time drying bottles before dividing them into lots. After a meeting with the operators and supervisors, we agreed to implement a pilot plan in which some of the freshly washed bottles were placed on a cart and left to dry at room temperature (1 h). After 1 h, all the bottles had dried out and operators could proceed with their lotting duties, reducing at least 17% of the entire process time.

Labeling Process: The process was moved to the start of the production line. Thus, bottles were labeled when they were still empty. When testing the hypothesis, this change facilitated labeling. Further, labels were less likely to detach because of the low label–bottle friction. This reduced label waste by more than 50%, which implies the requirement of less resources and better use of the operator's time because reprocessing was also reduced.

Bottling Process: The bottling process was inefficient because in addition to wasting soda pop, operators usually stopped the filling machine to finish filling the bottles. Therefore, the usage of funnels constituted an improvement for the company, and they have been using them ever since the trial run. This implementation reduced 70% of the waste generated in this process, which increased production to 134 bottles per month.

Spraying Process: In this process, the pilot was tested using carts and standard crates (until the purchase of custom crates is approved) facilitating bottle transfers. In addition, information signs were posted for the appropriate use of spray guns. This was complemented with training sessions on the proper use of spray guns and how to prevent unnecessary movements. After the pilot implementation, spraying times were reduced by 15% and even by 23% in case of experienced operators.

In addition, the described method and the instruction manual for the spray guns helped the operators to prevent overheating issues. This was proved because the number of discarded bags was reduced by 34% in the following months.

4.4 Component 4: Reliability-Centered Maintenance

After developing the maintenance plan and the list of spare parts that was required to be permanently in stock, the plant technician was requested to provide training to the operators in charge of the bottling process (when the specialized technician was not available). The training sessions focused on explaining the reason for the most common filling machine failures and how to replace spare parts, including valves, pistons, and outer pipes. These training sessions were developed in February after the standardization was implemented. Thus, the monthly liquid waste was observed to reduce by 7.4%.

In addition, machine stops were also reduced, as can be seen from the following table. The result was an overall 27% reduction in machine stops and 14% in stops because of technical reasons, which are the stops that generate the highest expenses for the company.

5 Results

After the validation of the project, the improvements, which were implemented from December to February, can be given as follows.

From the following table, it can be observed that the OEE of the company has changed from being unacceptable with 62% to being adjustable with 71% according to the OEE classification (Table 1).

Table 1. Validation

Tool	Indicator	AS IS	TO BE	Improvement
PROBLEM	Efficiency	75.2%	84.4%	+9.2%
5S Methodology	5S audit (Sort)	2.6	4	+50%
	5S audit (Set in Order)	0.8	2.8	+65%
	5S audit (Shine)	1.2	3.6	+75%
	5S audit (Standardize)	1.4	2.4	+71%
	5S audit (Sustain)	1.4	2.6	+86%
Workplace Design	OWAS	category 3	category 1	−100%
	Drop with dents	163	86	−50%
Process Standardization	Wasted labels	780 labels/month	337 labels/month	−56%
	Liquid liters wasted	766 l	204.5 l	−73%
	Bags wasted	83 bags	56 bags	−32%
	Lotting cycle time	23.2 min	19.258 min	−16%
	Spraying cycle time	5.495 min	4.7 min	−14%
RCM	Liters wasted per stop	1265.4 l	1193 l	−5%
	Machine stop Times	175 min	115 min	+9%
	Number of machine stops	37	25	−32%
	OEE	62%	71%	−34%

According to the validation, the OWAS tool, which assessed physical loads, decreased from 3 to 1. The process standardization tool will reduce waste and cycle times, with the most significant reduction of 73.30% observed in the bottling area.

6 Conclusions

Recently, the beverage industry has generated considerable market demand, but there is not much research on its production processes. The proposed design improved the company efficiency from 75.2% to 84.4%, representing process improvement and better use of resources.

Some of the resulting process improvements were as follows. Bag and label wastes were reduced to 72% and 74%, respectively, and liquid waste was reduced to 84%. The process time of the spraying process decreased by 14%. On the other hand, the RCM was increased by 9% of the OEE.

References

1. Instituto Nacional de Estadísticas e Informática (2019). Producto Bruto por Actividad Económica. Recuperado de: https://www.inei.gob.pe/
2. Ministerio de Producción. Boletín de Comercio Interno. Enero 2019. Grandes Almacenes, Supermercados y Tiendas Especializadas. Recuperado de (2019). http://ogeiee.produce.gob.pe/index.php/shortcode/oee-documentospublicaciones/boletines-comercio-interno
3. Retail, P.: Cuál es el panorama de las bebidas gaseosas en el mercado peruano?. Recuperado de (2016). https://www.peruretail.com/panorama-bebidas-gaseosas-mercado-peruano/
4. Ministerio de Producción. Comercio Interno. Enero 2019. Resultados de las Ventas de Grandes Almacenes, Supermercados y Tiendas Especializadas. Recuperado de (2019). http://ogeiee.produce.gob.pe/images/oee/CI/enero2019.pdf
5. Centro de Comercio Internacional. Perspectivas de competitividad de las pymes 2017 – Suplemento especial: Guía para las pymes sobre las cadenas de valor. Ginebra, Suiza: ITC (2017)
6. Ministerio de Producción. Relación de principales productos de empresas que participan en la muestra del índice de volumen físico, mes de enero 2019. Recuperado de (2019). http://ogeiee.produce.gob.pe/
7. Sahoo, S.: Assessing lean implementation and benefits within Indian automotive component manufacturing SMEs. India (2020)
8. Lucherini, F., Rapaccini, M.: Exploring the impact of Lean Manufacturing on Flexibility in SMEs. Italia (2017)
9. Solano, N.E.C, Llinas, G.AG., Torres, J.R. M: Towards the integration of lean principles and optimization for agricultural prouction systems: a conceptuaal review proposition, Colombia (2020)
10. Argüello: Application of the 5S methodology in line number # 1 of classification and packaging of a shrimp packing company located in Duran (2011)
11. Barcia, K., De Loor, C.: Metodología para mejorar un proceso de ensamble aplicando el mapeo de la cadena de valor (VSM). Guayaquil, Ecuador: ESPOL (2007)
12. Ncube, F., Kanda, A., Chirengendure, Y.: An evaluation of ergonomic risks associated with tailoring tasks using the rapid entire body assesment method (2019)
13. Iranzo, S., Piedrabuena, A., Iordanov, D., MartinezIranzo, U., Belda-Lois, J.M.: Ergonomic assesment of passive upper-limb exoskeletons in an automotive assembly plan. Spain (2020)
14. Tortorella, G.L., Vergara, L.G.L., Ferreira, E.P.: Lean manufacturing implementation: an assesment method with regards to socio-technical and ergonomics practices adoption. Brazil (2017). https://doi.org/10.1007/s00170-016-9227-7

15. Pourhosseini, O., Nasiri, F.: AFF.M.ASCE (2018) Availability-Based Reliability-Centered Maintance Shedduling: Case study of Domestic Building – Integrated Hot water sytems, India (2018)
16. Fuentes-Huerta, M., González-González, D., Cantú-Sifuentes, M., Praga-Alejo, R.: RCM implementation on plastic injection molding machine considering correlated failure modes and small size sample. Int. J. Adv. Manuf. Technol. **95**(9–12), 3465–3473 (2017). https://doi.org/10.1007/s00170-017-1402-y
17. Capristano Cueva, A.N.: Aplicación de la Metodologìa 5S para incrementar la Productividad de la Empresa Acadic S.R.L. SJL. Lima, Lima, Perú (2017)
18. Munfiáh, I.P., Fitriadi, R., Sufa, M.F.: Evaluation of work posture in Sohun Noodles Workes using OWAS and WERA method (2019)
19. Mikba, et al.: Estandarizacion Standardization – One of the Tools of Continuous Improvement (2016)

Lean Manufacturing Model for Production Management Based on SAP-LAP to Reduce Delays in the Production Line in Mypes of the Metalworking Sector

Adeir Leon-Huerta[1], Leslie Zamora-Huamancha[1], Ernesto Altamirano-Flores[1], and Carlos Raymundo[2](✉)

[1] Ingenieria Industrial, Universidad Peruana de Ciencias Aplicadas, Prolongación Primavera 2390, Lima 15023, Peru
{u201517150,u201517373,pcinealt}@upc.edu.pe
[2] Direccion de Investigacion, Universidad Peruana de Ciencias Aplicadas, Prolongación Primavera 2390, Lima 15023, Peru
carlos.raymundo@upc.edu.pe

Abstract. The purpose of this research is to present the repercussions generated by a deficient production process in a MYPE company in the metalworking sector. It is intended to implement a production management model based on the Lean manufacturing technique with SAP-LAP analysis to reduce delays in the production line. The objective of this project is to quantify the degree of improvement in the fulfillment of deliveries by implementing the improvement proposal. The model in this work consists of taking previous data in a metalworking MYPE company, in order to generate a work base based on a production line, to later be able to validate through a simulation with the implementation of Lean Manufacturing techniques such as 5s, SMED, Preventive Maintenance, Poka-yoke. With the application of the SAP-LAP analysis (Situation, Actors, Processes, Learning, Action and performance). The results obtained demonstrate the reduction of delays in the production line.

Keywords: Lean manufacturing · Metalworking · Production management · SAP-LAP

1 Introduction

In recent years, the metalworking sector has evolved worldwide. The most developed countries in the metalworking sector are: United States, Japan, China, Germany and Spain [1]. In Latin America, according to the IMF report, the growth of the economy is 1.2% at the end of 2017 and 1.9% in 2018 [2]. In the case of Colombia and Chile, the ECLAC expects growth in 2019 close to 3.3%, while Peru can reach 3.6% of its GDP [3]. In Peru, according to the minister of production, Raúl Pérez-Reyes, in 2018 production in the metalworking sector registered a growth of 6.1%. Finally, the minister

S. Trzcielinski et al. (Eds.): AHFE 2021, LNNS 274, pp. 473–481, 2021.
https://doi.org/10.1007/978-3-030-80462-6_57

of production indicated that more than 45 thousand formal companies operate in the metalworking sector, of which 98.7% (44,918) are MYPE [4]. However, the evolution also indicates the problem that MYPES have in the face of delivery delays.

According to Maximixe, in 2016 a fall of 5.6% was estimated, putting the metalworking industry in a state of business recession, which forces them to take refuge in short-term objectives to improve productivity, quality and customer loyalty [5, 6]. In a case study in India, it is confirmed that the use of lean manufacturing tools and SAP-LAP are efficient for the improvement of production in MYPIMES companies, emphasizing that they are of small investment and are easy to implement [7]. In another case study in Colombia in a metalworking company, lean manufacturing tools such as 5S and TPM are used where 37.1 kg of waste was eliminated corresponding to 22% and corrective maintenance was mitigated reducing to 20% from 47% [8]. Continuing with the case studies referring to POKA YOKE there is a study which demonstrates the effectiveness of this tool when applied in a company in the automotive sector which evidenced an improvement to reduce PPM's by 87.97%, validating it with a study repeatability and reproducibility [R&R] [9].

2 State of Art

2.1 Model for Managing Production in MSEs of the Metalworking Sector

Before creating a management model, it must be taken into account to try to cover as much as possible the activities or areas of the company to seek the synergy of these, it must not only be based on production, we also have to take into account the financial and environmental side, the current situation in which the company finds itself (maturity point), among other important points so as not to leave gaps in this [10, 11]. Good production management allows any company to be competitive in the market where it operates, but focused on the competitiveness of the product itself [10, 12, 13].

2.2 SAP-LAP Analysis in the Production Management Model in MSEs of the Metalworking Sector

SAP-LAP is an ideal parameter that helps to identify the vulnerability of the organization and is a flexible technique and offers high performance [14, 15]. To obtain a good product it is necessary to have good process practices to increase competitiveness [16, 17]. Other authors carry out a validation through a separate analysis, first the SAP Situation-Actor-Process model using a qualitative approach and then the LAP Learning-Action-Performance model using a quantitative approach [15, 16, 18].

2.3 LM Model of Production Management Based on SAP-LAP Analysis in Mypes of the Metalworking Sector

SAP-LAP analysis is used for the flexibility of resources in companies to improve productivity and reduce costs inclined to manufacturing. The Lean Manufacturing methodology helps the adoption of tools to face soft problems and thus return to more competitive companies. The application of both methodologies in a single model helps improve flexibility and lean manufacturing, benefiting stakeholders by suggesting actions that can be taken to improve the competitive priorities of MSEs [7, 19].

3 Input

A model is proposed to be able to fulfill the objective of the research, this model can be seen in Fig. 1. This model was proposed based on a study [19], based on a research and comparative analysis of different management models of production and Lean Manufacturing methodologies and SAP-LAP analysis to reduce delivery delays in the production line in MYPES of the metalworking sector.

3.1 Proposed Model

This Production Management model based on SAP-LAP and Lean Manufacturing is innovative since both methodologies are merged in order to mitigate delivery delays in manufacturing companies (Fig. 1).

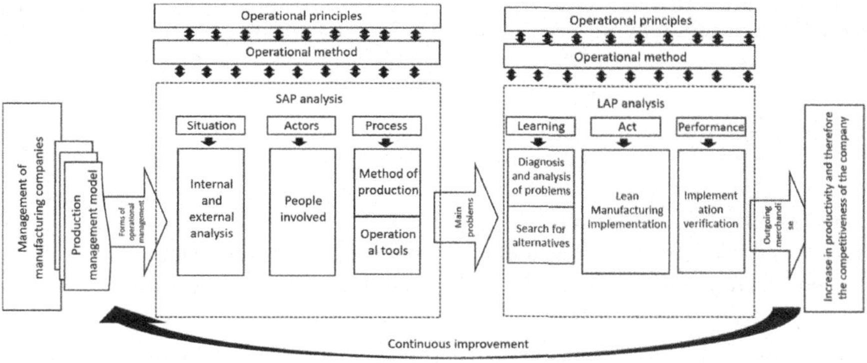

Fig. 1. Production Management Model based on SAP-LAP and Lean Manufacturing [19]

Advantage:

- It is aimed at industrial companies specifically in the metalworking machinery manufacturing sector.
- For micro and small businesses (MYPES) and by people with little experience.
- Its implementation and relationship with the production system.
- It offers a simple, agile and low-price alternative.
- Identifies deficiencies, which imply a negative impact on the competitiveness of the company.
- By having the SAP-LAP analysis immersed, it makes it a flexible model.
- It allows a synthesis of soft and hard problems.

3.2 Detail

SAP-LAP

- S (Situation): Identify the current situation of the company through an internal and external analysis.
- A (Actors): Establish the key actors, that is, the people involved in the production process.
- P (Process): Establish production methods, process management and operational tools.
- L (Learning): Carry out a diagnosis of and analysis of the main problems in the production process
- A (Act): Take action by implementing Lean Manufacturing.

Lean Manufacturing

- Start: Establish the work team
- Have: Establish which Lean Manufacturing tools are the right ones
- Implement: Implementation of the tools according to their processes.

- P (Performance): See the development and verification of the implementation through indicators.

3.3 Process

For the implementation of the proposed model, the steps shown in Fig. 2 are carried out. Of the entire model, the contribution focuses on the Action for the evaluation in general separately from SAP and LAP. Focusing on Process with the implementation of Lean Manufacturing tools.

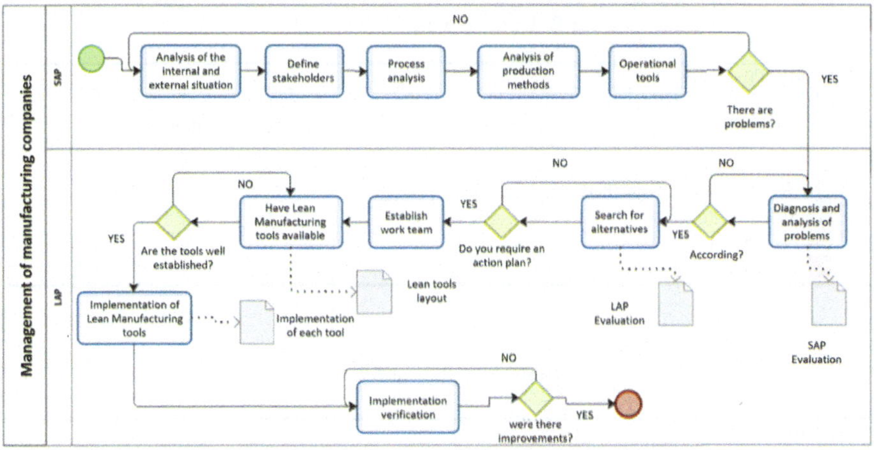

Fig. 2. Flow chart of the model

3.4 Indicators

Productivity:
This indicator aims to see an increase in productivity by reducing delays (Table 1).

Table 1. Productivity indicator.

1. NAME: Productivity	
1. OBJECTIVE OF THE INDICATOR: Achieve that the percentage of productivity is greater than 80%.	
2. CALCULATION: Percentage of productivity	$Pv = \frac{number\ of\ hours\ used}{number\ of\ hours\ available} * 100$
3. CHARACTERISTICS OF THE INDICATOR: (green) 80 < Pv < 90 (yellow) 70 < Pv < 80 (red) Pv < 70	

Production:
This indicator aims to increase production by reducing delays (Table 2).

Labor Utilization:
This indicator aims to decrease the production time after the implemented improvements (Table 3).

Table 2. Production indicator

1. NAME: Production
1. OBJECTIVE OF THE INDICATOR: Achieve that the percentage of production is greater than 90%.
2. CALCULATION: Percentage of production $P = \frac{number\ of\ orders\ placed}{quantity\ of\ orders} * 100$
3. CHARACTERISTICS OF THE INDICATOR: ■ (green) 80 < P < 90 ■ (yellow) 70 < P < 80 ■ (red) P < 70

Table 3. Labor utilization

1. NAME: Labor utilization
1. OBJECTIVE OF THE INDICATOR: Achieve that the percentage of this indicator is greater than 80%.
2. CALCULATION: $Tp = \frac{Productive\ hours\ of\ the\ operator}{Total\ operator\ hours} * 100$
3. CHARACTERISTICS OF THE INDICATOR: ■ (green) 80 < P < 95 ■ (yellow) 75 < P < 80 ■ (red) Pm < 75

4 Validation

4.1 Validation Scenario

To validate the proposed model, the company Maquiprocesos EIRL is taken as a case study, which is a metalworking Mype dedicated to the manufacture of machinery and equipment in stainless steel for the food industry, fishing sector, agro-industry and the like. This company is located in Cercado de Lima, Lima-Peru. This case study is accredited by the support of the production manager of the company Maquiprocesos EIRL.

Simulation

The program to use is the Arena. This program will allow us to simulate the current system and validate our information. Before this, the Input Analyzer program had to be used, which allows us to identify the type of distribution that each sample has, which will be placed in the Arena program.

The company is currently going through a problem which is identified with the delays in the production lines, it is shown that the company is in a decline in its production, as a percentage of approximately 42% with respect to what it should actually be. Producing this is demonstrated using Fig. 2, which shows us all the monthly productions that were delivered on time versus those that were delivered out of time and with that margin it is concluded that the company is working well below what should actually deliver.

4.2 Initial Diagnosis

The company is currently under a saturated production regime since it is working more with respect to its available hours. Regarding the current situation and the VSM of this situation, we have the main results based on the indicators shown before the numbers shown in the table (Table 4).

Table 4. Indicators Current situation

INDICATOR		
Productivity	Production	Using MDO
3% more than the established schedule (103%)	86%	76%

In this table we can see the results regarding the indicators raised of the current situation of the company.

4.3 Validation Design

In this current simulation it can be seen that there are 37 order orders where only 32 were executed. In addition, it should be noted that the system time was 413.71 h, showing that if there is a delay in the production line, this is due to the fact that the standard production time is 400 h.

Of 37 orders where 34 were executed, improving production to 5% with the current situation. In addition, the system time with this new simulation improved, reducing to 290.03 h, improving to 30% compared to the current situation. It becomes apparent that production line delivery delays were reduced (See Table 5).

Table 5. Simulation results

SIMULATION			
	Actual	Improved	
TC	413.71	290.03	**30%**
Orders	37		
Production	32	34	
	86%	92%	**5%**

As can be seen in the improved simulation, production time was reduced to 30% reducing delays in the production line and production was increased to 5% obtaining better performance. This shows that delays on the production line were reduced.

5 Conclusions

The proposed implementation helped reduce lead time on the production line by improving production time to 30% and increasing production to 5%. Good implementation of the model is essential to improve production processes. This model is aimed at industrial companies specifically in the metalworking sector for the manufacture of machinery for micro and small companies (MYPES) and by people with little experience.

The scenario that was validated, which is the production of steel tanks, shows a general improvement in the efficiency of the line, which allows the feasibility of the proposed model to be applied to other production lines with a similar condition.

The implementation and relationship of the model with the production system offers a simple, agile and low-price alternative. In addition, it identifies deficiencies, which imply a negative impact on the competitiveness of the company. Having the SAP-LAP analysis makes it a flexible model.

References

1. Metalworking beginning and evolution. http://metalmecanicainicioyevolucion.blogspot.com/. Accessed 28 June 2020
2. Investments of the Latin American metalworking industry for 2018 (2020). http://www.metalmecanica.com/temas/Hacia-donde-se-dirigiran-las-inversiones-de-la-industria-metalmecanica-latinoamericana-en-2018+123011. Accessed 28 June 2020
3. Economic panorama of the metalworking industry (2020). http://www.reporteroindustrial.com/temas/Panorama-economico-de-la-industria-metalmecanica+130855?pagina=1. Accessed 28 June 2020
4. Management, R.: Metalworking sector registered growth of 6.1% in the first four months of 2018 (2020). https://gestion.pe/economia/mercados/sector-metalmecanico-registro-crecimiento-6-1-primer-cuatrimestre-2018-237415-noticia/?ref=gesr(5). Accessed 28 June 2020
5. Peruvian metalworking industry grew 10.2% between January and October 2018. (2020). https://www.americaeconomia.com/negocios-industrias/industria-metalmecanica-peruana-crecio-102-entre-enero-y-octubre-2018#:~:text=%2D%20La%20producci%C3%B3n%20industrial%20metalmec%C3%A1nica%20creci%C3%B3,National%20de%20Industrias%20(SNI). Accessed 28 June 2020
6. (2020). http://repositorio.usil.edu.pe/bitstream/USIL/2779/1/2017_Jimenez_Reduccion-de-tiempo-de-entrega.pdf. Accessed 28 June 2020
7. Matharu, M.: Lean implementation in Indian manufacturing MSMES: a sap-lap analysis. Manage. Prod. Eng. Rev. **10**(1), 68–78 (2019). https://doi.org/10.24425/mper.2019.128245
8. Landazábal, M.S.C., Ruiz, C.G.A., Álvarez, Y.Y.M., Padilla, H.E.C.: Lean manufacturing: 5 s y TPM, herramientas de mejora de la calidad. Caso empresa metalmecánica en Cartagena, Colombia. SIGNOS - Investigación en sistemas de gestión **11**(1), 71–86 (2019). https://doi.org/10.15332/s2145-1389-4934
9. Ochoa, T.H., Bull, K.G., Mejía, G.I., Salgado, M.V., Guaderrama, A.M.: Poka-Yoke Implementation In Tools To Decrease Ppms In Assembly Station. Sci. Technol. Cult. 0(64) (2018)
10. Tikhonov, V.: Operational production management model of competitive products in mechanical engineering. Russia **8**(4), 1137–1142 (2019). https://doi.org/10.18421/TEM84-05
11. Kosieradzka, A.: Maturity model for production management. Procedia Eng. **182**, 342–349 (2017). https://doi.org/10.1016/j.proeng.2017.03.109

12. Klimecka-Tatar, D.: Context of production engineering in management model of value stream flow according to manufacturing industry. Prod. Eng. Arch. **21**(21), 32–35 (2018). https://doi.org/10.30657/pea.2018.21.07
13. Grabowska, S. : Industry 4.0 challenges for the business model, Poland. Silesian Univ. Technol. Publishing House. **138**, 137–144 (2019)
14. Gupta, M., Shri, C.: Modeling of changes: a case study on corrugated packaging firm. Int. J. Adv. Packag. Technol. **4**(1), 226–245 (2016)
15. Soni, V., Shukla, A.: Performance improvement of a service industry using sap-lap: a case study. Ind. Eng. J. **10**(12) (2018). https://doi.org/10.26488/iej.10.12.83
16. Gupta, M., Shri, C.: Developing a conceptual framework for performance evaluation of corrugated packaging firms in India. J. Packag. Technol. Res. **2**(1), 45–60 (2018). https://doi.org/10.1007/s41783-018-0028-3
17. Kaur, S.P., Kumar, J., Kumar, R.: Impact of competitiveness on business performance: a case study of manufacturing SMEs in Northern India. Int. J. Indian Cult. Bus. Manage. **13**(2), 205 (2016). https://doi.org/10.1504/ijicbm.2016.078039
18. Sharma, P., Sangal, A.L.: Building a hierarchical structure model of enablers that affect the software process improvement in software SMEs—a mixed method approach. Comput. Stand. Interfaces, 6.https://doi.org/10.1016/j.csi.2019.04.009
19. Chauhan, G.: An analysis of the status of resource flexibility and lean manufacturing in a textile machinery manufacturing company. Int. J. Organ. Anal. **24**(1), 107–122 (2016). https://doi.org/10.1108/IJOA-11-2012-0625

Author Index

S. Trzcielinski et al. (Eds.): AHFE 2021, LNNS 274, pp. 483–485, 2021.
https://doi.org/10.1007/978-3-030-80462-6

The manufacturer's authorised representative in the EU is Springer Nature Customer Service Centre GmbH, Europaplatz 3, 69115 Heidelberg, Germany. If you have any concerns regarding our products, please contact ProductSafety@springernature.com

Printed and bound by CPI Group (UK) Ltd, Croydon, CR0 4YY
30/09/2026
02225292-0012